Modern Concepts in *Penicillium* and *Aspergillus* Classification

NATO ASI Series

Advanced Science Institutes Series

A series presenting the results of activities sponsored by the NATO Science Committee, which aims at the dissemination of advanced scientific and technological knowledge, with a view to strengthening links between scientific communities.

The series is published by an international board of publishers in conjunction with the NATO Scientific Affairs Division

A	**Life Sciences**	Plenum Publishing Corporation
B	**Physics**	New York and London
C	**Mathematical and Physical Sciences**	Kluwer Academic Publishers Dordrecht, Boston, and London
D	**Behavioral and Social Sciences**	
E	**Applied Sciences**	
F	**Computer and Systems Sciences**	Springer-Verlag
G	**Ecological Sciences**	Berlin, Heidelberg, New York, London,
H	**Cell Biology**	Paris, and Tokyo

Recent Volumes in this Series

Series A: Life Sciences

Modern Concepts in *Penicillium* and *Aspergillus* Classification

Edited by

Robert A. Samson

Centraalbureau voor Schimmelcultures
Baarn, The Netherlands

and

John I. Pitt

CSIRO Division of Food Processing
North Ryde, New South Wales, Australia

Plenum Press
New York and London
Published in cooperation with NATO Scientific Affairs Division

Proceedings of the Second International *Penicillium*
and *Aspergillus* NATO Advanced Research Workshop,
held May 8–12, 1989,
in Baarn, The Netherlands

Library of Congress Cataloging-in-Publication Data

International Penicillium and Aspergillus NATO Advanced Research
  Workshop (2nd : 1989 : Baarn, Netherlands)
   Modern concepts in pencillium and aspergillus classification /
  edited by Robert A. Samson and John I. Pitt.
       p.   cm. -- (NATO ASI series. Series A, Life sciences ; v.
  185)
   "Proceedings of the Second International Penicillium and
  Aspergillus NATO Advanced Research Workshop, held May 8-12, 1989, in
  Baarn, The Netherlands"--Verso of t.p.
   "Published in cooperation with NATO Scientific Affairs Division."
   Includes bibliographical references.
   ISBN 0-306-43516-0
   1. Penicillium--Classification--Congresses. 2. Aspergillus-
  -Classification--Congresses.   I. Samson, Robert A. II. Pitt, John
  I. III. North Atlantic Treaty Organization. Scientific Affairs
  Division. IV. Title. V. Series.
  QK625.M7I57 1989
  589.2'3--dc20   90-7008
                                                                CIP

© 1990 Plenum Press, New York
A Division of Plenum Publishing Corporation
233 Spring Street, New York, N.Y. 10013

Printed in the United States of America

PREFACE

In our view, the First International *Penicillium* and *Aspergillus* Workshop held in Baarn and Amsterdam in May, 1985, was a great success. The assembly in one place of so many specialists in these two genera produced both interesting viewpoints and lively discussions. But more particularly, a remarkable cohesion of ideas emerged, borne primarily of the realisation that taxonomy has passed from the hands of the solitary morphologist. The future of taxonomy lay in collaborative and multidisciplinary studies embracing morphology, physiology and newer methodologies.

The Second International *Penicillium* and *Aspergillus* Workshop was borne logically from the first, and was held in Baarn on May 8-12, 1989. It was attended by 38 scientists from 16 countries. At this Workshop we have attempted to move further into new methods, especially by bringing together molecular biologists, medical and food mycologists and biochemists as well as more traditional taxonomists. We feel that the meeting contributed greatly to dialogue between taxonomists, and also fundamental and applied mycologists. At the meeting, we became aware that the approach to taxonomy of these genera is now becoming more pragmatic, with an increasing emphasis on consensus, and on stability of names. This is a noteworthy development, which we, as editors, welcome. So many species in *Penicillium* and *Aspergillus* are economically important in biotechnology, foods and medicine, and practical, stable taxonomy is of vital importance.

These Proceedings comprise 40 papers divided into 9 chapters. Discussions relating to each paper were taped and after careful editing, have been included following the relevant paper. Some general discussion is also included. Dr Keith Seifert kindly helped us with typing the discussions, and we are extremely grateful for his assistance. We also wish to thank Mr Guido van Reenen, who prepared the layout and camera ready copy of both the Abstracts and these Proceedings.

The Second International *Penicillium* and *Aspergillus* Workshop was sponsored by the NATO ARW programme. Cosponsors were the Dutch Programmebureau Biotechnology, the Royal Academy of Sciences and some industrial companies. We wish to sincerely thank all of these sponsors.

Local arrangements for the Workshop were very efficiently and smoothly organised with the assistance of Mrs Ans Spaapen-de Veer, Tineke van der Berg and Marjolein van der Horst; we wish to thank each of them, as well as many other colleagues from the Centraalbureau voor Schimmelcultures.

The Editors
Baarn, September 1989

CONTENTS

Chapter 6
COMPUTER-ASSISTED IDENTIFICATION OF PENICILLIA AND ASPERGILLIA

Chapter 7
NEW APPROACHES FOR PENICILLIUM AND ASPERGILLUS SYSTEMATICS: MOLECULAR BIOLOGICAL TECHNIQUES

Chapter 8
NEW APPROACHES FOR PENICILLIUM AND ASPERGILLUS SYSTEMATICS: BIOCHEMICAL AND IMMUNOLOGICAL TECHNIQUES

Chapter 9
TAXONOMIC STUDIES ON THE TELEOMORPHS OF PENICILLIUM AND ASPERGILLUS

1

INTRODUCTION

SYSTEMATICS OF *PENICILLIUM* AND *ASPERGILLUS* – PAST, PRESENT AND FUTURE

J.I. Pitt[1] and R.A. Samson[2]

[1]*CSIRO*
Division of Food Processing
North Ryde, NSW 2113, Australia

[2]*Centraalbureau voor Schimmelcultures*
3740 AG Baarn, The Netherlands

SUMMARY

The great pioneers in the systematics of *Penicillium* and *Aspergillus* were Charles Thom and Kenneth B. Raper, who together developed the first workable and widely accepted taxonomies of these genera. Both were instrumental also in the development of culture collections for these industrially important fungi. Raper made the invaluable contribution of developing freeze drying techniques suitable for fungi, and this has proved to be of great practical importance. However, they also left a legacy of nomenclatural and taxonomic problems which have been addressed only recently. Moreover, inherent difficulties in achieving consensus on species concepts and on identification of isolates have provided a challenge for modern students of these genera.

This paper describes the new approaches which have led to clarification of the systematics of these genera. On the one hand, increased awareness of the principles of priority, typification and type specimens has helped to bring order to nomenclature. On the other, the use of gross physiological characters, secondary metabolites and isoenzyme patterns has greatly assisted in clarifying taxonomy. Collaborative studies on difficult aspects of both genera have been initiated both by individuals and under the control of an international working group, the Subcommission on *Penicillium* and *Aspergillus* Systematics.

Great progress has been made. In particular, the taxonomy of the important and notoriously difficult subgenus *Penicillium* has now become firmly established, and criteria for differentiating aflatoxin producing and nontoxigenic *Aspergillus* species have been clarified.

The immediate end of all this work lies in a better understanding of relationships within these genera, and improved identifications. The wider benefits lie in a greater knowledge of the role of particular species in food spoilage, toxicology, biodeterioration, biotechnology and ecology.

INTRODUCTION

Penicillium and *Aspergillus* are two of the most economically important genera of fungi. Much of their economic impact is deleterious, with food spoilage, mycotoxin production and biodeterioration heading the list, but in fact their potential for economic utility is equally great. Penicillin has been the great success story in the utilisation of *Penicillium*, but the physiological and biochemical diversity of both genera is such that their potential for benefiting mankind has only just begun to be tapped. The future includes cheaper production of chemicals such as citric and other organic acids, biochemical syntheses and conversions of intricate molecules not amenable to chemical techniques, and a whole range of enzymes, proteins and other organic molecules of value in commercial applications. Genetic engineering will play a major role.

It is imperative, then, that the systematics of these genera be clarified and stabilised, as a matter of urgency. Ideally, identification should be unequivocal, accurate, simple and

Modern Concepts in Penicillium and Aspergillus Classification
Edited by R. A. Samson and J. I. Pitt
Plenum Press, New York, 1990

immutable. In some parts of *Penicillium* we are approaching this goal, a situation which even a few years ago would have been regarded as unlikely to ever be reached.

The majority of the significant recent work has dealt with *Penicillium*, the larger and more complex of these two genera, so it will be primarily discussed in this paper. *Aspergillus* will be considered also where appropriate.

THE PAST: A SHORT HISTORY

During the 19th century, the taxonomy of anamorphic genera such as *Penicillium* and *Aspergillus* was strictly botanical. Descriptions were based on observations of the fungi on natural substrates and were based mainly on microscopical observations. Because fruiting structures are ephemeral, especially in *Penicillium*, descriptions were usually rather meagre. Many such species subsequently went unrecognised.

With the advent of pure culture techniques around the turn of the century, colony and fruiting structure development began to be observed. As a result of improvements in microscopes and microscopic techniques, great advances occurred in descriptions. Recognition of different types of fruiting structure in both genera led to the splitting out of genera such as *Citromyces* and *Sterigmatocystis*, but most taxonomists retained *Penicillium* and *Aspergillus* as monolithic and broadly based, a situation still existing today. As they were less ephemeral than the anamorphs, teleomorphs of these genera were described early. The connection between *Eurotium* Link 1809 and *Aspergillus*, established by de Bary in 1854. Ludwig described *Eupenicillium* in 1892, though this teleomorph genus was not accepted until the 1970s.

The great pioneer in the study of *Penicillium* and *Aspergillus* was Charles Thom (1872-1956). Thom (1906) produced the first readily recognisable descriptions of *Penicillium* species and Thom (1910) the first taxonomy. He was author or coauthor of two further monumental works on *Penicillium*. Thom (1930) was a compendium of the 300 species described to that time, with keys and some indications of synonymy. Raper and Thom (1949) published "A Manual of the Penicillia", which was the authoritative taxonomy in almost exclusive use until 1980.

Thom and his colleagues also studied *Aspergillus* taxonomy. Thom and Church (1926) produced the first complete taxonomy of the genus, which was expanded by Thom and Raper (1945), and enlarged and refined again by Raper and Fennell (1965).

Other taxonomists.

Other taxonomists have also made significant contributions to the taxonomy of *Penicillium*, especially in Europe. Westling (1911) and Sopp (1912) both described a number of *Penicillium* species from Scandinavian soils. In France, Bainier and Sartory (e.g. Bainier, 1907; Bainier and Sartory, 1912) published a series of short papers on new species. Biourge (1923) published an extensive taxonomy, and revived and characterised some names of Dierckx (1901) which had been inadequately described.

In *Aspergillus*, description of new species appears to have proceeded more or less piecemeal, with no major taxonomies other than those mentioned above.

Culture collections.

At the same time as the taxonomy of *Penicillium* and *Aspergillus* developed, collections of fungal cultures began to be established. The role of culture collections in taxonomy, allowing the preservation and distribution of living ex type cultures and other important isolates, cannot be overemphasized. Their significance in the effective utilisation of fungi for beneficial purposes will prove equally great in the coming years.

The Centraalbureau voor Schimmelcultures in Baarn, and the Commonwealth Mycological Institute in Kew, Surrey, have played a major role in *Penicillium* and *Aspergillus* taxonomy by providing a vital continuity in cultures. Many of the important European taxonomists sent their cultures to CBS, including Westling and Biourge. Both Thom and Raper visited Baarn, and both obtained cultures which laid the foundation of their own collection, at the USDA Northern Regional Research Center in Peoria, Illinois, a collection which has played a pivotal role in *Penicillium* and *Aspergillus* taxonomy.

Culture preservation.
It is worth noting here also that K.B. Raper pioneered the technique of freeze drying for the preservation of fungal cultures. This has also been of great importance. Studies by Pitt (1979) were based on the Raper collection at Peoria, which in many cases provided him with better quality cultures than were available elsewhere, from collections where cultures were maintained by less effective preservation techniques, or had been freeze dried later.

Shortcomings of Raper and Thom's work.
Charles Thom and Kenneth Raper were taxonomic giants, whose perceptions of species and species relationships were without parallel among their contemporaries. Thom, Raper and their colleagues dominated the taxonomy of these two genera for half a century. However, as the number of species described in *Penicillium* and *Aspergillus* increased, so nomenclatural and taxonomic difficulties multiplied. Without any disrespect to their memories, it can be stated in the late 1980s that some of Thom and Raper's concepts have proved to be less than satisfactory and, equally importantly, that their very conservative approach to nomenclatural changes ultimately led to serious difficulties in the systematics of both genera: in *Penicillium*, especially, to a state little short of chaos.

Nomenclatural problems.
The nomenclatural problems which arose as a consequence of the Raper and Thom taxonomies can be summarised in three points: (a) failure to observe priority of names; (b) failure to typify species; and (c) failure to give precedence to teleomorphs in naming species.

(a) *Priority.*
In the sense used in the International Code of Botanical Nomenclature (ICBN; Greuter, 1988) "priority" has the specific meaning of "requiring use of the earliest validly published name". "Validity" is governed by a number of Articles, most of which are not relevant here. Suffice to say that Thom and Raper sometimes did not accept names which had priority, on various grounds which are not acceptable now. Ignoring Biourge's neotypifications, Thom and Raper rejected many names published by Dierckx (1901) on the grounds of inadequate descriptions, unless it suited them, as in the case of an apt name like *Penicillium brevicompactum*. Valid names of other authors were sometimes rejected for similar reasons. Name changes made much later (Samson *et al.*, 1976; Pitt, 1979), with the inevitable confusion that name changes cause, could have been avoided in many cases.

(b) *Failure to typify.*
Among the criteria for valid publication, the ICBN demands that species by "typified", i.e.
that a representative herbarium specimen be preserved from material used to produce the
original description. Raper and Thom failed to specify dried types, though in their
defense, it must be said that many other taxonomists were equally negligent. More
seriously, however, Raper and Thom frequently failed to identify even living type
material in their standard descriptions of species, and sometimes ignored existing type
material altogether.

(c) *The problem of naming sexual states.*
Thom and Raper, and to a lesser extent, Fennell, maintained that teleomorph species with
anamorphs in *Aspergillus* or *Penicillium*, including those with teleomorphs, should be
named in the anamorph genera, making new combinations as necessary (Thom and
Raper, 1946; Thom, 1954; Raper, 1957; Fennell, 1973). For example, *Eurotium amstelodami*
Mangin was renamed *Aspergillus amstelodami* (Mangin) Thom & Church (Thom and
Church, 1926) and maintained this way in their later taxonomies. However the ICBN, and
the weight of nomenclaturalists' opinions, dictate that teleomorph names have priority
over anamorph names.
     A further problem concerns species described in an anamorph genus but which
included the teleomorph, for example, *Penicillium javanicum* van Beyma. Raper and Thom
(1949) accepted such species as being correctly placed in *Penicillium*, contrary to the ICBN.
Hawksworth and Sutton (1974) showed that the ICBN then current was ambiguous in the
naming of a species such as *Penicillium javanicum*. As a consequence, the International
Botanical Congress in Sydney altered articles in the new ICBN (Voss, 1983) to make
naming of such species unequivocal. The price was that some well established names,
especially in *Aspergillus* and *Penicillium*, had to be changed. The nomenclatural problem
was overcome, but at the expense of introducing new and unfamiliar names. Pitt (1979)
made the necessary changes to *Penicillium*. For example, "*Penicillium javanicum* van
Beyma" is now known correctly as *Eupenicillium javanicum* (van Beyma) Stolk & Scott
when the teleomorph, or the whole fungus, is under consideration, and as *Penicillium
indonesiae* Pitt when the anamorph alone is being considered. Samson and Gams (1985)
carried out the necessary alterations in *Aspergillus*: "*Aspergillus amstelodami* (Mangin)
Thom & Church" is correctly called *Eurotium amstelodami* Mangin when the teleomorph is
present, and *Aspergillus hollandicus* Samson & Gams for the *Aspergillus* state alone. For a
more detailed discussion of this point see Pitt (1988).

"Groups".
Thom and his coworkers arranged clusters of species in *Aspergillus* and *Penicillium* into
"Groups", a subgeneric classification without nomenclatural status under the ICBN. This
has caused confusion, because most mycologists believe that if there is to be
nomenclatural stability, there must be a a code of nomenclature (the ICBN), and if there is
an ICBN, it must be accepted in all aspects. In industrially important genera, stability is
essential, and stability at the present time comes only from strict adherence to the ICBN.
     Pitt (1979) replaced the "Group" structure in *Penicillium* by a subgeneric and sectional
structure. Gams *et al.* (1985) carried out the same changes to *Aspergillus*.

Taxonomic problems.
The taxonomy of Penicillia has always been difficult. All taxonomists have agreed that the
genus has a lot of species separated by very little. Raper and Thom (1949) produced the

first definitive taxonomy, in its day a masterpiece, and in almost exclusive use for nearly 30 years. But it had shortcomings. First, the "taxonomic base", i.e. the range of characters used to distinguish species, was rather narrow, so that species were difficult to distinguish from each other, even to the expert. Also, Raper and Thom (1949) placed too much reliance on colony texture. For example, in the Raper and Thom classification terverticillate species were spread across Sect. *Asymmetrica* subsects. *Velutina, Fasciculata, Funiculosa* and *Lanata*, mainly on the basis of differences in colony texture. Without expert guidance, the classification was largely unworkable.

Samson *et al.* (1976) approached this problem by reducing a number of difficult to distinguish Raper and Thom species to the status of varieties. Pitt (1979) disagreed both with the Raper and Thom reliance on texture, and the varietal approach of Samson *et al.* (1976). Instead, he relied heavily on colony diameters and morphology, especially conidial colour. However, this approach was relatively ineffective in subgenus *Penicillium* where the majority of species respond similarly to temperature and reduced water activity (a_w), and differences in colony colours are small and rather subjective.

The competition among three classification systems with quite different approaches, and the shortcomings of all three, resulted in confusion (Onions *et al.*, 1984).

THE PRESENT: NEW APPROACHES

Clearly, new approaches were needed to improve the taxonomy of *Penicillium* and *Aspergillus*; changes which would broaden the taxonomic base, clarify species concepts and bring both nomenclature and taxonomy into line with the ICBN. Change started slowly, then gathered pace, on several fronts.

Nomenclatural changes.
The first developments came with nomenclatural changes relating to the use of teleomorph names. Benjamin (1955) introduced *Talaromyces* for *Penicillium* teleomorphs with gymnothecia, and took up *Carpenteles* Langeron for teleomorphs with cleistothecia. Raper (1957) resisted these changes. Stolk and Scott (1967) revived the earlier name *Eupenicillium* for *Carpenteles* and, for example, correctly renamed *Penicillium javanicum* van Beyma as *Eupenicillium javanicum* (van Beyma) Stolk & Scott. This approach won general approval over the next decade.

In search of a broader base.
A variety of attempts to improve taxonomy occurred more or less simultaneously. Abe (1956) published a modified version of Raper and Thom (1949) in which he made use of some physiological characters as an aid in *Penicillium* classification. These included the use of 37°C as a growth temperature and the use of nitrite as a selective agent for growth. Pitt (1973) developed this idea, making use of both 5°C and 37°C as measures of the effect of temperature on the growth of *Penicillium* isolates. He also introduced 25% glycerol nitrate agar (G25N, a_w 0.935) as a medium of reduced water activity, which provided some measure of physiological adaptation to low a_w. To make effective use of these concepts, Pitt (1973) used a carefully standardised set of media and incubation conditions. He proposed quantifying the influences of temperature and a_w by measurements of colony diameters grown for a standard incubation time of 7 days.

A variety of other approaches followed, all aimed at broadening the base of *Penicillium* taxonomy. Pyrolysis gas liquid chromatography (Kulik and Vincent, 1973; Burns *et al.*, 1976; Søderstrøm and Frisvad, 1984) and differences in long chain fatty acids

(Dart *et al.*, 1976) each showed the capacity to distinguish genera or species when relatively few isolates were examined, but became impossible to interpret when a large genus like *Penicillium* was studied. Moreover, the match of long chain fatty acids with morphological taxonomy was poor (Pitt, 1984). This was the case also with the production of nigeran, a macromolecular cell wall constituent studied by Bobbitt and Nordin (1978).

Secondary metabolite production.

The quest for taxonomically valuable physiological characters started along another line when Ciegler *et al.* (1973) suggested that secondary metabolites, particularly mycotoxins, might be of taxonomic value. The concept was applied on a limited scale to *Penicillium viridicatum* (Ciegler *et al.*, 1973, 1981). The development of simpler techniques followed. Of particular interest was the concept of directly spotting TLC plates with small samples of culture cut directly from Petri dishes with a cork borer, followed by chromatography (Filtenborg and Frisvad, 1980; Filtenborg *et al.*, 1983). Mycotoxin and secondary metabolite production can be assessed qualitatively much faster by this method than by conventional extraction, clean up and concentration techniques.

These approaches met with some early opposition because of the complexity of the concepts (Frisvad and Filtenborg, 1983; Frisvad, 1985, 1986). However, difficulties with misidentified cultures, and the relationship of mycotoxin profiles to morphological taxonomy have now been clarified, especially with the introduction of another new technique, electrophoretic comparison of certain enzymes (see below). Patterns of secondary metabolites have now become an effective taxonomic tool, especially when used in conjunction with traditional taxonomy as governed by the ICBN (Pitt, 1984).

Frisvad and Filtenborg (1983), Paterson (1986) and El-Banna *et al.* (1987) have published detailed thin layer chromatographic solvent systems and R_f values for a wide variety of *Penicillium* metabolites, so that metabolite profiles can now be used by the determinative taxonomist.

Isoenzyme electrophoresis.

Cruickshank (Cruickshank and Wade, 1980; Cruickshank, 1983) developed effective methods for separation of species of *Sclerotinia*, *Botrytis* and other genera by examining patterns of pectic enzymes after separation by gel electrophoresis. Small samples of culture fluid were subjected to electrophoresis at low temperatures, then the separated enzymes allowed to act on methoxy pectin incorporated into the gel, and the sites of enzyme action visualised by ruthenium red staining.

To enable the differentiation of the many species in subgenus *Penicillium*, Cruickshank's technique was broadened by including the examination of amylase and ribonuclease isoenzymes. For amylase production, soluble starch was incorporated in the gels as a substrate, and for ribonucleases, ribosomal RNA. Fungi were cultured on wheat grains for the production of both these sets of enzymes (Cruickshank and Pitt, 1987a).

A study of the isoenzyme patterns (zymograms) for all species accepted by Pitt (1979) in subgenus *Penicillium* has now been successfuly completed (Cruickshank and Pitt, 1987b). For the first time, an objective physiological test has been correlated to an acceptable degree with classical taxonomic methods. All isolates examined from the great majority of species accepted by Pitt (1979) and now agreed on by Samson and Pitt (1985) showed an absolute correlation with specific zymogram patterns, greatly reinforcing our confidence in current species concepts in this subgenus. In certain species, notably *P. aurantiogriseum* and *P. viridicatum*, correspondence was less clear; however, a lack of clarity in classical taxonomic concepts was already evident.

By 1989, the taxonomy of subgen. *Penicillium* has been greatly clarified. Close agreement on taxonomy by classical techniques, secondary metabolites and isoenzyme patterns has resulted in well defined species (Pitt and Cruickshank, 1990; Stolk *et al.*, 1990; Frisvad and Filtenborg, 1989).

Examination of herbarium specimens.

As was stated earlier in this paper, the name of any fungus is linked to a herbarium specimen. The fruiting structures in some genera, including *Penicillium* and *Aspergillus*, are of a relatively emphemeral nature, and taxonomists have frequently ignored herbarium specimens in reaching taxonomic conclusions. Recently Seifert and Samson (1985) undertook a study of herbarium specimens in an obsolete genus, *Coremium*, and were able to recognise some old, valid species which needed to be transferred to *Penicillium*. As a result, the names of a few species have been changed. These include *P. claviforme* Bainier to *P. vulpinum* (Cooke & Massee) Seifert & Samson, *P. granulatum* Bainier to *P. glandicola* (Oudem.) Seifert & Samson, and *P. concentricum* Samson *et al.* to *P. coprophilum* (Berk. & Curt.) Seifert & Samson.

Collaborative examination of cultures.

To assist in clarification of the taxonomy of aflatoxigenic fungi, Klich and Pitt (1985, 1988) carried out a collaborative study involving the detailed morphological and gross physiological examination of more than 200 isolates of *Aspergillus flavus* and related species. Cultures were examined in two different locations, with isolates identified only by code. Features distinguishing *A. flavus* from *A. parasiticus* and from the closely related food fermentation species, *A. flavus* and *A. sojae*, were documented.

A subcommission of the International Commission on Taxonomy of Fungi (IUMS) has been formed, which will carry out collaborative studies on the taxonomy of *Penicillium* and *Aspergillus*. Known as the Subcommission on *Penicillium* and *Aspergillus* Systematics (SPAS), this group has twelve members from seven countries, and comprises both morphological taxonomists and those skilled in the newer techniques which have been described here. A significant impact is expected.

Genetic studies.

Until recently, the genetic tools of DNA hybridisation and analysis of DNA and RNA sequences have been little used in the systematics of *Penicillium* and *Aspergillus*.

On the basis of studies of DNA homology among *A. flavus* and related species, Kurtzman *et al.* (1986) reduced several well known species to subspecies or varietal status. Specifically *A. parasiticus* became *A. flavus* subspecies *parasiticus* Kurtzman *et al.*; *A. oryzae* became *A. flavus* subspecies *flavus* variety *oryzae* Kurtzman *et al.*; and *A. sojae* became *A. flavus* subspecies *parasiticus* variety *sojae* Kurtzman *et al.*

Klich and Mullaney (1987) disagreed with this conclusion, arguing that DNA restriction enzyme fragment polymorphism showed differences between the DNA of *A. flavus* and *A. oryzae*. Klich and Pitt (1988) also did not accept this change, on the grounds that morphological differences existed between these two species, and on the more practical grounds that it was important that species used for food fermentations possess different species names from those which are known to be mycotoxigenic.

THE FUTURE

Finally, a look at the future. Some aspects of prediction are not difficult, because some techniques of the future are already in use, and are described more fully in the papers which follow in these Proceedings.

In the future, increased reliance will be placed on secondary metabolites and isoenzyme profiles as aids in taxonomy -not, perhaps, in routine identifications, but certainly in more specialist laboratories and those where fundamental taxonomic studies are carried out.

Care is required in interpreting results from both of these techniques. The fact that two isolates or sets of isolates have similar profiles by either technique does not prove a close relationship. However, in our view, similar profiles with both techniques, and close similarities in morphological characters, provide an excellent guide to species definition.

Second, genetic techniques will be increasingly employed -again, for basic studies, not as a determinative tool. Genetic studies will range from fundamental work on RNA, which will enable clarification of phylogeny on a generic scale, and the setting of evolutionary clocks, to comparisons of nuclear and mitochondrial DNA as an aid in delimiting species. DNA probes will be of great value, once suitable probes have been developed. Again it needs to be emphasised that genetic techniques, like other taxonomic methods, are tools, not the ultimate solution. Similarities or differences in genetic material, measured by whatever technique, must be assessed in conjunction with other taxonomic yardsticks.

No doubt other new and innovative genetic methods will appear in the next few years. Of particular interest is the new technique of "gene amplification", which promises to make it possible to identify the most limited herbarium material by copying DNA from even a single spore until there is suffcient to hybridise or sequence. The impact of this technique on current taxonomy is difficult to assess, but may be profound, unless carefully managed.

Third, computer based systems will be increasingly applied to *Penicillium* and *Aspergillus* taxonomy. Culture collections, the indispensible libraries of the taxonomist, will become computerised, in time enabling access for all interested scientists to much more data than is currently available, and much more rapidly. Networks of information of this kind will be established, though rather slowly.

At a different level, computers will be used to produce databases relating to individual genera, related series, or isolates within a species. Keys will be developed from this information, and will to some degree replace current written taxonomies.

Fourth, more rapid taxonomic methods will be devised. The trend will be towards systems which can be used directly on isolation plates, and may involve computer databases, recognition of specific or more general antibodies, or secondary metabolites (Filtenborg and Frisvad, 1990).

Fifth, a trend towards protection of the names of well recognised species will develop in the next few years. *Penicillium*, especially, will be in the forefront of these studies, because general agreement on species concepts throughout much of the genus is already close, or will become so, especially through international collaboration under SPAS. Such protection is becoming imperative (Hawksworth, 1990).

However, regardless of the introduction and use of new techniques, we will not, and indeed should not, see the replacement of current morphological and gross physiological identification methods, at least into the next century. The reason is simple, and compelling: it is most important that taxonomy, at the determinative level, be accessible, as far as possible, to those in Bangkok or Bogotá, Berlize or Bhutan, as well as those working in the favoured environments of Baarn, Berlin or Britain.

REFERENCES

ABE, S. 1956. Studies on the classification of the Penicillia. *Journal of General and Applied Microbiology, Tokyo* 2: 1-344.

BAINIER, G. 1907. Mycothece de l'Ecole de Pharmacie. Part IX. *Bulletin trimestriele de la Societe Mycologique de France* 23: 11-22.

BAINIER, G. and SARTORY, A. 1912. Etude d'un *Penicillium* nouveau, *Penicillium herquei. Bulletin trimestriele de la Societe Mycologique de France* 28: 121-126.

BENJAMIN, C.R. 1955. Ascocarps of *Aspergillus* and *Penicillium. Mycologia* 47: 669-687.

BIOURGE, P. 1923. Les moisissures du groupe *Penicillium* Link. *Cellule* 33: 1-322.

BOBBITT, T.F. and NORDIN, J.H. 1978. Hyphal nigeran as a potentialphylogenetic marker for *Aspergillus* and *Penicillium* species. *Mycologia* 70: 1201-1211.

BURNS, D.T., STRETTON, R.J. and JAYATILAKE, S.D.A.K. 1976. Pyrolysis gas chromatography as an aid to the identification of *Penicillium* species. *Journal of Chromatography* 116: 107-115.

CIEGLER, A., FENNELL, D.I., SANSING, G., DETROY, R.W. and BENNETT, G.A. 1973. Mycotoxin-producing strains of *Penicillium viridicatum*: classification into subgroups. *Applied Microbiology* 26: 271-278.

CIEGLER, A., LEE, L.S. and DUNN, J.J. 1981. Production of naphthoquinone mycotoxins and taxonomy of *Penicillium viridicatum. Applied and Environmental Microbiology* 42: 446-449.

CRUICKSHANK, R.H. 1983. Distinction between *Sclerotinia* species by their pectic zymograms. *Transactions of the British Mycological Society* 80: 117-119.

CRUICKSHANK, R.H. and PITT, J.I. 1987a. The zymogram technique: isoenzyme patterns as an aid in *Penicillium* classification. *Microbiological Sciences* 4: 14-17.

—— 1987b. Identification of species in *Penicillium* subgenus *Penicillium* by enzyme electrophoresis. *Mycologia* 79: 614-620.

CRUICKSHANK, R.H. and WADE, G.C. 1980. Detection of pectic enzymes in pectin-acrylamide gels. *Analytical Biochemistry* 107: 177-181.

DART, R.K., STRETTON, R.J. and LEE, J.D. 1976. Relationships of *Penicillium* species based on their long chain fatty acids. *Transactions of the British Mycological Society* 66: 525-529.

DIERCKX, R.P. 1901. Un essai de revision du genre *Penicillium* Link. *Annales de la Societe scientifique de Bruxelles* 25: 83-89.

EL-BANNA, A.A., PITT, J.I. and LEISTNER, L. 1987. Production of mycotoxins by *Penicillium* species. *Systematic and Applied Microbiology* 10: 42-46.

FENNELL, D.I. 1973. Review of The genus *Talaromyces*: Studies on *Talaromyces* and Related Genera. II. *Mycologia* 65: 1221-1223.

FILTENBORG, O. and FRISVAD, J.C. 1980. A simple screening method for toxigenic moulds in pure culture. *Lebensmittel Wissenschaft und Technologie* 13: 128-130.

—— 1990. Identification of *Penicillium* and *Aspergillus* species in mixed cultures without subculturing. *In* Modern Concepts in *Penicillium* and *Aspergillus* Classification, eds. R.A. Samson and J.I. Pitt, pp. 27-37. New York and London: Plenum Press.

FILTENBORG, O., FRISVAD, J.C. and SVENDSEN, J. 1983. Simple screening procedure for molds producing intracellular mycotoxins in pure culture. *Applied and Environmental Microbiology* 45: 581-585.

FRISVAD, J.C. 1985. Profiles of primary and secondary metabolites of value in classification of *Penicillium viridicatum* and related species. *In* Advances in *Penicillium* and *Aspergillus* Systematics, eds. R.A. Samson and J.I. Pitt, pp. 311-325. New York and London: Plenum Press.

—— 1986. Taxonomic approaches to mycotoxin identification. *In* Modern Methods in the Analysis and Structural Elucidation of Mycotoxins, ed. R.J. Cole, pp. 415-457. Orlando, Florida: Academic Press.

FRISVAD, J.C. and FILTENBORG, O. 1983. Classification of terverticillate Penicillia based on profiles of mycotoxins and other secondary metabolites. *Applied and Environmental Microbiology* 46: 1301-1310.

—— 1989. Chemotaxonomy of and mycotoxin production by terverticillate *Penicillium* species. *Mycologia* (in press).

GAMS, W., CHRISTENSEN, M., ONIONS, A.H.S., PITT, J.I. and SAMSON, R.A. 1985. Infrageneric taxa of *Aspergillus. In* Advances in *Penicillium* and *Aspergillus* Systematics, eds. R.A. Samson and J.I. Pitt, pp. 55-62. New York and London: Plenum Press.

GREUTER, W. *et al.* 1988. International Code of Botanical Nomenclature adopted by the Fourteenth International Botanical Congress, Berlin, July-August, 1987. Königstein, W. Germany: Koeltz Scientific Books.

HAWKSWORTH, D.L. 1990. Problems and prospects for improving the stability of names in *Aspergillus* and *Penicillium*. In Modern Concepts in *Penicillium* and *Aspergillus* Classification, eds. R.A. Samson and J.I. Pitt, pp. 75-82. New York and London: Plenum Press.

HAWKSWORTH, D.L. and SUTTON, B.C. 1974. Article 59 and names of perfect state taxa in imperfect genera. *Taxon* 23: 563-568.

KLICH, M.A. and MULLANEY, E.J. 1987. DNA restriction enzyme fragment polymorphism as a tool for rapid differentiation of *Aspergillus flavus* from *Aspergillus oryzae*. *Experimental Mycology* 11: 170-175.

KLICH, M.A. and PITT, J.I. 1985. The theory and practice of distinguishing species of the *Aspergillus flavus* group. In Advances in *Penicillium* and *Aspergillus* Systematics, eds. R.A. Samson and J.I. Pitt, pp. 211-220. New York and London: Plenum Press.

—— 1988. Differentiation of *Aspergillus flavus* from *A. parasiticus* and other closely related species. *Transactions of the British Mycological Society* 91: 99-108.

KULIK, M.M. and VINCENT, P.G. 1973. Pyrolysis gas-liquid chromatography of fungi: observations on variability among nine *Penicillium* species of the section *Asymmetrica*, subsection *Fasciculata*. *Mycopathologia et Mycologia Applicata* 51: 1-18.

KURTZMAN, C.P., SMILEY, M.J., ROBNETT, C.J. and WICKLOW, D.T. 1986. DNA relatedness among wild and domesticated species in the *Aspergillus flavus* group. *Mycologia* 78: 955-959.

ONIONS, A.H.S., BRIDGE, P.D. and PATERSON, R.R. 1984. Problems and prospects for the taxonomy of *Penicillium*. *Microbiological Sciences* 1: 185-189.

PATERSON, R.R.M. 1986. Standardized one- and two-dimensional chromatographic methods for the identification of secondary metabolites in *Penicillium* and other fungi. *Journal of Chromatography* 368: 249-264.

PITT, J.I. 1973. An appraisal of identification methods for *Penicillium* species: novel taxonomic criteria based on temperature and water relations. *Mycologia* 65: 1135-1157.

—— 1979. The Genus *Penicillium* and its Teleomorphic States *Eupenicillium* and *Talaromyces*. London: Academic Press.

—— 1984. The value of physiological characters in the taxonomy of *Penicillium*. In Toxigenic Fungi - Their Toxins and Health Hazard, eds. H. Kurata and Y. Ueno, pp. 107-118. Amsterdam: Elsevier.

—— 1988. A Laboratory Guide to Common *Penicillium* Species. 2nd ed. North Ryde, N.S.W.: CSIRO Division of Food Processing.

PITT, J.I. and CRUICKSHANK, R.H. 1990. Speciation and synonymy in *Penicillium* subgenus *Penicillium* - towards a definitive taxonomy. In Modern Concepts in *Penicillium* and *Aspergillus* Classification, eds. R.A. Samson and J.I. Pitt, pp. 103-119. New York and London: Plenum Press.

RAPER, K.B. 1957. Nomenclature in *Aspergillus* and *Penicillium*. *Mycologia* 49: 644-662.

RAPER, K.B. and FENNELL, D.I. 1965. The Genus *Aspergillus*. Baltimore, Maryland: Williams and Wilkins.

RAPER, K.B. and THOM, C. 1949. A Manual of the Penicillia. Baltimore, Maryland: Williams and Wilkins.

SAMSON, R.A. and GAMS, W. 1985. Typification of the species of *Aspergillus* and associated teleomorphs. In Advances in *Penicillium* and *Aspergillus* Systematics, eds. R.A. Samson and J.I. Pitt, pp. 31-54. New York and London: Plenum Press.

SAMSON, R.A. and PITT, J.I. 1985. Check list of common *Penicillium* species. In Advances in *Penicillium* and *Aspergillus* Systematics, eds. R.A. Samson and J.I. Pitt, pp. 461-463. New York and London: Plenum Press.

SAMSON, R.A., STOLK, A.C. and HADLOK, R. 1976. Revision of the Subsection *Fasciculata* of *Penicillium* and some allied species. *Studies in Mycology, Baarn* 11: 1-47.

SEIFERT, K.A. and SAMSON, R.A. 1985. The genus *Coremium* and the synnematous Penicillia. In Advances in *Penicillium* and *Aspergillus* Systematics, eds. R.A. Samson and J.I. Pitt, pp. 143-154. New York and London: Plenum Press.

SØDERSTRØM, B. and FRISVAD, J.C. 1984. Separation of closely related asymmetric Penicillia by pyrolysis gas chromatography and mycotoxin production. *Mycologia* 76: 408-419.

SOPP, O.J.-O. 1912. Monographie der Pilzgruppe *Penicillium* mit besonder erucksichtigung der in Norwegen gefungdenen Arten. *Skrifter udgivna af Videnskabsselskabet i Christiana* 11: 1-208.

STOLK, A.C and SCOTT, D.B. 1967. Studies on the genus *Eupenicillium* Ludwig 1. Taxonomy and nomenclature of Penicillia in relation to their sclerotioid ascocarpic states. *Persoonia* 4: 391-405.

STOLK, A.C., SAMSON, R.A., FRSIVAD, J.C. and FILTENBORG, O. 1990. The systematics of the terverticillate Penicillia. *In* Modern Concepts in *Penicillium* and *Aspergillus* Classification, eds. R.A. Samson and J.I. Pitt, pp. 121-137. New York and London: Plenum Press.

THOM, C. 1906. Fungi in cheese ripening: Camembert and Roquefort. *Bulletin of the Bureau of Animal Industry, U.S. Department of Agriculture* 82: 1-39.

—— 1910. Cultural studies of species of *Penicillium*. *Bulletin of the Bureau of Animal Industry, U.S. Department of Agriculture* 118: 1-109.

—— 1930. The Penicillia. Baltimore, Maryland: Williams and Wilkins.

—— 1954. The evolution of species concepts in *Aspergillus* and *Penicillium*. *Annals of the New York Academy of Sciences* 60: 24-34.

THOM, C. and CHURCH, M.B. 1926. The Aspergilli. Baltimore, Maryland: Williams and Wilkins.

THOM, C. and RAPER, K.B. 1945. A Manual of the Aspergilli. Baltimore, Maryland: Williams and Wilkins.

—— 1946. *Aspergillus* or what? *Science, N.Y.* 103: 735.

VOSS, E.G. *et al.*, eds. 1983. International Code of Botanical Nomenclature adopted by the Thirteenth International Botanical Congress, Sydney, August, 1981. Utrecht: Bohn, Scheltema and Holkema.

WESTLING, R. 1911. Über die grünen Spezies der Gattung *Penicillium*. *Arkiv für Botanik* 11: 1-156.

DIALOGUE FOLLOWING DR. PITT'S PRESENTATION

TAYLOR: It's interesting to see new methods for identification being developed for these genera. Alternative methods of identification may involve genetic methods or secondary metabolites. Right now, we're seeing some of those methods used in medical mycology. Later on, we may see some of them being applied to food.

SAMSON: How about identification of *Penicillium* and *Aspergillus* by serological techniques?

PITT: I don't mind what methods we use. Food and pharmaceutical people are desperate for faster methods. Presently, there are no simple methods that anyone can use. I intend to concentrate on developing some new ideas in this area over the next few years.

2

TECHNIQUES AND PRACTICAL ASPECTS FOR IDENTIFICATION OF PENICILLIUM AND ASPERGILLUS

STANDARDIZATION IN *PENICILLIUM* IDENTIFICATION

O. Constantinescu

The Herbarium
Uppsala University
751 21 Uppsala, Sweden

SUMMARY

The cultural and morphological characters of 36 species of *Penicillium* were examined after growth at 25°C on Czapek yeast extract agar (CYA), malt extract agar (MEA), and 25% glycerol nitrate agar (G25N), in glass and in unsealed and sealed plastic Petri dishes. As a result, the following suggestions for the standardization of *Penicillium* identification are made: a conidial suspension in sterile water supplemented with a wetting agent to be used as inoculum; addition of trace elements solution to the CYA; the possible use of unsealed plastic Petri dishes provided with ribs instead of glass ones for the cultivation on CYA and MEA, and the same but partially sealed for cultivation on G25N; the inoculation in one point instead of three as a better alternative for the fast growing species.

INTRODUCTION

The identification of Penicillia is still to a large extent based on the examination of macro- and micromorphological characters. Moreover, the results of the new approaches in *Penicillium* identification, such as the use of secondary metabolites and biochemical tests, are always checked against the traditional identification procedures. The interest in developing standardized media and growth conditions in order to guarantee reproducibility in *Penicillium* identification goes back to 1893 when Biourge made the first attempt to monograph the genus (Hennebert, 1985) and was continued until recently (Pitt, 1985; Samson and Pitt, 1985 b). However, in spite of almost one hundred years of efforts, it seems that a consensus is far from attained. It is symptomatic that no complete agreement has been reached even on the formulation of malt extract agar, one of the two most used media for *Penicillium* identification. This is rather unfortunate, particularly for the non-specialist who is confronted with controversial procedures and approaches in his attempt to name *Penicillium* isolates.

During routine identification of Penicillia with the aid of recent monographic and other studies (Pitt, 1979, 1985; Williams and Pitt, 1985), certain difficulties were encountered. Atypical micromorphology was present in most species when grown on Czapek yeast extract agar (CYA). The evaluation of the growth was difficult in fast growing species because interference between the colonies. When, for time-saving reasons, disposable, without ribs, plastic Petri dishes were used instead of glass ones, the diameter of the colonies on CYA and malt extract agar (MEA) was reduced by 25 to 40 mm by occasional tightening of the dishes during incubation. Similarly, the colony diameters on glycerol-nitrate agar (G25N) were consistently smaller than expected in almost all species. Wrapping the dishes in polyethylene film did not reduce variation in colony diameters, but increased the variation on both CYA and MEA.

In an attempt to critically evaluate and standardize the procedures for inoculum preparation, inoculation and growth conditions, a series of experiments were initiated.

Modern Concepts in Penicillium and Aspergillus Classification
Edited by R. A. Samson and J. I. Pitt
Plenum Press, New York, 1990

MATERIAL AND METHODS

Fungi.

Thirty six species of *Penicillium* preserved at UPSC were tested (Table 1).

Table 1. Variation in colony diameters (in mm) of cultures grown in glass Petri dishes, and occurrence (+) of some cultural and microscopical features in 36 species of *Penicillium*

number of species	strain	colony diameter CYA	MEA	G25N	A	B	C	D
P. aurantiogriseum	1297	31-38	32-39	14-18	+			+
P. brevicompactum	1267	21-25	14-19	15-16	+	+		
P. camemberti	1718	31-34	45-49	20-27				+
P. canescens	1536	32-33	1-35	19-23		+		+
P. chrysogenum	1503	29-37	*48-56	14-20		+		+
P. citreonigrum	2161	23-24	18-22	10-13	+		+	
P. citrinum	1831	22-25	15-21	10-14	+	+		
P. corylophilum	2495	27-32	*42-45	13-17			+	+
P. crustosum	1590	*41-50	*33-50	18-26		+		+
P. decumbens	1733	20-23	25-28	12-17			+	
P. digitatum	980	*55-56	*62-70	4-5				+
P. echinulatum	1005	33-38	31-38	20-22	+	+		+
P. expansum	1296	34-40	*43-50	19-21				+
P. glabrum	2736	*45-48	*50-58	17-19			+	
P. hirsutum	2697	26-36	24-31	19-25				
P. islandicum	2444	16-19	17-19	4-6	+			
P. italicum	1577	*40-48	*56-61	14-18			+	+
P. janczewskii	1828	27-33	28-35	12-16			+	
P. janthinellum	1036	32-43	40-50	13-15				+
P. jensenii	1354	27-32	13-18	13-17		+		+
P. lividum	2488	26-32	32-42	10-16			+	+
P. miczynskii	1974	20-27	20-26	9-16	+			
P. minioluteum	2178	34-39	*48-48	2-4			+	+
P. olsonii	2019	26-37	22-28	19-21			+	
P. puberulum	2737	22-25	23-24	13-14	+			
P. purpurogenum	2715	14-15	34-38	0	+	+		
P. restrictum	2040	27-29	15-25	14-18				
P. roqueforti	972	*59-65	*61-70	19-22	+			+
P. rugulosum	2492	7-10	13-15	3-5				
P. sclerotiorum	1597	35-41	*43-50	13-19			+	+
P. spinulosum	1695	*42-58	*55-57	15-23			+	+
P. variabile	1735	7-10	16-20	4-7				
P. verrucosum	1624	18-28	15-17	15-17			+	
P. viridicatum	2700	24-34	*33-40	16-20			+	
P. vulpinum	830	26-34	24-28	9-14				+

* Values affected by interference of the colonies

A = Abnormal microscopical structures on CYA; B = Reduction of exudate production in plastic Petri dishes; C = Additional colonies in plastic Petri dishes; D = Interface effect

Media.

MEA and CYA were prepared according to Samson and Pitt (1985b), and G25N according to Pitt (1979). In addition, CYA supplemented with the trace elements solution according to Smith (1949) was also used. Yeast extract, malt extract, peptone and agar (all Difco), sucrose (BDH Analar), and other components (all Merck) were used.

Culture vessels.

Glass Petri dishes 90 x 14 mm, and plastic dishes 87 x 15 mm provided with ribs, were used. Eighteen ml of medium was used for each Petri dish. Inoculum was prepared by

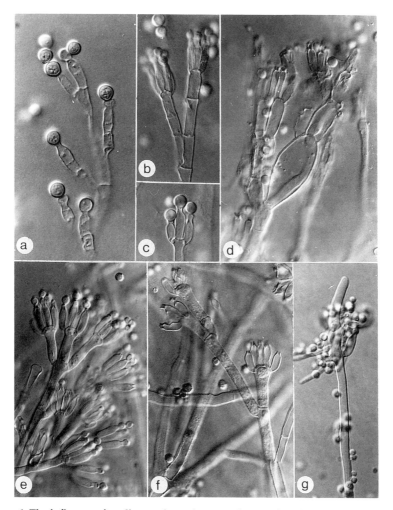

Figure 1. The influence of medium and growing procedure on the micromorphology of some Penicillia. a-c. Abnormal structures in *P. roqueforti*; a. from the interface zone on CYA; b-c. from the interface zone on MEA. d; *P. puberulum*, swollen branch on CYA; e-f. *P. canescens*; e. regular and smooth-walled penicilli on MEA; f. irregular and rugose penicilli on CYA; g. *P. miczynskii*. Abnormal, reduced penicilli on CYA.

mixing conidia with sterile distilled water containing 0.1% Tween 20 (KEBO), in a deep-well slide. A lid of a plastic dish, in which three holes 1.5 mm diam and 38 mm from each other had been drilled, was used to mark the places of inoculation on the back of the Petri dish. The plates were inoculated upside-down by dipping the inoculation needle into the conidial suspension and piercing the medium at these previously marked points. The plates were incubated upside-down, at 25°C for 7 days, in darkness, in an incubator provided with vertical ventilation. For some fast growing species (i.e. with colonies of 30 mm or more after 7 days) a series of cultures with single colonies were also made. One of the two series of cultures grown in plastic Petri dishes was kept during incubation in partially sealed polyethylene bags. Two diameters of each colony, one along the radius and one perpendicular to it, were measured by using a transparent sheet of plastic on which both these diameters and the inoculation points were traced. Two duplicates were made for each combination of medium, isolate and type of culture vessel. The experiments were carried out in duplicate.

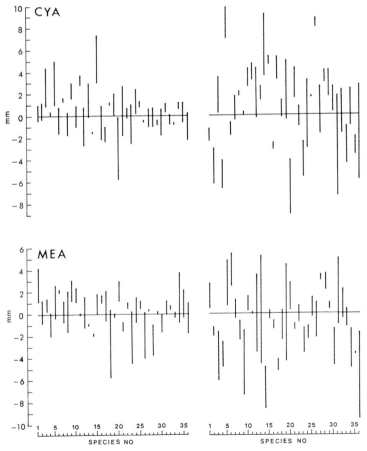

Figure 2. Variation in colony diameters of 36 species of *Penicillium* grown on CYA and MEA in unsealed (left) and partially sealed (right) plastic Petri dishes, as compared to cultures grown in glass dishes. Bars indicate the variation in the averages of two series of measurements.

RESULTS AND DISCUSSION

Preparation of inoculum and inoculation.

The addition of 0.1% Tween 20 to the sterile water for inoculum preparation prevented the formation of additional colonies from stray conidia, a frequent phenomenon when dry conidia or water alone (Raper and Thom, 1949) is used. This procedure, as well as the one described by Frisvad (1981), is simpler than the one employing molten agar as dilution agent and screw cap vials (Raper and Thom, 1949; Pitt, 1979). Moreover, it proved very suitable for rapid inoculation of a large number of Petri plates. Inoculating the plates by piercing the medium assures that a rather large number of conidia are placed on a minimum of surface.

Micromorphology on CYA.

Eleven out of 36 species tested showed frequent atypical microscopic structures when grown on CYA (Table 1). These include reduction in number of the metulae and/or phialides, swollen branches or metulae, presence of rudimentary, non developing metulae etc. (Fig. 1 d, f-g). By comparison, the micromorphology on MEA was far more regular (Fig. 1 e). This observation supports the claim that better development of Penicillia occurs on MEA than on CYA (Samson and Pitt, 1985a, p. 100). By adding trace elements solution of Smith (1949) to CYA, as recommended by Frisvad and Filtenborg (1983) and Pitt (1988), the presence of these abnormal structures was almost eliminated.

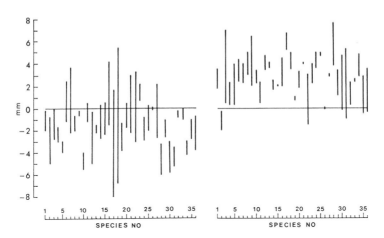

Figure 3. Variation in colony diameters of 36 species of *Penicillium* grown on G25N in unsealed (left) and partially sealed (right) plastic Petri dishes, as compared to cultures grown in glass dishes. Bars indicate the variation of the averages of two series of measurements.

Glass versus plastic Petri dishes.

No significant differences were observed in the colony shape, margin, colour and production of diffusible pigments when the fungi were grown in glass or plastic dishes. However, the production of exudate was diminished in 7 species when plastic dishes were used (Table 1). Although inoculated at three points, frequent additional colonies were formed in 14 species when grown in plastic dishes (Table 1). In some cases the whole surface of the plate was covered with colonies which made the measurement of an individual colony impossible. The presence of additional colonies is extremely rare when glass dishes are used. It seems that species belonging to subgenus *Aspergilloides* particularly have the tendency to form such additional colonie, as five out of the seven tested showed this problem. The remaining two, *P. sclerotiorum* and *P. restrictum*, showed late, and rather poor sporulation, which may explain why no additional colonies were formed. The static electricity is, most probably, responsible for this phenomenon. However, many heavily sporulating species do not form additional colonies. This different behaviour is certainly connected with characters of conidiogenesis, conidium size, weight and surface morphology.

The variation in colony diameters of cultures grown in unsealed or partially sealed plastic Petri dishes, compared with glass dishes are summarized in figures 2-4. Low variation (ca. 1 mm or less) on CYA occurred in 13 species when grown in unsealed dishes and in seven species when the dishes were sealed. The corresponding values for MEA are 16 and 10, and for G25N 6 and 8. Higher variation (4 mm or more) between the two series of measurements was present on CYA in four species when unsealed and in eight species when sealed. Corresponding values for MEA are 4 and 8, and for G25N 9 and 5. Only a few species showed higher than 8 mm variation between two measurements: two on each sealed CYA and MEA, and two on unsealed G25N. These figures indicate that sealing the dishes results in an increase of variation between two series of measurements on CYA and MEA, but not on G25N. In most of the species the cultivation in plastic dishes gives higher or lower values of the colony diam compared to the control (Fig. 4). Sealing the dishes induced an increase of these differences on both CYA and MEA. On G25N most of the species responded by a reduction of the diameter when grown in unsealed dishes, but an increase occurred in all but two species when the dishes were sealed. This is certainly caused by the rapid drying of this low water activity medium when plastic dishes provided with ribs are used.

The use of glass Petri dishes is recommended for *Penicillium* identification. According to Pitt (1979, p. 19) plastic dishes may be used if they are wrapped in polyethylene film. Our results show that unsealed plastic dishes provided with ribs, in which the aeration is relatively constant, provide more uniform results on CYA and MEA. A partial sealing is beneficial in the case of G25N medium.

Three inoculum points versus one.

When the colony diameter is ca. 30 mm, a mutual influence between adjacent colonies becomes noticeable. This effect mostly consists of the inhibition of growth and sporulation, presence of abnormal micromorphological structures (Table 1, Fig. 1 a-c), and alteration of the colony texture. The inhibitory effect on the growth was measured in eight fast or relatively fast growing species: *P. camemberti*, *P. chrysogenum*, *P. crustosum*, *P. digitatum*, *P. expansum*, *P. italicum*, *P. roqueforti* and *P. spinulosum*. These were grown in glass Petri dishes both as one or three colonies. Single colonies had larger diameters, in average by 9.5 mm in *P. roqueforti*, 5.8 mm in *P. italicum*, 4.5 mm in *P. spinulosum* and 1.7 in *P. crustosum*. The remaining species showed no significant differences in this respect. In *P. jensenii* a reduction in exudate production was noticed at the interference zone. The interface effect is clearly evident in *P. vulpinum* and *P. expansum* in which no synnemata are formed at the interference zone. Traditionally, the Penicillia are inoculated in three points, although at least one of the monographers, Abe (1956), used one colony in his study. According to Raper and Thom (1949: 71) "one can study the mature growth and fruiting habits of the mold best in the area where the colonies approach one another". However, our experience does not support this view. The only advantage of growing three colonies in each plate may be that three instead of one colony are examined and occasional variation may be noticed. The practice in cultivating Penicillia shows, however, that significant variation between colonies obtained in the same plate is extremely rare.

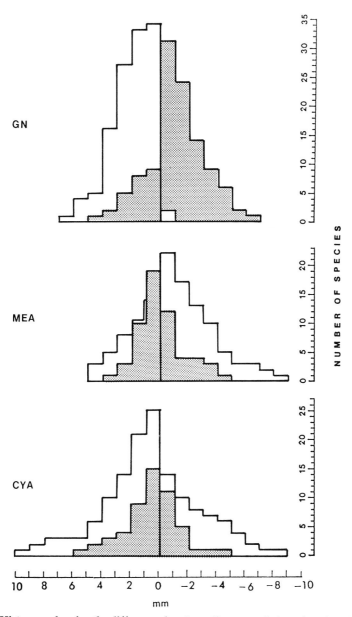

Figure 4. Histogram showing the differences in colony diameters of 36 species of *Penicillium* grown on CYA, MEA and G25N in unsealed (shaded area) and partially sealed (unfilled area) plastic Petri dishes as compared to cultures grown in glass dishes.

CONCLUSIONS

For a better standardization of the procedures for *Penicillium* identification the following suggestions are made: Preparation of inoculum in sterile distilled water supplemented with a wetting agent; addition of trace elements solution to the CYA; inoculation in one point, at least for fast growing species; if plastic Petri dishes are used, those provided with ribs should be preferred; cultures on G25N medium should be partially sealed if plastic dishes are used.

ACKNOWLEDGEMENTS

Thanks are due to Mrs Ulla-Britt Sahlstrøm for printing the photographs. This work was supported by grants from the Swedish Natural Science Research Council and Swedish Council for Forestry and Agricultural Research.

REFERENCES

ABE, S. 1956. Studies on the classification of the Penicillia. *Journal of General and Applied Microbiology* 2: 1-344.

FRISVAD, J.C 1981. Physiological criteria and mycotoxin production as aids in identification of common asymmetric Penicilia. *Applied and Environmental Microbiology* 41: 568-579.

FRISVAD, J.C. and FILTENBORG, O. 1983. Classification of terverticillate Penicillia based on profiles of mycotoxins and other secondary metabolites. *Applied and Environmental Microbiology* 46: 1301-1310.

HENNEBERT, G.L. 1985. Dierckx' contribution to the genus *Penicillium*. *In* Advances in *Penicillium* and Aspergillus Systematics, eds. R.A. Samson and J.I. Pitt, pp. 9-21. New York and London: Plenum Press.

PITT, J.I. 1979. The Genus *Penicillium* and its Teleomorphic States *Eupenicillium* and *Talaromyces*. London: Academic Press.

—— 1985. A Laboratory Guide to Common *Penicillium* Species. North Ryde, N.S.W.: CSIRO, Division of Food Research.

—— 1985. Media and incubation conditions for *Penicillium* and Aspergillus taxonomy. *In* Advances in *Penicillium* and Aspergillus Systematics, eds. R.A. Samson and J.I. Pitt, pp. 93-103. New York and London: Plenum Press.

—— 1988. A Laboratory Guide to Common Penicillium Species. Second edition. North Ryde, N.S.W.: CSIRO, Division of Food Processing.

RAPER, K.B. and THOM, C. 1949. A Manual of the Penicillia. Baltimore: Williams and Wilkins.

SAMSON, R.A. and PITT, J.I., eds. 1985a. Advances in *Penicillium* and Aspergillus Systematics. New York and London: Plenum Press.

—— 1985. General recommendations. *In* Advances in *Penicillium* and Aspergillus Systematics, eds. R.A. Samson and J. I. Pitt, pp. 455-460. New York and London: Plenum Press.

SMITH, G. 1949. The effect of adding trace elements to Czapek-Dox medium. *Transactions of the British Mycological Society* 32: 280-283.

WILLIAMS, A.P. and PITT, J.I. 1985. A revised key to *Penicillium* subgenus *Penicillium*. *In* Advances in *Penicillium* and *Aspergillus* Systematics, eds. R. A. Samson and J. I. Pitt, pp. 129-134. New York and London: Plenum Press.

DIALOGUE FOLLOWING DR. CONSTANTINESCU'S PAPER

PITT: I'd like to make some comments rather than ask questions about what you've done. Much of it is already known. I think any mycologist doing morphological work now uses Petri dishes with ribs. Fungi are sensitive to carbon dioxide, so ribbed Petri dishes are essential. Ones without ribs are not suitable for mycological work.

As for the use of CYA as the standard medium for morphological examination, this is based on what Dr. Raper did. I followed his lead in using Czapek as the primary medium. He used malt very little, and his descriptions on malt are often quite incomplete. I followed him with reasonable justification. If a consensus of people now feel that malt extract agar is preferable for morphological examination, then go ahead. It is impractical for me to change my system now. I still prefer CYA. You do get more morphological variation on CYA so you must examine more than one penicillus and find typical forms.

As for using three colonies rather than one, this gives you a replication of your colony sizes. If a culture is old or moribund, you will find an unacceptable variation in the size of colonies. This gives an indication of problems with the culture. If there is too much variation, the culture should be reinoculated. Certainly faster growth is possible from single colonies, but the three colony system is a standard system and is what is used in the literature. The fact that you can get faster growth with a single colony is irrelevant.

The differences in morphology between margin and the centre of the colonies is also unimportant. Normally, one works near the margins of the colony where the growth is youngest. On the other hand, if you have weakly sporulating colonies, as happens with some *Eupenicillium* species, spores may form near the colony centres. The system is designed to obtain results in one run. Industry wants results now.

One other point. In your presentation and your table 1 on variations of colony diameter. You have measured colonies both on the radial and tagential diameters and averaged them. The data in Pitt (1979), however, is based only on the largest diameter i.e. the tangential diameter, as it is evident that colonies must interfere with each other in the radial direction when they become large. The methodology you used seems therefore to be different.

CONSTANTINESCU: I was just reporting my observations, not intending to make a criticism of your work. In my paper, I state that the only advantage to having three colonies on one plate instead of only one, is that one can observe the morphological variaton between the three colonies. Significant variation in three colonies is extremely rare provided that the colonies are the same distance apart and have the same amount of inoculum. I would not recommend changing the system.

SAMSON: At CBS, we still use Czapek and 2% malt extract agar. We find that the 2% malt extract, without any additions, is excellent for morphology of *Penicillium* and the work that we will present later in this workshop is based on this medium. We particularly like 2% malt extract for micromorphology of *Penicillium* and also of *Aspergillus*.

CONSTANTINESCU: At Uppsala, we want to follow a single system, so we are following Pitt.

SAMSON: We use CYA only for growth diameters, not micromorphology.

GAMS: We are still in the phase of learning about taxonomic relations from molecular biology, genetics, biochemistry and so on. If we get clues from these disciplines that help us distinguish between entities, we will have to make all efforts to correlate these distinctions with morphology. In these cases, we may have to use not only malt extract agar, but also oatmeal agar, SNA and other weak media to get optimal expression of morphology. Only under such circumstances, will we be able to find the crucial distinguishing features.

IDENTIFICATION OF *PENICILLIUM* AND *ASPERGILLUS* SPECIES IN MIXED CULTURES IN PETRI DISHES USING SECONDARY METABOLITE PROFILE

O. Filtenborg and J.C. Frisvad

Department of Biotechnology
The Technical University of Denmark
2800 Lyngby, Denmark

SUMMARY

Identification of mould species is a time consuming procedure, which usually takes at least 14 days, including isolation and subculturing. We have investigated the possibility of performing identification directly on the isolation media, using TLC screening of secondary metabolite profiles. An identification procedure is proposed using standard isolation media, based on comparison of secondary metabolite profiles from a selected number of isolates. The procedure has been tested on a number of food samples, and a high degree of agreement was found with results from normal identification procedures with pure cultures. An international testing of this identification procedure is proposed and we are willing to provide test samples and photos of standard secondary metabolite profiles.

INTRODUCTION

Over the years a lot of effort has been devoted to the development of general purpose isolation and diagnostic media for mould species important in foods and feeds. Many isolation media have been devised, of which Dichloran 18% Glycerol nitrate agar (DG18: Hocking and Pitt, 1980) and Dichloran rose bengal chloramphenicol agar (DRBC: King *et al.*, 1979) are good examples. These media support the growth of all important species of *Penicillium*, *Aspergillus* and several other genera. Pure cultures for identification can be readily isolated from them (King *et al.*, 1986). However, this procedure is time consuming, requiring at least five to seven days for initial growth on the selective medium and seven days for identification on diagnostic media. An alternative approach is the development of selective media,. However, only two successfull media exist so far: AFPA for the screening of *Aspergillus flavus* and *A. parasiticus* (Pitt *et al.*, 1983) and PRYES for the screening of *P. verrucosum* chemotype II (Frisvad, 1983).

As another alternative to the procedure just described, it may be easier to identify the moulds directly on the isolation plates. One way may be the determination of secondary metabolite profiles of the fungi directly on these media, because these profiles are specific for particular fungal species (Frisvad and Filtenborg, 1983; Frisvad, 1989). In this paper we report on investigations on the ability of a number of food spoilage and myotoxinogenic moulds to produce secondary metabolites on the isolation media DRBC and DG18 and modifications of them. The aim is to determine if such fungi can be identified directly in mixed culture on isolation media, by using simple TLC screening procedures to detect their production of secondary metabolites.

Modern Concepts in Penicillium and Aspergillus Classification
Edited by R. A. Samson and J. I. Pitt
Plenum Press, New York, 1990

MATERIAL AND METHODS

Fungi. The investigation of metabolite production in pure culture was performed on isolates from commercial, not visibly mouldy, food products. They were: *Penicillium roqueforti* from artichoke, *P. expansum* from apple, *P. brevicompactum* from potatoes, *A. alutaceus* from chili, *A. versicolor* and *A. flavus* from pepper, *P. verrucosum* II, *P. chrysogenum*, *P. hordei* and *P. aurantiogriseum* from barley. Mixed cultures originating from commercial food products were tested as well.

Table 1. Influence of media on secondary metabolite production of selected *Penicillium* and *Aspergillus* species after incubation for seven days.

species	metabolite	medium			
		CYA	YES	DRBC	DRYES
P. roqueforti	PR-toxin	–	ec(a)	–	–
	Roquefortine C	ic	ic	ic	–
P. expansum	Patulin	ec>ic	ec>ic	ec>ic	ec>ic
	Citrinin	ec>ic	ec>ic	ec>ic	ec>ic
A. alutaceus	Xanthomegnin	ic	ic	ic	ic
	Viomellein	ic	ic	ic	ic
	Penicillic acid	ec>ic	ec>ic	ec>ic	ec>ic
P. aurantiogriseum	Xanthomegnin	ic	–	–	–
	Viomellein	ic	–	–	–
P. chrysogenum	Roquefortine C	ic	–	ic	–
P. hordei	Roquefortine C	ic	–	–	–
P. verrucosum	Citrinin	ec>ic	ec>ic	ec>ic	ec>ic
	Ochratoxin A	ec>ic	ec>ic	–	ec>ic
P. brevicompactum	Brevianamid A	ic	ec>ic	ec>ic	ec>ic
A. versicolor	Sterigmatocystin	ic	ic	ic	ic
A. flavus	Aflatoxin B1	ec>ic	ec>ic	ec>ic	ec>ic

(a) ec = extracellular detection, ic = intracellular detection; - = no detection, ec>ic = extracellular detection better than intracellular detection.

Media.

The media used were Dichloran rose bengal yeast extract sucrose agar (DRYES; Frisvad, 1983), DG18 and DRBC. DRYES was used instead of the similar medium PRYES, although it is less restrictive towards *Rhizopus*, since the pentachloronitrobenzene used in PRYES is considered a possible cancinogen (King *et al.*, 1979)

Inoculation.

Pure cultures were inoculated in triplicate, either as three point inoculations or spread plates. Samples of food products were homogenized, diluted and inoculated on the media by spread plates in triplicate. The TLC agar plug method (Filtenborg and Frisvad, 1980; Filtenborg *et al.*, 1983) was used for screening for metabolite production. Plug samples were taken from all three colonies. Variations in metabolite concentrations between triplicates were never significant. TLC conditions were adjusted to be optimal for individual metabolites during the pure culture investigations. When mixed cultures were used the following protocol was chosen: elution in chloroform/acetone/isopropanol

85:15:20 (CAP) followed by Ce(SO4)2-spray for DRBC and elution in toluene/ethylacetate/90% formic acid 4:4:1 (TEF) followed by anisaldehyde-spray for both DRBC and DRYES. For details see Filtenborg and Frisvad (1980) and Filtenborg *et al.* (1983).

RESULTS AND DISCUSSION

Production of secondary metabolites on isolation media by pure cultures.

The ability of the selected *Penicillium* species to produce secondary metabolites in pure culture was compared on isolation media, DRYES, DG18 and DRBC and media known to support production of secondary metabolites: Czapek yeastextract agar (CYA) and yeast extract sucrose agar (YES)(Frisvad and Filtenborg, 1983) (Fig.1).

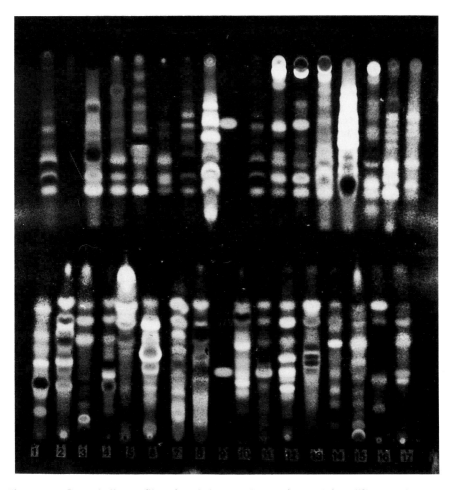

Figure 1. TLC-metabolite profiles of pooled pure cultures of several *Penicillium* species on CYA and YES, extracted with chloroform/methanol. The metabolite-extracts have been concentrated by evaporation. TLC-condition: Eluation in TEF, spraying with anisaldehyde and 365 nm UV-illumination.

Most of the metabolites investigated were produced on DRYES and/or DRBC within seven days (Table 1). However, none of the metabolites investigated were detected after growth on DG18 for seven days. This is probably due to the lower water activity level of this medium (0.95), compared to 0.995 or more for the other media. Reduced water activity is well known to inhibit metabolite production dramatically, so an extended incubation time is needed to detect secondary metabolites on DG18.

On DRYES and DRBC generally the metabolite concentrations were equal to or slightly lower than those on YES and CYA but a detection was still possible with the rather insensitive TLC screening method. Detection is significantly improved, if incubation time is extended to nine days. Extracellular metabolites were normally produced in greater amounts in DRYES than in DRBC while the reverse was true for intracellular metabolites.

Table 2. Production of secondary metabolites by *Penicillium* and *Aspergillus* colonies in mixed culture on DRYES and DRBC.

food	medium	identity metabolites[1]	secondary metabolites	identity morphology[2]
Walnut	DRYES	*P. crustosum* [2][3]	Terrestric acid, etc.	*P. crustosum*
		P. expansum [2]	Citrinin, etc.	*P. expansum*
		P. aurantio-griseum	Penicillic acid, etc.	*P. aurantio-griseum*
		P. echinulatum chemotype II	several	*P. echinulatum* chemotype II
		P. brevicompactum [2]	several	*P. brevicompactum*
		? [2]	weak profile	*P. solitum*
	DRBC	*P. aurantiogriseum*	Verrucosidin, etc.	*P. aurantio-griseum*
		P. crustosum [2]	Penitrem A, etc.	*P. crustosum*
Pepper (white)	DRYES	*P. citrinum* [4]	Citrinin, etc.	*P. citrinum*
		A. versicolor [2]	Sterigmatocystin, etc.	*A. versicolor*
		A. candidus [2]	several	*A. candidus*
Grapes	DRYES	*P. brevicompactum* [4]	Brevianamid A, etc.	*P. brevicompactum*
		P. glabrum [4]	several	*P. glabrum*
Barley	DRYES	*P. verrucosum* [5]	Citrinin, Ochratoxin A, etc.	*P. verrucosum*
		P. aurantio-griseum [3]	few	*P. aurantio-griseum*
		P. hordei	several	*P. hordei*
		?	weak profile	*P. solitum*
	DRBC	*P. aurantio-griseum* [2]	Xanthomegnin, viomellein, etc.	*P. aurantio-griseum*

[1]Frisvad and Filtenborg (1983); [2]Pitt, (1979), Raper and Fennell (1965), Frisvad (1981); [3]Number of isolates tested.

Besides the metabolites listed in Table 1, several other members of the expected metabolite-profile of each species were detected (Frisvad and Filtenborg, 1983), but this will not be discussed in detail here. Generally the profiles were almost complete on DRYES and/or DRBC, but at concentrations slightly lower than in YES and/or CYA.

Some metabolites appeared on DG18, especially in the mycelium. In *Aspergillus* species, DG18 sometimes supported the highest production of metabolites of all the media tested.

Some metabolites are mainly excreted in the medium, but usually can be detected into the mycelium too, although in lower concentrations. This is important in this context, as the screening method is intended to be used on mixed cultures, where extracellular metabolites from different isolates may mix in the medium, but probably not in the mycelium.

The results listed in Table 1 are based on inoculations with a great number of conidia. This is unlike the normal spread plate technique where conidia are present on countable plates in low numbers, 10-50 per plate. To test if the results with mass inocula could be repeated using this inoculation technique, the metabolite screening was repeated with three isolates of *P. aurantiogriseum*, *P. verrucosum* and *P. chrysogenum*. Generally the secondary metabolite profiles obtained were the same, but a few differences were seen. Xanthomegnin and viomellein was now detected in DRBC, while penicillic acid and roquefortine C were no longer observed in this medium.

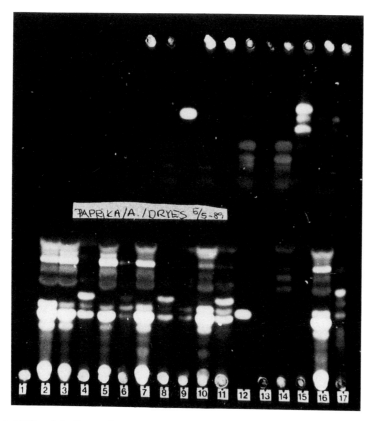

Figure 2. TLC-metabolite profiles of isolates from a paprika sample in a mixed culture on DRYES. TLC-conditions: Eluation in TEF, no spraying and 365 nm UV-illumination. Single metabolites were not identified, but profiles characteristic of *Eurotium* species (profile 2 and 4 representing different species) are dominating, and also *Aspergillus niger* (profile 17) is present.

Production of secondary metabolites in isolation media: mixed cultures.

Metabolite production may be influenced by a competition of different species. Reduced production of secondary metabolites could be the result. Hence, the screening method with mixed cultures may be expected to produce different results from those obtained in pure cultures. Therefore, the metabolite screening method was tested on the fungal flora of several food products, inoculated on DRYES and DRBC (Figs 2-6). The metabolite profiles of the mixed flora on DRYES and on DRBC were used to identify the fungi present according to Frisvad and Filteborg (1983) and by comparation with photos of metabolite profiles of well known isolates, based on culture extracts (Frisvad and Thrane, 1987). Verification of the identification was carried out on pure cultures according to Pitt (1979), Raper and Fennell (1965), and Frisvad (1981). Some of the results are listed in Table 2.

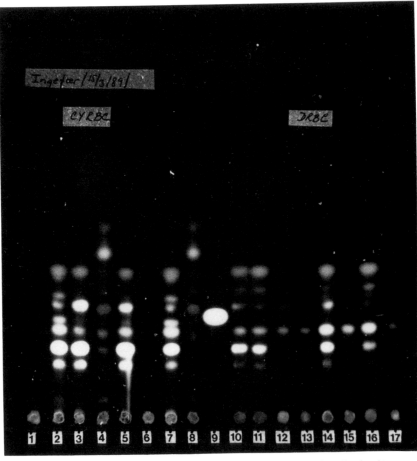

Figure 3. TLC-metabolite profiles of isolates from a ginger sample in mixed cultures on DRBC and CYRBC (DRBC with added 0.5% yeast extract). TLC-conditions: Eluation in TEF, no spraying and 254 nm UV-illumination. Metabolites are brevianamides and raistrick phenoles. A metabolite profiles characteristic of *Penicillium brevicompactum*.

Table 2 shows that even with a mixed flora on DRYES and DRBC, several secondary metabolites were produced detectable with the simple TLC screening method used in this investigation. Sometimes profiles were weak, for example with *P. solitum*, making identification impossible unless several colonies of the same species were present which had more distinct but comparable profiles. This was often the case. Some metabolites were not detected for example roquefortine C and patulin, due to unfavourable production or detection conditions or simply because the fungus did not have the ability to produce them. Nevertheless, it was often possible to correctly identity the important species in a sample.

The most significant metabolite profiles were often produced on DRYES (Fig. 2 and 5). When the intracellular metabolite profiles were tested, no interfering metabolites from adjacent colonies were observed. However when testing extracellular profiles we observed a certain degree of metabolite diffusion. However, valuable information was often obtained by comparing intra- and extracellular metabolites as well as profiles of the same sample from both DRYES and DRBC.

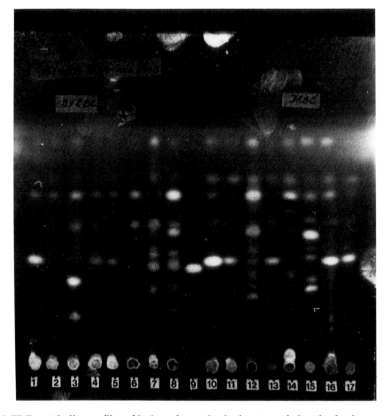

Figure 4. TLC-metabolite profiles of isolates from a buckwheat sample in mixed cultures on DRBC and CYRBC (DRBC with added 0.5% yeast extract). TLC-conditions: Eluation in TEF, spraying with anisaldehyde and 254 nm UV-illumination. A metabolite in profile 1 and several others is penicillic acid. A metabolite in profile 8 and 12 is verrucosidin. Profiles characteristic of *P. aurantiogriseum*, *P. roqueforti* and *P. brevicompactum* are present.

CONCLUSIONS

From these investigations it can be concluded that the production of several *Penicillium* and *Aspergillus* secondary metabolites can be detected in mixed cultures on the mould isolation media DRYES and DRBC, using a simple TLC screening method (Filtenborg and Frisvad, 1980; Filtenborg *et al.*, 1983). Not all of the metabolites in the profile of the individual species could be detected in the same analysis, as TLC conditions, media and competitive fungal flora plays an important role in this respect. Roquefortine C, xanthomegnin and viomellein, for example, were rarely detected under these conditions.

However, as only a few metabolites of the total profile, even two or three, can be sufficient, to identify the individual colony (Frisvad and Filtenborg, 1989), we believe that by screening for their secondary metabolite profiles on the media DRYES and /or DRBC will enable identification of several important moulds directly in the mixed flora. The most significant metabolite profiles were often observed on DRYES. The choice of medium and TLC conditions, however, must depend on the species and metabolites which are of interest in the particular situation.

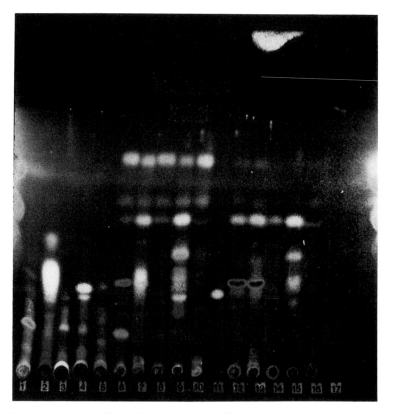

Figure 5. TLC-metabolite profiles of isolates from a wallnut sample in a mixed culture on DRYES. TLC-conditions: Eluation in TEF, spraying with anisaldehyde and 365 nm UV-illumination. The tale in profile 2 and 7 is citrinin. A metabolite in profile 5 and 10 is terrestric acid. Several other metabolites can be identified. Profiles characteristic of *P. expansum, P. aurantiogriseum, P. solitum* and *P. crustosum* are present.

This method was tested on several food products and the majority of the *Penicillium* and *Aspergillus* species could be identified. For example, the important species *A. versicolor*, which grows to a colony diameter of only a few millimeters on these media, produced a spectacular and very specific metabolite profile, including sterigmatocystin. Using this method we also often observed that colonies, with very different appearances especially on DRBC, had the same metabolite profile and thus represented the same species. *P. brevicompactum* and *P. aurantiogriseum* in particular showed this effect. This is time saving information when trying to separate and identify such species.

The application of this method of course depends on available metabolite standards and data on metabolite profiles. These standards and methods exists to a certain degree (Frisvad and Filtenborg, 1983, 1990) and the number can be expected to increase considerably in the future. We are also using extracts (Frisvad and Thrane, 1987) of well known isolates as standard profiles, and can recommend this as an interim solution until a sufficient number of standard metabolites are available.

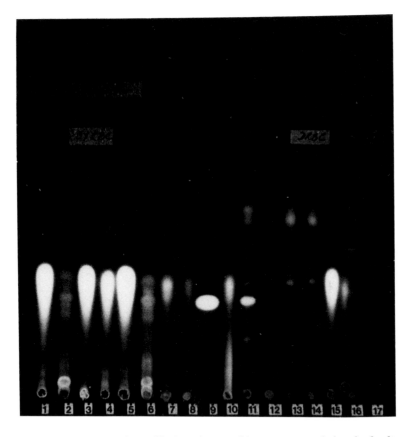

Figure 6. TLC-metabolite profiles of isolates from a white pepper sample in mixed cultures on DRBC and CYRBC (DRBC with added 0.5% yeast extract). TLC-conditions: Eluation in TEF, no spraying and 365 nm UV-illumination. The tale in profile 1 and several others is citrinin. One of the metabolites in profile 11 and others is sterigmatocystin. Profiles characteristic of *P. citrinum* and *A. versicolor* are present.

The described identification procedure needs further testing to prove its value. This may lead to the need for modification of isolation media to take account of the possibilities of enhance metaboliteproduction. The addition of $MgSO_4$ to DRYES is highly recommended. Investigations have proven (Filtenborg et al., 1990) that the production of secondary metabolites varies significantly with the brand of yeast extract used in YES on which DRYES is based, but that the addition of $MgSO_4$ compensates for this.

REFERENCES

FILTENBORG, O. and FRISVAD, J.C. 1980. A simple screening-method for toxigenic moulds in pure cultures. Lebensmittel Wissenschaft und Technologie 13: 128-130.

FILTENBORG, O., FRISVAD, J.C. and SVENDSEN, J.A. 1983. Simple screening method for molds producing intracellular mycotoxins in pure cultures. Applied and Environmental Microbiology 45: 581-585.

FILTENBORG, O., FRISVAD, J.C. and THRANE, U. 1990. The significance of yeast extract composition on metabolite production in Penicillium. In Modern Concepts in Penicillium and Aspergillus Classification, eds. R.A. Samson and J.I. Pitt, pp. 433-440. New York and London: Plenum Press.

FRISVAD, J.C. 1981. Physiological criteria and mycotoxin production as aids in identification of common asymmetric Penicillia. Applied and Environmental Microbiology 41: 568-579.

FRISVAD, J.C. 1983. A selective and indicative medium for groups of Penicillium viridicatum producing different mycotoxins in cereals. Journal of Applied Bacteriology 54: 409-416.

FRISVAD, J.C. 1989. The use of high-performance liquid chromatography and diode array detection in fungal chemotaxonomy based on profiles of secondary metabolites. Botanical Journal of the Linnaean Society 99: 81-95.

FRISVAD, J.C. and FILTENBORG, O. 1983. Classification of terverticillate Penicillia based on profiles of mycotoxins and other secondary metabolites. Applied and Environmental Microbiology 46: 1301-1310.

FRISVAD, J.C. and FILTENBORG, O. 1990. Secondary metabolites as consistent criteria in Penicillium taxonomy. In Modern Concepts in Penicillium and Aspergillus Classification, eds. R.A. Samson and J.I. Pitt, pp. 373-384. New York and London: Plenum Press.

FRISVAD, J.C. and THRANE, U. 1987. Standardized high-performance liquid chromatography of 182 mycotoxins and other fungal metabolites based on alkylphenone retention indices and UV-VIS spectra (diode array detection). Journal of Chromatography 404: 195-214.

HOCKING, A.D. and PITT, J.I. 1980. Dichloran-glycerol medium for enumeration of xerophilic fungi from low moisture foods. Applied and Environmental Microbiology 39: 480-492.

KING, A.D., HOCKING, A.D. and PITT, J.I. 1979. Dichloran-rose bengal medium for enumeration and isolation of molds from foods. Applied and Environmental Microbiology 37: 959-964.

KING, A.D., PITT, J.I., BEUCHAT, L.R. and CORRY, J.E.L., eds. 1986. Methods for the Mycological Examination of Foods. New York and London: Plenum Press,

PITT, J.I. 1979. The Genus Penicillium and its Teleomorphic States Eupenicillium and Talaromyces. London: Academic Press,

PITT, J.I., HOCKING, A.D. and GLENN, D.R. 1983. An improved medium for the detection of Aspergillus flavus and A. parasiticus. Journal of Applied Bacteriology 54: 109-114.

RAPER, K.B. and FENNELL, D.I. 1965. The genus Aspergillus. Baltimore: Williams and Wilkins.

DIALOGUE FOLLOWING DR. FILTENBORG'S PRESENTATION

SAMSON: I'd be interested to see what kind of results you would get on DG18. This has become a commonly used general purpose medium.

FILTENBORG: On DG18, the water activity is a little low, and the Penicillia do not produce secondary metabolites well within a week under these conditions. But Aspergilli, they are wonderful on DG18. You can separate a lot of Aspergilli very well on DG18.

SAMSON: What happens when you have other fungi, like *Cladosporium* or *Ulocladium*, on these plates?

FILTENBORG: They were there!

GAMS – As you are taking agar plugs from mixed cultures, isn't there a problem of diffusion of metabolites from one colony to another?

FILTENBORG: There is no problem, because we are taking intracellular metabolites from the mycelium, not diffusible ones from the agar. This could be a problem with some metabolites, such as citrinin, which diffuses easily. But in general we don't have many problems.

PITT: Do you turn the plug over so that the mycelium is in contact with the TLC plate, or do you use the substrate side?

FILTENBORG: We turn it over and put the mixture of chloroform and methanol on the plug, then turn it around and press it on the TLC plate.

PATERSON: On your slides, there was some variation in the penicillic acid spot between the isolates. Is this a particular problem with penicillic acid?

FILTENBORG: It is varying, yes. You see this with citrinin, as well.

PATERSON: Why do you see this variability?

FILTENBORG: It depends on the position of the plate, proximity of other colonies, competition and so on.

PATERSON: Some of the spots are very close together. Is there coelution of some compounds?

FILTENBORG: We can usually separate them.

HEARN: Does the brand of TLC plates have any effect on your analysis?

FILTENBORG: I hope not. We haven't tested this.

VARIATION IN *PENICILLIUM* AND *ASPERGILLUS* CONIDIA IN RELATION TO PREPARATORY TECHNIQUES FOR SCANNING ELECTRON AND LIGHT MICROSCOPY

P. Staugaard* , R.A. Samson and M.I. van der Horst

Centraalbureau voor Schimmelcultures
3740 AG, Baarn, The Netherlands

SUMMARY

Conidia and conidiophores of selected species of *Penicillium* and *Aspergillus* were examined by light microscopy and scanning electron microscopy. SEM micrographs were taken from frozen hydrated material and of critical point dried material after various chemical fixation procedures. Comparison between the techniques were made with emphasis on the conidial dimensions and surface detail of the conidia, conidiogenous cells and conidiophore elements. Frozen hydrated specimens reveal conidial dimensions closest to the living material and show significant shrinkage in chemical fixed and critical point dried specimens.

INTRODUCTION

The gross microscopic features of *Penicillium* and *Aspergillus* isolates can be readily observed by light microscopy, but in many species it is difficult to elucidate the shape and ornamentation of the conidia, which are especially small. Therefore scanning electron microscopy (SEM) has been increasingly used to investigate the pattern of ornamentation or to measure conidial dimensions (Ramírez, 1982, Bridge *et al.*, 1985; Kozakiewicz, 1989 a and b).

Beckett *et al.* (1984) and Read *et al.* (1983) have discussed the various preparation techniques for SEM and found significant differences in the dimensions of uredospores of *Uromyces viciae-fabae* (Pers.) Schroeter and ascospores of *Sordaria humana* (Fuckel) Winter. The authors concluded that low temperature SEM of frozen hydrated samples was the best technique for studying fungal spores under conditions as near to their natural environment.

This paper reports a comparative study of selected *Penicillium* and *Aspergillus* species using light microscopy and various preparation techniques for SEM.

MATERIAL AND METHODS

Fungi.

The following isolates were examined: *Penicillium aurantiogriseum* ex moulded food, *P. digitatum* ex Citrus, *P. echinulatum* CBS 317.48, *P. expansum* CBS 489.75, *P. griseofulvum* CBS 384.48, *P. italicum* CBS 278.58, *P. megasporum* CBS 529.64, *P. oxalicum* CBS 460.67, *P. variabile* CBS 385.48, *Aspergillus flavus* CBS 573.65, *A. niger* CBS 554.65, *A. phoenicis* CBS 135.48, *A. tamarii* CBS 104.13, *A. terreus* CBS 601.65.

---

* Present address: RIVM, Laboratory for Bacterial Vaccins, P.O. Box 1, 3720 BA Bilthoven, The Netherlands

Modern Concepts in Penicillium and Aspergillus Classification
Edited by R. A. Samson and J. I. Pitt
Plenum Press, New York, 1990

Cultures of the different species were grown on 2% maltextract agar at 25°C. Five to seven days after inoculation samples were taken for microscopical examination.

Light microscopy.
With a glass needle, a piece of the sporulating culture was suspended in a drop of 80% DL lactic acid on a microscope slide and covered with a coverslip. Samples were observed and photographed using an Olympus BH2 photo microscope equipped with Nomarski interference contrast optics.

Low temperature SEM of frozen hydrated material.
Blocks of approximately 5 x 5 mm were cut from the agar containing a sporulating section of a colony. The blocks were mounted in a cup specimen holder with cryo-glue, freshly made up as a 50/50 mixture of colloidal graphite (Emscope No A1533 Aquadag) and Tissue-Tek mounting medium (Miles Scientific).

The specimen holder was designed as a cylindrical cup made of copper. By providing radiation cooling this provided maximum protection to the sensitive hyphal material from damage during warming up. Samples were frozen in slushed nitrogen inside the slushing chamber of the Hexland CT1000A cryo system and quickly transferred under vacuum to the cold stage in the working chamber. In the working chamber the samples were sputtered with gold at low temperature (T= 170°C) for three to five minutes (thickness approximately 20 nm). After transfer to the cold stage (T= -140°C) of the Jeol 840A scanning electron microscope specimens were observed and photographed at 3 to 7 kV acceleration voltage.

Chemical fixation & critical point drying (CPD).
Blocks of approximately 3 x 7 mm, cut from a sporulating colony on the plate, werereplaced in a perforated beem capsule and immersed in fixing fluid. Fixation was carried out in a 5% aqueous solution of glutaraldehyde (GA) overnight (16 hours) at 4°C. Potassium permanganate (1%) was used for 30 minutes and osmium tetroxide (1% solution) for 30 minutes both at room temperature. After fixation, the samples were generally dehydrated in methoxy ethylene glycol in two steps of 20 min. followed by two changes of absolute acetone (Samson *et al.*, 1979). A few samples were slowly dehydrated (30 min. each step) in an ethanol series (15, 30, 45, 60, 70, 80, 90, 96, 100, 100% ethanol in distilled water. After dehydration was completed, the samples were critical point dried in a Balzers critical point drying apparatus using liquid carbon dioxide.

The blocks were mounted on specimen stubs with colloidal silver paint (Agar G302) and sputtered with gold for 150 s in a Polaron E5200 sputter coater. Specimens were observed and photographedin a Jeol 840A SEM at 7 to 10 kV accelerating voltage.

Measurements of conidial diameter.
Micrographs were enlarged and printed to a final magnification of 2000 x or in a few cases for very small conidia, 4000 x. From each series of photographs 100 to 300 conidia were measured and counted. The average diameter, measured at right angles to the growth axis, was taken as a quantitative measurement for shrinkage and swelling comparisons.

RESULTS AND DISCUSSION

Dimensions: conidial diameter measurements.

When *P. aurantiogriseum* conidia were relatively rapidly dehydrated in methoxy ethylene glycol as described by Samson *et al.* (1979) or by gentle dehydration in an ethanol series no significant differences in conidial size could be determined (Fig. 1).

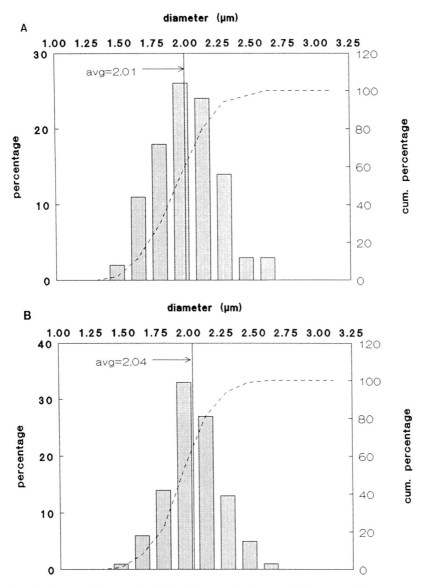

Figure 1. Average diameter of conidia of *P. aurantiogriseum* of critical point dried specimens after in (A) gentle dehydration in an ethanol series and (B) rapid dehydration in methoxy ethylene glycol.

As shown in Table 1, both *Aspergillus* and *Penicillium* conidial diameters measured after chemical fixation and critical point drying were smaller than in the frozen hydrated state (Table 1). Shrinkage ranged from 8 to 34%, and in most species an average of 20-30 % could be measured when comparing lightmicroscopy and SEM.

Table 1. Conidial diameters of various *Aspergillus* and *Penicillium* species.

Species	Light microscopy	Frozen hydrated	GA fix CPD
A. flavus	6.16	4.91	4.88
		20%	21%
A. tamarii	6.62	4.76	4.37
		28%	34%
P. digitatum	3.71	3.62	3.05
		2%	18%
P. expansum	3.09	2.92	2.55
		6%	18%
P. italicum	2.84	2.4	1.95
		15%	31%
P. megasporum	6.75	4.81	5.15
		29%	24%
P. oxalicum	3.27	2.99	2.88
		8%	12%
P. variabile	2.7	2.15	2.01
		20%	27%

Average diameters are expressed in μm. Shrinkage is expressed as a percentage compared with the light microscope value.

When conidia of *P. aurantogriseum, P. echinulatum* and *P. griseofulvum* were fixed in OsO_4, glutaraldehyde or potassium permanganate small differences in conidial size could be measured but they appear not to be consistent (Table 2). The three species also show a different degree of shrinkage.

Table 2. Conidial diameters after different SEM preparatoration techniques.

Species	Light microscopy	Frozen hydrated	GA fix CPD	OsO_4 fix CPD	$KMnO_4$ fix CPD
P. echinulatum	4.68	4.43	3.93	4.57	4.38
		5%	16%	2%	6%
P. aurantiogriseum	2.49	2.38	2.14	1.98	2.07
		4%	14%	20%	17%
P. griseofulvum	2.87	2.32	2.24	2.17	2.34
		19%	22%	24%	18%

Average diameters are expressed in μm. Shrinkage is expressed in percentage compared with the light microscopical value.

Conidial diameters of all examined *Aspergillus* and *Penicillium* isolates, measured in the light microscope after suspension in 80 % lactic acid were significantly larger than those measured in the SEM.

Ultrastructural detail.
The conidia of *P. echinulatum* and *P. megasporum* appeared more regular and spherical in light microscopy and frozen hydrated specimens than in chemically fixed samples. Frozen hydrated samples showed well preserved conidia with regular spines but in fixed and critical point dried samples the conidial wall was more wrinkled. This indicates that shrinkage leads to artificial ornamentation in fixed specimens. The conidia of *P. aurantiogriseum* and *P. digitatum* have a very fine surface structure in the frozen hydrated state. This structure can not been observed in fixed samples and is too small to be resolved in light microscopy.

Stipes and metulae show significant shrinkage in fixed samples as septa, which have relatively a more rigid structure, form protruding rings with lower shrinkage than elsewhere. These rings are rarely observed in frozen hydrated samples. In LM and Cryo SEM the stipes of *P. echinulatum* showed a smooth surface with clearly defined warts as ornamentation. In fixed samples the surface of the stipes is less smooth and the warts are deformed. Apart from shrinkage and deformation, a lot of debris is to be seen in OsO_4 and $KMnO_4$ fixed samples, probably material from the exudate which is fixed unto the surface during immersion in the fixing fluid.

Phialides of *P. echinulatum* and *P. aurantiogriseum* are smooth in LM and in frozen hydrated state; however in fixed and CPD specimens the phialides have a wrinkled surface, again indicating shrinkage.

DISCUSSION

The preparation of soft biological material to be observed in the dried state inevitably results in its shrinkage. Conidial size is determined during the final drying process, although the actual amount of shrinkage varies depending on the actual mode of drying and the specimen. Read *et al.* (1983) found that air-dried conidia probably showed most shrinkage, while frozen hydrated spores of *Sordaria humana* provided superior preservation. However, in some cases expansion of the fungal cell may occur due to the conversion to ice and this may explain the variation observed in the present study. In addition the structure of the conidial wall of Penicillia and Aspergillia can also vary, which may influence the degree of shrinkage.

In our SEM investigations conidial ornamentation was not significantly different than when observed in light microscopy. In some species with very small conidia, e.g. *Aspergillus terreus*, striations on the conidial wall were observed, which confirmed observations by Miegeville (1987) and Kozakiewicz (1989b). This ornamentation could not be seen in LM. The "microverrucate" and "reticulate" ornamentation reported by Kozakiewicz (1989a) for conidia of *P. corylophilum* and *P. hirsutum* have not been seen in many isolates representing these taxa (Samson, unpubl. data). However, she examined air-dried conidia from three weeks old cultures grown on Czapek agar, which probably had become dehydrated and collapsed naturally.

The measurements of conidial dimensions are not only dependant on the age and growth of the fungal material, but to a great extend to the preparatory techniques involved. Frozen hydrated samples will give dimensions closest to those of living

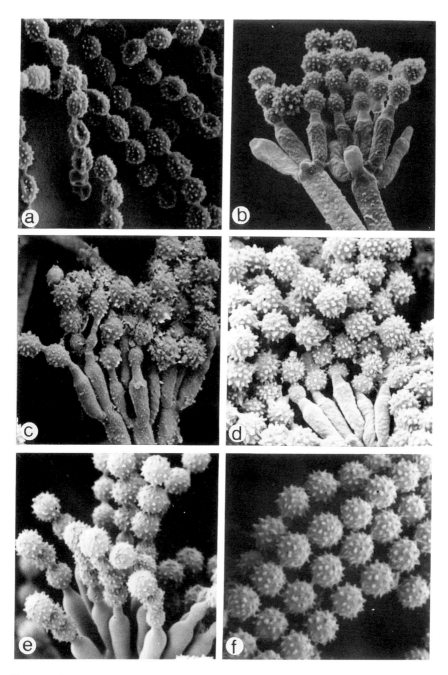

Figure 2. Scanning electronmicrographs of conidia and conidiogenous structures of *P. echinulatum* CBS 317.48. a. air-dried; b-d. critical point dried; b. fixed with glutaraldehyde, c. potassium permanganate, d. osmiumtetroxide; e-f. frozen hydrated (magnification all 2000x)

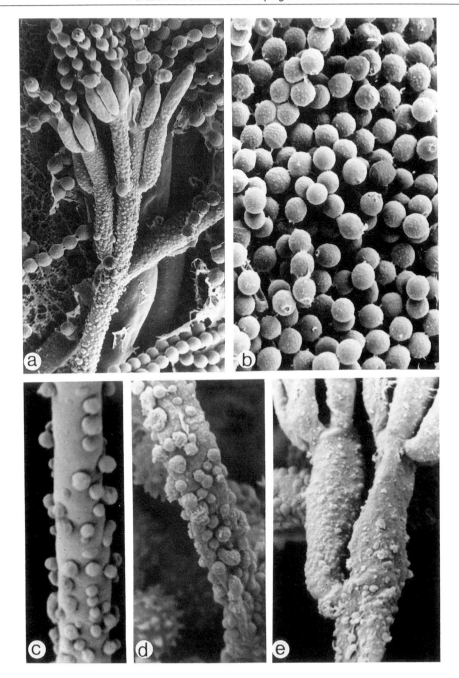

Figure 3. a-b. Scanning electronmicrographs of conidia and conidiogenous structures of *P. hirsutum* CBS 201.57, frozen hydrated, a. conidiophore (1500x), b. conidia (2500x). Note the finely roughened surface of the conidia; c-e. stipes of the penicillus of *P. echinulatum*, c. frozen hydrated, d. CPD with osmiumtetroxide fixation, f. CPD with potassium permanganate fixation (magnification all 2000x)

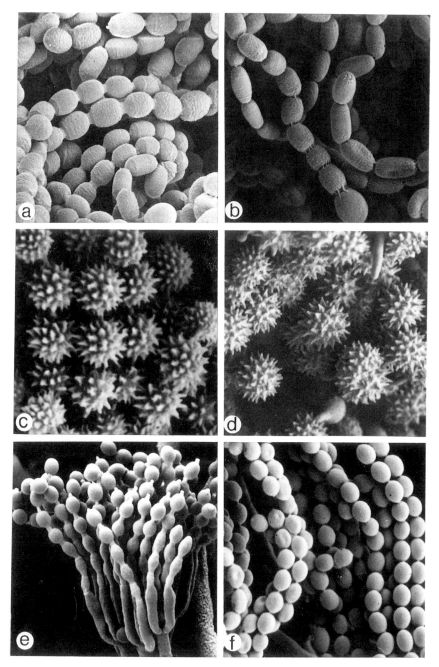

Figure 4. Scanning electronmicrographs of *Penicillium* conidia, a-b. *P. digitatum*, a. CPD fixed with glutaraldehyde, b. frozen hydrated; c-d. *P. megasporum* CBS 529.64, c. CPD fixed with glutaraldehyde, d. frozen hydrated; e-f. *P. expansum* CBS 486.75, e. CPD fixed with glutaraldehyde, f. frozen hydrated (magnification all 2000x).

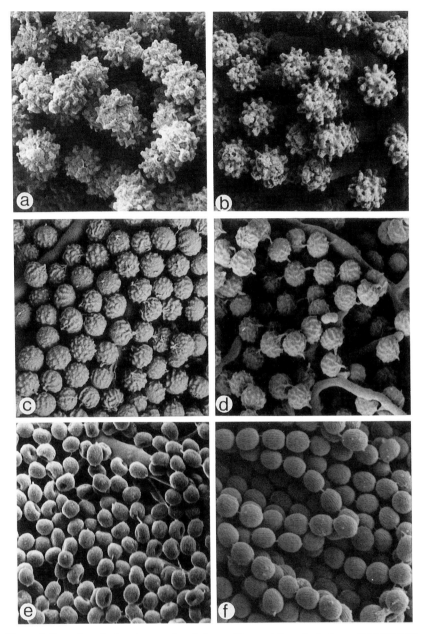

Figure 5. Scanning electromicrographs of *Aspergillus* conidia, a-b. *A. tamarii* CBS 104.13, a. frozen hydrated, b. CPD fixed with glutaraldehyde; c-d. *A. niger* CBS 554.65, c. frozen hydrated, d. CPD fixed with glutaraldehyde. Note that with both methods the conidia show longitudinal bars; e-f. *A. terreus* CBS 601.65, e. frozen hydrated, f. CPD fixed with glutaraldehyde. Note the fine striations(magnification all 2000x).

material. We agree with Read *et al.* (1983) that, if SEM is used, frozen hydrated material should be used as a base line, although it should be not considered as the panacea for morphological studies.

It is tempting to employ cryo SEM for the examination of fungal material, but particularly for the important and common species of Penicillia and Aspergillia a simple technique, usable in any well-equiped laboratory should be employed. For taxonomic studies light microscopical observations are preferable and in our view SEM has only a limited value in resolving taxonomic problems in *Penicillium* and *Aspergillus* anamorphs. It is especially emphasised that conventional SEM techniques should be used with great care, as dehydration may cause artefacts in ornamentation.

REFERENCES

BECKETT, A., READ, N.D. and PORTER, R. 1984. Variation in fungal spore dimensions in relation to preparatory techniques for light microscopy and scanning elctron microscopy. *Journal of Microscopy* 136: 87-95.

BRIDGE, P.D., HAWKSWORTH, D.L., KOZAKIEWICZ, Z., ONIONS, A.H.S., PATERSON, R.R. and SACKIN, M.J. 1985. An integrated approach to *Penicillium* systematics. *In* Advances in *Penicillium* and *Aspergillus* Systematics. eds. R.A. Samson & J.I. Pitt, pp. 281-309. New York and London: Plenum Press.

MIEGEVILLE, M. 1987. Apport de la microscopie electronique à balayage dans l'observation de divers *Aspergillus*. Bulletin Society France Mycologie Medicinal 16: 133-140

KOZAKIEWICZ, Z. 1989a. Ornamentation types of conidia and conidiogenous structures in fasciculate *Penicillium* species using scanning electron microscopy. *Botanical Journal of the Linnean Society* 99: 273-293.

—— 1989b. *Aspergillus* species on stored products. *Mycological Papers* 161: 1-188.

RAMIREZ, C. 1982. Manual and Atlas of the Penicillia. Amsterdam: Elsevier Biomedial Press.

READ, N.D., PORTER, R. and BECKETT, A. 1983. A comparison of preparative techniques for the examination of the external morphology of fungal material with the scanning electron microscope. *Canadian Journal of Botany* 61: 2059-2078.

SAMSON, R.A., STALPERS, J.A. and VERKERKE, W. 1979. A simplified technique to prepare fungal specimens for scanning electron microscopy. *Cytobios* 24: 7-11

VARIANTS OF *PENICILLIUM EXPANSUM*: AN ANALYSIS OF CULTURAL AND MICROSCOPIC CHARACTERS AS TAXONOMIC CRITERIA

J.F. Berny and G.L. Hennebert

Laboratoire de Mycologie Systématique et Appliquée, MUCL
Catholic University of Louvain
1348 Louvain-la-Neuve, Belgium

SUMMARY

The strain *Penicillium expansum* MUCL 29412, a monoconidial culture derivated from the wild type LCP 3384, after being submitted to divers physical factors, has bred ten variants. The variations are maintained from one generation to an other. They concern mainly the cultural characters and some microscopic characters. The implications for taxonomic studies and identification on a culture issued from a monospore are discussed. The value of cultural and microscopic characters, stable or not, as taxonomic criteria is investigated.

INTRODUCTION

Variants of *Penicillium notatum* NRRL 1249 B1 differing in cultural characters have been reported to occur spontaneously in successive single conidium transfers on malt agar under laboratory conditions (Sansome, 1947). Two distinctive variants resulted: one dark coloured and heavily sporulating, the other pale coloured and poorly sporulating. Morphological variants producing double sized conidia have also been obtained from the same strain on a mutagenic camphor culture medium (Sansome, 1949). In single conidium lines of *Penicilium viridicatum P. crustosum* and *P. aurantiocandidum*, two or three types of variation have been obtained on Czapek agar. These differed in growth rate, sporulation rate, colony reverse colour, presence and colour of diffusible pigment, presence of white mycelial outgrowth and conidium ornamentation. Two were reported as stable, the third one reverted partly into the two stable types (Bridge *et al.*, 1986). Samson *et al.* (1976), Pitt (1979) and Onions *et al.* (1984), have observed an intergradation between the fasciculate species of *Penicillium* subg. *Penicillium* and reported degradation of significant taxonomic characters in strains after long cultivation.

For taxonomic purposes, it is most important to assess the degree of variability within particular isolates (an isolate usually being a population), and the degree of variability of the isolates within the species and between close species. To check the stability of type strains during subculturing and preservation in culture collections, variations within the strain depending on conditions needs to be critically examined in order to facilitate the recognition of deviated elements amongst typical ones within the strain population.

Preservation techniques applied in culture collections and incidental X-ray exposure in customs have induced nine types of variants in subcultures of the single conidium strain *P. expansum* MUCL 29412. These variants differ from the original parent strain on some of their cultural and microscopic characters.

This study shows the extent of variability of those characters and investigates its implication in their taxonomy.

Modern Concepts in Penicillium and Aspergillus Classification
Edited by R. A. Samson and J. I. Pitt
Plenum Press, New York, 1990

MATERIAL AND METHODS

Strain.

P. expansum MUCL 29412 is a single conidium strain derived from the 1984 wild type isolate *P. expansum* LCP 3384.

Origin of variants.

In the course of a CEC research programme (BAP 0028) on the improvement of preservation techniques of fungal strains, subcultures of *P. expansum* MUCL 29412 were submitted to 19 different preservation techniques by freeze-drying and by freeze-storage at -20° to -196°C, 11 strains being exposed incidentally to uncontrolled X-ray exposure soon after being revived from preservation. Mass conidia (5000 conidia) and 200 single conidium transfers were prepared from each of the 19 revived cultures after treatment and from the parent strain MUCL 29412 preserved at 4°C. The mass conidia transfers and the 4000 single conidium isolates were grown on 2% malt yeast agar (MYA2: malt extract 2%, yeast extract 0.1%, agar 2%, pH 7) and on Czapek agar and incubated for 21 days in the laboratory under diffuse light. The mass conidia transfers showed numerous sectors on both culture media, while 9 variant types were recognized amongst the single conidium isolates on MYA2. One strain of each variant has been isolated. The variant strains and the parent strain have been preserved on MYA2 slants at 4°C.

Growth conditions for strain characterization.

Ten single conidium replicates of the original parent strain and of each variant type were inoculated on MYA2 in Petri dishes and incubated at 24°C in the dark and at 20°C in diffuse light in the laboratory. Observations of cultural and microscopic characters were recorded after 7 days.

Scanning electron microscopy.

Three day old conidia taken at the half radius of 7 day old colonies were mounted on double sided cellophane tape and examined by scanning electron microscopy (Jeol JSM-35, 20 KV, magnification 10000 X).

Phytopathogenicity testing.

Ten day old conidia from a single conidium isolate of each of the 8 sporulating variants and of the parent strain were inoculated into subepidermic scars made in apples of cultivar Gloster and incubated in a moist chamber for 4 weeks.

Nuclear staining.

Nuclei were visualized in conidia of the parent strain in 21 day old cultures according to the HCl-Giemsa method (Robinow, 1981).

Stability testing.

Three successive generations of the 9 variant strains and their parent strain were investigated by means of 20 single conidium isolates from each one. The colonies of each strain obtained were compared with each other macroscopically and microscopically to detect any further deviation of the variant.

RESULTS

Mass conidia cultures of the treated strains developed quite a number of sectors originated from the inoculum at the center of the plates, while no sectors were observed in the mass conidia cultures of the untreated parent strain. Sectoring being considered as the sign of a mixed culture, it was implied that sectors originated from inoculated variant conidia. Single conidia were therefore isolated from the strains.

Seven hundred and forty three variant colonies were obtained from the 3800 single conidium isolates, that is a variation rate of 19.55%, while no variation was detected at all in the untreated parent strain. Although it is not the purpose of this paper to analyse the mutagenic effect of the treatment, it can be said that the 1600 single conidium isolates derived from the 8 strains treated by preservation techniques alone and not subjected to X-ray have shown an average of 9.44% variant colonies including variants 1 to 6 and variant 8, while the 2200 single conidium isolates from the other 11 strains incidentally exposed after preservation to uncontrolled X-rays developed an average of 26.91% variants including variant 1 and variants 4 to 9.

The rates of occurrence of the nine variants and their relative frequency is given in Table 1.

Descriptions and illustrations (Figs. 1, 3-5) of the single conidium isolates of the parent strain and of the nine variants were made from MYA2 cultures grown at 24°C in the dark for 7 days. Differences observed on MYA2 cultures grown at 20°C in diffuse light and on Czapek agar cultures grown at 24°C in darkness (Fig. 2) are indicated as remarks.

Table 1. Occurrence and relative frequency of variants from untreated and treated *Penicillium expansum* MUCL 29412.

Variant	Occurrence number	Occurrence rate (%)	Relative frequency (%)
From untreated parent strain:			
Variant	0/200	0.0	0.0
From treated strains:			
Variant 1	288/3800	7.57	38.76
Variant 2	10/3800	0.26	1.34
Variant 3	6/3800	0.16	0.81
Variant 4	198/3800	5.21	26.65
Variant 5	7/3800	0.18	0.94
Variant 6	5/3800	0.13	0.67
Variant 7	66/3800	1.73	8.88
Variant 8	90/3800	2.37	12.11
Variant 9	73/3800	1.92	9.82

Parent strain (MUCL 29412)

Colony: diameter 44 mm; relief plane; depth 1 mm; immersed margin 2 mm wide, hyphae straw yellow, regular; texture coremiform; colour dull green; exudate colourless as small droplets; soluble pigment none; reverse yellow green.

Conidioma: stipes 500 x 3.5-4.5 µm, smooth; penicilli 3 branch points, 35-45 x 25-40 µm; rami 2-3, 15-20 x 2.5-3.5 µm, smooth, appressed; metulae 3-5, 10-17 x 2.5-3.5 µm, smooth, appressed; phialides 3-7, 7-10 x 2.5-3.5 µm, ampulliform, smooth, appressed; collula short; conidia mostly ellipsoidal, 3-4 x 2.5-3 µm, smooth.

Remarks. Colony grown at 20°C in diffuse light strongly coremial, coremia being distinctly zonated. On Czapek agar at 24°C in darkness, diffusely zonate, coremiform to fasciculate, abundant light brown exudate, reverse yellow to orange. Staining of nuclei in this strain has demonstrated that all conidia are conspicuously uninucleate.

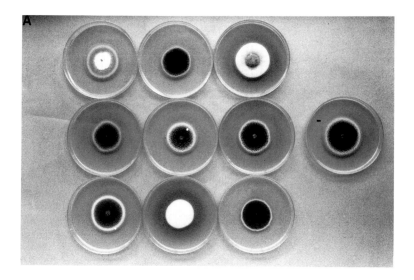

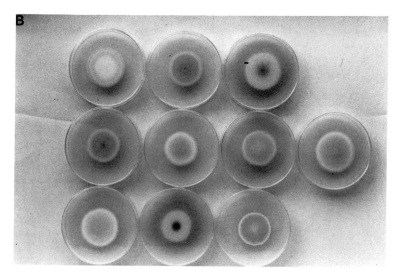

Figure 1. Colonies on 2 % malt agar at 24°C in darkness of variants 1 to 9 (from left to right, row after row) and the parent strain *P. expansum* MUCL 29412. A, upper surface and B, reverse

Variant 1 (MUCL 30223)

Colony: diameter 44 mm; relief plane; depth 0,5mm; immersed margin 2mm wide, pure yellow, regular; texture velutinous but with 2 cm diameter central area of white lanose mycelium with delayed sporulation; colour yellow green; exudate none; soluble pigment none; reverse pure yellow.

Conidioma: stipes 200 x 2-3 µm, smooth; penicilli 2-3 branch points, (20-)30-35(-45) x 20-25 µm; rami 1-2, 25 x 2.5-3 µm, smooth, appressed; metulae 3-5, 10-17 x 2.5-3 µm, smooth, appressed; phialides 3-7, 7-10 x 2.5-3 µm, cylindroidal, smooth, appressed; collula short; conidia mostly ellipsoidal, 3-4 x 2.5-3 µm, smooth.

Remarks. At 20°C in diffuse light, velutinous with no fasciculation, but slightly zonate, with a minute central outgrowth of white lanose mycelium. On Czapek agar, totally white, plane, furrowed, sterile, reverse white to pale yellow.

Distinctive characters. The colony is lower and the texture is velutinous. The sporulating area is yellow green, the colour of immersed mycelium at the margin is pure yellow. No exudate is produced. The reverse is pure yellow. Conidiophore stipes are shorter and narrower; some penicilli are biverticillate; terverticillate penicilli are far narrower than those in the parent strain and there are fewer rami on each stipe, and they are longer and narrower. Metulae and phialides are also narrower. The phialides are cylindroidal.

Variant 2 (MUCL 30224)

Colony: diameter 38 mm; relief plane; depth 1 mm; immersed margin 2mm wide, pure yellow, regular; texture fasciculate into small coremia; colour dull green; exudate colourless, as small droplets; soluble pigments none; reverse yellow green.

Conidioma: stipes 300 x 3.5-4.5 µm, smooth; penicilli 2-3 branch points; (20-)30-35(-45) x 25-40 µm; rami 1-2, 15-20 x 2.5-3.5 µm, smooth, appressed; metulae 3-5, 10-14 x 2.5-3.5 µm, smooth, appressed; phialides 3-7, 7-10 x 2.5-4 µm, ventricose, smooth, appressed; collula short; conidia mostly ellipsoidal, 3.5-4.5 x 2.5-3 µm, smooth.

Remarks. At 20°C in diffuse light, small coremia with distinct zonation. On Czapek agar, zonate and slightly furrowed, coremiform, reverse orange.

Distinctive characters. The diameter of the colony is smaller; the immersed mycelium at the margin is pure yellow. The conidiophores are fasciculate into smaller coremia, with conidiophore stipes shorter, penicilli sometimes biverticillate, rami fewer on each stipe, metulae shorter, phialides wider and ventricose and conidia somewhat larger.

Variant 3 (MUCL 30225)

Colony: diameter 41 mm; relief umbilicate; depth 2-3 mm; immersed margin 1 mm wide, pure yellow, regular; aerial margin of white lanose mycelium; texture fasciculate at the sporulating central area; colour dull green; exudate colourless as small droplets; soluble pigment red brown; reverse red brown.

Conidioma: stipes 500 x 3.5-4.5 µm, smooth; penicilli 3 branch points, 35-45 x 25-40 µm; rami 2-3, 15-20 x 2.5-3.5 µm smooth, appressed; metulae 3-5, 10-17 x 2.5-3.5 µm, smooth, appressed; phialides 3-7, 7-10 x 2.5-3.5 µm, ampulliform, smooth, appressed; collula short; conidia mostly subglobose, 2-3.5 x 2.5-3 µm, smooth.

Remarks. At 20°C in diffuse light, zonate but sporulating slowly from the center. No aerial white lanose mycelium. On Czapek agar, faster growing, zonate and furrowed, granular, sporulating well, reverse red brown to dark brown.

Distinctive characters. The colony is somewhat smaller in diameter; the relief is umbilicate and the colony is higher because of its lanose texture. Sporulation from the center of the colony is delayed, leaving a very wide margin of aerial lanose white mycelium. The immersed mycelium at the margin is wider and pure yellow. Conidiophores are grouped in fascicles from aerial hyphae. The colony produces a red brown diffusible pigment and the colour of the reverse is red brown. Conidia are more globose and their average length is shorter than those of the parent.

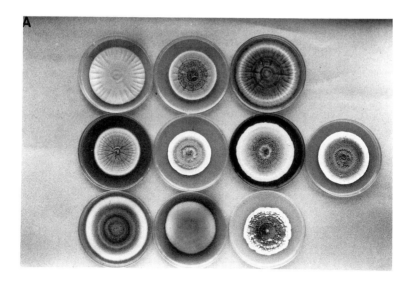

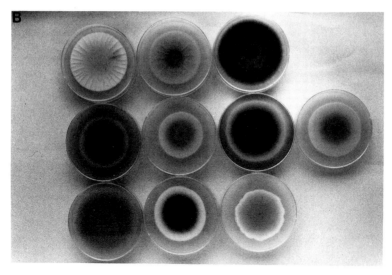

Figure 2. Colonies on Czapek agar at 24°C in dark of variants 1 to 9 (from left to right, row after row) and the parent strain *P. expansum* MUCL 29412. A, upper surface and B, reverse.

Variant 4 (MUCL 30226)

Colony: diameter 35 mm; relief plane; depth 0.5 mm; immersed margin 2mm wide, pure yellow, lobate; texture fasciculate; colour dull green; exudate brownish in droplets; soluble pigment red brown; reverse yellow green.

Conidioma: stipes 300 x 3.5-4.5 µm, smooth; penicilli 2-3 branch points, (20-)30-35(-45) x 25-40 µm; rami 2-3, 20-40 x 2.5-3.5 µm, smooth, appressed; metulae 3-5, 10-17 x 2.5-3.5 µm, smooth, appressed; phialides

3-7, 7-10 x 2.5-3.5 μm, ampulliform, smooth, appressed; collula short; conidia mostly ellipsoidal, 3-4 x 2.5-3 μm, smooth.

Remarks. At 20°C in diffuse light, zonate, fasciculate, more paler and slowly sporulating. On Czapek agar, colony texture granular, grey green, furrowed, red brown soluble pigment, reverse orange to dark brown.

Distinctive characters. The diameter of the colony is smaller and the colony lower. The outline is slightly lobate and the immersed mycelium at the margin is pure yellow. Brown exudate and red brown diffusible pigment are produced. Conidiophores are grouped in fascicules, with shorter stipes, some biverticillate penicilli and some longer rami.

Variant 5 (MUCL 30227)
Colony: diameter 37 mm; relief plane; depth 1 mm; immersed margin 2mm wide, straw yellow, regular; texture coremiform; colour dull green; exudate colourless as large droplets; soluble pigment none; reverse yellow green;
Conidioma: stipes 600 x 3.5-4.5 μm, smooth; penicilli 2-3 branch points, (20-)30-35(-45) x 25-40 μm; rami 2-3, 15-20 x 2.5-3.5 μm, smooth, appressed; metulae 3-5, 10-17 x 2.5-3.5 μm, smooth, appressed; phialides 3-7, 7-10 x 2.5-3.5 μm, cylindroidal, smooth, appressed; collula short; conidia mostly ellipsoidal, 3-4 x 2.5-3 μm, smooth.
Remarks. At 20°C in diffuse light, similar growth rate to the parent, strongly coremial and sporulating from the center to the margin. On Czapek agar, slowly growing, zonate and coremiform, ochre to yellow green, reverse yellow to orange.

Distinctive characters. The diameter of the colony is smaller with exudate in large droplets. Coremia are larger; the stipes are longer; some penicilli are biverticillate; the phialides are more cylindroidal.

Variant 6 (MUCL 30228)
Colony: diameter 40 mm; relief plane; depth 1 mm; immersed margin 2mm wide, pure yellow, regular; texture coremiform; colour dull green; exudate colourless, as small droplets; soluble pigment red brown; reverse yellow green.
Conidioma: stipes 600 x 3.5-4.5 μm, smooth; penicilli 2-3 branch points, (20-)30-35(-45) x 25-40 μm; rami 2-3, 15-20 x 2.5-3.5 μm, smooth, appressed; metulae 3-5, 10-14 x 2.5-3.5 μm, smooth, appressed; phialides 3-7, 7-10 x 2.5-3.5 μm, ampulliform, smooth, appressed; collula short; conidia mostly ellipsoidal, 3-4 x 2.5-3 μm, smooth.
Remarks. At 20°C in diffuse light, similar growth rate to parent, strongly coremial and sporulating throughout. On Czapek agar, fast growing, coremiform in central zones, pale yellow green, intense red brown soluble pigment, reverse orange red to dark brown.

Distinctive characters. The diameter of the colony is somewhat smaller. The immersed mycelium at the margin is pure yellow. A red brown soluble pigment is produced. Some penicilli are biverticillate; the stipes are longer, with shorter metulae.

Variant 7 (MUCL 30229)
Colony: diameter 44 mm; relief plane; depth 1 mm; immersed margin 2 mm wide, pale yellow, regular; texture fasciculate; colour dull green; exudate colorless as small droplets; soluble pigment none; reverse yellow green.
Conidioma: stipes 500 x 3-4 μm, smooth; penicilli 2-3 branch points, (20-)30-35(-50) x 20-25 μm; rami 1-2, 25 x 2.5-3 μm, smooth, appressed; metulae 3-5, 10-17 x 2.5-3 μm, smooth, appressed; phialides 3-7, 7-10 x 2.5-3 μm, cylindroidal, smooth, appressed; collula short; conidia mostly subglobose, 2.5-3.5 x 2.5-3.5 μm, smooth.
Remarks. At 20°C in diffuse light, colony smaller, heavily sporulating throughout and closely zonate. On Czapek agar, fast growing, strongly zonate, zones differently coloured ochre to yellow green, exudate light brown, reverse orange.

Distinctive characters. Immersed mycelium in the marginal area is pale yellow. The conidiophores are grouped in fascicles with some biverticillate and terverticillate penicilli; penicilli always reduced and sometimes somewhat longer; metulae are less numerous on each stipe, but narrower and eventually longer; metulae and phialides are also narrower. The phialides are cylindroidal; the conidia are mostly subspherical.

Variant 8 (MUCL 30230)

Colony: diameter 35 mm; relief umbonate; depth 6 mm; immersed margin 1 mm wide, pure yellow, regular; texture floccose; colour white; exudate none; soluble pigment red brown; reverse red brown.
Conidioma: none observed, although a very few conidia have been found.
Remarks. At 20°C in diffuse light, mycelium sterile. On Czapek agar, colony white, thick and lanose, reverse red brown.

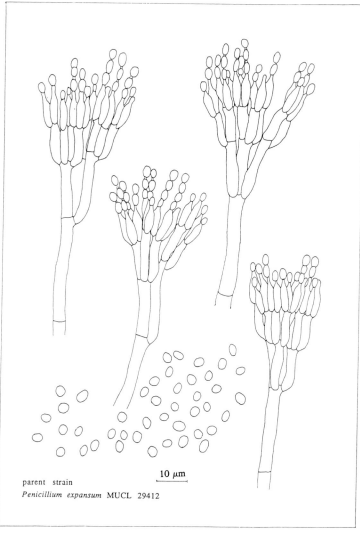

parent strain
Penicillium expansum MUCL 29412

10 μm

Figure 3. Conidiophores and conidia of parent strain *P. expansum* MUCL 29412, on 2% malt agar at 24°C.

Distinctive characters. This variant is distinct from the parent strain in all characters. The colony is smaller in diameter; the immersed mycelium at the margin narrower and pure yellow. The colony is umbonate and deeper, made of sterile, white, lanose mycelium. No exudate is produced, but red brown soluble pigment is. The reverse is red brown.

Variant 9 (MUCL 30231)

Colony: diameter 36 mm; relief plane; depth 0.5 mm; immersed margin 0.5 mm wide, pale yellow, regular; texture fasciculate; colour dull green; exudate colourless as small droplets; soluble pigment none; reverse yellow green.

Conidioma: stipes 500 x 3.5-4.5 µm, finely roughened; penicilli 2-3 branch points, (20-)30-35(-50) x 20-25 µm; rami 2-3, 15-25 x 2.5-3.5 µm, smooth or finely roughened, appressed; metulae 3-5, 10-17 x 2.5-3.5 µm, smooth, appressed; phialides 3-7, 7-10 x 2.5-3.5 µm, ampulliform, smooth, appressed; collula short; conidia globose, 3-4.5 x 3-4.5 µm, smooth.

Remarks. At 20°C in diffuse light, zonate and moderately sporulating Conidia globose. On Czapek agar, slowly growing, yellow green, colony texture very tufty, tufts of white mycelium becoming sporulating, reverse yellow.

Distinctive characters. The colony is smaller and lower; immersed margin is narrower and pale yellow. The conidiophores are grouped in fascicles; walls of the stipes and of some rami are finely roughened. Some penicilli are biverticillate; terverticillate penicilli are sometimes longer but they are always narrower; rami are also sometimes longer. Conidia are distinctly globose and larger.

Scanning electron microscopy examination of three day old conidia of the parent strain and its eigth sporulating variants shows that they are smooth in all cases but size and shape vary. Variants 2, 4, 5 and 6, all well sporulating from typically coremial to strongly fasciculate conidiophores, have acquired conidia somewhat larger than those of the parent strain but similar in shape (Fig. 6). The less fasciculate variants 7 and 3 have somewhat smaller conidia, but these are the same shape as the parent strain.

Two variants are noteworthy. Variant 9 shows larger, globose and smooth conidia; however it is conspicuously fasciculate and does not shed conidial crusts. Variant 1 is not fasciculate and appears velutinous. Its conidiophores are of a reduced complexity, a feature also seen in variant 7. It also shows a strong tendency to produce outgrowths of almost sterile white lanose mycelium similar to the wholly sterile variant 8 (Fig. 7). Subepidermic inoculations of conidia from the parent and eight sporulating variant made on Gloster apples has demonstrated the virulence of all the strains; a soft rot of the fruits soon develops and somewhat small to large coremia emerge at the surface of the rot. The size of those coremia is similar to their respective size observed on malt agar.

DISCUSSION

The parent strain *P. expansum* MUCL 29412 showed uninucleate conidia. The absence of sectors in mass conidia cultures of this isolate, the absence of variant colonies amongst the 200 single conidium isolates and the absence of variation in single conidium subcultures during three generations demonstrate the genetic homogeneity and stability of the parent strain cultivated on MYA2 and maintained at 4°C. A similar test has been carried out on 50 single conidium isolates from conidia of *P. expansum* collected on Boscop apple rot. All single conidium cultures obtained immediately from nature as well as all single conidium cultures derived from these were typical of the wild type in the first generation.

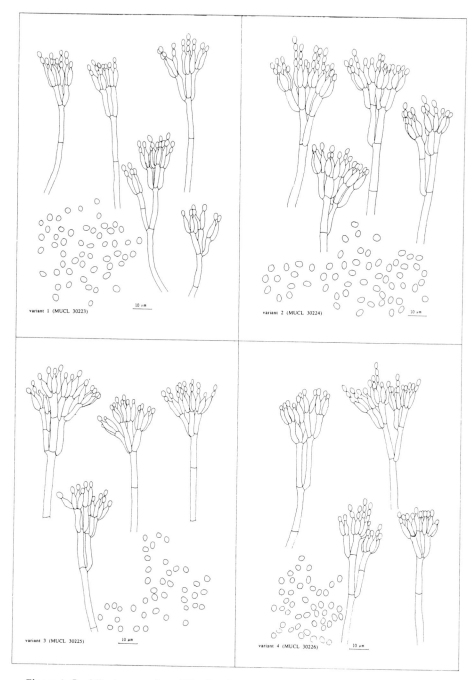

Figure 4. Conidiophores and conidia of variant 1 to 4 of *P. expansum* MUCL 29412, on 2% malt agar at 24°C.

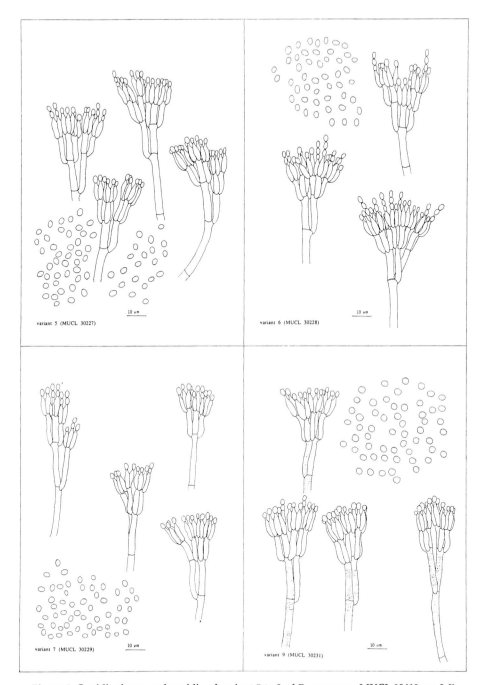

Figure 5. Conidiophores and conidia of variant 5 to 9 of *P. expansum* MUCL 29412, on 2 % malt agar at 24°C.

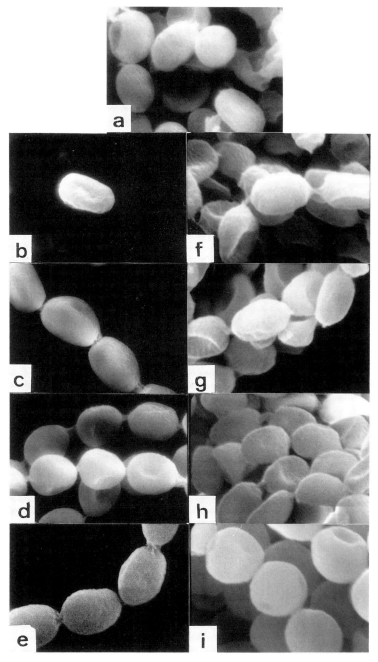

Figure 6. Conidia of the parent strain (a) and variants 1 to 9 (from b to i) of *P. expansum* MUCL 29412 as seen by scanning electron microscopy (x 3800).

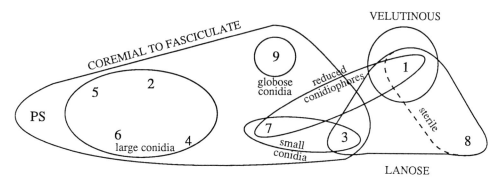

Figure 7. Schematic representation of the divergence of variants 1 to 9 from the parent strain *P. expansum* MUCL 29412 with indication of their common characters.

However it has been reported that large variation may naturally exist in strains from *Penicillium* subsection *Fasciculata* (Samson *et al.*, 1976), and that repeated cultivation on agar slants can be responsible for variation in *Penicillium* subgen. *Penicillium* (Bridge *et al.*, 1986, 1987). In the present case, some of the factors which having induced variation, i.e. preservation techniques such as freeze-drying and frozen-storage at -20° and -196°C, as well as freight control by X-ray in customs, are currently being used.

Nine types of variants have been induced. They differ from the original parent strain by their cultural characters and some of their microscopic characters. The growth rate can decrease up to 20% (variants 4, 8, 9). The texture varies from coremial to fasciculate to velutinous (variant 1) and to lanose (variants 3, 8). This change appears to be correlated to a delay and a decrease in conidiation (variant 8 sporulated after 7 months). At the same time the colour of the colony changes from dull green to yellow green (variant 1, 3) and even white in the lanose, asporogenous mycelium. The colour of the immersed marginal mycelium varies from pale straw yellow to bright yellow (variants 1, 2, 7). Exudates may be absent, when present they usually are colourless or brownish (variant 4). A red-brown soluble pigment can be produced (variants 3, 4, 6, 8).

There are variations in the structure and size of the elements of the conidioma. The most conspicuous variations occur in the length and width (variants 1, 5, 6) or the roughness of the conidiophore stipe (variant 9), the reduction of the penicillus (variants 1, 7), the shape of the phialides, that varying from ampulliform to either ventricose (variant 2) or cylindroidal (variants 1, 5, 7), and the size and shape of the conidia that can become large and globose (variant 9). A few characters are stable, such as the length of the phialides and the absence of ornamentation of the conidia. Also the pathogenic virulence to apples appears unchanged in the sporulating variants.

The characters shown to be variable are primarily important in the identification of the species (Samson *et al.*, 1976; Pitt, 1979; Ramírez, 1982). It also appears that different characters vary from one variant to another. These observations suggest that characters do not exist which are sufficiently discriminant either to identify all nine variants as one species or to differenciate these variants from the parent species and from any other related species. Indeed when one important character is changed, another may remain that makes species identification possible.

Variant 9, with rugose conidophore stipes and rami, globose smooth conidia, with a granular to floccose colony surface and a narrow margin, resembles *P. crustosum* Thom closely. However, variant 9 does not form any powdery or loose crusts of conidia on MYA2 and is very tufted on Czapek agar. Raper and Thom (1949) classified *P. crustosum* together with *P. expansum* in the same series *Expansa*. Fassatiová (1977), comparing authentic strains of both species, redisposed *P. crustosum* as *P. expansum* var. *crustosum*. Samson *et al.* (1976) considered *P. crustosum* as a synonym of *P. cyclopium* as *P. verrucosum* var. *cyclopium*, but Stolk and Samson (1985) reverted to maintaining *P. crustosum* as a distinct species in the series *Expansa*. Pitt (1979) also considered *P. crustosum* as a distinct species. These opinions indicate that *P. crustosum* is a taxon very close of *P. expansum* Variant 9 of *P. expansum* appears to corroborate this observation.

Variant 1 has conidiophores that are so short that it appears velutinous; its penicilli are also reduced and the strain has a strong tendency to form sterile mycelium. On Czapek agar, it is totally white and sterile, like the lanose variant 8. The character of colony texture indeed may vary considerably, not only within the species but within the strain.

Considering the possible variations of taxonomic characters within a species as has been demonstrated here within a strain, identification is therefore not helped by rigid dichotomous keys constructed on the basis of typical isolates. To encompass such a variability within an identification key, a probabilistic computerized matrix should be constructed on the basis of records of numerous pure strains. This method would allow consideration as to whether the observed variations are in the range of the natural variations or if new species or varieties have to be created.

The analysis of the extent of variability of an isolate that might appear to be an heterogenous population, can be made possible by the observation of single conidium cultures derivated from the original sample.

This prodedure is applicable to the study of morphological characters of an isolate and also its physiological or biochemical properties. The production of secondary metabolites of the nine morphological variants is now investigated by a similar method.

REFERENCES

BRIDGE, P.D., HAWKSWORTH, D.L., KOZAKIEWICS, Z., ONIONS, A.H.S. and PATERSON, R.R.M. 1986. Morphological and biochemical variation in single isolates of *Penicillium*. *Transactions of the British mycological Society* 87, 389-396.

BRIDGE, P.D., HUDSON, L., KOZAKIEWICS, Z., ONIONS, A.H.S. and PATERSON R.R.M. 1987. Investigation of variation in phenotype and DNA content between single-conidium isolates of single *Penicillium* strains. *Journal of General Microbiology* 133, 995-1004.

FASSATIOVA, O. 1977. A taxonomic study of *Penicillium* series *Expansa* Thom emend. Fassatiová. *Acta Univ. Carol. Biol.*, 1974: 283-335.

ONIONS, A.H.S., BRIDGE, P.D. and PATERSON, R.R.M. 1984. Problems and prospects for the taxonomy of *Penicillium*. *Microbiological Sciences* 1, 185-189.

PITT, J.I. 1979. The Genus *Penicillium* and its Teleomorphic States *Eupenicillium* and *Talaromyces*. London: Academic Press.

RAMIREZ, C. 1982. Manual and Atlas of the Penicillia. Amsterdam: Elsevier Biomedical Press.

RAPER, K.B. and THOM, C. 1949. A Manual of the Penicillia. Baltimore: Williams and Wilkins.

ROBINOW, C.F. 1981. Nuclear behaviour in conidial fungi. In Biology of conidial fungi, Vol. 2, eds. G.T. Cole and B. Kendrick. New York: Academic Press.

SAMSON, R.A., STOLK, A.C. and HADLOK, R. 1976. Revision of the subsection *Fasciculata* of *Penicillium* and some allied species. *Studies in Mycology, Baarn* 11: 1-47.

SANSOME, E.R. 1947. Spontaneous variation in *Penicillium notatum* strain NRRL 1249 B21. *Transactions of the British mycological Society* 31: 66-79.

—— 1949. Spontaneous mutation in standard and 'GIGAS' forms of *Penicillium notatum* strain 1249 B21. *Transactions of the British mycological society* 32: 305-314.

STOLK, A.C. and SAMSON, R.A. 1985. A new taxonomic scheme for *Penicillium* anamorphs. *In* Advances in *Penicillium* and *Aspergillus* systematics, eds. R.A. Samson R.A. and J.I. Pitt, pp. 163-192. New York: Plenum Press.

DIALOGUE FOLLOWING DR. BERNY'S PAPER

BRIDGE: The kind of variation that you have reported here have, in fact, been reported in *Penicillium* before, as well as in other fungi. They occur in many fungi without any X-ray, UV or other treatment whatever. In most of these cases, you are exerting some kind of selective pressure on the population when you do the isolation. It is difficult to say what is and what is not a mutant. Most colonies of *Penicillium*, for example, readily yield nutritional mutants when we grow them on substrates they do not normally encounter. More work is required.

PITT: Variant number 9 looks alot like *P. crustosum*. Is it possible that this might have been a contaminant?

HENNEBERT: This is not a contaminant. It comprised 10% (check paper) of the single conidium isolates. It does look similar to *P. crustosum* because it does have globose conidia and rough stipes, but it still rots apples, which *P. crustosum* does not.

SAMSON: Yes it does!

HENNEBERT: Well, maybe this character is not good enough. But these isolates are derived from a true, typical *P. expansum*. These are not contaminants. So, we have to believe that this "*P. crustosum*" is a true variant of *P. expansum*. So the question is, what is the species concept of *P. expansum* in this case? The same question arises regarding the velutinous colony texture. Is this a good enough character to exclude a strain from *P. expansum*? Maybe not. And how about the lanose variant? Everyone knows that lanose strains are degenerate and difficult to identify and classify by morphological methods. So we have to look at them with other methods, such as secondary metabolites.

But what is important here is that we have all these variants from a single conidium of *P. expansum*. It is a question of preservation methods, or perhaps the accidental X-ray treatment. Some preservation methods we used gave no variants.

My main conclusion is that the species concept should be based on a better study of variation. The only way to do this is with single conidium isolates.

FASSATIOVA: In some of your strains, you have found rough metulae and phialides, which are characters of *Geosmithia*. (Note of the editors: on the slides of the drawings, interrupted lines were showing background phialides giving an impression of a rough wall. They have been omitted in Figures 3-5 of this paper)

BERNY: No, the phialides were not rough.

SAMSON: However, we have also found several isolates of *P. expansum* with rough stipes. Otherwise, they are quite characteristic, and Dr. Frisvad has confirmed this with the secondary metabolites.

SAMSON: I don't agree that you should key out these deviating forms. When you isolate from foods or other materials, you will get the wild type. Deviating strains are very rare, I think. Therefore, if you go to a computer key with these deviating strains, it makes things quite difficult, but this is not the fault of the key.

PITT: I agree with Dr. Samson. Keys are normally written for fresh isolates. They should work for people who are isolating from nature. There is no way any key can take into account variation that is found in cultural conditions. One of the probelms with some of Raper's descriptions was that they were based on isolates that had been in collections for a long time. They were therefore difficult to use. Keys should be written for fresh isolates. They cannot be expected to work for isolates that have been manipulated.

TAYLOR: I wonder if there is correlation between the different kinds of variants and the X-ray treatment or the preservation methods?

BERNY: There is no correlation.

HENNEBERT: Of the nineteen different preservation methods studied, only eight procedures were used with this particular monoconidial strain. Up to forty percent of the first generation single conidium isolates were variants. Some of the cultures received from other culture collections stayed in customs for about one month and were subjected to X-ray treatments. In these cultures, up to seventy-five percent of the single conidium isolates were variants. Therefore, one pure culture, stored using different methods over time, can yield many different variants. This could also happen with extype cultures.

SAMSON: If you start making monoconidial isolates, then you impose another selection on your type culture.

HENNBERT: You can reselect the original one.

SAMSON: You can also lose the original one.

HENNBERT: Or have lost it.

PITT: Four years ago, we agreed to use a mixed spore inoculum for identifying and handling Penicillia. This is an important point. The variability in individual conidia of a haploid genus like *Penicillium* is quite high. I suspect there may be more than one genome present in an ordinary *Penicillium* colony. The way that *Penicillium* has adapted so rapidly into man-made environments such a cereals, relates to it having a lot of variants available. If we make single spore cultures, we are seeing a correct expression of a part of the genome, but these may be variants that would not normally survive in nature. *Penicillium* has the capability of shedding a lot of variants into the environment, with the possibility that this might allow it to exploit new habitats. Some of the variants that Dr. Berny has shown us, which consist mostly of mycelium, are most unlikely to survive in nature. With mass inocula, anastomoses of the germ hyphae result in the production of a mixed genome representing the sum of the variation in those conidia.

PETERSON: In our lab, we routinely compare colony morphology before and after lyophilization. Changes in colony morphology are very rare in my experience.

SEIFERT: Is it possible that these different variants may represent different vegetative compatability groups (VCGs)?

TAYLOR: VCG's do occur in anamorphic fungi, as judged by hyphal anastomosis, such as *Fusarium oxysporum*. But you wouldn't see these if you made many single conidium isolates from a single isolate from nature. You would only see them if you made isolates from different habitats or geographical localities.

CROFT: Regarding the problem of generating variance from a single isolate, I suspect that in some cases chromosomal rearrangement may be the cause.

PENICILLIUM AND *ASPERGILLUS* IN THE FOOD MICROBIOLOGY LABORATORY

A.P. Williams

Leatherhead Food R.A.
Randalls Road, Leatherhead, Surrey, KT22 7RY, UK

SUMMARY

Species of these two genera have long been dominant among the spoilage flora of a range of foods, especially those with reduced water activity and low pH. As various lobbies seek to reduce levels of humectant preservatives and acids, such spoilage is likely to increase and, where controlled chill is used as a substitute, psychrotrophic Penicillia have a particular advantage. Although some species are ubiquitous, the presence of others can be strongly indicative of particular types of problem (e.g. *Eurotium* spp. and marginal moisture abuse). The concept of recognising Indicator and Index organisms is not unknown in Bacteriology, yet has been slow to be adopted in Food Mycology, mainly because of the difficulty in identification.

Mycology is advancing rapidly, with clarifications of taxonomy, simplification of keying and new analytical procedures (e.g. E.L.I.S.A. and lectin binding). It is now time to consolidate those achievements so that recognition of important species, especially in *Penicillium* and *Aspergillus*, and assessment of their significance, are made available to laboratories that lack the services of an expert mycologist.

INTRODUCTION

During the last 10 years, *Penicillium* and *Aspergillus* have dominated the spoilage mycoflora of mouldy samples examined in our laboratory. Williams and Bialkowska (1985) reported that, from 294 samples with visible growth of one or more mould species, *Penicillium* accounted for 53.1% of isolates, *Aspergillus* and associated teleomorphs 15.1%, *Cladosporium* 8.9%, *Mucor* 8.5%, *Rhizopus* 4.8% and other genera 9.6%. Of the penicillia, subgenus *Penicillium* (90.1%) was dominant.

The following species lists give details of spoiled foods from which species of *Penicillium* and *Aspergillus* have been isolated, including isolations more recent than 1985. It is hoped that other laboratories will be encouraged to publish similar lists, so that further information becomes available on the relationship between species and pattern of spoilage. This would not only supplement information already available about foods at risk from particular mould species, but would also highlight those ubiquitous species whose presence gives no information on likely contamination source.

SOURCES OF *PENICILLIUM* AND *ASPERGILLUS*

Penicillia and Aspergilli described below were all isolated either by direct subculture of visible mould growth on food surfaces or by direct plating of visible mouldy material from within a food sample. In the following lists, a standard sequence has been adopted for food types: meat pies and pasties; meat products; cheese (British hard cheese unless otherwise stated); bread; cakes; jams; nuts; cereals; fruit and vegetables and others. A tabular form has not been used because the variety of foods is too large. Nomenclature

Modern Concepts in Penicillium and Aspergillus Classification
Edited by R. A. Samson and J. I. Pitt
Plenum Press, New York, 1990

follows the conventions of Williams (1990) and, where more than one isolate was obtained in any category, the number of isolates is given in parenthesis.

Subgenus *Penicillium*

P. atramentosum: Cheese

P. aurantiogriseum: Pork sausages; slated anchovies; cheese (3); bread (2); palm kernels; maize (2); muesli; animal feed; peach; apple juice; pickled gherkin; milk shake mix; pepper.

P. brevicompactum: Cottage pie; meat pie (2); sausage roll; salted anchovies (2); cheese (2); bread (3); jams (5); maize; grapes; pear; tomato; celeriac; coconut; cabbage; tomato puree; milk shake mix; pet food; dairy spread.

P. camemberti: Cheese (4).

P. chrysogenum: Cottage pie; Scotch egg; meat pie (3); cooked sausages (2); salted anchovies; cheese; cottage cheese; bread (5); cake (6); jam (2); cabbage; apple; pet food (2); milk shake mix; whole egg; mustard; fruit sweets; chocolate biscuit; paprika; tartrazine solution; suet; dairy spread.

P. commune: Cottage pie; meat pie (9); Scotch egg; sausage roll (3); chicken; sausages (3); cheese (18); bread (6); Madeira cake (7); jam (2); marrow; banana; orange; lemon; pickled onion; bay leaves; yoghurt (2); milk shake mix; tartrazine solution (2); dairy spread.

P. coprophilum: Muesli; pet food.

P. crustosum: Sausage; cheese (3); bread (4); cake; jam (3); chestnut; maize (2); raw pastry; apricot; pear; nectarine; raspberries; peach; tomato (2); apple juice; celeriac; carrot; baked beans; pickled onions; fruit jelly sweets; ground coffee; pectin; dairy spread.

P. cyclopium: Meat pies (9); sausages (2); cheese (3); bread (6); cake; mincemeat; maize; rusk; raw pastry (2); canned peaches; suet.

P. digitatum: Lemon (2); orange; tangerine.

P. echinulatum: Sausage roll; bread; jam (2).

P. expansum: Meat pie (2); sausage; cheese (2); Brie cheese; cake (4); walnut; soya meal; tomato; tomato puree; apple (2); pear (2); pet food; spaghetti; cream; dairy spread.

P. glandicola: Processed potato; pet food.

P. griseofulvum: Rice

P. hirsutum: Pork pie; cheese; bread.

P. italicum: Meat pie; bread; cake; satsuma; lemon; mustard.

P. olsonii: Bread; jam; rusk; tomato (3); orange juice; suet; ground coffee.

P. roqueforti: Fish pate; cheese (15); bread (5); jam (2); biscuit; fruit drink; gherkin; cucumber; pickled onions (2); fruit wine; pet food.

P. solitum: Meat pie (2); cheese (6); bread (2); processed potato; pet food.

P. verrucosum: Cake.

P. viridicatum: Meat pie (2); cheese (4); raw peanuts; maize (2); banana; suet.

Other *Penicillia*

Talaromyces flavus: Apple juice.

Eupenicillium cinnamopurpureum: Jam.

Penicillium citrinum: Bread; raw peanuts; maize; muesli.

P. corylophilum: Bread (2); cake; jam (many); animal feed; pickled onions (2); pepper.

P. fellutanum: Cake; suet.

P. glabrum: Mixed cereal; marrow.

P. islandicum: Suet; pepper.

P. janczweskii: Jam.

P. jensenii: Flour.

P. minioluteum: Pickled onions.

P. miczynskii: Cheese.

P. pinophilum: Animal feed.

P. purpurogenum: Soya meal; onion.

P. rugulosum: Flour.

P. simplicissimum: Pickled onion; food thickener.

P. spinulosum: Meat pie; bread; cake; spaghetti; carrot; pickled onion; onion; fruit drink; ground coffee.

P. variabile: Chicken paste; onion.

Aspergillus and its teleomorphs

Eurotium amstelodami: Cake (4); jam (3); raw peanuts.

E. chevalieri: Cake; fruit jelly; coconut cream.

E. herbariorum: Cake.

E. manginii: Cake (2).

E. medius: Malt extract.

E. repens: Bread; cake (5); jam (many); rice; paprika.

E. rubrum: Jam; raw peanuts; wheat; fruit drink; malt extract.

Emericella nidulans: Pepper.

Neosartorya fischeri: Raspberry puree.

Aspergillus candidus: Rice; processed potato; pepper.

A. flavus: Raw peanuts; palm kernels; maize (2); wheat; soya (2); rice; rusk; pet food; ground coffee; pepper (2); suet; margarine.

A. fumigatus: Bread; nuts; muesli; animal feed; marinated cherries; pepper.

A. niger: Cheese; bread (3); raw peanuts; pistache nuts; cocoa beans; sesame seed; pepper.

A. ochraceus: Fish food.

A. parasiticus: Animal feed.

A. restrictus: Jam (3); rice; malt extract.

A. tamarii: Muesli; pet food; pepper.

A. terreus: Maize; processed potato; animal feed; pepper.

A. ustus: Animal feed; dairy spread.

A. versicolor: Pepper; tartrazine solution.

A. wentii: Raw peanuts; cocoa beans.

IMPLICATIONS OF THE RANGE OF MOLDS ISOLATED

Although the moulds listed above were isolated from food samples that had visually spoiled, it is inappropriate to comment on their relative incidence, because of the wide variation in the size of the investigations. For example, although a considerable number of species have been found growing on pepper, they all came from one episode in which ground black pepper from a single source had been packed after moisture abuse. The pickled onion isolates were obtained in a similar manner. On the other hand, isolates from jams, cheese, bread and meat products were obtained from many samples over a number of years and are much more representative of the natural spoilage pattern in the U.K.

For practical identification purposes, the number of different species to be separated presents some difficulty. Of 59 different species identified, 40 were found five times or less and, of those, 17 were found only once. This means that a microbiologists may well lack the opportunity to gain familiarity with a considerable number of the species that can occasionally cause a spoilage problem. In this context, the single isolations of *Talaromyces flavus* and of *Neosartorya fischeri* are noteworthy. Both had caused substantial problems associated with heat processing and the recognition of these heat-resistant species played an important part in controlling the spoilage. Where species were isolated frequently, from different sources, certain trends are evident. Although *Eurotium repens* continues to be a major spoilage fungus in jams, *Penicillium corylophilum* and *P. brevicompactum* occur nearly as often. Both of these Penicillia are xerophilic. Perhaps the trend towards reduction in jam solids, as well as moisture migration during unsatisfactory storage, account for their regular isolation. Many other Penicillia are ubiquitous e.g. *P. aurantiogriseum*, *P. commune* and *P. cyclopium*. *P. commune* appears to have a particular affinity for protein such as meat and cheese. *P. expansum* and *P. italicum* are usually considered as agents of decay of fruit, yet both were isolated from various sources, *P. digitatum* in contrasts was only found on citrus fruit. Among the remaining Penicillia, few other clear trends were apparent, although *P. crustosum* was often isolated from fruit and vegetables, *P. olsonii* was less rare than the literature would suggest and *P. roqueforti* was the frequent spoilage agent of non mould-ripened hard cheeses, as well as growing on many other foods.

Aspergilli were, in general, found less often than Penicillia. This is undoubtedly because the majority of foods submitted to our laboratory have been of high water activity and intended for chill storage. *Aspergillus* spoilage usually occurred after accidental wetting of stable, low water activity foods. The only clear cut affinity was between *Eurotium* spp. and spoilage of jams and cakes.

For a food microbiology laboratory that has to deal with mould spoilage, it is apparent that some species may be expected to occur frequently, others rarely. The necessity remains to identify them accurately, in order to assess their significance, to trace their origins and thus to control and prevent the recurrence of spoilage. Now that mycology is advancing rapidly, with clarification of taxonomy, simplification of keying and new analytical procedures (e.g. ELISA and lectin binding) it should be possible to devise a strategy for investigating mould spoilage of foods. Such an approach might include a primary screening of isolates to select genera (by ELISA for example); a secondary range of tests to identify isolates as far as necessary, for example by using computer-assisted data-sorting and finally a library of information on the physiological and ecological characteristics of the fungi, to assist in problem solving. Most of the ingredients needed for this strategy now exist and only await assembly.

REFERENCES

WILLIAMS, A.P. 1990. Identification of *Penicillium* and *Aspergillus*: computer-assisted keying. In Modern Concepts in *Penicillium* and *Aspergillus* classification, eds. R.A. Samson and J.I. Pitt, pp. 287-292, New York and London: Plenum Press.

WILLIAMS, A.P. & BIALKOWSKA, A. 1985. Moulds in mould spoiled foods and food products. *Leatherhead Food R.A. Research Report* No. 527.

3

NOMENCLATURE: CONSERVATION AND STABILITY OF NAMES OF ECONOMICALLY IMPORTANT SPECIES

PROBLEMS AND PROSPECTS FOR IMPROVING THE STABILITY OF NAMES IN *ASPERGILLUS* AND *PENICILLIUM*

D.L. Hawksworth

CAB International Mycological Institute
Ferry Lane, Kew, Surrey TW9 3AF, UK

SUMMARY

The main reasons for instability in names in *Aspergillus* and *Penicillium* are reviewed, emphasizing the differences between those due to nomenclatural and taxonomic changes. Attention is drawn to the options now available under the Code for the conservation or rejection of names in the rank of species, including the conservation of types. Recent international initiatives to improve the stability of names are summarized, but in the final analysis taxonomic responsibility towards users is crucial. Rapidly published peer reviews by internationally established panels of specialists may assume increasing importance in the future.

INTRODUCTION

The stability of names is an emotive subject for both applied biologists and taxonomists, but for opposing reasons. Applied biologists are becoming increasingly frustrated and impatient, often refusing to accept changes (Bennett, 1985; Rossmore, 1988), whereas taxonomists regard any mention of restriction on name changes as a threat to their right to undertake objective scientific research and publish the results. With increased emphasis on the relevance of research to practical applications, systematic work which leads to instability in the vocabulary of applied biology finds difficulty in obtaining support. Indeed, stability in names is one of the primary demands of the consumers of the products of taxonomy, that is all users of scientific names. The need for a common language has been accentuated in recent years by the requirements of computerized bibliographies and culture information retrieval services, quarantine, and health and safety regulations. Lack of attention to consumer requirements by taxonomists has been identified as the main reason why the massive demand for the practical products of systematics is not matched by the level of resources allocated to the subject (Hawksworth and Bisby, 1988). Perhaps the greates challenge facing systematists into the next century is the adaptation of their practices to better satisfy consumer demands and so regain the confidence and respect of their contemporaries.

In the case of *Penicillium* Link, only 64 (47%) of the names accepted by Raper and Thom (1949) were adopted in the monograph of Pitt (1979); the real level of stability will be even less than these figures indicate as the circumscriptions and typifications of some of the species names still employed were also changed. The current position is confusing to users and ways must be found to alleviate this situation (Onions *et al.*, 1984; Samson and Gams, 1984).

This contribution endeavours to identify the main reasons for such substantial levels of change, and ways in which these can be minimized under the current International Code of Botanical Nomenclature (Greuter *et al.*, 1988). Recent international initiatives to increased nomenclatural and taxonomic stability are reviewed, and action the Subcommision on *Penicillium* and *Aspergillus* Systematics (SPAS) of the International

Modern Concepts in Penicillium and Aspergillus Classification
Edited by R. A. Samson and J. I. Pitt
Plenum Press, New York, 1990

Commission on the Taxonomy of Fungi (ICTF) could initiate to limit the possibilities for future changes in these genera is discussed.

Changes in names can arise in one of two ways: either from the application of the International Code of Botanical Nomenclature (i.e. nomenclatural instability), or from new scientific evidence and the differing interpretations of taxonomists (i.e. taxonomic instability). These different situations require disparate solutions, and the problems and prospects for each category are considered in turn below.

NOMENCLATURAL INSTABILITY

Problems

The International Code of Botanical Nomenclature provides a system of 76 mandatory rules (Articles), and also Recommendations, aimed at promoting nomenclatural stability. The Code lays down criteria for effective and valid publication, the ways in which names are formed, typification procedures, the choice of name when several compete, conditions under which names can be automatically rejected, etc.

Proposals to modify any provisions in the Code can be made to each International Botanical Congress; proposals are voted on in a mail ballot, debated and voted on at the Nomenclature Session of the Congress, and if they gain the necessary substantial majority (two-thirds is normally required) are incorporated into the new edition of the Code published after each Congress. Even though the Code was approved at the First International Botanical Congress held in 1867 (De Candolle, 1867), the number of proposals continues to be substantial; 349 were made to the 1987 Berlin Congress (Greuter and McNeill, 1987). Changes in the Code itself are sometimes operative only after the Congress, but in many cases they are retrospective. Such alterations can consequently endanger names which were correct under previous editions of the Code.

A major cause of name changes in both *Aspergillus* Micheli ex Link and *Penicillium* was the fundamental retrospective revision of the rules concerning fungi with pleomorphic life cycles, Art. 59, adopted by the 1981 Congress in Sydney (Voss *et al.*, 1983). Even prior to this, students of these genera had been reluctant to consider that the names could not be used for all states (Raper and Thom, 1949; Raper and Fennell, 1965). I quote Bennett (1959, p. 584): "I have been calling, and will continue to call, both the perfect (teleomorphic) and imperfect (anamorphic) state of this fungus *Aspergillus*". Against this background particularly destabilizing was the decision that even if a species name was proposed under an anamorphic generic name, if the description and the type included the sexual ascosporic stage then the name had to be applied to the teleomorph and was no longer available for the anamorph, the conidial state. The effect this was forecast to have in genera such as *Aspergillus* and *Penicillium* (Hawksworth and Sutton, 1974a,b) was passed over in favour of simplification of the Code (Weresub, 1979). Samson and Gams (1985) and Pitt (1979) had to introduce 24 and 15 new anamorph names respectively as a direct result of this single change; many were for well-known species. While such a massive number of changes are unlikely to be necessary for this reason again, when a "sclerotium" proves to be an immature non-ostiolate ascoma in a *Penicillium* type, the name will no longer be usable in that genus if the structure was included in the original description.

Changes due to other nomenclatural provisions have had substantially less impact than those due to the revision of Art. 59. For example, the revision of the starting point for the nomenclature of conidial fungi from 1 January 1821 back to 1 May 1753 (Demoulin *et al.*, 1981), only led to the simplification of some author citations in these two genera (e.g. Hawksworth, 1985).

Mycologists still all too frequently fail to deposit dried cultures as holotypes for the names of species described from culture. Living cultures are specifically excluded as acceptable as the types of names under Art. 9.5, and names not complying with this provision published after 1 January 1958 are not valid (Art. 37.1). Most important pre-1958 names in *Aspergillus* and *Penicillium* have now been lectotypified by Samson and Gams (1985) and Pitt (1979) respectively so this should not now threaten names currently in use. Proposals to change this provision by the International Mycological Association (van Warmelo, 1979) were not accepted. This matter is also of concern to phytologists and yeast specialists particularly. The debate continues, although a proposal to establish a Special Committee on living types was unfortunately rejected at the 1987 Congress (Greuter *et al.*, 1989, pp. 59-60). The value of dried material should not, however, be underestimated; it is amenable to scanning electron microscopy, analysis of secondary metabolites (Paterson and Hawksworth, 1985), and DNA extraction and amplification from minute samples can be expected to be possible shortly.

A greater nomenclatural threat is the discovery of earlier names for accepted species. In both *Aspergillus* and *Penicillium* considerable numbers of names are of uncertain application; Raper and Fennell (1965) listed 70 in *Aspergillus* and Pitt (1979) 175 in *Penicillium*. Some of these epithets are among the earliest in the genera and could potentially threaten common species: *A. griseus* Link, *A. virens* Link, *A. flavescens* Wreden, *P. fasciculatum* Sommerf., *P. glaucum* Link, and *P. nigrescens* Jungh. These could potentially be resurrected by neotypification, and the First International *Penicillium* and *Aspergillus* Workshop recommended: "That old and unfamiliar names for which no holotype or material suitable for lectotypification exists are not reintroduced by neotypification other than in exceptional circumstance" (Samson and Pitt, 1985, p. 455). A proposal to incorporate this into the Berlin Code received some support but was regretably rejected (Greuter *et al.*, 1989, p. 46).

Prospects

While the nomenclatural problems surrounding the issue of fungi with pleomorphic life cycles and occasioned by later starting points have now been clarified and embodied in the Code, nomenclatural changes due to the taking up of older names (Art. 11.2) remain a threat. Weresub (1979) counselled that mycologists should not seek too many special provisions in the Code, and I concur; name changes due to the rule of priority at species rank concern all groups.

Disadvantageous changes in wellknown family and generic names due to the strict application of the Code have long been avoidable by invoking the conservation procedures (Art. 14). At the Sydney Congress in 1981 this possibility was opened up for the first time for specific names of "species of major economic importance" (Art. 14.2; Voss et al., 1983). While "case law" needs to develop to define "major economic importance", species such as *Aspergillus fumigatus* Fres., *A. niger* van Tieghem, *Penicillium chrysogenum* Thom and *P. roqueforti* Thom would clearly be expected to fulfil this requirement.

In addition to overcoming threats to well-known names due to priority of publication, under Art. 14.8 names can be conserved with different types. This provision has not yet been adopted for any fungal name, but appears to provide a possible mechanism for avoiding particularly unfortunate changes occasioned by the revised Art. 59, as in the case of *A. nidulans* (Frisvad *et al.*, see p. 83-89).

The possibility of widening the scope for the conservation of specific names was introduced at the 1987 Berlin Congress where it was agreed that cases of rejected misapplied names previously handled under Art. 69 could be treated in a similar manner to those being conserved (Art. 69.3). Art. 69 applies to taxa where significant confusion might result, but is not confined to species of "major economic importance". This procedure, and that of conservation on grounds where names are endangered by a strict adherance to priority, merit greater use for specific names by mycologists than has so far been the case. Before a name is changed, a responsible taxonomist should always determine whether the conservation procedures could be invoked. Proposals backed by an international group of specialists such as the present Workshop or SPAS are more likely to be favourably received than ones from individuals.

Conservation procedures are lengthy and time consuming. Proposers are required to publish formal proposals which can then be considered and voted on by the appropriate Committee. Approval has to await the subsequent International Botanical Congress and the process can consequently take up to six years, or even longer if the Committee does not reach a clear majority decision quickly.

The current priority rule and restricted use of conservation procedures for species names fails to achieve the primary aim of the Code: "... the provision of a stable method of naming taxonomic groups, avoiding and rejecting the use of names which may cause error or ambiguity or throw science into confusion" (Greuter *et al.*, 1988, Preface, p. 1). Since 1985 the matter has been addressed at a series of international meetings and special working groups. Resultant proposals (Hawksworth, 1988a; Hawksworth and Greuter, 1989) aim at preparing Lists of Names in Current Use for all groups of organisms covered by the Code, and subject to decisions at the next International Botanical Congress in Tokyo in 1993, when being given a specially protected nomenclatural status at that time. This initiative was accepted as a part of the International Union of Biological Sciences (IUBS) scientific programme in 1988 and the Commission on Botanical Nomenclature (International Association for Plant Taxonomy General Committee) appointed a Special Committee in March 1989 to initiate the preparation of lists and prepare proposals for the 1993 Congress. The 36,500 accepted generic names are being listed first, together with trials at species rank for different groups.

This move has been partly inspired by the success of the "Approved Lists of Bacterial Names" (Skerman *et al.*, 1980). By that single step and the devalidation of all omitted names the number of species epithets available for use by bacterial taxonomists was reduced from about 21,000 to 1,800. Approved names were linked to a revised Code and procedures established for the registration of newly published names (Sneath, 1986). In the case of botanical nomenclature, it is not envisaged that names omitted from the Lists would be devalidated; they would still be available for use so long as existing requirements for validity and legitimacy were fulfilled, but could not be taken up in preference to a listed name even if of an earlier date. If adopted, that procedure would in practice parallel that for the "sanctioning" of names of fungi embodied in the Code since 1981 (Demoulin *et al.*, 1981; Korf, 1982).

It is important to emphasize that the Lists initiatives are concerned with nomenclature and not taxonomy. Names listed are envisaged as including those necessary to enable alternative taxonomic schemes to be employed. They can be expected to promote productive taxonomic work through taxonomists having to spend less time investigating obscure long-neglected literature and more time collecting data and providing soundly based schemes. What Cowan said of the former state of bacterial taxonomy applies also to fungi: "too much attention has been given to nomenclature and too little to the bacteria themselves, their characters and what they do" (Cowan, 1970, p. 145).

Procedures for the treatment of newly published names are being addressed separately by a complementary Special Committee on Registration charged with examining various options (Greuter, 1986; Brummitt *et al.*, 1986) also charged with reporting to the 1993 Congress.

Aspergillus and *Penicillium* species names are now some of the best catalogued and typified in the fungi. In the light of the work of the current series of workshops, they are now approaching consensus systems. They would now be ideal for a pilot study for the Lists of Names in Current Use initiative, especially in view of their particular economic importance. The resultant Lists could also be commended for adoption if "sanctioned" by the International Commission on the Taxonomy of Fungi in the period prior to such procedures formally being embodied in the Code.

TAXONOMIC INSTABILITY

Problems

Science is progressive, concepts are tested as new data become available, and where necessary modified to better explain the new observations. Objective taxonomic changes are therefore to be expected and indeed welcomed by users, Revisions should lead to clearer understandings of species limits, evolutionary relationships and, more importantly, increased levels of predictability of functional attributes (e.g. toxin production, pathogenicity, enzyme and metabolite production, growth parameters).

Inevitably, differences in interpretation of the same data set will occur and often several stances will be logically justifiable. Variant systems then need testing by new or more extensive data to see which better fits the evidence. In taxonomy it should not pass unnoticed that substantial monographs based on large data sets of both specimens and characters tend to be the most rapidly accepted by peers and also to stand the rigorous test of time.

Taxonomic instability, in the sense of repeated switching and modification, is generally associated with poor standards of taxonomic research or independent data sets. This is especially so when changes are based on studies of few isolates or specimens (sometimes even wrongly named), and consider very few characters. It is a common failing of the pioneers of new approaches to consider that their novel techniques or data sets provides *the* answer. History generally proves otherwise, and it is prudent to consider the experience of taxonomists in other groups where particular techniques have already been employed more extensively (Hawksworth, 1988b).

It would be invidious to single out particular studies in *Aspergillus* and *Penicillium* here, but as a cautionary pertinent example the lichen-forming *Ramalina siliquosa* (Huds.) A.L. Sm. may be mentioned. Culberson (1967) recognized six species on the basis of secondary metabolites. Comprehensive numerical taxonomic studies (Sheard, 1978), and protein patterns (Mattsson and Kärnefelt, 1986) have shown two species to be present, as already recognized by field observations since the 1860s.

Information on the applicability of some of the newer techniques to mycology is provided in Jury and Cannon (1989) and Hawksworth and Bridge (1988).

Prospects

In the long-term, the main need for the production of greater taxonomic stability is funding for major multidisciplinary integrated studies. In the case of the filamentous fungi, these approaches are being pioneered in *Aspergillus* (Chapter 5, this volume) and *Penicillium* (Bridge *et al.*, 1990), and further through the initiatives of SPAS.

Users are often uncertain as to which of two or more differing taxonomies to follow, rarely being in a position to assess the merits of each objectively. A case for guidance from international peer groups exists and the ICTF has made a start in this direction. Published changes are reviewed by the international panel of 11 members that make up the Commission, and their views published in the series "Name changes in fungi of microbiological, industrial, and medical importance" (Cannon, 1986). Authors are less likely to rush into print if peer reviews can be published widely and rapidly. There have also been proposals to produce standard lists of names reflecting a single taxonomy reviewed by internationally appointed peer groups on perhaps a five-yearly basis (Barnett, 1987). While of great value to users, such proposals remain unpopular as they could tend to stifle the development of science and delay sound improvements being widely or quickly adopted. In the case of the generic names of the Ascomycotina, outline classifications produced since 1982 by an interactive process and revised annually are proving to be of value (Eriksson, 1989; Hawksworth, 1989; Reynolds, 1989). The twice yearly journal "Systema Ascomycetum" is devoted to that project, and is being considered as a model that might be applicable in other groups of organisms.

Most importantly, the standards of training of taxonomists need to be improved. Mycologists can learn from treatises primarily written for other groups (e.g. Austin and Priest, 1986; Davis and Heywood, 1963; Heywood and Moore, 1984; Jeffrey, 1982; Stace, 1982) as well as those directed at mycologists (e.g. Hawksworth, 1974). In the case of *Aspergillus* and *Penicillium*, the Recommendations produced by the First International *Penicillium* and *Aspergillus* Workshop (Samson and Pitt, 1985) are of particular value in promoting good practic; the Second Workshop may wish to consider taking this further. That Workshop also encouraged the ICTF to prepare a Code of Practice for Systematic Mycologists which was subsequently drafted, agreed and issued (Sigler and Hawksworth, 1987); Chinese and Russian versions of the ICTF Code have also been prepared in the final analysis, taxonomic responsibility towards users will remain crucial.

ACKNOWLEDGEMENTS

The proposals made in this paper partly arose from the results of Science and Engineering Research Counci (SERC) contract no. SO/17/84.

REFERENCES

AUSTIN, B. and PRIEST, F. 1986. Modern Bacterial Taxonomy. 145 pp. Workingham: Van Norstrand Reinhold.
BARNETT, J.A. 1987. Classification. *Nature, London* 325: 384.
BENNETT, J.W. 1985. Taxonomy of fungi and biology of the Aspergilli. *In* Biology of Industrial Microorganisms eds. A.L. Demain and N.A. Solomon, pp. 359-406. Melno Park, Calif.: Benjamin/Cummings Publishing.
BRIDGE, P.D., HAWKSWORTH, D.L., KOZAKIEWICZ, Z., ONIONS, A.H.S., PATERSON, R.R.M., SACKIN, M.I. and SNEATH, P.H.A. 1990. A reappraisal of the terverticillate Penicillia using biochemical, physiological and morphological features. *In* Modern Concepts in *Penicillium* and *Aspergillus* classification, eds. R.A. Samson and J.I. Pitt, pp. 139-147. New York and London: Plenum Press.
BRUMMITT, R.K., HAWKSWORTH, D.L. and McNEILL, J. 1986. Proposals for a method of defining effective publication by means of approved publications. *Taxon* 35: 823-826.
CANNON, P.F. 1986. Name changes in fungi of microbiological, industrial, and medical importance. *Microbiological Sciences* 3: 168-171, 285-287.
COWAN, S.T. 1970. Heretical taxonomy for bacteriologists. *Journal of General Microbiology* 61: 145-154.
CULBERTON, W.L. 1967. Analysis of chemical and morphological variation in the *Ramalina siliquosa* species complex. *Brittonia* 19: 333-352.

DAVIS, P.H. and HEYWOOD, V.H. 1963. Principes of Angiosperm Taxonomy. 556 pp. Edinburgh and London: Oliver and Boyd.

DE CANDOLLE, A. 1867. Lois de la Nomenclature Botanique. Paris: Masson and fils.

DEMOULIN, V., HAWKSWORTH, D.L., KORF, R.P. and POUZAR, Z. 1981. A solution to the starting point problem in the nomenclature of fungi. *Taxon* 30: 52-63.

ERIKSSON, O. 1989. Eclectic mycology: a response to Reynolds. *Taxon* 38: 64-67.

GREUTER,W. 1986. Proposals on registration of plant names, a new concept for the nomenclature of the future. *Taxon* 35: 816-819.

GREUTER, W. et al., 1988. International Code of Botanical Nomenclature adopted by the Fourteenth International Botanical Congress, Berlin, July-August 1987. Regnum Vegetabile 118: 1-328.

GREUTER, W. and McNEILL, J. 1987. Synopsis of proposals on botanical nomenclature Berlin 1987. *Taxon* 36: 174-281.

GREUTER, W., McNEILL, J. and NICOLSON, D.H. 1989. Report on botanical nomenclature - Berlin 1987. *Englera* 9: 1-228.

HAWKSWORTH, D.L. 1974. Mycologists' Handbook. 231 pp. Kew, Surrey: Commonwealth Mycological Institute.

—— 1985. The typification and citation of the generic name *Penicillium. In* Advances in *Penicillium* and *Aspergillus* Systematics, eds. R.A. Samson and J.I. Pitt, pp. 3-7. New York and London: Plenum Press.

—— 1988a. Improved stability for biological nomenclature. *Nature, London* 334: 301.

——, ed. 1988b. Prospects in Systematics. Oxford: Clarendon Press.

—— 1989. Taxonomic stability. *Nature, London* 337: 416.

HAWKSWORTH, D.L. and BISBY, F.A. 1988. Systematics: the keystone of biology. *In* Prospects in Systematics, ed. D.L. Hawksworth, pp. 3-30. Oxford: Clarendon Press.

HAWKSWORTH, D.L. and BRIDGE, P.D. 1988. Recent and future developments in techniques of value in the systematics of fungi. *Mycosystema* 1: 5-19.

HAWKSWORTH, D.L. and GREUTER, W. 1989. Report of the first meeting of a working group on Lists of Names in Current Use. *Taxon* 38: 142-149.

HAWKSWORTH, D.L. and SUTTON, B.C. 1974a. Article 59 and names of perfect state taxa in imperfect state genera. *Taxon* 23: 563- 568.

—— 1974b. Comments on Weresub, Malloch and Pirozynski's proposal for Article 59. *Taxon* 23: 659-661.

HEYWOOD, V.H. and MOORE, D.M., eds. 1984. Current concepts in plant taxonomy. 432 pp. London: Academic Press

JEFFREY, C. 1982. An introduction to plant taxonomy. 154 pp. Second edition. Cambridge: Cambridge University Press.

JURY, S.L. and CANNON, P.F., eds. 1989. Novel approaches to the systematics of fungi. *Botanical Journal of the Linnean Society* 99: 1-95.

KORF, R.P. 1982. Mycological and lichenological implications of changes in the Code of Nomenclature enacted in 1981. *Mycotaxon* 14: 476-490.

MATTSSON, J.E. and KÄRNEFELT, I. 1986. Protein banding patterns in the *Ramalina siliquosa* group. *Lichenologist* 18: 231-240.

ONIONS, A.H.S., BRIDGE, P.D. and PATERSON, R.R.M. 1984. Problems and prospects for the taxonomy of *Penicillium. Microbiological Sciences* 1: 185-189.

PATERSON, R.R.M. and HAWKSWORTH, D.L. 1985. Detection of secondary metabolites in dried cultures of *Penicillium* preserved in herbaria. *Transactions of the British Mycological Society* 85: 95-100.

PITT, J.I. 1979. The Genus *Penicillium* and its teleomorpic states *Eupenicillium* and *Talaromyces*. 634 pp. London: Academic Press.

RAPER, K.B. and FENNELL, D.I. 1965. The Genus *Aspergillus*. 686 pp. Baltimore: Williams and Wilkins.

RAPER, K.B. and THOM, C. 1949. A Manual for the Penicillia. 875 pp. Baltimore: Williams and Wilkins.

REYNOLDS, D.R. 1989. Consensus mycology. *Taxon* 38: 62-63.

ROSSMORE, H.W. 1988. A rose by any other name. *International Biodeterioration* 23: 325-327.

SAMSON, R.A. and GAMS, W. 1984. The taxonomic situation in the hyphomycete genera *Penicillium, Aspergillus,* and *Fusarium. Antonie van Leeuwenhoek* 50: 815-824.

SAMSON, R.A. and GAMS, W. 1985. Typification of the species of *Aspergillus* and associated teleomorphs. *In* Advances in *Penicillium* and *Aspergillus* Systematics, eds. R.A. Samson and J.I. Pitt, pp. 31-54. New York and London: Plenum Press.

SAMSON, R.A. and PITT, J.I., eds. 1985. Advances in *Penicillium* and *Aspergillus* Systematics. 438 pp. New York and London: Plenum Press.

SHEARD, J.W. 1978. The taxonomy of the *Ramalina siliquosa* species aggregate (lichenized Ascomycetes). *Canadian Journal of Botany* 56: 915-938.

SIGLER, L. and HAWKSWORTH, D.L. 1987. International Commission on the Taxonomy of Fungi: Code of practice for systematic mycologists. *Microbiological Sciences* 4: 82-86; *Mycopathologia* 99: 3-7; *Mycologist* 21: 101-105.

SKERMAN, V.B.D., McGOWAN, V. and SNEATH, P.H.A., eds. 1980. Approved lists of bacterial names. *International Journal of Systematic Bacteriology* 30: 225-420.

SNEATH, P.H.A. 1986. Nomenclature of bacteria. In Biological Nomenclature Today, eds. W.D.L. Ride and T. Younes, pp. 36-48. Eynesham, Oxford: IRL Press.

STACE, C.A. 1989. Plant Taxonomy and Biosystematics. Second edition. 264 pp. London: Edward Arnold.

VOSS, E.G. et al., 1983. International Code of Botanical Nomenclature adopted at the Thirteenth International Botanical Congress, Sydney, August 1981. *Regnum Vegetabile* 111: 1-472.

van WARMELO, K.T. 1979. Proposals for modifications of the Code of Botanical Nomenclature: IMC2 proposals. *Taxon* 28: 424-431.

WERESUB, L.K. 1979. Mycological nomenclature: reflections on its future in the light of its past. *Sydowia Beihefte* 8: 416-431.

PROPOSALS TO CONSERVE IMPORTANT SPECIES NAMES IN *ASPERGILLUS* AND *PENICILLIUM*

J.C. Frisvad[1], D.L. Hawksworth[2], Z. Kozakiewicz[2], J.I. Pitt[3],
R.A. Samson[4] and A. C. Stolk[4]

[1]*Dept of Biotechnology*
Technical University of Denmark
2800 Lyngby, Denmark

[2]*CAB International Mycological Institute*
Kew, Surrey, TW9 3AF, United Kingdom

[3]*CSIRO*
Division of Food Processing
North Ryde, N.S.W. 2113, Australia

[4]*Centraalbureau voor Schimmelcultures*
3740 AG Baarn, The Netherlands

Although the Botanical Code has allowed for conservation of generic names for a number of years, until recently no mechanism existed for protecting important species names. For example, under the Botanical Code prevailing at the time, Samson *et al.* (1976) and Pitt (1979) had no alternative to taking up earlier names for currently accepted species, at least in cases where it could be shown that well documented neotypes existed. In particular, Samson *et al.* (1976) revived *P. verrucosum* Dierckx for *P. viridicatum* Westling, while Pitt (1979) took up *P. aurantiogriseum* Dierckx 1901 for *P. cyclopium* Westling 1911 and *P. glabrum* (Wehmer) Westling 1893 for *P. frequentans* Westling 1911. Both Samson *et al.* (1976; 1977) and Pitt (1979) evaded the issue of the relationship of *P. griseoroseum* Dierckx 1901 with *P. chrysogenum* Thom 1910, clearly in the hope that a method might eventuate for saving the obviously threatened *P. chrysogenum*. More recent work (Cruickshank and Pitt, 1987) has clearly demonstrated the synonymy of these two species.

In the meantime, a mechanism for conserving threatened specific names has been added to the Botanical Code. The Sydney Botanical Congress voted to allow conservation of species where they were "of major economic importance" (Art. 14, Voss *et al.*, 1983). Consequent on that change, this paper foreshadows a formal proposal to conserve *P. chrysogenum* against *P. griseoroseum* at the next Botanical Congress. The case for conservation rests on the importance of *P. chrysogenum* as the one species which produces penicillin, the antibiotic which has revolutionised the treatment of gram-positive bacterial infections in man and animals. To permit the synonymising of *P. chrysogenum* is unacceptable, indeed unthinkable.

Although recent work has threatened other *Penicillium* names in current use, e.g. *P. granulatum* Bainier 1907, *P. claviforme* Bainier and *P. concentricum* Samson *et al.* 1976 (Seifert and Samson, 1985), these species are of limited importance, and it has been judged unnecessary, and probably unrewarding, to attempt conservation. Recent taxonomies have accepted the older names *P. glandicola* (Oudem.) Seifert & Samson, *P. vulpinum* (Cooke & Massee) Seifert & Samson and *P. coprophilum* (Berk. & Curt.) Seifert & Samson respectively (Pitt, 1988; Frisvad and Filtenborg, 1989).

Modern Concepts in Penicillium and Aspergillus Classification
Edited by R. A. Samson and J. I. Pitt
Plenum Press, New York, 1990

The taxonomy of *Aspergillus* has not been so thoroughly revised as that of *Penicillium* in recent years. However, two obviously threatened names of great importance exist, *A. niger* van Tieghem and *A. nidulans* (Eidam) Winter 1884.

Al-Musallam (1980) pointed out that two older species, *A. phoenicis* (Corda) Thom 1840 and *A. ficuum* (Reichardt) Hennings 1867, accepted by Thom and Raper (1945) and Raper and Fennell (1965), were synonymous with *A. niger*, or at most *A. niger* represented a variety of a single earlier species. A clear case exists for conserving *A. niger* against these two species, as *A. niger* is the source of commercial production of citric and other organic acids around the world, and clearly of major economic importance.

A. nidulans is threatened for a different reason, outlined more fully in the section below, which relates to the use of holomorphic names in anamorphic genera. The case for conservation is equally compelling: *A. nidulans* has been used in a wide range of significant genetic studies, and adoption of the legitimate name for the anamorph, *A. nidulellus* Samson & Gams (1985) would lead only to long term confusion.

Samson and Gams (1985) suggested revival of *A. alutaceus* Berk. & Curt. 1875 for *A. ochraceus* Wilhelm 1877, but later authors (Klich and Pitt, 1988) have not accepted this change, on the grounds that synonymy of these two species has not been clearly established. If synonymy becomes accepted, a case for conservation of *A. ochraceus* may have to be developed, but the chances of success under the present Botanical Code appear slim.

Stability in names of *Penicillium* and *Aspergillus* may, in the slightly longer term, come from current proposals to develop lists of protected names (Hawksworth, 1990; see also the final discussion in these proceedings). However, at this time, conservation of *P. chrysogenum*, *A. niger* and *A. nidulans* appears to be essential.

The following sections outline proposals which will be placed before the relevant committees for consideration at the next Botanical Congress in 1993. The formal proposals will be prepared and submitted for publication in *Taxon*.

Proposal to conserve the name *Aspergillus niger* van Tieghem against *Aspergillus phoenicis* (Corda) Thom and *A. ficuum* (Reichardt) Hennings

The name of the very common and industrially important species *Aspergillus niger* is threatened by the older names *A. phoenicis* (Corda) Thom and *A. ficuum* (Reichardt) Hennings (Al-Musallam, 1980). Conservation of *A. niger* is proposed because of its great industrial and economic importance. The earlier names have been used only rarely in modern publications.

Ustilago phoenicis Corda, Icon. Fung. 4: 9, 1840 = *Aspergillus phoenicis* (Corda) Thom, J. Agric. Res. 7: 14, 1916 = *Aspergillus niger* var. *phoenicis* (Corda) Al-Musallam, Rev. Black *Aspergillus* Species: 56, 1980.

Holotype material of *U. phoenicis*, on fruit of *Phoenix dactylifera* in PRM. Representative culture: CBS 126.49 = ATCC 10698 = NRRL 363.

Ustilago ficuum Reichardt, Verh. Zool. Bot. Ges. Wien 17: 335, 1867 (issue 1 of volume; month not determinable exactly). = *Aspergillus ficuum* (Reichardt) Hennings, Hedwigia 34: 86, 1985.

No type material is preserved at W. The identity of this taxon is established by CBS 115.34, chosen by Raper & Fennell (1965) as a representative isolate. Though the date of the original publication could not be established with certainty, it appears probable that this name predates the publication of *A. niger* by a few months.

Aspergillus niger van Tieghem, Annls Sci. nat. Bot., Sér. 5, 8: 240, 1867 (issue 4 of the volume; month not determinable exactly).

Holotype material is preserved in PC. Representative culture: CBS 554.65 = ATCC 16888 (designated as "neotype" by Al-Musallam, 1980).
Full descriptions in Raper & Fennell (1965), Al-Musallam (1980), and Klich and Pitt (1988).

A number of additional, later synonyms of *A. niger* have been compiled by Al-Musallam (1980).

A. phoenicis was distinguished from *A. niger* at specific level by Raper & Fennell (1965) and at varietal level by Al- Musallam (1980), because of smaller, longitudinally striate, slightly prolate conidia. Recent cryo-SEM pictures of numerous isolates from *Aspergillus* Sect. *Nigri* have shown that striation and tubercles intergrade freely and that in CBS 554.65 a considerable amount of striation is visible. Thus this distinction represents only a gradation. Even if the two taxa were separated as varieties within a single species, as Al-Musallam (1980) proposed, the nomenclatural consequence is that *A. niger* must become a variety of *A. phoenicis*.

Raper & Fennell (1965) distinguished *A. ficuum* from *A. niger* by more rapidly spreading colonies on Czapek agar and from *A. phoenicis* by perfectly globose conidia. Al-Musallam (1980) included *A. ficuum* as a synonym of *A. niger* var. *niger*.

The taxonomic distances between representative isolates of these taxa were assessed by Al-Musallam (1980), using cluster analysis involving all available morphological parameters, after both equal and iterative weighting of characters. Frisvad (unpublished data) has indicated that *A. niger* and *A. phoenicis* possess numerous secondary metabolites in common, suggesting a lack of separation even at varietal level.

Polonelli *et al.* (1985) found that neither immunodiffusion nor two dimensional immunoelectrophoresis (TDIE) permitted an antigenic differentiation of any of the several isolates of *A. niger*, *A. phoenicis* and *A. pulverulentus* studied.

Frisvad (unpublished data) also found identical pectinase isoenzyme profiles in the isolates concerned.

Economic importance.
A. niger is commonly used for the commercial production of citric and other organic acids. In previous decades it has been used frequently as an indicator for trace elements and in numerous assays (Domsch *et al.*, 1980). Its economic value in citric acid fermentation can be estimated from the fact that world-wide 5850 million litres are produced annually.

Test searches in the BIOSIS databank have shown that in the last decade annually some 300 references deal with *A. niger*, 2 with *A. phoenicis* and 2 with *A. ficuum*.

Proposal to retain the name *Aspergillus nidulans* (Eidam) Winter by conservation of *Sterigmatocystis nidulans* Eidam with a conserved type specimen

Sterigmatocystis nidulans Eidam, in Cohn, Beitr. Biol. Pfl. 3: 392, 1883; type:- ? France, sine loc., ex coll. G. Bainier (IMI 86806 - stat. anam.); typ. cons. prop. = *Aspergillus nidulans* (Eidam) Winter, Rabenh. Krypt. - Fl. 1: 62, 1884.

Aspergillus nidulellus Samson & W. Gams, in Samson and Pitt, Adv. *Penicillium* and *Aspergillus* Syst.: 44, 1985 (1986).

The revision of Art. 59.1 adopted by the XIII International Botanical Congress in 1981 (Voss *et al.*, 1983), endangered many well known species names of anamorphs in *Aspergillus* and *Penicillium*, as already anticipated by Hawksworth and Sutton (1974). However, that same Congress also amended Art. 14.1 to permit the conservation of epithets at species rank. Under Art. 14.8, it is further possible to conserve a name with a type different from that designated by the author or determined by application of the Code; this provision is not restricted by rank. The familiar fungus *Aspergillus nidulans* was first described as *Sterigmatocystis nidulans* by Eidam (1883). His description included both the anamorphic and teleomorphic states. His excellent illustrations clearly indicate an *Aspergillus*, and the biseriate columnar conidial heads point unequivocally to *Aspergillus nidulans* as identified by later authors. Winter (1884) transferred the name *S. nidulans* to *Aspergillus* as *A. nidulans* (Eidam) Winter, and again the description included both the anamorphic and teleomorphic states. Later Vuillemin (1927) transferred *S. nidulans* to *Emericella* Berk. & Broome, as *Emericella nidulans* (Eidam) Vuill. Thus the name *Sterigmatocystis nidulans* functions as the basionym for binomials in both *Emericella* and *Aspergillus*.

Under the current Art. 59, names introduced inclusive of both teleomorph and anamorph are automatically typified by the teleomorph (Art. 59.6) and do not cover the anamorph alone. Samson and Gams (1985) in addition to the epithet *nidulans* for the teleomorph, introduced the new species name *Aspergillus nidulellus* for the anamorph, quite correctly under this provision of the Code (Greuter et al., 1988).

However, *Aspergillus nidulans* is one of the most familiar names of filamentous fungi to biologists and in the field of fungal genetics is second only in prominence to *Neurospora crassa*. Numerous papers and reviews (approximately 100 per annum) on production, detection and isolation of mutants, mechanisms of mitotic recombination, arrangement of alleles and development of chromosome maps, have been published using *A. nidulans* as the research tool (e.g. Gossop *et al.*, 1940; Pontecorvo, 1949; Jinks, 1952; Pontecorvo *et al.*, 1953; Käfer, 1958; Grindle, 1963; Clutterbuck, 1974; Smith and Pateman, 1977). More recently it has been used as one of the best model systems for the study of eukaryotic gene expression and its control (e.g. Cove, 1977; Ballance and Turner, 1985; Arst and Scazzochio, 1985).

Culture collections throughout the world maintain mutant strains of *A. nidulans* (IMI, ATCC, CBS, IFO, JCM, NRRL, IMUR, BKMF). Indeed there are many genetics laboratories which maintain their own *A. nidulans* genotypes (see Clutterbuck, 1974; 1986), and the Fungal Genetics Stock Center has 616 genetic variants of this species (Fungal Genetics Stock Center, 1988).

It has recently been shown that *A. nidulans* can be genetically engineered to secrete proteins of bacterial or mammalian origin (Gwynne *et al.*, 1987; Upshall *et al.*, 1987), potentially a first step towards the use of filamentous fungi for the production of many important pharmaceutical and industrial processes. It poses an attractive proposition for the efficient, inexpensive production of valuable and useful proteins. Thus, the name

Aspergillus nidulans has been widely used and accepted for more than 100 years by geneticists and biochemists, as well as by mycologists. The change to *A. nidulellus* in this context is highly undesirable, most unlikely to be followed, and will cause confusion and difficulties in on-line retrieval of data. The above proposal will safeguard this name.

With the proposed typification, the name of the teleomorph cannot be based on Eidam's epithet. However, in using the epithet in *Emericella*, Vuillemin (1927: 137) was clearly basing his decision on the character of the teleomorph. Under Art. 59.6, however, the name can be regarded as attributed to Vuillemin alone provided that it is neotypified by the material of the teleomorph, as follows:

Emericella nidulans Vuillemin, C. r. Acad. Sci. Paris 184: 137, 1927; neotype:- ? France, sine loc., ex coll. G. Bainier (IMI 86806 - stat. teleo.).

The Bainier strain used for the typification of both morphs is that upon which the diagnosis of this taxon in both its morphs was based primarily by Raper and Fennell (1965: 497). Further, this same strain was accepted in part as neotype for Eidam's teleomorphic taxon and in part as holotype for *A. nidulellus* by Samson and Gams (1985: 44).

Proposal to conserve *Penicillium chrysogenum* Thom as the species name for the principal producer of penicillin.

Penicillium chrysogenum Thom, Bull. Bur. Anim. Ind. US Dep. Agric. 118: 58, 1910.
    Penicillium griseoroseum Dierckx, Annls Soc. Scient. Brux. 25: 86, 1901.
    Penicillium brunneorubrum Dierckx, *op. cit.* 25: 88, 1901.
    Penicillium citreoroseum Dierckx, *op. cit.* 25: 86, 1901
    Penicillium notatum Westling, Ark. Bot. 11: 95, 1911

Raper and Thom (1949) accepted *P. chrysogenum* and *P. notatum* as separate species, both being regarded as penicillin producers. *P. citreoseum* Dierckx was regarded as a synonym of *P. cyaneofulvum* Biourge 1923, and placed in synonymy with *P. griseoroseum* by Pitt (1979). Subsequent authors (Samson *et al.*, 1976; Pitt, 1979; Pitt, 1988) have all accepted *P. chrysogenum*, but have placed the other species mentioned, with the exception of *P. griseoroseum*, in synonymy.

Raper and Thom (1949) regarded the description of *P. griseoroseum* as too meagre to establish identity and therefore rejected the species; this species, along with *P. brunneorubrum* and *P. citreoroseum*, was also considered doubtful by Samson *et al.* (1977). Pitt (1979) recognised that cultures derived from Dierckx' type of *P. griseoroseum* still existed, and on the basis of examination of IMI 92220i retained this species as distinct from *P. chrysogenum*, with *P. griseoroseum* and *P. citreoroseum* as synonyms. More recently, the identity of *P. griseoroseum* with *P. chrysogenum* has been firmly established by morphological comparisons (Pitt, unpublished), similar isoenzyme patterns (Cruickshank and Pitt, 1987) and by identical patterns of secondary metabolites including penicillin, roquefortine C and meleagrin (Frisvad and Filtenborg, 1989).

Consequently, *P. griseoroseum* threatens the well-established and economically important name *P. chrysogenum*, necessitating conservation against *P. griseoroseum*. It will also be necessary to conserve *P. chrysogenum* against the other Dierckxian names *P. brunneorubrum* and *P. citreoroseum*, considered by Pitt (1979) to be synonyms of *P. griseoroseum*.

Economic significance.

P. chrysogenum is the only name for a penicillin-producing species which is in common use. In past decades the synonymous name *P. notatum* has also been used extensively in the penicillin literature, but it has been little used in recent years. Together with a variety of biological and chemical derivatives, penicillin is still the major fungal compound used in medicine. A sample search in the BIOSIS data base still produced annual numbers of 56 publications. In addition, *P. chrysogenum* has been used for the commercial production of single cell protein and enzyme production. As one of the few common species in *Penicillium* subgenus *Penicillium* which produces no significant mycotoxins, and in view of its past history of industrial use, further industrial and biotechnological applications can be anticipated. Stabilisation of the name *P. chrysogenum* is of great importance.

REFERENCES

AL-MUSALLAM, A. 1980. Revision of the black *Aspergillus* species. Dissertation, University of Utrecht. 92 pp.

ARST, H.N. and SCAZZOCHIO, C. 1985. Formal genetics and molecular biology of the control of gene expression in *Aspergillus nidulans*. In Gene Manipulation in Fungi, eds J.W. Bennett and L.L. Lesure, pp. 309-343. New York: Academic Press.

BALLANCE, D.J.B. and TURNER, G. 1985. Development of a high-frequency transforming vector for *Aspergillus nidulans*. Gene 36: 321-331.

CLUTTERBUCK, A.J. 1974. *Aspergillus nidulans*. In Handbook of Genetics, Vol. 1, ed. R.C. King, pp. 77-510. New York: Plenum Press.

——— 1986. Glasgow stock list of *Aspergillus nidulans* strains 1986. Fungal Genetics Newsletter 33: 59-69.

COVE, D.J. 1977. The genetics of *Aspergillus nidulans*. In Genetics and Physiology of Aspergillus, eds J.E. Smith and J.A. Pateman, pp. 81-95. London: Academic Press.

CRUICKSHANK, R.H. and PITT, J.I. 1987. Identification of species in *Penicillium* subgenus *Penicillium* by enzyme electrophoresis. *Mycologia* 79: 614-620.

DOMSCH, K. H., GAMS, W. and ANDERSON, T.-H. 1980. Compendium of Soil Fungi. London: Academic Press.

EIDAM, E. 1883. Zur Kenntniss der Entwicklung bei den Ascomyceten III. *Sterigmatocystis nidulans* n.sp. In Beiträge zur Biologie der Pflanzen, Vol. 3, ed. F.S. Cohn, pp. 392-411.

FRISVAD, J.C. and FILTENBORG, O. 1989. Chemotaxonomy of and mycotoxin production by the terverticillate Penicillia. Mycologia (in press).

FUNGAL GENETICS STOCK CENTER. 1988. Catalogue of Strains, Supplement 35: 12-23. Kansas: Fungal genetics stolk center.

GOSSOP, G.H., YUILL, E. and YUILL, J.L. 1940. Heterogeneous fructifications in species of *Aspergillus*. *Transactions of the British Mycological Society* 24: 337-344.

GREUTER, W. et al. 1988. International Code of Botanical Nomenclature adopted by the Fourteenth International Botanical Congress, Berlin, July-August, 1987. Königstein, W. Germany: Koeltz Scientific Press.

GRINDLE, M. 1963. Heterokaryon compatibility of closely related wild isolates of *Aspergillus nidulans*. Heredity, London 18: 397-405.

GWYNNE, D.I., BUXTON, F.P., WILLIAMS, S.A., GARVEN, S. and DAVIES R.W. 1987. Genetically engineered secretion of active human interferon and a bacterial endoglucanase from *Aspergillus nidulans*. Biotechnology 5: 713-719.

HAWKSWORTH, D.L. and SUTTON, B.C. 1974. Article 59 and names of perfect state taxa in imperfect genera. *Taxon* 23: 563-568.

HAWKSWORTH, D.L. 1990. Problems and prospects for improving the stability of names in *Aspergillus* and *Penicillium*. In Modern Concepts in *Penicillium* and *Aspergillus* Classification, eds R.A. Samson and J.I. Pitt, pp. 75-82. New York and London: Plenum Press.

JINKS, J.L. 1952. Heterokaryosis; a system of adaptation in wild fungi. *Proceedings of the Royal Society, London, Series B,* 140: 83.

KLICH, M.A. and PITT, J.I. 1988. A Laboratory Guide to Common *Aspergillus* Species and Their Teleomorphs. North Ryde, N.S.W.: CSIRO Division of Food Processing.

KÄFER, E. 1958. An 8-chromosome map of *Aspergillus nidulans*. *Advances in Genetics* 9: 105-145.

PITT, J.I. 1979. The Genus *Penicillium* and its Teleomorphic States *Eupenicillium* and *Talaromyces*. London: Academic Press.

—— 1988. A Laboratory Guide to Common *Penicillium* Species. North Ryde, N.S.W.: CSIRO Division of Food Processing.

POLONELLI, L., CASTAGNOLA, M., D'URSO, C. and MORACE, G. 1986. Serological approaches for identification of Aspergillus and Penicillium species. *In* Advances in *Penicillium* and *Aspergillus* Systematics, eds R.A. Samson and J. I. Pitt, pp. 267-280. New York and London: Plenum Press.

PONTECORVO, G. 1949. Auxanographic techniques in biochemical genetics. *Journal of General Microbiology* 3: 122-126.

PONTECORVO, G., ROPER, J.A., HEMMONS, L.M., MACDONALD, K.D. and BUFTON, A.W.J. 1953. The genetics of *Aspergillus nidulans*. *Advances in Genetics* 5: 141-238.

RAPER, K.B. and FENNELL, D.I. 1965. The Genus *Aspergillus*. Baltimore, Maryland: Williams and Wilkins.

RAPER, K.B. and THOM, C. 1949. A Manual of the Penicillia. Baltimore: Williams and Wilkins.

SAMSON, R.A. and GAMS, W. 1985. Typification of the species of *Aspergillus* and associated teleomorphs. *In* Advances in *Penicillium* and *Aspergillus* Systematics, eds R.A. Samson and J.I. Pitt, pp. 31-54. New York and London: Plenum Press.

SAMSON, R.A., STOLK, A.C. and HADLOK, R. 1976. Revision of the Subsection *Fasciculata* of *Penicillium* and allied species. *Studies in Mycology, Baarn* 11: 1-47.

SAMSON, R.A., HADLOK, R. and STOLK, A.C. 1977. A taxonomic study of the *Penicillium chrysogenum* series. *Antonie van Leeuwenhoek* 43: 169-175.

SEIFERT, K.A. and SAMSON, R.A. 1985. The genus *Coremium* and the synnematous Penicillia. *In* Advances in *Penicillium* and *Aspergillus* Systematics, eds R.A. Samson and J. I. Pitt, pp. 143-154. New York and London: Plenum Press.

SMITH, J.E. and PATEMAN, J.A. 1977. Genetics and Physiology of *Aspergillus*. London: Academic Press.

THOM, C. and RAPER, K.B. 1945. A Manual of the Aspergilli. Baltimore, Maryland: Williams and Wilkins.

UPSHALL, A., KUMAR, A.A., BAILEY, M.C., PARKER, M.D., FAVREAU, M.A., LEWISON, K.P., JOSEPH, M.J., MARAGANORE, J.M. and McKNIGHT, G.L. 1987. Secretion of active human tissue plasminogen activator from the filamentous fungus *Aspergillus nidulans*. *Biotechnology* 5: 1301-1304.

VOSS, E.G. et al., eds. 1983. International Code of Botanical Nomenclature adopted by the Thirteenth International Botanical Congress, Sydney, August, 1981. Utrecht: Bohn, Scheltema and Holkema.

VUILLEMIN, P. 1927. *Sartorya*, nouveau genre de plectascinées angiocarpes. *Compte rendu hebdomadaire des Séances de l'Académie des Sciences* 184: 136-137.

WINTER, G. 1884. Rabenhorst Kryptogamen Flora. Die Pilze Deutschlands, Oesterreichs und der Schweiz 1(2): 62.

NOMENCLATURE STABILITY IN *PENICILLIUM* AND *ASPERGILLUS*, AN ALTERNATIVE VIEW

W. Gams

Centraalbureau voor Schimmelcultures
3740 AG Baarn, The Netherlands

The situation of nomenclatural instability is less deplorable than Hawksworth (1990) has suggested. As he pointed out, the mere fact that *Penicillium* and *Aspergillus* experts meet and discuss the taxonomy of these genera at regular intervals is the best warranty for eventual stability.

Hawksworth's main concern is nomenclatural instability. Major changes occurred at the Sydney Botanical Congress in 1981. The rules of Art. 59 concerning teleomorph-anamorph nomenclature were clarified for the first time. Thanks to the efforts of Luella K. Weresub and G. L. Hennebert, the meaning of the article has become transparent. Liberal redispositions have become possible and the situation is straight forward. Until 1981 it was not legitimate to make combinations from anamorphic to teleomorphic genera and vice versa; now the correctness of such a procedure is only dictated by the kind of fungus involved. It is exaggerated to state that "teleomorph names are no longer available for anamorphs", as Article 59 primarily states that teleomorph names are holomorph names, i.e. they cover the whole fungus, thus also the anamorph. The new ruling implies a small sacrifice of some *Aspergillus* and *Penicillium* names in those cases where ascomata have been observed from the beginning, if a separate name is found desirable to cover the anamorph only. In these cases, the expert will use the teleomorph name anyhow in preference to indicate natural relationships. Use of the same epithets in *Aspergillus* and *Penicillium* is not illegitimate, it is just incorrect, as it comprises the whole fungus and not only the anamorph, and that is exactly what Thom, Raper and Fennell had intended to do. A separate name to cover the anamorph only is meaningful in such cases as *P. janthinellum* that rarely shows an *Eupenicillium javanicum* teleomorph, but it has little importance in other cases where ascomata are regularly found. It is quite likely that the anamorph epithets introduced by Pitt (1979) and Samson and Gams (1985) will not become popular, but this is of little concern as the stable teleomorph names are to be used preferentially anyhow.

Of greater effect on stability are old obscure names. A suggestion that it is good mycological practice not to resurrect them has been incorporated in the conclusions of the previous *Penicillium* workshop, but unfortunately not in the ICTF Code of Practice (Sigler and Hawksworth, 1987). It is regrettable that even now no ruling exists that efficiently prevents the resurrection of obscure old names by neotypification. At the Berlin Congress considerable discussion was spent on Art. 69, that allows rejection of names that have been widely and persistently used for a taxon or taxa not including the type. A proposal to remove it from the Code was defeated. While Art. 14.2, newly formulated at Sydney,

Modern Concepts in Penicillium and Aspergillus Classification
Edited by R. A. Samson and J. I. Pitt
Plenum Press, New York, 1990

(conservation of an important name against others) is intended only for species of "major economic importance", Art. 69 can and should be invoked for many more species. Long debates have been held at Berlin (Greuter *et al.*, 1989) on the desirability of indexing of recognized (listing) and/or registration of newly published names. The term Indexing is used for lists like the one in the "Index of Fungi" that are of tremendous value to the practicing taxonomist. But standard lists now envisaged additionally imply a form of recognition, or comment on the availability of names for use (comparable to sanctioning). At Berlin it was quite evident that the great majority of botanists were strongly opposed to the introduction of any kind of censorship by a registring authority. In the meantime a special committee appointed by IAPT (Hawksworth and Greuter, 1989) have concluded that listing of all botanical names is technically feasible by international collaboration and indeed highly desirable. But there is very little chance that any protected status will be assigned to registered names at the next Botanical Congress. At present the rules on fungal nomenclature are made by the International Association of Plant Taxonomy (IAPT), with a strong input from its Committee for Fungi and Lichens. The IUMS International Commission on Taxonomy of Fungi (ICTF) is also entering the scene of taxonomy, but it would be detrimental to taxonomy if the two organizations developed divergent nomenclatural ideas.

Indexing is very useful, and should be supported on a worldwide basis. On-line accessible databases are the technical solution to the problem of complete documentation on all available names. It will also serve nomenclatural stability if the status of all names, including obscure ones, is elucidated once and for ever. But freezing nomenclature to the status quo for the sake of stability does no justice to the historical development of our science; in fact this would mean abolishing all of the philosophy of nomenclature so far prevailing in taxonomy.

At a *Fusarium* workshop at Sydney in 1983, Nelson and his colleagues (see Nelson *et al.*, 1983) proposed recognizing Wollenweber and Reinking (1935) as the starting point for *Fusarium* nomenclature. This proposal did not find the support of the community of experts. Similarly in *Penicillium* and *Aspergillus* it would not make sense to sanction the names used by Thom, Raper, and Fennell at the present day, where dramatic progress is made in understanding taxonomic structures with a whole range of modern experimental techniques.

At present the taxonomy of many taxa is under review at both the specific and infraspecific level. I would like to recommend that in this situation it is better not to formalise immediately every discovery of taxonomic relevance, such as creation of infraspecific taxa, new combinations etc., but to wait for support from other related studies. Self-restraint, especially in the context of resurrecting doubtful old names, cannot be recommended strongly enough.

Great pressure, some of which appears justified, is now being exerted on taxonomists to stabilize names in use. But in the present situation it is more important to tell outsiders patiently what is going on and why all this is necessary rather than to yield to pressure by adopting standardized names. It is quite likely that considerable consensus on many Aspergilli and Penicillia will emerge in the next 10-20 years that cannot be anticipated now. For the time being we must make use of the provisions of the Botanical Code to conserve important names, as is being considered elsewhere in this Workshop.

With thanks to Prof. D. L. Hawksworth for some improvements, though opinions still diverge.

REFERENCES

GREUTER, W., MCNEIL, J. and NICOLSON, D. H. 1989. Report on botanical nomenclature - Berlin 1989. *Englera* 9: 1-228.

HAWKSWORTH, D.L. 1990. Problems and prospects for improving the stability of names in *Aspergilus* and *Penicillium*. *In* Modern Concepts in *Penicillium* and *Aspergillus* Classification, eds. R.A. Samson and J.I. Pitt, pp. 75-82. New York and London: Plenum Press.

HAWKSWORTH, D. L. and GREUTER, W. 1989. Report of the first meeting of a working group on lists of names in current use. *Taxon* 38: 142-148.

NELSON, P. E., TOUSSOUN, T. A. and MARASAS, W. F. O. 1983. *Fusarium* Species, an Illustrated Manual for Identification. University, Park, Pennsylvania: Pennsylvania State University Press.

PITT, J. I. 1979. The genus *Penicillium* and its teleomorphic states *Eupenicillium* and *Talaromyces*. London: Academic Press.

SAMSON, R. A. and GAMS, W. 1985. Typification of the species of *Aspergillus* and associated teleomorphs. *In* Advances in *Penicillium* and *Aspergillus* Systematics, eds. R.A. Samson and J.I. Pitt, pp. 31-54. New York and London: Plenum Press.

SAMSON, R. A., STOLK, A. C. and HADLOK, R. 1976. Revision of the subsection Fasciculata of *Penicillium* and some allied species. *Studies in Mycology, Baarn* 11: 1-47.

SIGLER, L. and HAWKSWORTH, D. L. 1987. International Commission on the Taxonomy of Fungi (ICTF) code of practice for systematic mycologists. *Mycopathologia* 99: 3-7.

WOLLENWEBER, H. W. and REINKING, O. A. 1935. Die Fusarien. Berlin: P. Parey.

DIALOGUE FOLLOWING THE PRESENTATIONS BY PROFESSORS HAWKSWORTH AND GAMS

HAWKSWORTH: The movement for stabilization of names comes from the International Union of Biological Sciences (IUBS). It's not really a matter of mycologists isolating themselves from botany, it's a matter of botanists isolating themselves from biology, both for nomenclaturalists and users of taxonomy. The work for some botanical groups is very advanced, for example with the flowering plants. A draft list of generic names in use for flowering plants will go out this autumn. The moss and hepatic lists will be being debated by the International Biological Congress in St. Louis later on this summer. The draft algal list should be finished by the end of the year. The list of fungal genera should be available at the Fourth International Mycological Congress in Regensberg in 1990.

GAMS: I have no word against the importance of the name *Aspergillus nidulans*. In fact, we were reluctant to propose the new name *Aspergillus nidulellus*. However, I am uneasy about that conservation proposal. Is there a need to split the anamorph and the teleomorph in this fungus? All the authors who are talking about *Aspergillus nidulans* are in fact talking about the teleomorph, *Emericella nidulans*. It is not as if the name "*nidulans*" is unavailable for the fungus. It's still the right epithet for the holomorph. I doubt that this proposal will be successful.

NIRENBERG: I think that we are all concerned about what nonspecialists feel about all the name changes with fungi. Students, plant pathologists, and technical assistants have problems understanding why we have two different species epithets in one fungus when we talk about the teleomorph and the anamorph. Why can't we change the rules in the Code? It would be a lot easier to use the same epithet for the anamorph in the teleomorph, and just use the oldest one. This would solve many problems.

HAWKSWORTH: The fact is that mycologists went off in their own direction a long time ago. Phycologists don't have this problem. We have to live with history, or it becomes too destabilizing now. If we were starting again I would agree with you. Is anyone, apart from Dr Gams, against the proposal on *Aspergillus nidulans*?

SAMSON: Do the geneticists have any feeling about this? Is it necessary to conserve *Aspergillus nidulans*? You still have the epithet *nidulans* for the teleomorph. It's not quite correct according to the rules to use it for the anamorph, but who cares?

MULLANEY: For geneticists, this fungus is just a tool, and I don't think they are too concerned about why the fungus has this name.

SAMSON: French cheese makers still call *Penicillium camemberti* by an old name, *P. candidum*. Everybody knows what is meant by that name. So, it may be unnecessary to make such a complicated proposal to protect this name. *Penicillium chrysogenum* and *Aspergillus niger* are different because all mycologists use these names and both are clear synonyms of earlier names. The formal proposals are necessary for these two cases.

PITT: I agree with Dr Samson. If we do not conserve these names, we are bound to run into problems later on. But it may also be worth trying *A. nidulans* as a test case.

(At this point, the proposals were voted upon by the participants. The proposals for *Aspergillus niger* and *P. chrysogenum* received unanimous support. There was one vote against the *A. nidulans* proposal).

HAWKSWORTH: Does the group wish to make any comments on the problem of resurrecting old names in *Penicillium* and *Aspergillus*? Does the group wish to produce a list of names that might be later granted protected status, because we are producing a more stable taxonomy? Without making any final taxonomic judgements, we can make a list and suggest that these are the names that people should be considering.

SAMSON: You can't just reject a name when there is a very good holotype specimen available. In *Penicillium*, there probably are not too many older names that need to be considered, but in *Aspergillus* there may be quite a number.

HAWKSWORTH: Linnaeus did this, Fries did this, and there is a general feeling that it may be time to do this again.

GAMS: This really is relevant to Article 69 cases about which there has been a lot of misunderstanding. The cases that really concern me are those where no holotype exists. We should confine our discussion to these cases.

HAWKSWORTH: What is there to discuss? The reality is that people do sometimes take up old names by neotypification when they should be trying to keep the established name. This hasn't happened so much with *Penicillium* and *Aspergillus* in recent times. Particularly with generic names, it's clear you should follow the conservation route. The danger is that if taxonomists and nomenclaturalists don't take this matter into hand themselves, they will find that other bodies will have done it. For example, a major list being produced of arthropods of economic importance will be used by all the indexing services and people preparing reference work worldwide. It's happening with seed testing, it's happening with plant conservation, and it's the way of the world. The pressure is very much from the information scientists and legislators, not the taxonomists and nomenclaturalists. We can't risk isolating ourselves continually from the users if we want to be funded.

CHRISTENSEN: The main advantage is that it protects against the loss of valuable information now on hand. If the name changes, we may lose this information when people forget about the synonymy in years to come. As an ecologist, this worries me.

HAWKSWORTH: This is true.

CHRISTENSEN: As a body of responsible biologists, we should consider this as the primary rationale for our proposals.

SAMSON: But how much value can you give to this old information? Many of these reports on particular *Penicillium* and *Aspergillus* species are undoubtedly based on misidentificatons. For many years, the species were not well characterized. There is a burden of literature with incorrect identifications and this causes a lot of confusion.

HAWKSWORTH: Perhaps we in mycology should think about rejecting a lot more names.

PITT: I have divided loyalties. With *Aspergillus* and *Penicillium*, that the best thing we can do right now is to do the groundwork to stabilize the names. We have made extraordinary progress in the past four or five years, so perhaps we should be thinking about producing a list of acceptable names if we have another workshop. There may very well be consensus on many of these names in four years' time. At that time, I would like to see this kind of proposal seriously considered by a group such as this.

HAWKSWORTH: Who should have the responsibility of compiling such a list that could be ratified by a third workshop? Perhaps this is the function of the Subcommission on *Penicillium* and *Aspergillus* Systematics. By the time it next meets, we will know what decisions have been made at the 1993 Botanical Congress concerning such lists.

GAMS: What Dr Pitt may envisage is a publication in 4 or 5 years, perhaps not an authoritative monograph, but a list of all published names with remarks on their status, that people will use, without having any nomenclatural status of sanctioning. *Penicillium* and *Aspergillus* taxonomy is obviously more advanced than that of many other fungal genera. The list concept will at that time not be suitable for many other genera.

HAWKSWORTH: There are large numbers of fungal names that are not catalogued anywhere, and I suspect that there may also be such names in *Aspergillus* and *Penicillium*. Recently, a zoologist colleague sent me a publication on insects from Greenland with a list of fungal genera in the appendix, and requested a list of current equivalent names. There were about 12 genera, none of which had ever been indexed in any fungal publication.

Final Discussion on the conservation of names.

On Aspergillus nidulans.

GAMS: I agree with the idea of the original CMI proposal that something must be done to save the use of *Aspergillus nidulans*. Certainly, geneticists will continue to use this name whether or not it is formally conserved. Neither the present situation nor conservation is satisfactory. The name *Aspergillus nidulans* is used by many authors; this is not quite correct because it is a holomorph that belongs to *Emericella*. Editors of reputable journals are uncertain whether they should impose the use of *Emericella nidulans* on authors or not. If *Aspergillus nidulans* is conserved as proposed, it will have to be done with the explicit exclusion of the teleomorph. Then, the future user would have an available and correct name, *Aspergillus nidulans*, but using this binomial explicitly excludes the teleomorph. This is not usually the intention. The original description of Eidam, under the *Aspergillus* name, included a description of the ascomata. It is difficult to see how we can now typify this described ascomycete only by its anamorph.

HAWKSWORTH: We're well aware of these problems, but the reality is that people will continue to use the name *Aspergillus nidulans*. Therefore, there will still be a nomenclaturally correct name available for the teleomorph if people wish to use one. The proposal protects that.

GAMS: I'm not too concerned about a slightly incorrect usage. Databases will have to be constructed to cope with synonyms anyway. But perhaps editors of journals will be quite concerned about what to do with such names. Editors of most mycological journals do not usually impose the use of correct names on their authors.

HENNEBERT: I think it is like a nickname and a proper name. *Aspergillus nidulans* can be like a nickname for this anamorph and it won't cause any confusion.

GAMS: But to have a nickname, we can stay with the status quo.

PITT: Even if the proposal is not strictly in accordance with the Code, I think we have to do something to save this name. This appears to be the simplest route. Several years ago, Dr Hawksworth and Dr Sutton unsuccessfully attempted to save some of the common names in *Aspergillus* and *Penicillium* that included teleomorphs in the protologue. There is a case here for trying a different approach.

HAWKSWORTH: It would be useful to have this debated in the larger community. After all, we still can't get people to use names like *Acremonium*; users still call it *Cephalosporium*. The *Fusarium*-people also have similar problems. *Fusarium* has teleomorphs in *Gibberella*, but these names are not consistently cited either.

Subcommission on *Penicillium* and *Aspergillus* Systematics (SPAS).

HENNEBERT: I would be interested to learn more about the Subcommission on *Penicillium* and *Aspergillus* Systematics (SPAS). Who are the members of this commission? What are their aims and what are the results? Are there special publications produced?

PITT: SPAS is an organization independent from this workshop. It is currently a subcommission of the ICTF, the International Commission on the Taxonomy of the Fungi, which is under the auspices of the Mycology Division of the International Union of Microbiological Sciences (IUMS). SPAS is a collection of experts governed by the rules of ICTF and the Mycology Division of IUMS. The aim is to promote the improvement in and dissemination of information on the nomenclature and taxonomy of *Penicillium* and *Aspergillus*. Selection of members is by the subcommission itself. The present members are myself as chairman, Dr Klich as secretary, and Dr Samson, Dr Frisvad, Dr Cruikshank, Dr Mullaney, Mr. Williams, and Dr Onions who is retiring. The subcommision was formed after the first *Penicillium* and *Aspergillus* workshop when it became clear that some kind of multidisciplinary working group was necessary. Our work so far has related to clarifying the taxonomy of certain groups of species in *Penicillium* where a concensus approach is desirable. Two projects are presently underway. The first is a study of *P. glabrum, P. spinulosum* and some closely related taxa. The second concerns some species in subgenus *Biverticillium* where both taxonomic and nomenclatural problems existed. Our work will obviously expand as a result of the resolutions proposed at this meeting.

Recommendations.

1. That SPAS be encouraged to produce a list of names in *Aspergillus* and *Penicillium* in current use, together with place of publication, dried types or ex-type isolates, for confirmation at the next *Penicillium* and *Aspergillus* workshop, with a view to these being incorporated into the IUBS/IAPT "List of Names in Current Use" project and granted some special, protected status subject to the decision of the next International Botanical Congress.

GAMS: I follow your proposal, until the protected status is mentioned. This idea divides biologists. Would such a list ever be officially recognized? If we include this aim in the proposal, it may hamper the progress of the project. Personally, I cannot support the idea of sanctioning names from such a list.

SAMSON: You don't have to sanction anything for this list. We would consider the list at the next workshop.

HAWKSWORTH: Perhaps this list could be published as part of the proceedings of the next *Penicillium* and *Aspergillus* workshop?

PITT: And if we don't like it at that time, it can just be discarded, after open discussion.

HENNEBERT: Is there any precedent for such a list being granted protected status?

HAWKSWORTH: Not yet. Other sample lists are being prepared at the behest of the IUBS that will be proposed for protected status at the International Botanical Congress in 1993. One is being made for the Leguminosae on a regional basis around the world. In fact, ICSU, the International Council of Scientific Unions, the parent body of IUBS, has a body called CODATA which has a project MSDN, the Microbial Strain Data Network. Last September, they set up a committee to examine standardization of terminology and nomenclature in biology. These are mainly information scientists and they have the funding to do this work. Either the scientists who understand organisms do the work, or they may get left behind. The CABI Thesaurus includes numerous plant pathogen names, for example, that is used as a standard by the National Agricultural Library in the US.

HENNEBERT: Do these lists actually exclude names, such as synonyms? Do you exclude names of species found only once? What do you call current use?

HAWKSWORTH: The idea is to include any name that people feel might be useful. It isn't a taxonomic exercise, it is a nomenclatural exercise. If there were differing views on particular species complexes, you would list all the names that people might choose to use. Names that had been considered synonyms and were long accepted as such would not be included. The effect of this would be to reduce the number of names that people have to consider. We do not intend to follow the bacteriologists in devalidating names that already exist. You could use names not in the list provided they do not conflict with names included in the list. It would protect us from the problem of people resurrecting names by neotypification. It would also save us the problem we now face of having to conserve names such as *Penicillium chrysogenum* and *Aspergillus niger*.

HENNEBERT: Is the rank of the name protected? For example, could the varietal names in *Neosartorya fischeri* be raised to species?

HAWKSWORTH: Yes, of course. It is not intended to fossilize taxonomy.

PITT: Would anyone care to second Dr Gams' proposal that the "protected status" clause be dropped?

There were no seconders to Dr Gams' proposal. The proposal as given above was accepted by all the participants; with two dissenters.

2. That SPAS prepare a list of isolates available from service culture collections that can be used as standards for the TLC detection of secondary metabolites of value in species separation in *Penicillium*.

This resolution was unanimously accepted.

Miscellaneous.

HAWKSWORTH: I think we should encourage work on the genetics of species in these two genera so that we will be able to better understand the basis of variation. We don't know anything about chromosome numbers in these genera, for example. Electrophoretic methods are now available for separating chromosomes.

CHRISTENSEN: I strongly support this. Genetic variation is basic.

SAMSON: We taxonomists should really try to help the geneticists, and make sure they are working with properly identified isolates.

These sentiments were endorsed by the workshop, but no formal wording was proposed.

4

TAXONOMIC SCHEMES OF PENICILLIUM

SPECIATION AND SYNONYMY IN *PENICILLIUM* SUBGENUS *PENICILLIUM* – TOWARDS A DEFINITIVE TAXONOMY

J.I. Pitt[1] and R.H. Cruickshank[2]

[1]*CSIRO Division of Food Processing*
North Ryde, N.S.W. 2113, Australia

[2]*Department of Agricultural Science*
University of Tasmania,
Hobart, Tas. 7000, Australia

SUMMARY

Until a few years ago, taxonomy of asexual, haploid fungi such as *Penicillium* relied primarily on morphological or gross physiological characters. Although, in *Penicillium*, a large measure of agreement existed, in the final analysis speciation depended on the concepts of the individual taxonomist. The resulting lack of agreement was nowhere more obvious than in *Penicillium* subgenus *Penicillium*. Recent developments have changed all this. In particular, the introduction of simple techniques for studying secondary metabolites and improved methods for distinguishing isoenzymes by gel electrophoresis have resulted in independent methods for assessing species concepts.

Backed by accurate identifications using traditional methods, careful studies using secondary metabolite profiles and isoenzyme patterns have produced remarkably consistent results both within and between species. Correlations are so clear that it can now be stated confidently that we are approaching a definitive taxonomy for subgenus *Penicillium*.

Of course some new species can be expected to be discovered as unfamiliar habitats are explored, for example, new species and varieties described from seed stores of marsupials on the U.S. prairies. Leaving these latter aside because their status has not yet been fully assessed, some 25 well defined species can be distinguished in subgenus *Penicillium*.

This paper sets out species in subgenus *Penicillium* as currently conceived by the authors. For each species a diagnosis is given, and a list of synonyms where these have been confirmed by studies on living cultures. Notes on species concepts, ecology and mycotoxin production are also given.

INTRODUCTION

In introducing subgenus *Penicillium*, Pitt (1979) defined it as including species producing terverticillate penicilli, and typified it by *P. expansum*, the type species of the genus. In this subgenus, he brought together species previously classified by Raper and Thom (1949) in sect. *Divaricata* subsects. *Lanata, Fasciculata, Funiculosa* (in part) and *Velutina* (in part). This appeared to provide for a more natural classification. In so doing, he placed in synonymy, or renamed, 22 species previously recognised by Raper and Thom (1949). However, 17 of Raper's species remained essentially unaltered, although three of these had to be renamed to conform with the Botanical Code.

The system devised by Pitt (1979) contrasted rather sharply with the speciation introduced by Samson *et al.* (1976) in which 17 species of *Penicillium* subsect. *Fasciculata* as defined by Raper and Thom (1949) were reduced to 8 species and 5 varieties.

Ramírez (1982) reclassification of the genus closely followed the taxonomy of Raper and Thom (1949) and added little to the picture. However, the differences between Pitt (1979) and Samson *et al.* (1976) led to considerable confusion (e.g. Onions *et al.*, 1984). The major problem lay with the differences in circumscription of species, and that in turn was

Modern Concepts in Penicillium and Aspergillus Classification
Edited by R. A. Samson and J. I. Pitt
Plenum Press, New York, 1990

due to the absence of objective methods for judging species concepts in this asexual, haploid genus.

During the decade since, the situation has changed dramatically. New techniques have emerged which have enabled the delimitation of species by techniques quite unrelated to the morphological and gross physiological criteria used by Pitt (1979) or other contemporary taxonomists. Two techniques in particular emerged: the use of secondary metabolites and of isoenzymes as taxonomic criteria. Following on initial work by Ciegler *et al.* (1973, 1981), Filtenborg and Frisvad (1980) developed the study of secondary metabolites in *Penicillium*, and in subgen. *Penicillium* in particular (Frisvad and Filtenborg, 1983). Although in early work a high degree of correlation between species and secondary metabolites was often not apparent, refinements in species identifications (e.g. El-Banna *et al.* ,1987) helped to overcome this problem.

At the same time, studies on electrophoretic patterns (zymograms) of certain enzymes were being undertaken: again a high correlation between species and specific zymograms was discovered in *Penicillium* subgen. *Penicillium* (Cruickshank and Pitt, 1987a, b).

Table 1. Classification of *Penicillium* Subgenus *Penicillium*

Section *Penicillium*	Series *Urticicola* Fassatiová
	P. *brevicompactum* Dierckx
Series *Expansa* Raper & Thom ex	P. *glandicola* (Oudemans) Seifert &
Fassatiová	Samson
P. *atramentosum* Thom	P. *griseofulvum* Dierckx
P. *chrysogenum* Thom	P. *hordei* Stolk
P. *coprophilum* (Berk. & Curt.) Seifert &	P. *solitum* Westling
Samson	P. *verrucosum* Dierckx
P. *expansum* Link	
	Section *Cylindrospora* Pitt
Series *Viridicata* Raper & Thom ex Pitt	
P. *aethiopicum* Frisvad *et al.*	Series *Italica* Raper & Thom ex Pitt
P. *allii* Vincent & Pitt	P. *italicum* Wehmer
P. *aurantiogriseum* Dierckx	P. *digitatum* (Pers.:Fr) Sacc.
P. *commune* Thom	P. *fennelliae* Stolk
P. *crustosum* Thom	
P. *echinulatum* Raper & Thom ex	**Section *Coronatum* Pitt**
Fassatiová	
P. *hirsutum* Dierckx	Series *Olsonii* Pitt
P. *roqueforti* Thom	P. *olsonii* Bainier & Sartory
P. *viridicatum* Westling	
	Section *Inordinate* Pitt
Series *Camemberti* Raper & Thom ex Pitt	
P. *camemberti* Thom	Series *Arenicola* Pitt
	P. *arenicola* Chalabuda

The superimposition of the information obtained from secondary metabolite patterns and zymograms on the more traditional taxonomic bases derived from morphology and gross physiology have shown a dramatic correlation: a correlation so clear cut that it can now be confidently stated that we are very close to a total comprehension of the biological species currently known to exist in subgen. *Penicillium*. This situation would have been unthinkable at the beginning of this decade. At the same time, this high correlation showed that morphological and gross physiological properties were effective taxonomic features. The major species conceived by Raper and Thom (1949) and emended slightly, or where necessary renamed, by Pitt (1979) have been sustained by the more recent evidence from unrelated sources.

This paper is designed to clarify the taxonomy of subgen. *Penicillium*. Table 1 provides a summary of our current thinking, showing sectional classification and accepted species. The text which follows lists species in alphabetical order, and a list of synonyms. Apart from a few obligate synonyms, all those included have been grown in culture and identification established by enzyme electrophoresis. The majority are as listed by Pitt (1979), but a number of corrections are included. A diagnosis of each accepted species follows, together with supplementary information on changes in species concepts, ecology, important secondary metabolites and mycotoxins. The listing of the latter is intended to cover major metabolites only.

It is believed that the listing in Table 1 and below provides a definitive listing of the currently known and accepted species. Subdivision of a number of these species into varieties or chemotypes (Pitt and Hawksworth, 1985) for specialist purposes is possible, especially if secondary metabolite production is taken as a primary criterion for speciation. The discovery of further species in *Penicillium* subgen. *Penicillium* is to be expected as more diverse habitats are examined, e.g. the isolates from desert kangaroo rat and other habitats in the U.S.A. (Frisvad *et al.*, 1987) which have not yet been fully evaluated. Nevertheless the major speciation now appears to be in place.

MATERIALS AND METHODS

In the course of this study, cultures of types or neotypes of all species of subgen. *Penicillium* known to exist in the world's major culture collections were examined. Other "authentic" cultures from major studies were included. The methods used were those of Pitt (1979) and Cruickshank and Pitt (1987a, b). A feature of the zymogram technique is that it was able to effectively classify most old and morphologically deteriorated cultures, including a number which Pitt (1979) and, earlier, Raper and Thom (1949) had found great difficulty in assigning to species. Such a result had been predicted by Paterson and Hawksworth (1985).

Other major sources of data used to draw the taxonomic conclusions given below included the studies of species mycotoxin relationships reported briefly by El-Banna *et al.* (1987) and published data of Frisvad (Frisvad and Filtenborg, 1983; Frisvad, 1985, 1986).

Penicillium aethiopicum Frisvad et al. 1989

DIAGNOSIS. Colonies growing quite rapidly, texture velutinous to fasciculate; conidia blue on CYA, dark green on MEA; exudate clear to pale brown, soluble pigment brown; reverse on CYA orange yellow to yellow brown, on MEA yellow grey. Stipes on CYA smooth, on MEA usually rough; penicilli terverticillate; conidia spherical to ellipsoidal, smooth walled. Sometimes growth at 37°C. The most distinctive feature of this species is the presence of definite yellow to brown reverse colours on CYA and MEA. Morphologically it is best described as intermediate between *P. aurantiogriseum* and *P. viridicatum*.

CONCEPT. Originally considered to be an unusual *P. viridicatum*, *P. aethiopicum* possesses distinctive secondary metabolite profiles (Frisvad, 1986, as *P. viridicatum* IV; El-Banna *et al.*, 1987, as Sp 1448). The species is also culturally distinct.

MYCOTOXINS. *P. aethiopicum* produces viridicatumtoxin and griseofulvin (Frisvad, 1986; El-Banna *et al.*, 1987).

ECOLOGY. According to Frisvad (unpublished data), this species is primarily found in African soils.

DESCRIPTION. Frisvad *et al.* (to be published).

Penicillium allii Vincent & Pitt 1989

DIAGNOSIS. Colonies relatively large, velutinous to distinctly coremial; conidia dull green; exudate and soluble pigment pale yellow; reverse dark brown. Stipes very rough, usually short, penicilli usually terverticillate, often with rami and metulae rough walled; conidia spherical and smooth walled.

CONCEPT. *P. allii* is a recently described species (Vincent and Pitt, 1989). It was recognised as distinct both morphologically and because of its unique enzyme electrophoretic patterns (Cruickshank and Pitt, 1987b, as *Penicillium* sp.).

MYCOTOXINS. Secondary metabolite production has not yet been studied.

ECOLOGY. The known isolates of *P. allii* have come from the Middle East, and are associated with garlic (*Allium* species).

DESCRIPTION. Vincent and Pitt (1989)

Penicillium arenicola Chalabuda 1950
  Penicillium canadense G. Smith 1956

DIAGNOSIS. Colonies of variable size, texture velutinous to floccose; conidia pale brown; soluble pigment red brown; reverse brown to dark brown. Stipes smooth to rough, penicilli terverticillate to irregular, sometimes with 4 or 5 branch levels; conidia spherical to subspheroidal, smooth to rough.

CONCEPT. The concept of this species remains unaltered from the original.

MYCOTOXINS. Secondary metabolite production is unreported.

ECOLOGY. *P. arenicola* has been isolated only from soil. It has been reported to be highly tolerant of copper.

DESCRIPTIONS. Pitt (1979); Pitt (1988).

Penicillium atramentosum Thom 1910

DIAGNOSIS. Colonies growing moderately rapidly, texture velutinous; exudate, soluble pigment and reverse colours on CYA reddish brown. Stipes smooth, penicilli terverticillate, delicate; conidia spherical, smooth walled.

CONCEPT. The concept of *P. atramentosum* has remained unaltered since first described.

MYCOTOXINS. No significant secondary metabolites have been reported.

ECOLOGY. This is a rarely reported species, without known significance.

DESCRIPTION. Pitt (1979).

Penicillium aurantiogriseum Dierckx 1901
  Penicillium aurantiocandidum Dierckx 1901
  Penicillium puberulum Bainier 1907
  Penicillium cyclopium Westling 1911
  Penicillium brunnioviolaceum Biourge 1923
  Penicillium porraceum Biourge 1923
  Penicillium martensii Biourge 1923
  Penicillium lanoso-coeruleum Thom 1930
  Penicillium carneolutescens G. Smith 1939
  Penicillium viridicyclopium Abe 1956

Penicillium verrucosum var. *cyclopium* (Westling) Samson *et al.* 1976
Penicillium cyclopium var. *aurantiovirens* (Biourge) Fassatiová 1977

OTHER SYNONYMS. *Penicillium solitum*, regarded as a synonym of *P. aurantiogriseum* by Pitt (1979), is now accepted as a distinct species (Cruickshank and Pitt, 1987b; Pitt and Spotts, in prep.) *P. puberulum*, regarded as a separate species by Pitt (1979), has been placed in synonymy (Cruickshank and Pitt, 1987a, b); many isolates placed in *P. puberulum* by Pitt (1979) are now assigned to *P. commune* (Pitt *et al.*, 1986).

DIAGNOSIS. Colonies of moderate growth rate, texture velutinous or granular to fasciculate; conidia blue or grey blue on both CYA and MEA; exudate clear to pale brown, soluble pigment brown to reddish brown. Stipes smooth or at most very finely roughened on both CYA and MEA; penicilli terverticillate, a proportion biverticillate in some isolates; conidia subspheroidal to ellipsoidal, smooth walled. The primary distinguishing features of this species are the distinctly blue conidial colour and smooth stipes, on both CYA and MEA.

CONCEPT. The concept of this species, established under the name *P. cyclopium* by Raper and Thom (1949) has been reestablished in recent years (Cruickshank and Pitt, 1987b; Pitt 1988), though in a slightly broader sense. Samson *et al.* (1976) considered this taxon to be a variety within their broad species *P. verrucosum*. Pitt (1979) rejected this, accepted *P. cyclopium* as a concept, but took up the earlier name *P. aurantiogriseum*, and broadened it by the inclusion of *P. martensii*, *P. commune* and *P. solitum*. Later studies (Cruickshank and Pitt, 1987a, b; El-Banna *et al.*, 1987; Polonelli *et al.*, 1987) showed that Pitt (1979) had confused *P. commune* with *P. puberulum*, and that both *P. commune* and *P. solitum* were separate species. Removal of these species from the concept of Pitt (1979) and Williams and Pitt (1985) has produced a well defined species (Pitt *et al.*, 1986; Pitt, 1988).

MYCOTOXINS. The principal toxins produced by *P. aurantiogriseum* are penicillic acid and verrucosidin (S-toxin; Frisvad, 1986; El-Banna *et al.*, 1987).

PROBLEMS. The interface between *P. aurantiogriseum* and *P. viridicatum* is unclear. Morphological differences do not tally completely with differences in mycotoxin production. However, this is a minor difficulty which can be resolved by further study.

ECOLOGY. This is a ubiquitous species, especially in cereals, but also in other foods of a very wide variety, including fresh and stored fruits and vegetables and fresh and processed meats (Pitt and Hocking, 1985). *P. aurantiogriseum* is also the cause of "blue-eye" disease of corn in the U.S.A. (Ciegler and Kurtzman, 1970). It is much less commonly isolated from sources other than foods.

DESCRIPTIONS. Pitt *et al.* (1986); Pitt (1988).

Penicillium brevicompactum Dierckx 1901

Penicillium griseobrunneum Dierckx 1901
Penicillium stoloniferum Thom 1910
Penicillium hagemii Zaleski 1927
Penicillium patris-mei Zaleski 1927
Penicillium brunneostoloniferum Abe ex Ramírez 1982
Penicillium volgaense Beljakova & Mil'ko 1972

DIAGNOSIS. Colonies slowly growing, especially on MEA, texture velutinous; conidia green; exudate clear to reddish brown; soluble pigment reddish brown; reverse yellowish to reddish brown. Stipes long and broad, smooth to finely roughened, penicilli short and broad, mostly terverticillate, sometimes quaterverticillate, biverticillate or irregular; metulae usually apically inflated; conidia usually ellipsoidal, smooth to finely roughened. *P. brevicompactum* is a distinctive species, with slower growth on MEA than

G25N, velutinous texture, green conidia, large usually smooth walled stipes, inflated metulae and very broad penicilli.

CONCEPT. Pitt (1979) placed *P. stoloniferum*, a species recognised by Raper and Thom (1949) primarily because of "arachnoid" margins, in synonymy with *P. brevicompactum*. Otherwise the concept of this species remains as Raper and Thom (1949) established it.

MYCOTOXINS. The major metabolite produced by this species is brevianamide A, not a compound of significant toxicity, but a very useful taxonomic marker.

ECOLOGY. Both a food spoilage fungus and an agent of biodeterioration, *P. brevicompactum* has been isolated from a very wide variety of habitats, especially dried foods and decaying vegetation. It is also a weak pathogen on fruits and vegetables (Pitt and Hocking, 1985).

DESCRIPTIONS. Pitt (1979); Domsch *et al.* (1980); Pitt and Hocking (1985); Pitt (1988).

Penicillium camemberti Thom 1906
Penicillium biforme Thom 1910
Penicillium candidum Roger *apud* Biourge 1923.

DIAGNOSIS. Colonies slowly growing, deeply floccose; conidia white or pale grey green; exudate clear; reverse pale, yellow, reddish brown or purple. Stipes smooth or rough, penicilli terverticillate or irregular; conidia subspheroidal to spherical, smooth walled. The primary distinguishing features of this species are the persistently white to pale grey conidia, deeply floccose texture, and recovery only from cheese and cheese related habitats.

CONCEPT. Raper and Thom (1949) distinguished *P. camemberti* from *P. caseicola* by the production of grey rather than white conidia. Samson *et al.* (1977b) and Pitt (1979) placed *P. caseicola* in synonymy. Otherwise the concept of *P. camemberti* is much as first described.

MYCOTOXINS. It is remarkable that all tested strains of *P. camemberti* consistently produce cyclopiazonic acid (El-Banna *et al.*, 1987). It is equally remarkable that many isolates of *Aspergillus oryzae*, another domesticated species, also produce cyclopiazonic acid.

ECOLOGY. As pointed out by previous authors, *P. camemberti* is a domesticated fungus, used in cheese manufacture, and is virtually unknown from other sources.

DESCRIPTIONS. Samson *et al.* (1977b); Pitt (1979); Pitt and Hocking (1985); Pitt (1988).

Penicillium chrysogenum Thom 1910
Penicillium notatum Westling 1911
Penicillium flavidomarginatum Biourge 1923
Penicillium rubens Biourge 1923
Penicillium meleagrinum Biourge 1923
Penicillium chlorophaeum Biourge 1923
Penicillium camerunense Heim 1949
Penicillium aromaticum forma *microsporum* Romankova 1955
Penicillium harmonense Baghdadi 1968

DIAGNOSIS. Growth at 25°C rapid, absent or occasionally slight at 37°C; texture velutinous to floccose; conidia blue or blue green; exudate, soluble pigment and/or reverse usually yellow. Stipes smooth, penicilli usually terverticillate; conidia smooth walled, usually ellipsoidal.

CONCEPT. This species has always been accepted. The concept was broadened marginally from that of Raper and Thom (1949) by Samson *et al.* (1977a) and Pitt (1979) to include *P. notatum* and *P. meleagrinum*.

PROBLEMS. Cruickshank and Pitt (1987b) showed that *P. griseoroseum* Dierckx 1901 is probably identical with *P. chrysogenum*. If this is confirmed by other techniques, *P. chrysogenum* will need to be conserved.

MYCOTOXINS. *P. chrysogenum* is the producer of penicillin. It has been reported to produce mycotoxins occasionally, e.g. by Frisvad (1986) and El-Banna *et al.* (1987), but the significance of these reports remains uncertain.

ECOLOGY. This is an ubiquitous species.

DESCRIPTIONS. Samson *et al.* (1977a); Pitt (1979); Domsch *et al.* (1980); Pitt and Hocking (1985); Pitt (1988).

Penicillium commune Thom 1910

Penicillium palitans Westling 1911
Penicillium flavoglaucum Biourge 1923
Penicillium aurantiogriseum var. *poznaniense* Zaleski 1927
Penicillium lanoso-coeruleum Thom 1930
Penicillium lanosogriseum Thom 1930
Penicillium lanosoviride Thom 1930
Penicillium ochraceum var. *macrosporum* Thom 1930
Penicillium australicum Sopp ex van Beyma 1944
Penicillium roqueforti var. *punctatum* Abe 1956 (nom. inval.)

DIAGNOSIS. Colonies of medium size, texture velutinous to floccose; conidia bluish grey to dull green on CYA, dull green on MEA; exudate clear to pale brown; reverse usually pale, occasionally yellow, brown or purple. Stipes finely to conspicuously roughened on both CYA and MEA, penicilli terverticillate; conidia spherical to subspheroidal, smooth walled. *P. commune* resembles *P. aurantiogriseum*: it differs by consistently green conidial colours on MEA and having stipes at least finely roughened.

CONCEPT. The concept of *P. commune* has been badly confused for a long period. Thom (1910) regarded it as a common species, but Raper and Thom (1949) as quite rare, a floccose species of little significance. Pitt (1979) placed it in synonymy with *P. puberulum*, and confused it with *P. aurantiogriseum* as well. Recent studies on secondary metabolites (Frisvad, 1983; Frisvad and Filtenborg, 1983; both as *P. camemberti II*), by enzyme electrophoresis (Cruickshank and Pitt, 1987b) and a reappraisal of morphology (Pitt *et al.*, 1987) have clearly shown the validity of Thom's original concept, and that this is a common and distinct species.

MYCOTOXINS. The principal mycotoxin produced by *P. commune* is cyclopiazonic acid (Frisvad, 1986; El-Banna *et al.*, 1987). It does not produce penicillic acid or verrucosidin, metabolites of *P. aurantiogriseum* as currently circumscribed.

ECOLOGY. Recent studies have shown that this species, which was unrecognised for so many years, is a common spoilage fungus in foods, feeds and decaying vegetation. It is also clear now that *P. commune* is the wild type ancestor of the domesticated species *P. camemberti*, used in cheese manufacture (Pitt *et al.*, 1987; Polonelli *et al.*, 1987).

DESCRIPTION. Pitt (1988).

Penicillium coprophilum (Berk. & Curt.) Seifert & Samson

Penicillium concentricum Samson *et al.* 1976

DIAGNOSIS. Colonies of moderate size, texture velutinous to fasciculate; conidia dull green. Stipes smooth walled; penicilli terverticillate; conidia borne as ellipsoids and remaining so.

CONCEPT. *P. concentricum* was regarded as a synonym of *P. italicum* by Pitt (1979). Later (Seifert and Samson, 1985; see discussion) Pitt agreed that the two species are distinct. Seifert and Samson (1985) took up the earlier name *P. coprophilum* for this species.

MYCOTOXINS. Griseofulvin is the major metabolite produced by *P. coprophilum*.

ECOLOGY. This species is found principally in dung: according to Seifert and Samson (1985) it is extremely common in rabbit dung in the Netherlands.

DESCRIPTION. Samson *et al.* (1976).

Penicillium crustosum Thom 1930

> *Penicillium crustosum* var. *spinulosporum* Sasaki 1950
> *Penicillium pseudocasei* Abe ex G. Smith 1963
> *Penicillium farinosum* Novobranova 1974
> *Penicillium terrestre* sensu Raper and Thom (1949)

OTHER SYNONYMS. *P. aurantiogriseum* var. *poznaniense* v. Szilvinyi and *P. australicum* Sopp ex v. Beyma, regarded as synonyms of *P. crustosum* by Pitt (1979), are now considered by the authors to be synonyms of *P. commune*. *P. verrucosum* var. *melanochlorum* Samson *et al.* is a synonym of *P. solitum*.

DIAGNOSIS. Colonies growing rapidly, fasciculate on CYA, granular to crustose on MEA; conidia on CYA blue at the margins, dull green elsewhere, on MEA uniformly dull green. Stipes rough, penicilli large, terverticillate to quaterverticillate; conidia smooth walled, usually spherical, occasionally ellipsoidal. The most striking diagnostic feature of this species is the tendency for masses of conidia to be shed by mature colonies on MEA when jarred.

CONCEPT. The concept accepted here is the same as that of Pitt (1979), and essentially that of Thom (1930). Samson *et al.* (1976) placed *P. crustosum* in synonymy with *P. aurantiogriseum*. Söderström and Frisvad (1984) discussed the secondary metabolites of *P. crustosum*, but their study included *P. commune* as accepted here. *P. crustosum* is now well accepted as a distinct species (Samson and Pitt, 1985; Frisvad, 1986; Cruickshank and Pitt, 1987b).

ECOLOGY. Raper and Thom (1949) regarded *P. crustosum* as a weak pathogen of fruit, and uncommon. However Pitt (1979) and Pitt and Hocking (1985) reported that *P. crustosum* is a ubiquitous species in foods, compounded feeds and probably also decaying vegetation in temperate climates.

MYCOTOXINS. *P. crustosum* is the major producer of the neurotoxins known as penitrems. Penitrem A is a very powerful toxin, responsible for death or brain damage in a variety of domestic animals (Pitt and Leistner, 1989). Toxicity to man remains uncertain.

DESCRIPTIONS. Pitt (1979); Pitt and Hocking (1985); Pitt (1988).

Penicillium digitatum (Pers.:Fr.) Sacc. 1832

SYNONYMS. None of the 10 synonyms of this species listed by Pitt (1979) are known in culture.

DIAGNOSIS. Colonies moderately large on CYA, rapidly growing on MEA, texture velutinous; conidia olive; reverse sometimes brownish. Stipes smooth walled, penicilli at their most complex terverticillate, but often biverticillate or irregular; conidia borne as cylinders, in maturity ellipsoidal to cylindroidal, exceptionally large, smooth walled. This is a distinctive species with olive coloured, very large, elongate conidia.

CONCEPT. The concept of Raper and Thom (1949) remains unaltered.

MYCOTOXINS. No significant secondary metabolites have been reported.

ECOLOGY. Primarily a pathogen of citrus fruits, *P. digitatum* has been isolated occasionally from a variety of other habitats.

DESCRIPTIONS. Pitt (1979); Domsch *et al.* (1980); Pitt and Hocking (1985); Pitt (1988).

Penicillium echinulatum Raper & Thom ex Fassatiová 1977

Penicillium cyclopium var. *echinulatum* Raper & Thom 1949 (basionym)
Penicillium palitans var. *echinoconidium* Abe 1956 (nom. invalid.)

DIAGNOSIS. Colonies of moderate size, texture granular to fasciculate; conidia deep green. Stipes rough, penicilli terverticillate and undistinguished. The primary feature distinguishing *P. echinulatum* from other species in sect. *Penicillium* is the production of conidia which are spherical and distinctly spinose.

CONCEPT. The concept of *P. echinulatum* has not changed since the original description by Raper and Thom (1949).

MYCOTOXINS. No major metabolites are produced.

ECOLOGY. This is a rarely encountered species.

DESCRIPTIONS. Pitt (1979); Pitt (1988).

Penicillium expansum Link :Fr. 1832

Penicillium aurantiovirens Biourge 1923
Penicillium resticulosum Birkinshaw et al. 19421930

OTHER SYNONYMS. Pitt (1979) listed 14 other synonyms, most previously cited by Raper and Thom (1949). However, none of these is known in culture. Pitt (1979) regarded *P. aurantiovirens* as a synonym of *P. aurantiogriseum*, an error corrected here.

DIAGNOSIS. Colonies rapidly growing at 25°C, fasciculate to coremial; conidia dull green; exudate and soluble pigment sometimes produced, brown. Stipes smooth, penicilli typically terverticillate; conidia ellipsoidal, smooth walled.

CONCEPT. Thom (1910) clearly established the identity of Link's original concept, and this has subsequently remained unaltered. The name *P. expansum* has been in almost exclusive use for this species since that time also.

MYCOTOXINS. *P. expansum* produces patulin, an acutely toxic mycotoxin of some concern in apple juice, especially that prepared in small operations (Brackett and Marth, 1979).

ECOLOGY. Although known primarily as the cause of a destructive rot of apple and pear fruits, *P. expansum* has been isolated from a wide variety of other food and plant materials. Indeed it is a broad spectrum plant pathogen. It has been found less commonly away from living tissue.

DESCRIPTIONS. Pitt (1979); Domsch *et al.* (1980); Pitt and Hocking (1985); Pitt (1988).

Penicillium fennelliae Stolk 1969

DIAGNOSIS. Colonies of moderate size, velutinous to floccose; conidia dull green; mycelium salmon or near on both CYA and MEA; soluble pigment salmon or brown; reverse orange brown on both CYA and MEA. Stipes smooth, penicilli biverticillate to terverticillate; conidia borne as cylinders, ellipsoidal to cylindroidal, with walls spinulose. Colony pigmentation and elongate conidia with spinulose walls make this a distinctive species.

CONCEPT. The concept is as described by Stolk (1969).

MYCOTOXINS. No information on secondary metabolites is available.

ECOLOGY. This species has been found only from soil, from Zaire.

DESCRIPTIONS. Stolk (1969); Pitt (1979).

Penicillium glandicola (Oudemans) Seifert & Samson 1903
    Penicillium granulatum Bainier 1905

DIAGNOSIS. Colonies slowly growing, especially on MEA, texture fasciculate to coremiform; conidia dull green; exudate clear to pale yellow; soluble pigment yellow to deep brown; reverse usually deep brown. Stipes rough, penicilli terverticillate to quaterverticillate; metulae sometimes apically inflated; conidia ellipsoidal, smooth walled. Similar to *P. brevicompactum* in colony diameters, *P. glandicola* is readily distinguished by fasciculate texture and rough stipes.

CONCEPT. The concept of *P. glandicola* (under the name *P. granulatum*) as established by Raper and Thom (1949) remains unchanged. The name was changed to the earlier *P. glandicola* after studies by Seifert and Samson (1985).

MYCOTOXINS. Penitrem A and patulin are sometimes produced (Frisvad, 1986).

ECOLOGY. *P. glandicola* has been isolated from a wide range of habitats, but rather infrequently.

DESCRIPTIONS. Pitt (1979, as *P. granulatum*); Domsch *et al.* (1980, as *P. granulatum*); Pitt (1988).

Penicillium griseofulvum Dierckx 1901
    Penicillium patulum Bainier 1906
    Penicillium urticae Bainier 1907
    Penicillium flexuosum Dale *apud* Biourge 1923
    Penicillium griseofulvum var. *dipodomyicola* Frisvad *et al.* 1987

DIAGNOSIS. Colonies slowly growing, fasciculate to coremiform; exudate clear to pale yellow, soluble pigment reddish brown. Stipes smooth, penicilli terverticillate to quaterverticillate; phialides very short, conidia subspheroidal to spherical, smooth walled.

CONCEPT. Raper and Thom (1949) called this species *P. urticae*; Pitt (1979) took up the earlier valid epithet *P. griseofulvum*. The concept of the species has been unchanged for 40 years or more.

MYCOTOXINS. The principal metabolite produced by *P. griseofulvum* is the toxic antibiotic griseofulvin. Patulin and cyclopiazonic acid are produced by some isolates (Frisvad, 1986; El-Banna *et al.*, 1987).

ECOLOGY. A widely distributed species, *P. griseofulvum* is most commonly encountered as an agent of biodeterioration. No preferred habitat is obvious.

DESCRIPTIONS. Pitt (1979); Domsch *et al.* (1980); Pitt and Hocking (1985); Pitt (1988).

Penicillium hirsutum Dierckx 1901

OTHER SYNONYMS. *P. hordei,* considered a synonym of *P. hirsutum* by Pitt (1979), is now regarded as distinct (Frisvad, 1986; Cruickshank and Pitt, 1987a, b)

DIAGNOSIS. Growth relatively rapid, colonies usually deep and fasciculate; conidia green; exudate reddish or violet brown; soluble pigment yellow to orange brown. Stipes rough, penicilli terverticillate to quaterverticillate; conidia spherical to ellipsoidal, walls smooth to finely roughened.

CONCEPT. This species was called *P. corymbiferum* by Raper and Thom (1949); the earlier valid name *P. hirsutum* was taken up by Pitt (1979) without significant change in concept.

MYCOTOXINS. Only minor compounds are known to be produced.

ECOLOGY. A relatively uncommon though widespread species. One specific habitat is as a pathogen on the bulbs of Liliaceae.

DESCRIPTIONS. Pitt (1979); Pitt and Hocking (1985); Pitt (1988).

Penicillium hordei Stolk 1969

DIAGNOSIS. Colonies growing relatively slowly, texture sometimes floccose on CYA, but usually with fasciculate areas, often definitely coremial on MEA; mycelium yellow; exudate orange brown; conidia green. Stipes smooth, penicilli terverticillate; conidia spherical, small, rough walled.

CONCEPT. The concept of *P. hordei* erected by Stolk (1969) was not accepted by Pitt (1979), who regarded this species as a synonym of *P. hirsutum*. However, he later accepted the species (Samson and Pitt, 1985).

MYCOTOXINS. No significant mycotoxins have been reported.

ECOLOGY. The main reported habitat for this species has been barley seeds in Europe.

DESCRIPTIONS. Stolk (1969); Samson *et al.* (1976).

Penicillium italicum Wehmer 1894
    Penicillium japonicum G. Smith 1963

DIAGNOSIS. Colonies growing moderately rapidly on CYA, rapidly on MEA, texture usually fasciculate to coremial, occasionally velutinous; conidia greyish green; soluble pigment brown; reverse on CYA usually brownish orange to greyish brown, on MEA dark brown. Stipes smooth, penicilli terverticillate, sometimes irregular; conidia borne as cylinders, in maturity ellipsoidal to short cylindroidal, smooth walled. The main diagnostic feature of *P. italicum* is the formation of conidia as cylinders from the phialides. It also produces dense, fasciculate colonies with green conidia.

CONCEPT. The concept of *P. italicum* established by Raper and Thom (1949) was broadened by Pitt (1979) to include the newly described species *P. concentricum* Samson *et al. P. concentricum* was restored to species status (as *P. coprophilum*) in later publications (Samson and Pitt, 1985; Cruickshank and Pitt, 1987a), and the concept of *P. italicum* restored to that of Raper and Thom (1949).

MYCOTOXINS. No important secondary metabolites are produced.

ECOLOGY. The primary habitat for this species is as a pathogen of citrus fruits. It has been reported infrequently from other habitats. Some evidence exists that *P. italicum* has evolved from the coprophilic species *P. coprophilum*.

DESCRIPTIONS. Pitt (1979); Domsch *et al.* (1980); Pitt and Hocking (1985); Pitt (1988).

Penicillium olsonii Bainier & Sartory 1912.

DIAGNOSIS. Colonies of moderate size, deep but velutinous; conidia dull green; mycelium white to pale brown; reverse pale yellow to yellow brown. Stipes very long and broad, smooth walled; penicilli usually terverticillate, multiramulate; conidia ellipsoidal, smooth walled. *P. olsonii* is readily distinguished by large stipes bearing up to 6 rami in well ordered verticils.

CONCEPT. Although this species has sometimes been considered a synonym of *P. brevicompactum*, it is quite distinct morphologically. The concept clearly remains as described and illustrated by the original authors, although now greatly amplified.

MYCOTOXINS. No secondary metabolites of significance have been reported.

ECOLOGY. *P. olsonii* remains an uncommonly isolated species, although by no means as rare as reported by Pitt (1979). It is widely distributed.

DESCRIPTIONS. Pitt (1979); Pitt (1988).

Penicillium roqueforti Thom 1906
    Penicilium roqueforti var. *weidemannii* Westling 1911
    Penicillium gorgonzolae Biourge 1923

Penicillium roqueforti var. *viride* Dattilo-Rubbo 1938
Penicillium conservandi Novobranova 1974

DIAGNOSIS. Colonies rapidly growing, strictly velutinous; conidia at the margins bluish, elsewhere green; reverse on CYA sometimes dark green. Stipes usually very rough, penicilli large and terverticillate; conidia large, spherical, smooth walled.

CONCEPT. *P. roqueforti* has been regarded as a distinct species since it was described. No consistent differences appear to distinguish strains used in cheese manufacture from spoilage isolates.

MYCOTOXINS. Two distinct subspecies occur: one typically produces roquefortines and PR toxin, the other patulin and sometimes penicillic acid (Frisvad, 1986; El-Banna *et al.*, 1987). The two subspecies are indistinguishable morphologically.

ECOLOGY. Although more widely known for its role in cheese manufacture, *P. roqueforti* is in fact a very widely distributed spoilage fungus, especially on refrigerated products, and products of low oxygen tension (Pitt and Hocking, 1985).

DESCRIPTIONS. Samson *et al.* (1977b); Pitt (1979); Pitt and Hocking (1985); Pitt (1988).

Penicillium solitum Westling 1911
Penicillium psittacinum Thom 1930
Penicillium casei var. *compactum* Abe 1956 (nom. inval.)
Penicillium verrucosum var. *melanochlorum* Samson *et al.* 1976
Penicillium mali Novobranova 1972 = *Penicillium mali* Gorlenko & Novobranova 1983.

DIAGNOSIS. Colonies small to medium sized, usually velutinous; conidia dark bluish green to dark green; exudate sometimes produced, clear; reverse on CYA usually pale, on MEA usually a distinct greyish or brownish orange. Stipes smooth to rough, penicilli usually terverticillate, but sometimes biverticillate or quaterverticillate also; conidia quite large, spherical to subspheroidal, smooth walled.

CONCEPT. Considered to be an uncommon, funiculose species by Raper and Thom (1949), *P. solitum* was given recognition as a new variety *P. verrucosum* var. *melanochlorum* by Samson *et al.* (1976). It was placed in synonymy with *P. aurantiogriseum* and unrecognised by Pitt (1979); *P. verrucosum* var. *melanochlorum* was incorrectly placed in synonymy with *P. crustosum*. Cruickshank and Pitt (1987a, b) revived *P. solitum* on the basis of differences in enzyme electrophoretic patterns and morphology. Westling's concept is now placed on a sound basis.

MYCOTOXINS. Unlike most other common species in *Penicillium* subgen. *Penicillium*, *P. solitum* does not appear to produce distinctive metabolites or mycotoxins.

ECOLOGY. Because of lack of recognition, the ecology of *P. solitum* remains largely unknown. However, unpublished data (L. Leistner and J.I. Pitt) shows that this species is common in European processed meats. It also has a major niche as a pathogen of pomaceous fruit (Pitt and Spotts, in prep.).

DESCRIPTIONS. Pitt (1988); Pitt and Spotts (in prep.).

Penicillium verrucosum Dierckx 1901

DIAGNOSIS. Colonies slowly growing, especially on MEA, texture velutinous, floccose or fasciculate; conidia yellow green; exudate clear to pale yellow; reverse on CYA yellow brown to deep brown. Stipes robust, rough walled, penicilli usually terverticillate, in some isolates also quaterverticillate, in others also biverticillate or irregular; conidia usually spherical, smooth walled. Colony appearance is similar to *P. viridicatum*, but diameters on CYA and MEA are smaller. Stipes and penicilli are more robust than in *P. viridicatum*.

CONCEPT. Ignored by Raper and Thom (1949), *P. verrucosum* was taken up by Samson *et al.* (1976) in a very broad sense. Pitt (1979) accepted the species, but with a much narrower concept. Discussion over the validity of Pitt's concept has now declined, with corroboration coming from reexamination (Pitt, 1987), enzyme electrophoretic studies (Cruickshank and Pitt, 1987b) and secondary metabolite profiles (Frisvad, 1986; Pitt, 1987; El-Banna *et al.*, 1987).

MYCOTOXINS. *P. verrucosum* is the major source of ochratoxin A among the Penicillia. This toxin is not produced by *P. viridicatum* (Frisvad, 1986; Pitt, 1987), although this has been commonly reported in the literature. Some isolates of *P. verrucosum* also produce citrinin (Frisvad, 1986; El-Banna *et al.*, 1987).

ECOLOGY. *P. verrucosum* is of common occurrence in Scandinavian cereals (Frisvad and Vuif, 1986, as *P. viridicatum* Group II), and is the cause of ochratoxin poisoning of both animals and man in that region. It also occurs in European processed meats (Pitt and Hocking, 1985). It has not been isolated commonly elsewhere.

DESCRIPTIONS. Pitt (1979); Pitt (1988).

Penicillium viridicatum Westling 1911

> *Penicillium olivinoviride* Biourge 1923
> *Penicillium olivicolor* Pitt 1979
> *Penicillium ochraceum* Bain. ex Thom 1930

OTHER SYNONYMS. *P. palitans* Westling and *P. lanosoviride* Thom, placed in synonymy with *P. viridicatum* by Pitt (1979), are now regarded as synonyms of *P. commune*.

DIAGNOSIS. Colonies of moderate size, texture velutinous to fasciculate; conidia yellow green; exudate clear to pale yellow; sometimes soluble pigment, orange to reddish brown. Stipes rough, penicilli mostly terverticillate; conidia subspheroidal to ellipsoidal, smooth walled.

CONCEPT. *P. viridicatum* became established as a distinct species after acceptance by Raper and Thom (1949), and became well recognised. However, Raper and Thom (1949) incorrectly placed an earlier species, *P. verrucosum*, in synonymy with *P. viridicatum*. Samson *et al.* (1976) recognised *P. verrucosum*, but reduced several other species accepted by Raper and Thom (1949) to the status of varieties of it. Pitt (1979) also recognised *P. verrucosum*, but in a much narrower sense, and reinstated *P. viridicatum* in a sense similar to that of Raper and Thom (1949). This concept is now gaining acceptance (Frisvad, 1986; Pitt, 1987).

MYCOTOXINS. As now accepted, *P. viridicatum* produces the naphthoquinone toxins viomellein and xanthomegnin. The earlier reports that *P. viridicatum* produces ochratoxin A have been shown to be incorrect (Pitt, 1987).

PROBLEMS. A degree of overlap still exists between *P. aurantiogriseum* and *P. viridicatum*. See the discussion under *P. aurantiogriseum*.

ECOLOGY. Pitt (1979) regarded *P. viridicatum* as a ubiquitous species. However, although it is relatively common in cereals in temperate zones, it occurs at quite a low frequency elsewhere (Domsch *et al.*, 1980; Pitt and Hocking (1985)).

DESCRIPTIONS. Pitt (1979); Pitt and Hocking (1985); Pitt (1988).

Analytical Key to *Penicillium* Subgenus *Penicillium*

1. Conidia borne as cylinders, with at least a proportion remaining so at maturity.............................4
    Conidia borne as ellipsoids or spheroids and remaining so at maturity..2

2. Conidia *en masse* brown, penicilli often irregular.. *P. arenicola*
    Conidia *en masse* green or olive; penicilli usually well-defined... 3

3. Penicilli often with more than 3 rami... *P. olsonii*
    Penicilli with no more than 2 rami..6

4. Conidia *en masse* olive.. *P. digitatum*
    Conidia *en masse* green.. 5

5. Mycelium white; conidia smooth walled .... *P. italicum*
    Mycelium salmon or orange; conidia spinose ....*P. fennelliae*

6. Conidia *en masse* persistently white or pale grey; isolated from cheese or cheese factory. .... *P. camemberti*
    Conidia *en masse* blue, green or grey; source inconsequential... 7

7. Colonies on CYA at 25° exceeding 30 mm diam. 8
    Colonies on CYA at 25° not exceeding 30 mm diam... 22

8. Stipes on CYA and MEA smooth walled..9
    Stipes often rough-walled, especially on MEA..15

9. Soluble pigment and/or reverse yellow; sometimes growth at 37° ....*P. chrysogenum*
    Soluble pigment and reverse not yellow; no growth at 37°..10

10. Conidia blue on both CYA and MEA...*P. aurantiogriseum*
    Conidia green on both CYA and MEA 11

11. Exudate and soluble pigment red to reddish brown; conidia spherical. ....*P. atramentosum*
    Exudate and soluble pigment not red or reddish brown; conidia usually ellipsoidal................................12

12. Conidia dark blue green to dark green, reverse on CYA pale, on MEA greyish to brownish orange ....
    ....*P. solitum*
    Conidia yellow green or dull green, reverse on CYA pale or brown, on MEA pale or dull brown..........13

13. Colonies dense and compact, fascicles inconspicuous, coremia absent, conidia yellow green..............
    .... *P. viridicatum*
    Colonies fasciculate to coremial, conidia dull green or dark green..14

14. Conidia dull green, colonies deep and fasciculate to coremial, reverse on CYA pale to orange brown........
    ....*P. expansum*
    Conidia dark green, colonies granular to fasciculate, usually with small coremia, reverse on CYA dark
    brown ....*P. coprophilum*

15. Conidial walls rough or spinose. ....*P. echinulatum*
    Conidial walls smooth..16

16. Conidial colour on CYA uniformly blue or greyish blue 17
    Overall conidial colour on CYA green, although sometimes with bluish marginal areas........................18

17. Conidia dull green on MEA; reverse on CYA pale to dull brown...................................... *P. commune*
    Conidia dark green on MEA; reverse on CYA orange yellow to yellow brown .... *P. aethiopicum*

18. Colony reverse on CYA dark brown ....*P. allii*
    Colony reverse on CYA not dark brown..19

19. Conidia on CYA green with bluish margins; reverse on CYA pale or green; conidia usually spherical... 20
    Conidia on CYA uniformly green; reverse on CYA pale or yellow brown; conidia usually not spherical...
   21

20. Reverse on CYA pale or brownish; conidia dull green; texture fasciculate on CYA, granular on MEA; conidia on MEA usually detaching in masses when jarred .... *P. crustosum*
    Reverse on CYA often green; conidia dark green; texture velutinous; conidia on MEA not detaching when jarred .... *P. roqueforti*

21. Conidia yellow green; exudate clear to pale yellow; penicilli strictly terverticillate. .... *P. viridicatum*
    Conidia green; exudate reddish to violet brown; penicilli sometimes quaterverticillate............ *P. hirsutum*

22. Penicilli broad, often 40 mm or more across the metulae apices; metulae often apically enlarged ....
    .... *P. brevicompactum*
    Penicilli narrow, commonly 30 mm or less across the metulae apices; metulae cylindrical 23

23. Stipe walls smooth to finely roughened 24
    Stipe walls finely to conspicuously roughened 26

24. Phialides commonly 4.5 to 6 µm long .... *P. griseofulvum*
    Phialides exceeding 6 µm long 25

25. Conidia green; mycelium yellow; soluble pigment orange; conidia often rough walled .... *P. hordei*
    Conidia dark bluish green to dark green; mycelium white; soluble pigment clear; conidia smooth walled .... *P. solitum*

26. Conidia yellow green 27
    Conidia bluish or dull green to dark green 28

27. Colonies on CYA and MEA both exceeding 25 mm in diameter .... *P. viridicatum*
    Colonies on CYA and MEA not exceeding 25 mm in diameter. .... *P. verrucosum*

28. Texture fasciculate to coremial; soluble pigment yellow brown to deep brown; stipes conspicuously roughened; conidia ellipsoidal. .... *P. glandicola*
    Texture velutinous to floccose; soluble pigment absent; stipes smooth to roughened; conidia spherical to subspheroidal. 29

29. Conidia blue grey to dull green; reverse usually pale on both CYA and MEA .... *P. commune*
    Conidia dark blue green to dark green; reverse pale on CYA, characteristically greyish or brownish orange on MEA .... *P. solitum*

REFERENCES

BRACKETT, R.E. and MARTH, E.H. 1979. Patulin in apple juice from roadside stands in Wisconsin. *Journal of Food Protection* 42: 862-863.

CIEGLER, A. and KURTZMAN, C.P. 1970. Penicillic acid production by blue-eye fungi on various agricultural commodities. *Applied Microbiology* 20: 761-764.

CIEGLER, A., FENNELL, D.I., SANSING, G.A., DETROY, R.W. and BENNETT, G.A. 1973. Mycotoxin-producing strains of *Penicillium viridicatum*: classification into subgroups. *Applied Microbiology* 26: 271-278.

CIEGLER, A., LEE, L.S. and DUNN, J.J. 1981. Production of naphthoquinone mycotoxins and taxonomy of *Penicillium viridicatum*. *Applied and Environmental Microbiology* 42: 446-449.

CRUICKSHANK, R.H. and PITT, J.I. 1987a. The zymogram technique: isoenzyme patterns as an aid in *Penicillium* classification. *Microbiological Sciences* 4: 14-17.

—— 1987b. Identification of species in *Penicillium* subgenus *Penicillium* by enzyme electrophoresis. *Mycologia* 79: 614-620.

DOMSCH, K.H., GAMS, W. and ANDERSON, T.-H. 1980. Compendium of Soil Fungi. London: Academic Press. 2 vols.

EL-BANNA, A.A., LEISTNER, L. and PITT, J.I. 1987. Production of mycotoxins by *Penicillium* species. *Systematic and Applied Microbiology* 10: 42-46.

FILTENBORG, O. and FRISVAD, J.C. 1980. A simple screening method for toxigenic moulds in pure culture. *Lebensmittel Wissenschaft und Technologie* 13: 128-130.

FRISVAD, J.C. 1983. A selective and indicative medium for groups of *Penicillium viridicatum* producing different mycotoxins in cereals. *Journal of Applied Bacteriology* 54: 409-416.

FRISVAD, J.C. 1985. Profiles of primary and secondary metabolites of value in classification of *Penicillium viridicatum* and related species. *In* Advances in *Penicillium* and *Aspergillus* Systematics, eds. R.A. Samson and J. I. Pitt, pp. 311-325. New York: Plenum Press.

——— 1986. Taxonomic approaches to mycotoxin identification (taxonomic indication of mycotoxin content of foods). *In* Modern Methods in the Analysis and Structural Elucidation of Mycotoxins, ed. R.J. Cole, pp. 415-457. New York: Academic Press.

FRISVAD, J.C. and FILTENBORG, O. 1983. Classification of terverticillate Penicillia based on profiles of mycotoxins and other secondary metabolites. *Applied and Environmental Microbiology* 46: 1301-1310.

FRISVAD, J.C. and VUIF, B.T. 1986. Comparison of direct and dilution plating for detecting *Penicillium viridicatum* in barley containing ochratoxin. *In* Methods for the Mycological Examination of Food, eds. A.D. King, J.I. Pitt, L.R. Beuchat and J.E.L. Corry, pp. 45-47. New York: Plenum Press.

FRISVAD, J.C., FILTENBORG, O. and WICKLOW, D.T. 1987. Terverticillate Penicillia isolated from underground seed caches and cheek pouches of banner-tailed kangaroo rats (*Dipodomys spectabilis*). *Canadian Journal of Botany* 65: 765-773.

ONIONS, A.H.S., BRIDGE, P.D. and PATERSON, R.R. 1984. Problems and prospects for the taxonomy of *Penicillium*. *Micobiological Sciences* 1: 185-189.

PATERSON, R.R.M. and HAWKSWORTH, D.L. 1985. Detection of secondary metabolites in dried cultures of *Penicillium* preserved in herbaria. *Transactions of the British Mycological Society* 85: 95-100.

PITT, J.I. 1979. The Genus *Penicillium* and its Teleomorphic States *Eupenicillium* and *Talaromyces*. London: Academic Press.

——— 1987. *Penicillium viridicatum, Penicillium verrucosum*, and production of ochratoxin A. *Applied and Environmental Microbiology* 53: 266-269.

——— 1988. A Laboratory Guide to Common *Penicillium* Species. North Ryde, N.S.W.: CSIRO Division of Food Processing.

PITT, J.I. and HAWKSWORTH, D.L. 1985. The naming of chemical variants in *Penicillium* and *Aspergillus. In* Advances in *Penicillium* and *Aspergillus* Systematics, eds. R.A. Samson and J. I. Pitt, pp. 89-91. New York: Plenum Press.

PITT, J.I. and HOCKING, A.D. 1985. Fungi and Food Spoilage. Sydney: Academic Press.

PITT, J.I. and LEISTNER, L. 1989. Toxigenic *Penicillium* species. *In* Mycotoxins and Animal feedingstuffs: natural occurrence, toxicity and control. I. Toxigenic Fungi, J. E. Smith, ed. Boca Raton, Florida: CRC Press (in press).

PITT, J.I., CRUICKSHANK, R.H. and LEISTNER, L. 1986. *Penicillium commune, P. camemberti*, the origin of white cheese moulds, and the production of cyclopiazonic acid. *Food Microbiology* 3: 363-371.

POLONELLI, L., MORACE, G., ROSA, R., CASTAGNOLA, M. and FRISVAD, J.C. 1987. Antigenic characterization of *Penicillium camemberti* and related common cheese contaminants. *Applied and Environmental Microbiology* 53: 872-878.

RAPER, K.B. and THOM, C. 1949. A Manual of the Penicillia. Baltimore, Maryland. Williams and Wilkins.

RAMIREZ, C. 1982. Manual and Atlas of the Penicillia. Amsterdam: Elsevier Biomedical.

SAMSON, R.A. and PITT, J.I. 1985. Check list of common *Penicillium* species. *In* Advances in *Penicillium* and *Aspergillus* Systematics, eds. R.A. Samson and J. I. Pitt, pp. 461-463. New York: Plenum Press.

SAMSON, R.A., STOLK, A.C. and HADLOK, R. 1976. Revision of the Subsection *Fasciculata* of *Penicillium* and allied species. *Studies in Mycology*, Baarn 11: 1-47.

SAMSON, R.A., HADLOK, R. and STOLK, A.C. 1977a. A taxonomic study of the *Penicillium chrysogenum* series. *Antonie van Leeuwenhoek* 43: 169-175.

SAMSON, R.A., ECKARDT, C. and ORTH, R. 1977b. The taxonomy of *Penicillium* species from fermented cheeses. *Antonie van Leeuwenhoek* 43: 341-350.

SEIFERT, K.A. and SAMSON, R.A. 1985. The genus *Coremium* and the synnematous Penicillia. *In* Advances in *Penicillium* and *Aspergillus* Systematics, eds. R.A. Samson and J. I. Pitt, pp. 143-154. New York: Plenum Press.

SÖDERSTRÖM, B. and FRISVAD, J.C. 1984. Separation of closely related asymmetric Penicillia by pyrolysis gas chromatography and mycotoxin production. *Mycologia* 76: 408-419.

STOLK, A.C. 1969. Four new species of *Penicillium. Antonie van Leeuwenhoek* 35: 261-274.

THOM, C. 1910. Cultural studies on species of *Penicillium. Bulletin of the Bureau of Animal Industries of the United States Department of Agriculture* 118: 1-109.

——— 1930. The Penicillia. Baltimore, Maryland: Williams and Wilkins.

VINCENT, M.A. and PITT, J.I. 1989. *Penicillium allii*, a new species from Egyptian garlic. *Mycologia* 81: 300-303

WILLIAMS, A.P. and PITT, J.I. 1985. A revised key to *Penicillium* subgenus *Penicillium*. *In* Advances in *Penicillium* and *Aspergillus* Systematics, eds. R.A. Samson and J. I. Pitt, pp. 129-134. New York: Plenum Press.

DIALOGUE FOLLOWING DR. PITT'S PRESENTATION

GAMS: In cultures that have degenerated morphologically, do the isoenzyme patterns also change?

PITT: Occasionally a band disappears, but in general terms the isoenzyme patterns remain unchanged. The concentrations of the enzymes may decrease.

THE SYSTEMATICS OF THE TERVERTICILLATE PENICILLIA

A.C. Stolk[1], R.A. Samson[1], J.C. Frisvad[2] and O. Filtenborg[2]

[1]Centraalbureau voor Schimmelcultures
3740 AG Baarn, The Netherlands

[2]Department of Biotechnology
The Technical University of Denmark
2800 Lyngby, Denmark

SUMMARY

The species of the terverticillate Penicillia are re-investigated and delimited on the basis of the morphology of the conidiophores, phialides and conidia. In addition, growth characters and profiles of secondary metabolites were taken into account for definition of the taxa. In general, the terverticillate Penicillia represent a biologically homogenous group, but on the basis of their morphology subdivision into series is proposed. The series and the accepted taxa are discussed briefly and keyed out dichotomously. A list of the principal mycotoxins produced by each taxon is presented. Most species can be identified using morphological and cultural characters as observed on Czapek and 2% malt extract agar, while a more detailed speciation requires the use of a standardized medium regime including Czapek yeast extract agar, creatine sucrose agar and 5% acetic acid agar.

INTRODUCTION

The Penicillia included in the section Asymmetrica-Fasciculata by Raper and Thom (1949) are very important components of the mycoflora of foods (Pitt and Hocking, 1985; Samson and Van Reenen-Hoekstra, 1988). Many of these species produce potent mycotoxins, and because of the clear connection between taxa and mycotoxins (Frisvad, 1988), accurate identifications are most relevant and important (Frisvad, 1989). Three different approaches to the systematics of Raper and Thom's sect. *Asymmetrica*, subsect. *Fasciculata* (Samson *et al.*, 1976; Pitt, 1979; Frisvad and Filtenborg, 1983) led to confusion both nomenclature and species concepts, although Samson and Pitt (1985) have more recently agreed in may of these. Furthermore, recent biochemical and morphological reinvestigations in these species (Cruickshank and Pitt, 1987; Frisvad, 1988; Samson and Van Reenen-Hoekstra, 1988) have resulted in modified species circumscriptions.

In this paper we present a classification of the terverticillate Penicillia based on a reinvestigation of morphological criteria of cultures mainly grown on Czapek and 2 % MEA and supplemented with mycotoxin profiles. A dichotomous key to the taxa and a list of toxins produced by each taxon is provided. This paper includes only the taxa formerly classified as asymmetric-biverticillate Penicillia, although several terverticillate taxa can also be found in other subgenera. The taxa described by Frisvad *et al.* (1987) have not been treated as they are probably found in very specific habitats, while their proper status is not yet fully elucidated.

Modern Concepts in Penicillium and Aspergillus Classification
Edited by R. A. Samson and J. I. Pitt
Plenum Press, New York, 1990

Table 1. Disposition of series of subgenera *Furcatum* and *Penicillium* with terverticillate species

Subgenus *Furcatum*	Subgenus *Penicillium*
Series *Lanosa*	Series *Digitata*
Series *Chrysogena*	Series *Italica*
Series *Urticicola*	Series *Oxalica*
	Series *Gladiolii*
	Series *Viridicata*
	Series *Expansa*
	Series *Granulata*
	Series *Olsonii*
	Series *Claviformae*

MATERIAL AND METHODS

Ex type, authentic and fresh cultures of each species considered were examined on Czapek agar (Cz) and 2% malt agar (MA) (Samson *et al.*, 1976) and/or malt extract agar (MEA) (Raper and Thom, 1949) for morphological examinations and on Czapek yeast autolysate agar (CYA) (Pitt, 1979), Difco yeast extract sucrose agar (YES) (Filtenborg *et al.*, 1983) and creatine-sucrose agar (CREA) (Frisvad, 1985) for physiological and chemical examinations. Furthermore the isolates were grown on CYA at 37 C and inoculated into apples or lemons to test their pathogenicity. Cultures grown on CYA and YES were examined for profiles of secondary metabolites using the agar plug method (Filtenborg *et al.*, 1983). The production of particular secondary metabolites was confirmed by high-performance liquid chromatography (HPLC) with diode array detection (Frisvad and Thrane, 1987).

RESULTS AND DISCUSSION

Pitt (1979) classified the terverticillate Penicillia in subgenus *Penicillium*, including taxa previously accommodated by Raper and Thom (1949) as the *Asymmetrica* sects. *Fasciculata* and *Funiculosa*. We believe that the morphology of the conidiophore and shape of phialides and conidia justify a more detailed subdivision of the species. (Table 1). A more detailed subdivision of subgen. *Penicillium* into varieties new species and chemotypes, based on physiological characters and profiles of secondary metabolites is briefly presented by Frisvad and Filtenborg (1990).

Table 2 lists the specific mycotoxins of each accepted species. However, note that in some species many unknown, – but specific – compounds are present, e.g. in *P. olsonii*

Table 2. Accepted species of the terverticillate Penicillia and the production of mycotoxins.

P. atramentosum Thom	Oxaline Roquefortine C Rugulovasine A
P. aurantiogriseum Dierckx var. *aurantiogriseum*	Penicillic acid Verrucosidin Aurantiamin Xanthomegnin Viomellein Verrucofortine Cyclopenin Cyclopenol Viridicatin Viridicatol
P. aurantiogriseum var. *viridicatum* (Westling) Frisvad et Filtenborg	Penicillic acid Viridamine (Viridic acid)[a] Brevianamide A Verrucofortine Xanthomegnin Viomellein Cyclopenin, -ol Viridicatin, -ol
P. brevicompactum Dierckx	Mycophenolic acid Raistrick phenols Brevianamide A (Botryodiploidin)
P. camemberti Thom	Cyclopiazonic acid
P. chrysogenum Thom	Penicillin Roquefortine C Meleagrin (Xanthocillin) (emodic acid) (sorbicillin)
P. clavigerum Demelius	Penitrem A (Roquefortine A)
P. commune Thom	Cyclopiazonic acid Palitantin Cyclopaldic acid Rugulovasine A Roquefortine A
P. coprophilum (Berk. et Curt.) Seifert et Samson	Griseofulvin Roquefortine C Meleagrin Oxaline
P. crustosum Thom	Penitrem A Terrestric acid Cyclopenin, -ol Viridicatin, -ol Roquefortine C
P. digitatum (Pers.: Fr.) Sacc.	Tryptoquivalins

(cont.)

Table 2 (cont.)

P. echinulatum (Raper et Thom) Fassatiová	Cyclopenin, -ol Viridicatin, -ol Penechins
P. expansum Link	Patulin Citrinin Roquefortine C Chaetoglobosin C
P. fennelliae Stolk	Penicillic acid
P. glandicola (Oud.) Seifert et Samson	Penitrem A Patulin Roquefortine C Meleagrin Oxaline
P. griseofulvum Dierckx	Patulin Griseofulvin Cyclopiazonic acid Roquefortine C
P. hirsutum Dierckx	Roquefortine C Terrestric acis Compactin
P. hordei Stolk	Roquefortine C Terrestric acid
P. italicum Wehmer	Deoxybrevianamide E 5,6-dihydroxy-4-methoxy- 2H-pyran-2-one
P. lanosum Westling	Griseofulvin Kojic acid
P. olsonii Bain. et Sartory	–
P. roqueforti Thom	PR-toxin[b] Roquefortine C Mycophenolic acid Roquefortine A[b] Marcfortines[b] Patulin[c] Penicillic acid[c] Botryodiploidin[c]
P. solitum Westling	Compactin Cyclopenin, -ol Viridicatin, -ol
P. verrucosum Dierckx	Ochratoxin A Verrucolone Citrinin
P. vulpinum (Cooke et Massee) Seifert et Samson	Patulin Roquefortine C Oxaline Cyclopenin Viridicatin

[a] Secondary metabolites listed in parenthesis: freqeuncy of isolates producing this metabolite is unknown
[b] *P. roqueforti* chemotype I (dark green reverse on Czapek); [c] *P. roqueforti* chemotype II (light brown reverse on Czapek)

LIST OF ACCEPTED TAXA

The following list outlines the accepted taxa with a short discussion of their placement:

Series LANOSA Stolk et Samson

Adv. *Penicillium* and *Aspergilllus* Syst.: 180, 1985

Type species: *Penicillium lanosum* Westling

P. lanosum was placed as a synonym of *P. puberulum* by Pitt (1979), while *P. kojigenum* was placed as a synonym of *P. jensenii*. Pitt (1988) placed *P. lanosum* as a synonym of *P. commune*, but we consider it to be a distinct species based both on morphology (Samson *et al.*, 1976) and secondary metabolites (Table 2). Characteristically *P. lanosum* produces terverticillate penicilli with a quite divergent branch. Apart from its morphological characters, it is the only species in *Penicillium* producing both kojic acid and griseofulvin.

Penicillium lanosum Westling 1911

Penicillium kojigenum G. Smith 1961

Series CHRYSOGENA Raper & Thom ex Stolk & Samson

Series *P. chrysogenum* Raper & Thom - Man. Penicillia: 355, 1949 (nom. inval. Arts 21, 36) = Series *Chrysogena* Raper & Thom ex Stolk & Samson in Adv. *Penicillium* and *Aspergillus*

Raper & Thom (1949) proposed the series *P. chrysogenum*-series which included four species: *P. chrysogenum* Thom, *P. meleagrinum* Biourge, *P. notatum* Westling and *P. cyaneofulvum* Biourge. Samson *et al.* (1977a) observed no significient differences between the four species, as represented by their type strains and authentic cultures. Consequently they placed the latter three species in synonymy with *P. chrysogenum*.

There is general agreement that the ex type of *P. griseoroseum* Dierckx agrees very well with the ex type culture of *P. chrysogenum* Thom. It differs only in producing predominantly biverticillate penicilli and appears to have deteriorated in culture. The description of Biourge's (1923) *P. griseoroseum* was characterized by bi- to terverticillate conidiophores like those of *P. chrysogenum*. The species should be regarded as synonyms: the correct name is *P. griseoroseum*. However, as at present the name *P. chrysogenum* is commonly used in the literature and it is of great industrial significance, so it is proposed to conserve the name *P. chrysogenum* (see Chapter 2 on Nomenclature).

The type cultures of *P. chrysogenum*, and starter cultures of *P. nalgiovense* used for mould fermented salami resemble one another closely in morphological respects. Furthermore they both produce penicillin. *P. nalgiovense* as it is now used is therefore regarded as a domesticated form of *P. chrysogenum*. The ex type culture of *P. nalgiovense*, however is a different from the strains used for mould fermented products; the profiles of secondary metabolites are different: *P. nalgiovense* NRRL 911 produces nalgiovensin and nalgiolaxin (Birch and Stapleton, 1967) and other unique compounds, while the domesticated strains of *P. chrysogenum* produce penicillin and other specific coumpounds.

Series *Chrysogena* is classified in the subgenus *Furcatum* because of included species produce bi- to quaterverticillate conidiophores with divergent, subterminal and intercalary branches (metulae). As some *P. chrysogenum* isolates may develop up to five or more-stage-branched conidiophores it shows some affinities with the series *Urticicola*. However, conidiophores of in the Series *Chrysogena* are less complicated and the phialides and metulae are different.

Penicillium atramentosum Thom 1910

Penicillium chrysogenum Thom 1910
  Penicillium griseoroseum Dierckx 1901
  Penicillium citreoroseum Dierckx 1901
  Penicillium brunneorubrum Dierckx 1901
  Penicillium notatum Westling 1911
  Penicillium baculatum Westling 1911
  Penicillium meleagrinum Biourge 1923
  Penicillium cyaneofulvum Biourge 1923
  Penicillium roseocitreum Biourge 1923
  Penicillium flavidomarginatum Biourge 1923
  Penicillium rubens Biourge 1923
  Penicillium chlorophaeum Biourge 1923
  Penicillium camerunense Heim 1949
  Penicillium aromaticum f. *microsporum* Romankova
  Penicillium harmonense Baghdadi 1968
  Penicillium verrucosum var. *cyclopium* strain *ananas-olens* Ramírez 1982

Series URTICICOLA Fassatiová

Series *P. urticae* Raper & Thom - Man. Penicillia: 531, 1949 (nom. inval. Arts 21, 36) = Series *Urticicola* Fassatiová - Acta Univ. Carol. Biol. 12: 324. 1977

Fassatiová (1977) proposed the series *Urticicola* to accomodate a single species: *P. urticae* (= *P. griseofulvum*). Pitt (1979) emended Fassatiová's concept of the series by adding five species which like *P. griseofulvum* are characterized by relatively slow growth on CYA at 25°C. However, the added species produce regular, ter- to quaterverticillate penicilli and thus differ markedly from the complicated, loosely arranged, three- to five-stage-branched conidial structures, characteristic of *P. griseofulvum*. Moreover the phialides are quite different. Most authors (Raper & Thom, 1949; Fassatiová, 1977; Ramírez, 1982) have classified *P. griseofulvum*, because of its branched, synnematous conidiophores in Raper & Thom's subsection *Fasciculata* Raper and Thom or subgen.*Penicillium* (Pitt, 1979). However, in many respects, such as the complicated conidiophores, the divergent branches and the small metulae and phialides *P. griseofulvum* differs from species in subgen.*Penicillium*. It is preferred, here to include it in subgen. *Furcatum*. In fact *P. griseofulvum* morphologically is an unique species without close relationships in *Penicillium*. Biologically, however, *P. griseofulvum* is related to the coprophilic, synnemateous species *P. coprophilum, P. glandicola* and *P. vulpinum*. This relationship is also supported by the similar secondary metabolites produced by all four species: roquefortine C and patulin are produced by each species, except *P. coprophilum* which only produce roquefortine C (Frisvad and Filtenborg, 1989 a). The three-to five-stage-branched or sometimes even more complex conidiophores are slightly reminiscent of *P. chrysogenum*, however. *P. griseofulvum* differs from species in series *Chrysogena* in producing synnematous conidiophores, which are more complicated and irregular in structure, as well as by the very small phialides and metulae.

In agreement with Raper & Thom (1949) and Fassatiová (1977) *Urticicola*, as delimited here, is restricted to only a single species, *P. griseofulvum*.

Penicillium griseofulvum Dierckx 1901
  Penicillium patulum Bain. 1906
  Penicillium urticae Bain. 1907
  Penicillium flexuosum Dale apud Biourge 1923

? *Penicillium maltum* Hori et Yamamoto 1954
? *Penicillium duninii* Sidibe 1974

Series DIGITATA Raper et Thom ex Stolk et Samson

Series *P. digitatum* Raper et Thom - Man. Penicillia: 385, 1949 (nom. inval. Arts 21, 36) = Series *Digitata* Raper et Thom ex Stolk et Samson - Adv. *Penicillium* and *Aspergillus* Syst.: 183, 1985.

Type species: *Penicillium digitatum*

P. digitatum is a very distinctive species, related to *P. italicum* only by its cylindrical conidia and the ability to produce a destructive rot in citrus fruits. It shares no secondary metabolites with *P. italicum*. It is excluded from the key provided in this paper, since its conidiophores are not really terverticillate.

Penicillium digitatum (Pers.: Fr.) Sacc. 1881

Penicillium olivaceum Wehmer 1895
Penicillium olivaceum Sopp 1912
Penicillium olivaceum var. *norvegivum* Sopp 1912
Penicillium olivaceum var. *italicum* Sopp 1912
Penicillium digitatoides Peyronel 1913
Penicillium lanosogrisellum Biourge 1923
Penicillium terraconense Ramírez et Martinez 1980

Series ITALICA Raper et Thom *ex* Fassatiová

Series *P. italicum* Raper et Thom - Man. Penicillia: 523, 1949 (nom. inval. Arts 21, 36) = Series *Italica* Raper et Thom ex Fassatiová - Acta Univ. Carol., Biol.: 324, 1977 = Series *Italica* Raper et Thom ex Pitt - Gen. *Penicillium*: 381, 1979.

Type species: *Penicillium italicum*

Series *Italica* is represented by one distinctive species, *P. italicum*. It causes a characteristic blue rot of citrus fruits. *P. italicum* is distinguished from *P. digitatum* by the synnematous conidiophores, the better developed terverticillate penicilli and greyish, blue green colonies. White or avellaneous mutants of *P. italicum* may occasionally be encountered. *P. japonicum* G. Smith (= *P. digitatum* var. *latum* Abe) is a typical *P. italicum*, although strongly coloured and was incorrectly placed in synonymy with *P. resticulosum* Raistrick *et al.* by Pitt (1979).

Penicillium italicum Wehmer 1894

Penicillium aeruginosum Dierckx 1901
Penicillium ventruosum Westling 1911
Penicillium japonicum G. Smith 1963 (= *Penicillium digitatum* var. *latum* Abe 1956, nom. inval.)
Penicillium italicum var. *album* Wei 1940
Penicillium italicum var. *avellaneum* Samson et Gutter 1976

Series OXALICA Raper & Thom *ex* Pitt

Series *P. oxalicum* Raper & Thom - Manual Penicillia: 376, 1949 (nom. inval. Arts 21, 36) = series *Oxalica* Raper & Thom ex Pitt - Gen. Penicil: 273, 1979

Species of series *Oxalica* are characterized by well developed penicilli, robust, large phialides, large cylindrical to ellipsoidal conidia and rapidly growing, velutinous cultures. They are separated from species from series *Digitata* by much better developed conidiophores and from those in series *Italica* by veluti nous colonies. *P. oxalicum* is excluded from the key provided below because the conidiophores are predominantly biverticillate. The two species *P. oxalicum* and *P. fennelliae* are closely related morphologically and consequently are classified together in series *Oxalica*. The latter series is placed in the subgenus *Penicillium* because of the appressed structure of the penicilli.

Ramírez and Martinez (1981) described *P. aragonense* in terms suggesting a possible synonymy with *P. oxalicum*. The type culture of *P. aragonense* proved to be badly contaminated, it contained only *P. glabrum*.

P. sclerotigenum is another species, which is closely related to both *P. italicum* and *P. oxalicum*. Similarities to species in subgenus *Penicillium* include the ability to produce a destructive rot in yams, production of roquefortine C and griseofulvin and the appressed occasionally terverticillate penicilli.

P. oxalicum Currie et Thom 1915
> *P. aragonense* Ramírez et Martinez 1981
> *P. asturianum* Ramírez et Martinez 1981

P. fennelliae Stolk 1969

Series GLADIOLII Raper et Thom *ex* Stolk et Samson
> Series *P. gladioli* Raper et Thom - Man. Penicillia: 471, 1949 (nom. inval.); ex Stolk et Samson - Adv. *Penicillium* and *Aspergillus* Syst. p. 183. 1985

Penicillium gladioli is the only species classified in subgenus *Penicillium* which produces sclerotia. It is closely related to the anamorphs of *Eupenicillium crustaceum*. Up till now, no ascospores has been observed in any *P. gladioli* isolates.

P. gladioli McCulloch & Thom 1928

Series VIRIDICATA Raper et Thom *ex* Pitt
> Series *P. viridicatum* Raper et Thom - Man. Penicillia: 481, 1979
> Series *P. roqueforti* Raper et Thom - Man. Penicillia: 392, 1949
> Series *P. camemberti* Raper et Thom - Man. Penicillia: 421, 1949
> Series *P. commune* Raper et Thom - Man. Penicillia: 429, 1949
> Series *P. terrestre* Raper et Thom - Man. Penicillia: 446, 1949
> Series *P. ochraceum* Raper et Thom - Man. Penicillia: 475, 1949
> Series *P. cyclopium* Raper et Thom - Man. Penicillia: 490, 1949
> (all nom. inval., Arts 21, 36)

All species included in series *Viridicata* produce similar terverticillate, ocasionally quaterverticillate conidiophores, as well as robust phialides. The conidia of most species range from globose to subglobose, less commonly ellipsoidal, with walls smooth or nearly so, except in P. echinulatum, which has echinulate conidia.

Species classfied in series *Viridicata* are very important spoilage and mycotoxin producing fungi in food and feedstuffs. They have often been cited in the literature and consequently a correct identification is important (Frisvad, 1989). Unfortunately the identification of these species is often difficult, because of great variation within the species and the presence of morphologically intergrading strains. Raper and Thom (1949) considerably reduced the number of species accepted by Thom (1930) by regarding many of them as synonyms. However, they still maintained a great number of closely related species, which they distinguished from one another by minor differences, such a colony texture and colour. In an attempt to simplify species determination in these species, Samson *et al.* (1976) revived *P. verrucosum* Dierckx and emended the description, creating a large variable species, which they divided into six varieties, mainly based on conidial colour. By the use of colony diameters together with conidial and colony pigmentation to separate species in subgenus *Penicillium*, Pitt (1979) reinstated some varieties as species. However, recently novel characters and techniques have been introduced to provide a more accurate delimitation of these closely related species. Morphological differences between the species in series *Viridicata* are limited so more recently, the taxonomic importance of profiles of secondary metabolites (SMPs) have assisted in delimiting species in series *Viridicata*.

A comparative study of morphological and biochemical characters (SMPs) often proved a correlation between them to be present. For instance even though the morphological difference between *P. verrucosum* and *P. aurantiogriseum* is minute (*P. verrucosum* has more distinctly roughened stipes), physiological and biochemical differences between these two species are significant.

The morphology and physological characters of *P. aurantiogriseum* and *P. viridicatum* are very similar and both taxa can only be distinguished by their secondary metabolites and conidium colour. Frisvad and Filtenborg (1989) considered them as varieties.

P. aurantiogriseum Dierckx 1901 var. *aurantiogriseum*

P. aurantiocandidum Dierckx 1901
P. puberulum Bain. 1907
P. conditaneum Westling 1911
P. cyclopium Westling 1911
P. aurantiovirens Biourge 1923
P. brunneoviolaceum Biourge 1923
P. martensii Biourge 1923
P. porraceum Biourge 1923
P. aurantioalbidum Biourge 1923
P. johanniolii Zaleski 1927
P. polonicum Zaleski 1927
P. ochraceum Bain. apud Thom 1930
P. carneolutescens G. Smith 1939
P. viridicyclopium Abe 1956
P. verrucosum Dierckx var. *cyclopium* (Westling) Samson *et al.* 1976
P. verrucosum var. *ochraceum* (Bain.) Samson *et al.* 1976
P. cyclopium var. *aurantiovirens* (Biourge) Fassatiová 1977
P. cordubense Ramírez et Martinez 1981

P. aurantiogriseum var. *viridicatum* (Westling) Frisvad and Filtenborg 1989

P. olivinoviride Biourge 1923
? *P. blakeslei* Zaleski 1927
? *P. stephaniae* Zaleski 1927
P. olivicolor Pitt 1979

P. camemberti Thom 1906

? *P. album* Epstein 1902
? *P. epsteinii* Lindau 1904
P. rogeri Wehmer apud Lafar 1906
P. caseicola Bain. 1907
P. biforme Thom 1910
P. camemberti var. *rogeri* Thom 1910
P. camemberti Sopp 1912
P. candidum Roger *apud* Biourge 1923
P. paecilomyceforme von Szilvinyi 1949

P. commune Thom 1910

P. palitans Westling 1911
P. fuscoglaucum Biourge 1923
P. flavoglaucum Biourge 1923
P. ochraceum var. *macrosporum* Thom 1930

P. lanosoviride Thom 1930
P. psittacinum Thom 1930
P. australicum Sopp *ex* van Beyma 1944
P. cyclopium var. *album* G. Smith
P. roqueforti var. *punctatum* Abe 1956
P. caseiperdens Frank 1966
P. verrucosum var. *album* (G. Smith) Samson *et al.* 1976
P. album (G. Smith) Samson *et al.* 1976

P. crustosum Thom 1930

P. lanosogriseum Thom 1930
P. pseudocasei Abe 1956 (*ex* G. Smith 1963)
P. terrestre Jensen sensu Raper and Thom 1949
P. farinosum Novobranova 1974

P. echinulatum (Raper et Thom) Fassatiová 1977

P. cyclopum var. *echinulatum* Raper et Thom 1949
P. crustosum var. *spinulosporum* Sasaki 1950
P. palitans var. *echinoconidium* Abe 195

P. solitum Westling 1911

P. majusculum Westling 1911
P. casei Staub var. *compactum* Abe
P. mali Novobranova 1972
P. verrucosum var. *melanochlorum* Samson *et al.* 1976 = *P. melanochlorum* (Samson *et al*) Frisvad 1985
P. mali Gorlenko et Novobranova 1983

P. roqueforti Thom 1906

P. aromaticum-casei Sopp 1898
P. vesiculosum Bain. 1907
P. roqueforti var. *weidemannii* Westling 1911
P. atroviride Sopp 1912
P. roqueforti Sopp 1912
P. virescens Sopp 1912
P. aromaticum Sopp 1912
P. aromaticum-casei Sopp *ex* Sacc. 1913
P. suavolens Biourge 1923
P. gorgonzolae Weidemann apud Biourge 1923
P. weidemannii (Westling) Biourge 1923
P. stilton Biourge 1923
P. weidemannii var. *fuscum* Arnaudi 1928
P. biourgei Arnaudi 1928
P. roqueforti var. *viride* Datilo-Rubbo 1938
P. conservandi Novobranova 1074

P. verrucosum Dierckx 1901

P. casei Staub 1911
P. mediolanense Dragoni et Cantoni 1979
P. nordicum Dragoni et Cantoni 1979; ex Ramírez 1985 (atypically big (and white) conidia)

Series EXPANSA Raper et Thom ex Fassatiová

Series *P. expansum* Raper et Thom - Man. Penicillia: 508. 1949.

Type species: *P. expansum*

Series *Expansa* contains only one species: *P. expansum*, which develops ter- to quaterverticillate penicilli like those of species in series *Viridicata*, but differs by predominantly smooth stipes and ellipsoidal conidia. In addition, cultures differ also in growth rates, zonation and colour.

P. expansum causes a destructive rot of pomaceous fruit, on which it produces conspicuous synnemata. In fresh isolates,synnemata or strongly fasciculate growth can also be observed on agar media. *P. crustosum* and *P. solitum* in series *Viridicata* can also can cause rot of pomaceous fruit, but this is less destructive and limited.

Penicillium expansum Link 1809
   P. expansum Link ex S.F. Gray 1821
   P. glaucum Link 1822
   P. elongatum Dierckx 1901
   P. musae Weidemann 1907
   P. variabile Wehmer 1913
   P. plumiferum Demelius 1923
   P. aeruginosum Demelius 1923
   P. leucopus (Pers.) Biourge 1919
   P. kap-laboratorium Sopp apud Biourge 1925
   P. janthogenum Biourge 1923
   P. resticulosum Birkinshaw *et al.* 1942

Series GRANULATA Raper et Thom ex Fassatiová
Series *P. granulatum* Raper et Thom - Man. Penicillia: 539. 1949

Type species: *P. granulatum* Bain.

Series *Granulata* contains now five species which may easily be recognized by their conspicuous synnemata, consisting of a distinct stalk and apical somewhat cylindrical to subglobose feather-like capitulum. Conidiophores can be mono- or synnematous and are ter- to quaterverticillate. Species of series *Granulata* differ from those in series *Gladioli* by the presence of well-defined feather-like synnemata and the absence of ascomata and/or sclerotia. The production of synnemata is also the distinguishing character from the taxa included in series *Viridicata*.

P. coprophilum (Berk. et Curt.) Seifert et Samson 1985
   Coremium coprophilum Berk. et Curt. 1869
   Stilbum humanum P. Karst. 1888
   Pritzeliella caerulea Henn. 1903
   Penicillium concentricum Samson *et al.* 1976

P. glandicola (Oud.) Seifert et Samson 1985
   Coremium glandicola Oud. 1903
   P. granulatum Bain. 1905
   P. divergens Bain. et Sartory 1912
   P. schneggii Boas 1914

P. hirsutum Dierckx 1901
   P. corymbiferum Westling 1911
   P. verrucosum var. *corymbiferum* (Westling) Samson *et al.* 1976

P. hordei Stolk

P. allii Vincent et Pitt 1989

Series OLSONII Pitt
Series Olsonii Pitt - Gen. Penicillium: 392. 1979
Series *P. brevicompactum* Raper & Thom - Man. Penicillia: 404, 1949 (nom. inval. Arts. 21, 36)

Type species: *P. olsonii* Bain. et Sartory

Series Olsonii contains only two species, morphologically closely related: *P. olsonii* and *P. brevicompactum*. They differ mainly in the complexity of their penicilli. In *P. brevicompactum* branches are borne singly, though occasionally two or three of them per branching point may occur, whereas typical penicilli of *P. olsonii* produce a compact verticil of up to to six branches developing apically and sometimes also subapically on the stipe However, detioriorated strains of *P. olsonii* , such as the ex type of *P. volgaense* produce smaller verticils of branches. Penicilli of *P. olsonii* are sometimes suggestive of the conidiophores found in section *Inordinate,* but species of the latter section are distinguished by phialide shape and brown colony colours.

P. brevicompactum Dierckx 1901
P. griseobrunneum Dierckx 1901
P. stoloniferum Thom 1910
? *P. tabescens* Westling 1911
? *P. monstrosum* Sopp 1912
P. szaferi Zaleski 1927
P. hagemii Zaleski 1927
P. bialowiezense Zaleski 1927
? *P. biourgeanum* Zaleski 1927
P. patris-mei Zaleski 1927
P. brunneostoloniferum Abe 1956; *ex* Ramírez 1982

P. olsonii Bain. et Sartory 1912
P. volgaense Beljakova et Mil'ko
P. brevicompactum var. *magnum* Ramírez 1982

Series CLAVIFORMAE Raper and Thom ex Stolk, Samson et Frisvad
Series *P. claviforme* - Man. Penicillia: 548, 1949

Series in subgenus *Penicillium* cum speciebus coremiis vel synnematibus portatis vel stipitites parietibus levibus.

Species typica: *P. vulpinum* (Cooke et Massee) Seifert et Samson (= *P. claviforme* Bain. 1905))

Raper and Thom (1949) placed *P. claviforme* (= *P. vulpinum*) and *P. clavigerum* in *P. claviforme* series. Since both species produce well defined synnemata and branched conidiophores, they included the series in section *Asymmetrica* subsection *Fasciculata*. This assignment was accepted by Samson *et al.* (1976), Ramírez (1982) and Frisvad and Filtenborg (1983). Pitt (1979) classified both species in series *Duclauxii* of subgenus *Biverticillium* and regarded *P. clavigerum* as a synonym of *P. duclauxii*.

Acuminate phialide necks as occur in subgenus *Biverticillium* were not observed by us in *P. vulpinum* and *P. clavigerum*. *P. vulpinum*, characterized by long synnemata with irregularily branched penicilli, occupies an isolated position in *Penicillium* just as *P. griseofulvum*, and interestingly they are both coprophilous, synnematous, with smooth stipes and ellipsoid smooth conidia. Furthermore they both produce patulin and roquefortine C. They are clearly different in the structure of the conidiophores and the phialide morphology, however, and are therefore placed in different series. Synnematous species in subgenus *Biverticillium* typically occur on wood or feathers and they produce a completely different profile of secondary metabolites (Samson *et al.,* 1989).

P. vulpinum (Cooke et Massee) Seifert et Samson 1985
Coremium vulpinum Cooke et Massee 1888

P. claviforme Bain. 1905
Coremium silvaticum Wehmer 1914

P. clavigerum Demelius 1923

KEY TO THE TERVERTICILLATE PENICILLIA

The following key is based on the morphology of fresh isolates grown on MEA and Czapek agars. For the recognition of some species CREA and 0.5% acetic acid agar (Engel and Teuber, 1978; Frisvad, 1981) can be very useful. Only in some cases e.g. *P. aurantiogriseum*, an analysis of secondary metabolites can be conclusive for the identification of the taxa. For a more detailed synoptic key, using physiological and chemical data see Frisvad and Filtenborg (1990).

For the detailed description of the terverticillate species, the reader is referred to Samson *et al.* (1976, 1977 a and b), Pitt (1979), Samson and Van Reenen-Hoekstra (1988), Pitt and Hocking (1985) and Frisvad and Filtenborg (1989).

1. Sclerotia and/or ascomata present 2
   Sclerotia and ascomata absent 3

2. Sclerotia present, no ascospores, stipes strongly roughened, conidia globose to subglobose, 2.2-3.5 µm, smooth-walled .... *.P. gladioli*
   Sclerotia and fertile ascomata present, stipes with walls smooth or nearly so .... Genus *Eupenicillium*

3. Conidiophores strictly mononematous or both mononematous and loosely synnematous on MEA 4
   Conidiophores strongly synnematous on MEA 23

4. Stipes on MEA with walls smooth or nearly so, conidiophores usually strictly mononematous, but in a few species both mono- and loosely synnematous 5
   Stipes on MEA with walls finely to conspicously roughened, conidiophores mononematous and/or loosely synnematous..13

5. Phialides less than 6 µm long, with a very short, sometimes inconspicuous neck, conidia grey green, conidiophore branches strongly divergent, ...*P. griseofulvum*
   Phialides more than 6 µm long, mostly with a distinct neck, conidia in different shades of green, conidiophore branches slightly appressed or divergent..6

6. Conidiophores, loosely bi- or terverticillate, sometimes more, branches when terminal slightly appressed or divergent, when subterminal strongly divergent..7
   Conidiophores bi- to quaterverticillate with all elements strongly appressed 9

7. Conidia globose to subglobose, 2-2.8 µm diam., with walls finely but distinctly roughened, conidiophores bi- to terverticillate, occasionally with one or more single subterminal and/or intercalary branches, stipes with walls smooth or finely roughened ....*P. lanosum*
   Conidia broadly ellipsoidal to subglobose or globose, 2.5-4 x 2.5-3.5 µm, smooth-walled, conidiophores (bi-) ter to quaterverticillate, stipes smooth 8

8. Cultures blue-green to greyish blue green, reverse on Cz typically yellow, sometimes uncoloured, production of penicillin, creatine negative ....*P. chrysogenum*
   Cultures greyish green to dark green, reverse on Cz orange-red to brownish orange, no production of penicillin, creatine positive ....*P. atramentosum*

9. Penicilli short, compact, basically terverticillate, sometimes more complicated because of the septation of metulae and branches, branches 1-4(6) per verticil, appressed 10
   Penicilli bi- to ter- or quaterverticillate, elongate, not short and compact, branches when present, usually single 11

10. Conidiophores very large, with a terminal verticil of up to 6 appressed branches, developing on the apex but sometimes also on the subapical part of the stipe .... *P. olsonii*
    Conidiophores smaller, usually with one asymetrcial ramus and branches short, often simply, sometimes in a terminal verticil of 2-3(4) ....*P. brevicompactum*

11. Conidia ellipsoidal, 3-4 x 2.2-3 µm, smooth-walled, penicilli ter-(quater) verticillate; causing a rot of pomaceous fruits ....*P. expansum*
    Conidia cylindrical to strongly ellipsoidal, occasionally to pyriform, more than 3.5 µm in length, penicilli bi-to terverticillate **12**

12. Conidiophores mainly mononematous, in marginal areas often synnematous, forming sometimes well-developed loose synnemata, conidia (3.5-)4-6.5(-9) x 2.2-3.5 µm, smooth-walled, causing a rot of *Citrus* fruits ....*P. italicum*
    Conidiophores strictly mononematous, conidia 3.5-5.5(-7) x 2-2.5(-3), finely roughened, sometimes appearing slightly striate, not causing a *Citrus* rot ....*P. fennelliae*

13. Conidiophores mononematous, (bi-) terverticillate, with stipes smooth or finely roughened and branches divergent, colonies lanose ....*P. lanosum*
    Conidiophores predominantly mononematous, sometimes also loosely synnematous, ter- (quater) verticillate with branches usually appressed, especially in fresh cultures, colonies of fresh isolates at the margin, often granular, sometimes tufted...**14**

14. Conidia ellipsoidal, 3-4 x 2.3-3 µm, smooth-walled, conidiophores in marginal areas synnematous, especially in fresh isolates, stipes on MEA smooth (sometimes finely roughened); causing a distinct rot of pomaceous fruits...*P. expansum*
    Conidia globose and subglobose (sometimes slightly ellipsoidal), stipes on MEA finely to definitely roughened, odour usually mouldy, sometimes causing a limited rot of pomaceous fruit.........................**15**

15. Conidia small, (2.5-) 2.8-3.2(-3.5) µm diam., poor growth on creatine agar **16**
    Conidia larger, 3-4(-5) µm diam, heavy growth on creatine agar and production of base **18**

16. Colonies on Cz and MEA not exceeding 20 mm diam. in seven days at 25°C; no acid on creatine agar, growth on nitrite agar, producing ochratoxins.. *P. verrucosum*
    Colonies on Cz and MEA exceeding 20 mm diam. in seven days at 25°C, acid production on creatine agar, no growth on nitrite agar, not producing ochratoxins ....**17 (*P. aurantiogriseum*)**

17. Conidial areas dull blue-green or bright blue-green, metabolites: penicillic acid, xanthomegnin, viomellein, and viridicatin..*P. aurantiogriseum* var. *aurantiogriseum*
    Conidial areas green, metabolites as var. *aurantiogriseum*, but with viridamin and occasionally brevianamide ....*P. aurantiogriseum* var. *viridicatum*

18. Conidiophores strictly mononematous, with walls of stipes, branches and metulae usually strongly granular from encrustments, conidia 3.5 µm diam, smooth-walled; colonies on Cz and MEA spreading broadly, more than 40 mm diam, at 25°C in seven days, green, reverse usually dark green, growth on 0.5 % acetic acid .... *P. roqueforti*
    Conidiophores predominantly mononematous, marginal areas often also loosely synnematous, with stipes finely to strongly roughened on MEA, conidia 3-4(-5) µm diam.; colonies on Cz and MEA usually not exceeding 40 mm diam. at 25°C in seven days, reverse not dark green, not growth on 0.5% acetic acid **19**

19. Conidia with walls strongly roughened ....*P. echinulatum*
    Conidia with walls smooth or nearly so **20**

20. Conidial areas white or pale grey green, conidiophores mononematous, colonies often floccose.................
    ....*P. camemberti*
    Conidial areas in shades of blue-green, green or greyish green, colonies not floccose **21**

21. Conidial areas dull green to greyish green, forming crusts .... *P. crustosum*
    Conidia areas blue-green, green or dark green, no crust formation...**22**

22. Conidial areas greyish blue, blue-green or greenish; metabolites: cyclopiazonic acid, cyclopaldic acid, rugulovasine, palitantin, not causing apple rot .... *P. commune*
    Conidial areas dark green or dark blue-green, metabolites: viridicatin, compactins; causing restricted apple rot .... *P. solitum*

23. Colonies strictly synnematous, 2-4 mm in length, sometimes longer, with a pinkish to reddish stalk and a clavate green head .... *P. vulpinum*
    Conidiophores synnematous, but retaining their individuality, mononematous structures also present **24**

24. Synnemata 2-4 mm high, not differentiated into a distinct stalk and an enlarged head, conidiophores somewhat irregular, bi to ter-(quater)verticillate borne over nearly the entire length of the synnema, branches singly, usually divergent, sometimes appressed ....*P. clavigerum*
    Conidiophores aggregated into distinct, but loosely constructed coremia, consiting of a short, sometimes inconspicuous stalk and a feathery conidial head, regularly ter- to quaterverticillate, synnemata interspersed with abundant mononematous conidiophores **25**

25. Conidia ellipsoidal, smooth-walled... **26**
    Conidia globose, with walls smooth or roughened **27**

26. Synnemata on MEA, developing in concentric zones with a short white stalk and a green head, conidiophore elements smooth .... *P. coprophilum*
    Synnemata on MEA sometimes in concentric zones, strongly feathery, conidiophore elements strongly roughened .... *P. glandicola*

27. Conidia distinctly roughened, stipes with walls smooth or finely roughened, mostly associated with cereals and soil... *P. hordei*
    Conidia smooth, stipes, branches and metulae strongly roughened, mostly associated with flower bulbs, onions and vegetables ....*P. hirsutum* (compare also *P. allii*)

ACKNOWLEDGEMENTS

The authors thank the NATO Scientific Affairs Division (Brussel, Belgium) for the research grant (0216/86) for international collaboration.

REFERENCES

BIOURGE, P. 1923. Les moississures du groupe *Penicillium* Link. *Cellule* 33: 7-331.
BIRCH, A.J. and STAPLEFORD, K.S.J. 1967. The structure of nalgiolaxin. *Journal of the Chemical Society C* 1967: 2570-2571.
CRUICKSHANK, R.H. and PITT, J.I. 1987. Identification of species in *Penicillium* subgenus *Penicillium* by enzyme electrophoresis. *Mycologia* 79: 614-620.
ENGEL, G. and TEUBER, M. 1978. Simple aid in the identification of *Penicillium roqueforti* Thom. Growth in acetic acid. *European Journal of Applied Microbiology and Biotechnology* 6: 107-111.
FASSATIOVA, O. 1977. A taxonomic study of *Penicillium* series *Expansa* Thom emend. Fassatiová. *Acta of the University Caroliana, Biology.* 12: 283-335.
FILTENBORG, O., J.C. FRISVAD and SVENDSEN, J.A. 1983. Simple screening method for molds producing intracellular mycotoxins in pure culture. *Applied and Environmental Microbiology* 45: 581-585.
FRISVAD, J.C. 1981. Physiological criteria and mycotoxin production as aids in identification of common asymmetric Penicillia. *Applied and Environmental Microbiology* 41: 568-579.
—— 1985. Creatine-sucrose agar, a differential medium for mycotoxin producing terverticillate *Penicillium* species. *Letters in Applied Microbiology* 1: 109-113.
—— 1986. Taxonomic approaches to mycotoxin identification (taxonomic indication of mycotoxin content of foods). *In* Modern Methods in the Analysis and Structural Elucidation of Mycotoxins, ed. R.J. Cole, pp. 415-457. New York: Academic Press.
—— 1988. Fungal species and their specific production of mycotoxins. *In* Introduction to food-borne fungi, eds. R.A. Samson and E. van Reenen-Hoekstra, pp. 239-249. Baarn: Centraalbureau voor Schimmelcultures.

—— 1989a. The connection between the Penicillia and Aspergilli and mycotoxins with special emphasis on misidentified isolates. *Archives of Environmental Contamination and Toxicology* 18: 452-467.

—— 1990. The use of high-performance liquid chromatography and diode array detection in fungal chemotaxonomy based on profiles of secondary metabolites. *Botanical Journal of the Linnean Society* 99: 81-95.

FRISVAD, J.C. and FILTENBORG, O. 1983 Classification of terverticillate Penicillia based on profiles of mycotoxins and other secondary metabolites. *Applied and Environmental Microbiology* 46: 1301-1310.

—— 1989. Terverticillate Penicillia: chemotaxonomy and mycotoxin production. *Mycologia* 81 (in press)

—— 1990. Secondary metabolies as consistent criteria in *Penicillium* taxonomy and a synoptic key to *Penicillium* subgenus *Penicillium*. In Modern Concepts of *Penicillium* and *Aspergillus* Classification, eds. R.A. Samson and J.I. Pitt, pp. 373-384. New York and London: Plenum Press.

FRISVAD, J.C., FILTENBORG, O. and WICKLOW, D.T. 1987. Terverticillate Penicillia isolated from underground seed caches and cheek pouches of banner-tailed kangaroo rats (*Dipodomys spectabilis*). *Canadian Journal of Botany* 65: 765-773.

FRISVAD, J.C. and THRANE, U. 1987. Standardized high-performance liquid chromatography of 182 mycotoxins and other fungal secondary metabolites based on alkylphenone retention indices and UV-VIS spectra (diode array detection). *Journal of Chromatography* 404: 195-214.

PITT, J.I. 1979. The Genus *Penicillium* and its Teleomorphic States Eupenicillium and *Talaromyces*. London: Academic Press.

—— 1988. A Laboratory Guide to Common *Penicillium* Species. Second edition. North Ryde, N.S.W.: CSIRO Division of Food Processing.

PITT, J.I. and HOCKING, A.D. 1985. Fungi and food spoilage. Sydney: Academic Press.

RAPER, K.B. and THOM, C. 1949. A Manual of the Penicillia. Baltimore: Williams and Wilkins.

RAMIREZ, C. 1982. Manual and Atlas of the Penicillia. Amsterdam: Elsevier Biomedical.

SAMSON, R.A. and PITT, J.I. 1985. Check list of common *Penicillium* species. In Advances in *Penicillium* and *Aspergillus* Systematics, eds. R.A. Samson and J. I. Pitt, pp. 461-463. New York and London: Plenum Press.

SAMSON, R.A. and REENEN-HOEKSTRA, E.S. van. 1988. Introduction to food-borne fungi. Third edition. Baarn: Centraalbureau voor Schimmelcultures.

SAMSON, R.A., STOLK, A.C. and HADLOK, R. 1976. Revision of the Subsection *Fasciculata* of *Penicillium* and allied species. *Studies in Mycology, Baarn* 11: 1-47.

SAMSON, R.A., HADLOK, R. and STOLK, A.C. 1977a. A taxonomic study of the *Penicillium chrysogenum* series. *Antonie van Leeuwenhoek* 43: 169-175.

SAMSON, R.A., ECKARDT, C. and ORTH, R. 1977b. The taxonomy of *Penicillium* species from fermented cheeses. *Antonie van Leeuwenhoek* 43: 341-350.

SEIFERT, K.A. and SAMSON, R.A. 1985. The genus *Coremium* and the synnematous Penicillia. In Advances in *Penicillium* and *Aspergillus* Systematics, eds. R.A. Samson and J. I. Pitt, pp. 143-154. New York and Londen: Plenum Press.

SAMSON, R. A., STOLK, A.C. and FRISVAD, J.C. 1989. Two new synnematous species of *Penicillium*. *Studies in Mycology, Baarn* 31: 133-143.

THOM, C. 1930. The Penicillia. Baltimore: Williams and Wilkins.

DIALOGUE FOLLOWING DR. SAMSON'S PRESENTATION

PITT: In defining and delimiting subgen. *Penicillium,* I tried to avoid phylogenetic considerations. The taxonomy is directed towards the user rather than the systematist. A species like *P. oxalicum* should not be placed in subgen. *Penicillium* because it only rarely produces terverticillate penicilli and so a user would not look for it there. Regarding the possibility of using the name *P. majusculum* for what we have been calling *P. solitum,* the Code does not indicate priority when two species are described in the same paper. It's a matter of choosing which species best matches the adopted concept, or of maintaining usage of a name that is already being used. Raper used the name *P.*

solitum and so this name should be used, especially as the species concept has not changed.

SAMSON: But the description of *P. majusculum* fits the concept we are discussing here better.

STOLK: The original description of *P. majusculum* states that conidia are a dark green to black colour, like those of *P. melanochlorum*. *P. solitum* was described as having blue-green conidia. I think *P. majusculum* is a better name, but you can, of course, use *P. solitum*, because both names are valid.

SAMSON: Yes, I agree, we can use the epithet *solitum*. It is an established name.

A REAPPRAISAL OF THE TERVERTICILLATE PENICILLIA USING BIOCHEMICAL, PHYSIOLOGICAL AND MORPHOLOGICAL FEATURES

P.D. Bridge[1], D.L. Hawksworth[1], Z. Kozakiewicz[1], A.H.S. Onions[1], R.R.M. Paterson[1], M.J. Sackin[2] and P.H.A. Sneath[2]

[1]CAB International Mycological Institute
Kew, Surrey, TW9 3AF, UK

[2]Department of Microbiology
University of Leicester
Leicester, LE1 9HN, UK

SUMMARY

Three hundred and forty eight strains of terverticillate Penicillia were examined for 100 physiological, biochemical and morphological characters. These characters included assessing growth on specific carbon and nitrogen sources, screening for enzyme production, thin layer chromatography of secondary metabolites, and scanning electron microscopy of conidia. Test reproducibility and strain variation were critically considered. Resemblances between each pair of strains were calculated with Gower's coefficient and the Pattern difference, and dendrograms plotted. Overlap between groups was calculated and differences between dendrograms were considered in determining 37 species or species complex clusters. In addition the dendrogram contained a number of ungrouped strains, mainly single representatives of species included for comparison. In total, 91% of the strains were considered grouped.

Additional studies on strain variation and pectinase and amylase isoenzymes were undertaken by electrophoresis to test the homogeneity of the clusters. This showed that unique banding patterns were obtained for only two species, and in general a considerable range of patterns were produced for each species. High resolution melting curves of the DNBA of 14 of the test strains indicated considerable similarities in the DNA base distributions. Extensive studies on strain variation for all characters studied showed that although considerable phenotypic variation may occur within both species and strains, sufficient reliable characters were available to define taxa. Variation within and between strains could be accounted for by gene or partial chromosome duplication; this hypothesis was supported by some of the electrophoretic patterns and variations in conidial DNA contents.

This study showed the terverticillate Penicillia to be a group of very similar fungi. Many of the commonly used taxonomic characters are subjective and can vary significantly within single cultures. Satisfactory species concepts can be defined by using an integrated multidisciplinary approach and reliable quantitative identification schemes produced.

INTRODUCTION

The taxonomy of the terverticillate Penicillia has been considerably revised in recent years (e.g. Samson *et al.*, 1976; Pitt, 1979), but nevertheless areas of uncertainty remains (Onions *et al.*, 1984; Samson and Gams, 1984). To clarify the systematics and species concepts in the terverticillate Penicillia, a major integrated multidisciplinary study was undertaken. This involved selecting and developing taxonomic characters from morphology, physiology, biochemistry, and scanning electron microscopy. These characters were then critically examined for reproducibility and reliability. Additional studies were also carried out on intra-strain variation and stability (Bridge *et al.*, 1986b). The data were used in a numerical taxonomy to produce species level clusters in dendrograms (Bridge *et al.*, 1989a). The final

Modern Concepts in Penicillium and Aspergillus Classification
Edited by R. A. Samson and J. I. Pitt
Plenum Press, New York, 1990

clusters were further tested by the construction of a percent positive identification matrix (Bridge, 1990) and by comparison of electrophoretic amylase and pectinase isoenzyme patterns. The full methods and results from these studies are published elsewhere (Paterson, 1986; Bridge *et al.*, 1987; 1989a, b; Paterson *et al.*, 1989a; Kozakiewicz, 1989). This paper presents an overview of the completed study.

NUMERICAL TAXONOMY

One hundred characters from morphology, physiology, biochemistry and scanning electron microscopy were selected from over 200 assessed during the study (Bridge, 1985; Paterson, 1986; Bridge *et al.*, 1986a, 1987; Kozakiewicz, 1989). Results from these 100 characters were recorded for 348 strains of *Penicillium*, consisting mainly of terverticillate species, with some representatives of other subgenera for comparison. Test reproducibility and strain variation were assessed by including duplicate cultures in the study, performing tests in duplicate, and repeating tests for up to 80% of the strains at a later date. Resemblances between strains were calculated with Gower's coefficient (Sneath and Sokal, 1973) and the Pattern difference (D_{p2}; Sackin, 1981). The results were clustered by unweighted average linkage (UPGMA) and single linkage (Sneath and Sokal, 1973) and presented as dendrograms.

The UPGMA dendrogram from Gower's coefficient was used as the basis of the taxonomy. This represented the strains in 37 clusters that could be recognised as species or species complexes at approximately 70% similarity. These clusters contained 80% of the total number of strains. When single representatives of species and strains included for comparison were excluded, 91% of the strains were grouped. A simplified version of this dendrogram is given in Figure 1. Few of the groupings recovered were entirely surprising or new, and so only selected taxa will be discussed here.

The *P. expansum* strains were recovered in three small clusters separately at the top of the dendrogram. While some differences existed between these groups, they were overall very similar and so are considered as representing variants of a single species. The *P. aurantiogriseum* strains were mainly recovered in one large cluster, although five strains including the ex-type culture of *P. puberulum* were recovered as a separate cluster from the main *P. aurantiogriseum* cluster. There are no clear-cut differences between this small cluster and the main *P. aurantiogriseum* cluster; the name *P. puberulum* must therefore be considered synonymous with *P. aurantiogriseum*. The major area of taxonomic confusion, including strains named as *P. crustosum*, *P. ochraceum*, *P. commune* or *P. palitans*, was recovered as one large cluster. Considerable further work on this area of the study, including overlap calculations and principal coordinate analysis, suggested that there may be two very similar taxa here. The oldest reliable name recovered in this cluster was *P. solitum* and the main cluster was named as this, with a variety *P. solitum* var. *crustosum*.

Strains received as *P. chrysogenum* or its synonyms were recovered in two separate clusters. Most of the differences between these clusters were a matter of degree and it was considered that one cluster represented deteriorated lines of *P. chrysogenum*. Included in one of these clusters was the ex-type culture of *P. griseoroseum*, a name that pre-dates *P. chrysogenum*; this placement is in agreement with the findings of Samson *et al.* (1977) and Cruickshank and Pitt (1987). However, this strain was in a deteriorated condition and no original herbarium type material was available, therefore any name change is inappropriate (see chapter 2, this volume).

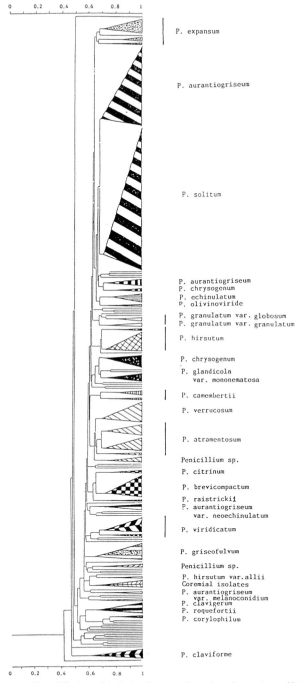

Figure 1. Simplified UPGMA dendrogram based on Gower's coefficient

The species *P. granulatum* and *P. camemberti* were both recovered as two clusters. In the case of *P. camemberti*, these two clusters remained linked together in all of the dendrograms, differing from each other in their conidial ornamentation and in their vigour and overall physiological activity. One of these clusters also contained a natural white-spored mutant of *P. solitum*. While the conidial ornamentation suggested a distinct separation into two taxa, the overlap and other data suggested that they were very similar. These cultures may have been assigned to the single species *P. camemberti* previously due to their production of white conidia. The presence of the white mutant further complicated the interpretation and these two groups have been tentatively retained as a single taxon. The only other white-spored culture included in the study, *P. aurantiogriseum* var. *album*, was recovered in the smaller of the two *P. aurantiogriseum* clusters. In the case of *P. granulatum* there were sufficient differences between the two clusters to warrant the creation of a variety, *P. granulatum* var. *globosum*.

Strains received as *P. atramentosum* and *P. meleagrinum* clustered as three groups in the UPGMA dendrogram based on Gower's coefficient and merged to form two groups in the Pattern difference dendrogram. Further investigations of a line of the ex-type culture of *P. atramentosum* not included in the study placed this in a different cluster from one regarded as typical of *P. atramentosum* by Pitt (1979). Overall, these clusters appear to represent variants of a single species correctly named as *P. atramentosum*.

The strains received as *P. brevicompactum* and *P. olsonii* clustered together as a single group. These species could not be separated on any characters and further work (see later; isoenzyme studies) also failed to separate all but one *P. brevicompactum* from *P. olsonii*. These species should therefore be considered synonymous.

Overall, the numerical taxonomy resolved the strains into satisfactory clusters. However, additional information can be gained from this approach. The cophenetic correlation for the UPGMA Gower's coefficient dendrogram is relatively low (0.78) and the overall test reproducibility was also low (0.82 for repeated cultures). The overlap calculations showed that although clusters could be regarded as distinct, there was less confidence in the distinctions between groups of clusters. Many species showed significant intra-specific variation and in some cases, such as conidial ornamentation in *P. solitum*, gradations in characters within the same species were observed. One possible interpretation of all of these factors would be that the species groups may represent distinct points within or on a semi-continuous spectrum. Regrettably there is very little data available on that situation for reliable comparisons to be made here.

VARIATION AND TEST REPRODUCIBILITY

The influence of basal media and additions to basal media were examined as part of the overall assessment of test reproducibility. The production of both secondary metabolites and extracellular enzymes can be profoundly influenced by relatively minor changes in basal medium such as the addition of copper and zinc (Smith, 1949) or the manufacturer and type of organic ingredients such as peptone (Odds *et al.*, 1978). Recently, the production of synnemata in the non-fasciculate species *P. funiculosum* has been demonstrated on a medium containing tributyltin compounds (Newby and Gadd, 1987). True synnemata can, however, be induced in many of the fasciculate Penicillia by growth on a variety of compounds including low levels of pentachlorophenol (see Figure 2). This demonstrates that the difference between the synnematal species *P. claviforme* and *P. clavigerum* and the fasciculate Penicillia is probably not based on a large structural difference but on a small difference in the control of gene expression; these taxa may therefore be more closely related than previously recognized.

Variations in test results between different workers in different laboratories is often due to small differences in the methods used. The ability of Penicillia to grow on sodium nitrite is an example of this, where different workers have used different concentrations of nitrite. Sodium nitrite can have a toxic effect which is dependent on pH and concentration, so variations in these parameters can affect the apparent ability of strains to grow on the medium (Bridge, 1988).

As a result of the relatively low levels of test reproducibility experienced during the numerical taxonomy, particularly with certain strains, intra-strain variation was studied. A number of lines of the same cultures were isolated from single conidia. These lines were maintained separately and differences in characteristics noted. Overall, minor differences were apparent in most cases, but in certain strains significant differences in phenotypic properties were found (Bridge *et al.*, 1986b; Bridge *et al.*, 1987). One possible explanation for this phenomenon is some kind of genetic rearrangement, perhaps involving aneuploidy or chromosome duplication and transposition. Further support for this is the variation in DNA content between conidia of these lines (Bridge *et al.*, 1986c; 1987) and the description of similar occurrences in other fungi where differences in chromosome size and number have been demonstrated, such as in *Candida* (Suzuki *et al.*, 1989).

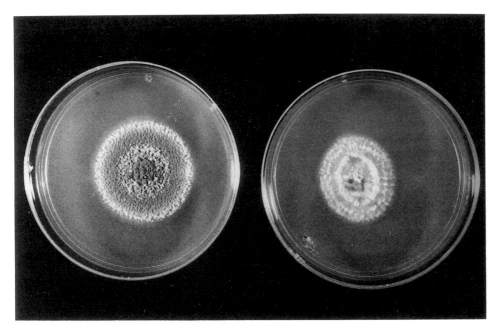

Figure 2. Induction of synnemata of fasciculate Penicillia by pentachlorophenol (right).

ISOENZYME STUDIES

The electrophoretic patterns of pectinase and amylase isoenzymes were determined for 170 representative strains from the main study using the methods of Cruickshank and Wade (1980) and Cruickshank and Pitt (1987). There was considerable variation in the intensity of the pectinase patterns from some strains in repeat runs. In many cases considerable variation was observed in the patterns shown by members of the same species. The results from the electrophoretic study were analyzed by numerical methods and the final data set was represented as a dendrogram. The clusters formed in this way represented strains with the same or similar enzyme banding patterns. These clusters did not in general correspond to species. All but one of the strains received as *P. brevicompactum* gave very similar patterns to those received as *P. olsonii*, a result similar to that observed by Cruickshank and Pitt (1987). The single strains of *P. funiculosum* and *P. roqueforti* clustered separately. The other species were represented in more than one cluster, although some clusters contained only one species (Paterson *et al.*, 1989a). Interpretation of the differences in isoenzyme patterns between members of the same species was carried out, and several of the observed differences could be explained by chromosomal reorganization; this adds further support to the possibility mentioned above.

DNA BASE CONTRIBUTIONS

High resolution DNA melting curves were obtained for 14 of the strains tested in the main study, 13 of which were also analysed for isoenzymes (Paterson *et al.*, 1989a, b). The DNA base distributions obtained indicated considerable similarity between strains at this fundamental level.

CONCLUSIONS

The application of an integrated multidisciplinary approach combined with numerical taxonomy techniques and critical studies on test reproducibility has proved to be a very powerful tool in the systematics of the filamentous fungi. The Penicillia studied here can be classified in species level groups. Although the fasciculate Penicillia can show considerable phenotypic variation within both species and strains, species concepts were firmly established. More detailed further studies have supported the numerical taxonomy in showing the overall similarity of these fungi, and the suggestion of a genetic basis for some phenotypic variation raises possibilities for future studies.

ACKNOWLEDGEMENTS

This work was supported by Science and Engineering Research Council contract SO/17/84 "Systematics of microfungi of biotechnological and industrial importance". We would like to thank E.M. Oliver, P. Farnell and D. Kavishe for their excellent technical assistance, and Dr. L.E. Fellows and the Royal Botanic Gardens, Kew, for providing certain biochemical facilities.

REFERENCES

BRIDGE, P.D. 1985. An evaluation of some physiological and biochemical methods as an aid to the characterization of species of *Penicillium* subsection *Fasciculata*. *Journal of General Microbiology* 131: 1887-1895.
—— 1988. A note on use and interpretation of nitrite assimilation tests in *Penicillium* systematics. *Transactions of the British Mycological Society* 88: 569-571.

—— 1990. Identification of terverticillate Penicillia from a matrix of percent positive test results. In Modern concepts in *Penicillium* and *Aspergillus* classification, eds. R.A. Samson and J.I. Pitt, pp.00-00. New York and London: Plenum Press.

BRIDGE, P.D., HAWKSWORTH, D.L., KOZAKIEWICZ, Z., ONIONS, A.H.S., PATERSON, R.R.M and SACKIN, M.J. 1986a. An integrated approach to *Penicillium* systematics. In Advances in *Penicillium* and *Aspergillus* Systematics, eds R.A. Samson and J.I. Pitt pp. 281-307. New York and London: Plenum Press.

BRIDGE, P.D., HAWKSWORTH, D.L., KOZAKIEWICZ, Z., ONIONS A.H.S. and PATERSON, R.R.M. 1986b. Morphological and biochemical variation in single isolates of *Penicillium*. *Transactions of the British Mycological Society* 87: 389-396

BRIDGE, P.D., HUDSON, L., HAWKSWORTH, D.L. and BRIDGE, D.A.. 1986c. Variation in nuclear DNA content in an ex-type isolate of *Penicillium* measured by continuous flow microfluorimetry. *FEMS Microbiology Letters* 37: 241-244

BRIDGE, P.D., HUDSON, L., KOZAKIEWICZ, Z., ONIONS, A.H.S. and PATERSON, R.R.M. 1987. Investigation of variation in phenotype and DNA content between single-conidium isolates of single *Penicillium* strains. *Journal of General Microbiology* 133: 995-1004

BRIDGE, P.D., HAWKSWORTH, D.L., KOZAKIEWICZ, Z., ONIONS, A.H.S., PATERSON R.R.M., SACKIN, M.J. and SNEATH, P.H.A. 1989a. A reappraisal of terverticillate Penicillia using biochemical, physiological and morphological features. I. Numerical taxonomy. *Journal of General Microbiology* (in press).

—— 1989b. A reappraisal of terverticillate Penicillia using biochemical, physiological and morphological features. II. Identification. *Journal of General* Microbiology (in press).

CRUICKSHANK, R.H. and WADE, G.C. 1980. Detection of pectin enzymes in pectin-acrylamide gels. *Analytical Biochemistry* 107: 177-181.

CRUICKSHANK, R.H. and PITT, J.I. 1987. Identification of species in *Penicillium* subgenus *Penicillium* by enzyme electrophoresis. *Mycologia* 79: 614-620.

KOZAKIEWICZ, Z. 1989. Ornamentation types of conidia and conidiogenous structures in fasciculate *Penicillium* species, using scanning electron microscopy. *Botanical Journal of the Linnean Society* 99: 273-293.

NEWBY, P.J. and GADD, G.M. 1987. Synnema induction in *Penicillium funiculosum* by tributyltin compounds. *Transactions of the British Mycological Society* 89: 381-384.

ODDS, F.C., HALL, C.A. and ABBOTT, A.B. 1978. Peptones and mycological reproducibility. *Sabouraudia* 16: 237-246.

ONIONS, A.H.S., BRIDGE, P.D. and PATERSON, R.R.M. 1984. Problems and prospects for the taxonomy of *Penicillium*. *Microbiological Sciences* 1: 185-189.

PATERSON, R.R.M. 1986. Standardized one and two-dimensional thin-layer chromatographic methods for the identification of secondary metabolites in *Penicillium* and other fungi. *Journal of Chromatography* 368: 249-264.

PATERSON, R.R.M.., BRIDGE, P.D., CROSSWAITE, M.J. and HAWKSWORTH, D.L. 1989a. A reappraisal of the terverticillate Penicillia using biochemical physiological and morphological features. III. An evaluation of pectinase and amylase isozymes for species characterization. *Journal of General Microbiology* (in press).

PATERSON, R.R.M., KING, G.J. and BRIDGE, P.D. 1989b. High resolution thermal denaturation studies on DNA from 14 *Penicillium* isolates. *Mycological Research* (in press).

PITT, J.I. 1979. The genus *Penicillium* and its Teleomorphic States *Eupenicillium* and *Talaromyces*. London: Academic Press.

SACKIN, M.J. 1981. Vigour and pattern as applied to multistate quantitative characters in taxonomy. *Journal of General Microbiology* 122: 247-254.

SAMSON, R.A. and GAMS, W. 1984. The taxonomic situation in the Hyphomycete genera *Penicillium*, *Aspergillus* and *Fusarium*. *Antonie van Leeuwenhoek* 50: 815-824.

SAMSON, R.A., STOLK, A.C. and HADLOK, R. 1976. Revision of the subsection *Fasciculata* of *Penicillium* and some allied species. *Studies in Mycology, Baarn* 11: 1-47

SAMSON., R.A., HADLOK, R. and STOLK, A.C. 1977. A taxonomic study of the *Penicillium chrysogenum* series. *Antonie van Leeuwenhoek* 43: 169-175.

SMITH, G. 1949. The effect of adding trace elements to Czapek Dox medium. *Transactions of the British Mycological Society* 32: 280-283.

SNEATH, P.H.A. and SOKAL, R.R. 1973. Numerical Taxonomy. San Francisco: W.H. Freeman and Sons.

SUZUKI, T., KOBAYASHI, I., KANBE, T. and TANAKA, K. 1989. High frequency variation of colony morphology and chromosome reorganization in the pathogenic yeast *Candida albicans*. *Journal of General Microbiology* 135: 425-434.

DIALOGUE FOLLOWING THE PAPER BY DR. BRIDGE AND COLLEAGUES

CHRISTENSEN: I have also been concerned about the problem of degenerated cultures. Did you try your Principal Components Analysis (PCA) using just the newly isolated cultures.

BRIDGE: We tried to avoid the situation of working with only freshly isolated cultures. The strains that we receive for identification at CMI vary enormously in quality. In our work, we tried to use a representative mix of cultures. If you are worried about things like deteriorated cultures, you can use coefficients such as the pattern coefficient, which has a component in it that allows you to compensate for the vigour of the strains. For example, the two *P. chrysogenum* strains came together easily using the pattern coefficient. Again, *P. puberulum* and *P. aurantiogriseum* came together with the pattern coefficient.

CHRISTENSEN: What percent of the variation was represented in the three axes of the PCA?

BRIDGE: Not an awful lot. About 28%.

CHRISTENSEN: That seems low to me. How do you account for that low figure?

BRIDGE: You are dealing with a reduction of 100 characters down to three dimensions. This is in effect a large compression of data to fit it into three dimensions, and it is inevitable that some information will be lost. In this case, it suggests that there is a lot of variation in a lot of characters. This is a complex situation. If you can reduce a multidimensional system to two or three dimensions, via PCA, and get 60-70% of the variability explained, you probably have a simple situation. It is a way of looking at the data without imposing a hierarchy.

CHRISTENSEN: In ecological studies often 70-80% of the variation is represented in the first three axes.

BRIDGE: But in ecological studies, one often has nice distinct groups.

PITT: Most of us consider *P. crustosum* to be a very well-defined species. It surprises me to see several other strains clustering with it. What did you do with *P. italicum*?

BRIDGE: We didn't really deal with it in this study, but we had a couple of marker strains of *P. italicum*, and they clustered together at the bottom.

SAMSON: But *P. italicum* is a typical terverticillate species.

PITT: Were some species from subgenus *Penicillium* left out of the analysis?

BRIDGE: Yes.

PITT: My other concern is your new variety, *P. granulatum* var. *globosum*. This species has ellipsoidal conidia, and it is difficult for me to imagine a variety with spherical conidia. It must be a different species.

BRIDGE: Here we have the problem of having a difference of only one or two characters, in this case mainly conidial characters. I don't wish to raise something like that to species status until we are more confident of these differences.

SAMSON: *P. glandicola* (= *P. granulatum*) typically has ellipsoidal conidia. We should not include a variety with globose conidia in the same species.

BRIDGE: But as we see it, the other morphological characters suggest that this is *P. granulatum*.

FRISVAD: A lot of your clusters fit nicely with our secondary metabolite groups. But I am concerned that *P. olsonii* and *P. brevicompactum* clustered together, especially as you included secondary metabolites. We have examined about 200 strains, and *P. brevicompactum* always produces mycophenolic acid and the five or so Raistrick phenols. Brevianamide A is sometimes, but not always, produced. *P. olsonii* produces some mystical unknown compounds, but does not produce any of those made by *P. brevicompactum*.

PATERSON: A difference of one metabolite may not make much difference in the numerical analysis.

FRISVAD: Even degenerated strains of *P. brevicompactum* produce some of these secondary metabolites.

BRIDGE: If these metabolites were consistently produced, I would expect to see some difference between the two species, even if it might be within one cluster. However, there was no evidence of breakdown within this cluster.

FRISVAD: In 1980, I checked all the *P. brevicompactum* strains listed in the CMI catalogue, including the type of *P. griseobrunneum*, and they all produced mycophenolic acid.

FILTENBORG: What yeast extract did you use for growing your fungi for analysis of secondary metabolites?

PATERSON: Difco in YES and CYA.

SAMSON: Was all the scanning electron microscopy done with cryofixation?

KOZAKIEWICZ: Cryofixation was used for examining the penicillus. Conidium preparations were air dried.

PITT: What temperature did you use for doing the isoenzyme patterns.

PATERSON: We used water cooled electrophoresis at 10°C.

PITT: This concerns me because Dr. Cruikshank uses 4°C. He gets good separation at that temperature. Perhaps 10° C does not give such good separation.

BRIDGE: We did get good band patterns, but they weren't reproducible within a species.

PATERSON: Well, some strains were reproducible, but others weren't. Proteins don't denature at 10°C.

EVALUATION OF THE DIAGNOSTIC FEATURES OF SOME SPECIES OF PENICILLIUM SECTION DIVARICATUM

O. Fassatiová and A. Kubatová

Department of Botany
Charles University
12801 Praha, Czechoslovakia

SUMMARY

The study was carried out with about 200 isolates mainly from Czechoslovakia and belonging to *Penicillium* sect. *Divaricatum*: *P. janczewskii, P. canescens, P. melinii, P. radulatum, P. janthinellum* and *P. simplicissimum*. The type isolates were compared according to their micro- and macromorphology. The following are recommended as diagnostical features: the branching of conidiophores, divergence of metulae and phialides and their length and the character of conidia. The length of phialides is accurately measured omitting the neck, as neck length can be variable in a single isolate. *P. janczewskii* and *P. daleae* can be distinguished satisfactory on the basis of their micromorphology and colonies. In both species colonies are very variable in degree of sporulation, in conidial colour and in exudate and reverse colours. *P. granatense* is a synonym of *P. janczewskii* based on our observation of the type isolate. Isolates of *P. canescens* have good micromorphological distinguishing features. Isolates of *P. melinii* are very similar in micromorphology and colony character to *P. radulatum*, which has been only found once in Czechoslovakia. The type isolate of *P. melinii* studied during our investigation is different from the original description and also from our own isolates. A great number of soil isolates could be identified as either *P. janthinellum* or *P. simplicissimum* because their micromorphology is very similar. They probably represent one species.

INTRODUCTION

During soil isolation of microfungi in Czechoslovakia and in Poland, the more abundantly occurring representatives of *Penicillium* sect. *Divaricatum* were selected on the basis of colony and conidiophore morphology. We present our own descriptions on colonies grown on Czapek-Dox agar (CZ) and Czapek yeast extract agar (CYA), incubated at 25°C. Strains of the species *P. janczewskii, P. daleae, P. canescens, P. simplicissimum, P. melinii, P. radulatum* and *P. janthinellum* from our region have been deposited in a provisional collection as D15-D220 and will gradually be transferred into the Culture Collection of Fungi, CCF, in the Department of Botany, Charles University, Prague. We were able to compare our isolates with some type cultures from foreign collections.

Penicillium simplicissimum (Oud.) Thom
Syn. *Penicillium janthinellum* Biourge

CZ: Colonies 30-60 mm diam in two weeks. White, lanose, floccose, raised at centre or depressed. The colonies becoming in centre pink, salmon, at the margin sometimes yellow. Sporulation in the submarginal area blue-green or grey-green or inconspicuous. Other colonies velvety or feltlike with yellowish, beige or ochraceous beige mycelium. Sporulation grey-green and gradually proceeding to the centre though sometimes only in the submarginal area. Colonies sometimes radially or irregularly sulcate with sparse shallow or deep furrows. Reverse colourless, intensive pink, orange-ochraceous, yellowish, light green, salmon, sometimes coral in the centre and at the margin brick coloured. Soluble pigment salmon, light

Modern Concepts in Penicillium and Aspergillus Classification
Edited by R. A. Samson and J. I. Pitt
Plenum Press, New York, 1990

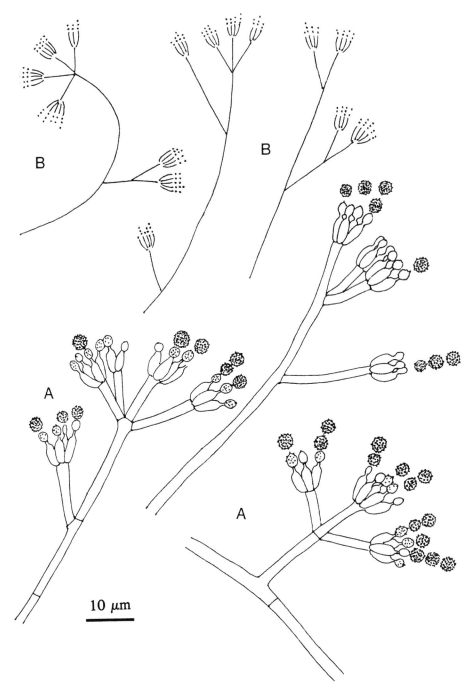

Figure 1. *Penicillium janszewkii* Zaleski. A. conidiophores, B. branching patterns of the penicilli.

ochraceous, yellowish, pink, light vinaceous, apricot, scarlet or absent. Unripened cream and soft ascomata were observed in six isolates after transfer from 37°C to 25°C.

CYA: Colonies 20-58 mm diam in two weeks, lanose, feltlike or velvety. Mycelium white, rosy buff or pinkish. Sterile margins 2-10 mm wide, white or yellowish. Colonies without radially furrows or radial sulcate. Sporulation heavy to weak, predominantly in the submarginal area or in the centre, light grey-green, blue-green, grey-brown, yellow-orange, sometimes pinkish or brownish in the centre. Exudate clear, lemon coloured, yellowish, pinkish brown, yellowish brown, green or absent. Reverse colourless, or rosy-buff, golden-yellow at the margin, deep yellow with brown or red centre, yellowish brown, cream-arange. Soluble pigment pinkish brown, pinkish red or not present. Conidiophores (50-)400-600 x 1.8-3 µm, with stipes smooth or rough. Branching variable, monoverticillate, biverticillate, slightly divergent, with a cluster of 2-4 metulae terminally and with one or two metulae or branches subterminally. Metulae smooth or rough 8-23 x 2.3-3 µm, sometimes proliferating in a new verticil of phialides. Phialides in clusters of 4-8, abruptly tapering into a long neck (1.5-3 µm). Dimensions of phialides 6.5-9 x 2.8-3.2 µm. Conidia ellipsoidal or subglobose, very often with a papilla on one end, rough to slightly echinulate, rarely also smooth, 2.5-3.9 x 2.8-3.2 µm.

P. simplicissimum has a very broad range in colony habit, but these did not differ substantially in microstructure. We have compared 53 isolates from Czechoslovakia and 2 isolates from Poland with cultures ex type of *P. janthinellum* (IMI 40238) and *P. simplicissimum* (CBS 280.39). Our results agree with the concept and figures of Stolk and Samson (1983), who considered *P. simplicissimum* to be the anamorph of *Eupenicillium javanicum* var. *javanicum*.

The following diagnostic features can be used for this species: a broad spectrum of colony colour, predominantly in pastel tones in reverse; smooth conidiophores with terminal divergent verticils of metulae and divergent metulae in subterminal positions. Conidia ellipsoidal to subglobose, distinctly rough or finely echinulate, often with a papilla on one end.

Penicillium janczewskii Zaleski – Fig. 1

   = *P. nigricans* Bainier ex Thom
   = *P. granatense* Ramírez

CZ: Colonies 26-50 mm diam in two weeks, grey-white, slightly raised or depressed in the centre; velutinous colonies grey-green with sporulation in the centre in dark grey-green or blue-green, at the margin lighter or vice versa, later becoming dark grey to black. Surface of the colonies smooth or radially sulcate with furrows shallows or deep. Margin white to yellowish or creamy 1-3 mm wide. Exudate clear, yellowish or absent. Reverse sometimes lightly sulcate, yellowish, yellowish green, brownish to violet-brown. Sometimes yellowish, creamish or orange-brown in the centre, olive brown or yellowish at the margin.

CYA: Colonies 30-57 mm diam in two weeks, lanose or velvety, sometimes lanose in the centre and velvety at the margin. Sporulation weak, colonies white, grey-white, sporulation dense, colonies grey, dark grey, green, olive-grey, olive-green, black-grey, sometimes with sterile white sectors. White margins 2.5 mm wide. Radial furrows shallow or deeper. Exudate clear, beige, yellow, creamy, creamy-orange or absent. Reverse rosy-buff with yellow margin, brownish red with ochraceous margins, brownish orange with yellowish margin, yellow-brown or pinkish. Sometimes sparse or dense furrows. Conidiophores branched, 30-500 x 1.6-3 µm, with smooth surfaces. Metulae numbering 2-6 in terminal whorls, very divergent, apically swollen, 7-15 x 2-2.5 µm, often also borne subterminally. Phialides numbering 6-8, in compact whorls, ampulliform, 5.4-7.8 x 1.8-2.3 µm with a distinct neck, 1-2.3 µm in length. Conidia globose to subglobose, echinulate, 2.3-3.2 µm.

We compared 42 isolates from Czechoslovakia and three isolates from Poland with an ex type culture of *P. janczewskii* (CBS 221.28). *P. janczewskii* is both in colonies and microstructure a very distinctive species. Important diagnostic features are: more or less a

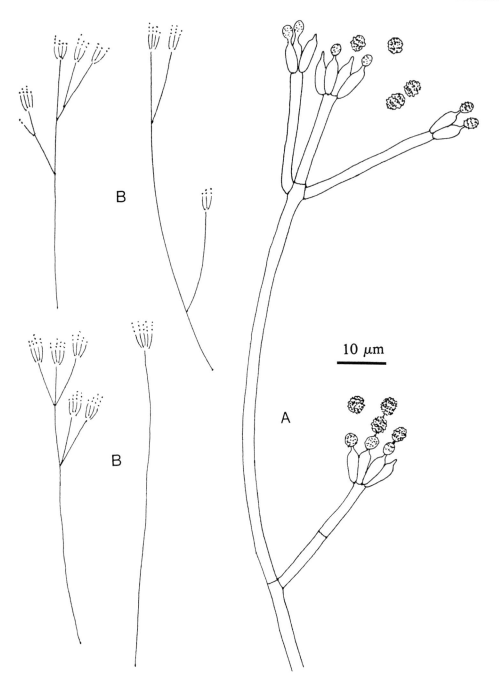

Figure 2. *Penicillium daleae* Zaleski. A. conidiophores, B. branching patterns of the penicilli.

lanose colonies often with sterile sectors, sporulation from light grey, grey-green, olive-grey to dark grey or grey-brown; exudate sometimes present, clear; reverse yellow-green, yellow-brown to brown. Conidiophores have very conspicuous divergent whorls of metulae (often at right angles), and often further metulae in subterminal positions. Phialides are very short (without neck 4-5 μm), abruptly tapering to shorter or longer necks. Conidia globose, dark, typically echinulate. We also examined *P. granatense* Ramírez et al. (IJFM 5965) and found that it was identical with *P. janczewkii* in all characters examined.

Penicillium daleae Zaleski – Fig. 2.

CZ: Colonies 32-48 mm diam in two weeks, lanose, velutinous, pinkish creamy, grey-green, brown-green. Sporulation weak or heavy in centre or at the margin, sometimes in concentric zones, with or without radial or irregular furrows. Margins with fine furrows white to creamy, 2-2.5 mm wide, later olive-green. Exudate clear, yellowish, brownish or brown-orange. Reverse light yellowish brown, pinkish brown, violet-brown, in the centre pinkish creamy, sometimes soluble pigment present, lightly brownish violet or brown.

CYA: Colonies 40-45 mm diam in 2 weeks, felted, low, dense, sometimes velvety or with velutinous sectors. Sporulation greyish olive or greyish creamy, grey or blue-green at the margin. Often lighter or darker concentric zones. Furrows thin, when present. Exudate clear, amber, ochraceous-brown to dark brown. Reverse cocoa-brown, dark flesh coloured, maroon or cinnamon brown, exceptionally colourless. Conidiophores little branched, often with 2-3 metulae terminally and 1-2 metulae subterminally, sometimes also with 1-2 branches, often monoverticillate. Conidiophores 20-150 x 1.6-3 μm, with smooth stipes. Metulae not very divergent, often unequal length, sometimes concurrent with phialides. Metulae 7.8-23 x 1.8-2.8 μm. Phialides in verticils of 2-7, weakly divergent or parallel, ampulliform 5.2-8.6 x 2.3-2.5 μm, with a distinct neck (1.5-3 μm). Conidia subglobose, with surface rough, with protuberances or spines and arranged in bands, 3-3.9 x 2.9-3.2 μm.

We compared 40 isolates from Czechoslovakia and 6 from Poland with an ex type of *P. daleae* (IMI 89338) and with IFJM 3004. *P. daleae* is very close to *P. janczewskii*. When mature it has brown-grey to creamy-brown colouration of the colonies, exudate is brown or amber-brown and the reverse has shades of brown or violet-brown. Conidiophores are less branched and divergent than in *P. janczewskii*. Conidia are subglobose and have protuberances or spines arranged in bands.

Penicillium canescens Sopp – Fig. 3.

CZ: colonies 34-50 mm diam in two weeks, lanose with very weak or heavier sporulation, grey-green, greyish yellow-green, sometimes with greyish yellow to creamy sectors. Margin white, yellowish, 1-3 mm wide. Radial furrows rare. Exudate clear, yellowish or absent. Reverse yellowish, orange, yellow-brown, yellow-orange, orange-red or middle brown. Soluble pigment sparse, yellowish, yellowish brown or light brown.

CYA: Colonies 43-60 mm diam in two weeks, lanose, compact, sporulation weak or heavy, grey-green. Margins white or yellowish, 1-3 mm wide. Radial furrows dense. Exudate brightly yellow to clear. Reverse brownish orange, maroon or light brown. Conidiophores very irregularly branched, 2-3 x 3.5 μm wide. Stipes are slightly sinuous, rough or granular. Metulae terminal, 2-5 in the verticils, also subterminally 9.3-19 x 1.6-3.1 μm, with apices somewhat enlarged, and sometimes concurrent with branches. Monoverticillate conidiophores also present. Phialides 3-8 in the verticils, ampulliform 6.3-8.6 x 1.8-2.3 μm, with distinct necks (1.6-2.5 μm). Conidia globose, subglobose, smooth or distinctly rough, 1.9-2.8 μm in diam.

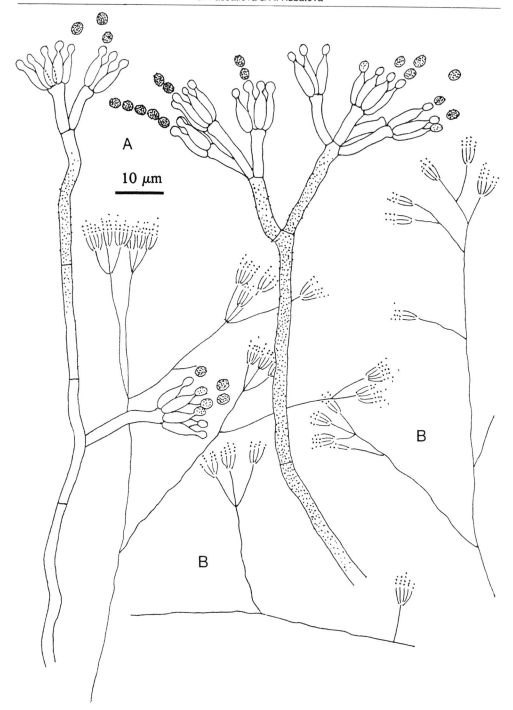

Figure 3. *Penicillium canescens* Sopp. A. conidiophores, B. branching patterns of the penicilli.

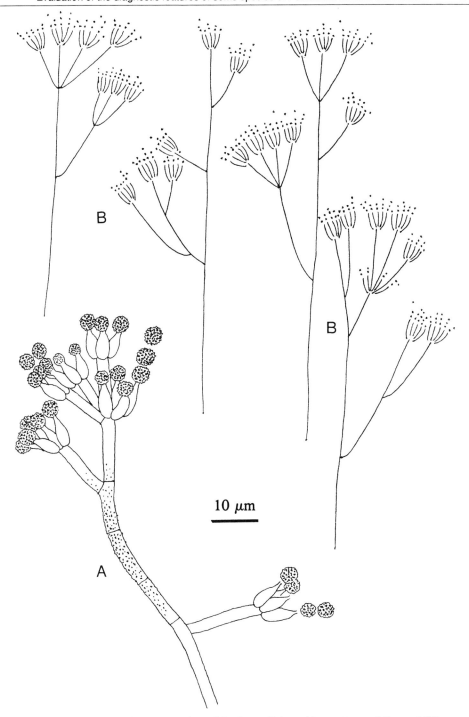

Figure 4. *Penicillium melinii* Thom. A. conidiophores, B. branching patterns of the penicilli.

We compared 8 isolates from Czechslovakia with an ex type culture of *P. canescens* (IMI 28260). *P. canescens* has the following important diagnostic features: colonies on CZ lanose, relatively sparsely sporulating in grey-green or yellow-green; reverse yellow to orange-red; conidiophores sinuous, unequally wide, mostly roughened with irregular whorls of metulae. Conidia are relatively small, ellipsoidal to subglobose, smooth or finely roughened.

Penicillium melinii Thom – Fig 4.

CZ: Colonies 30-60 mm diam in two weeks, velvety or powdery, with heavy sporulation, grey-blue-green, blue-green, in centre brownish grey, margins white, 1-1.8 mm wide, often arachnoid. Radial furrows dense or sparse, shallow, or absent. Exudate yellow, ochraceous or yellow-orange. Reverse orange, brown, violet-brown, intens yellow, pinkish-orange in the centre, yellow-green, brownish red or greenish brown at the margin. Soluble pigment orange-yellow, yellowish, brick-red, salmon or orange-green.
CYA: Colonies 33-57 mm diam in two weeks, velvety, heavy sporulating in dark greyish green, dark green, grey-blue-green or dark olive-green. Margins white, 0.5 mm wide. Sometimes sectors or radial furrows present. Exudate clear, amber, lightly ochraceous, yellowish or absent. Reverse orange, yellowish orange, orange-brown or not coloured in the centre, grey-green, grey-brown, yellow-beige at the margin. Soluble pigment golden-yellow, yellowish brown or rarely absent. Conidiophores 1.9-3.1 μm wide, finely roughened or rarely smooth, irregularly branched, with terminal whorls of 2-5 metulae, often of unequal length, not very divergent, with further subterminal metulae or branches or solitary phialides. Metulae rough, walled 7.5-18 x 2.2-2.5 μm. Phialides 5-8 in whorls, ampulliform, 5.6-8 x 1.9-2.5 μm with or without a distinct neck (0.6-1.5 μm). Conidia globose, echinulate or spinose, 2.8-3.8 μm in diam.

Thirty-nine isolates from Czechoslovakia were compared with an ex type culture of *P. melinii* (IMI 40126). *P. radulatum* G.Smith (CCF 1990), not type, isolated from cysts of *Globodera rostochienses* in Czechoslovakia (Novotná and Fassatiová, 1988) showed the same features in colonies and morphology as *P. melinii*.

The diagnostic features of *P. melinii* are as following: typical velvety or powdery colonies with dark blue-green or brown-grey sporulation and a narrow white margins; frequent radial furrows, exudate in various tones of yellow, orange to ochre; reverse orange-brown, also diffusing pigment in similar shades. Conidiophores are monoverticillate, or with a terminal whorl of metulae, sometimes also with lateral branches. Metulae are less divergent. Stipes and metulae are often roughened. Metulae are sometimes concurrent with a whorl of phialides. Phialides taper to short necks. Conidia are globose, finely echinulate or spinose.

DISCUSSION

In the present study, a number of morphological features were examined mostly in fresh isolates. The most typical features for diagnosis of the species were selected. The following microscopical features were preferred: branching of the conidiophore, surface texture of stipe and penicillus, divergence of metulae and phialides, the length of the phialide neck and length of the phialides without the neck and the shape, surface and size of the conidia. The colony character is best described on CZ, but the shades of colonies and other pigments are more distinctive on CYA.

Our isolates were compared not only with the ex type cultures but also with descriptions in Raper and Thom (1949), Pitt (1979), Ramírez (1982) and Stolk and Samson (1983). From its colony and conidiophore characters *P. janthinellum* was shown to be a synonym of *P. simplicissimum*, as reported by Stolk and Samson (1983). The ex type culture of *P. melinii* examined has deteriorated, and does not correspond in colony characters or especially in microscopical features either with the original description or to our isolates. Pitt (1979) has previously noted this. The conidiophores are very poorly branched and the dimensions of phialides and conidia are larger.

Further the isolate originally determined as *P. radulatum*, which was isolated from cysts of *Globodera rostochiensis* in Czechslovakia, was compared with our isolates of *P. melinii*, and we agree with Pitt (1979) that *P. radulatum* is a synonym of *P. melinii*. The type isolate of *P. granatense* Ramírez et al. is considered to be a synonym of *P. janczewskii*.

We assume that this study does not duplicitate the descriptions of the mentioned species in published monographies. We wanted to verify in a larger set of isolates from one region the utilisability of selected features for diagnostical purposes.

ACKNOWLEDGEMENTS

The authors wish to their thank Dr. Z. Lawrence (CAB International Mycological Institute, Kew, UK) and Drs. E. van Reenen-Hoekstra (Centraalbureau voor Schimmelcultures, Baarn, the Netherlands) for providing us type cultures. They also thank Prof. C. Ramírez (Thailand Institute of Science and Technological Research, Bangkok) for the isolates of *P. granatense* and *P. daleae*.

REFERENCES

NOVOTNA, J. and FASSATIOV´A O. 1988. Three species of the genus *Penicillium* Link isolated from the cysts of *Globodera rostochiensis* Wll. in Czechoslovakia. *Ceska Mykologie* 42: 90-96.
PITT, J.I. 1979. The genus *Penicillium* and its teleomorphic states *Eupenicillium* and *Talaromyces*. London; Academic Press.
RAMIREZ, C. 1981. Manual and Atlas of the Penicillia. Amsterdam; Elsevier Biomedical.
RAPER, K.B. and THOM, C. 1949. Manual of the Penicillia. Baltimore: Williams and Wilkins.
STOLK, A.C. and SAMSON R.A. 1983. The Ascomycete genus *Eupenicillium* and related *Penicillium* anamorphs. *Studies in Mycology, Baarn* 23: 1-149.

DIALOGUE FOLLOWING DR. FASSATIOVA'S PRESENTATION

CHRISTENSEN: You mentioned a species with very short phialides, only 4-5 µm long. Do your measurements for phialides include the neck or not?

FASSATIOVA: It is very difficult to measure a phialide without the neck, so our measurements include the neck.

SAMSON: Have you found any connection with *Eupenicillium*? Do any of the species you worked with produce sclerotia? Do any of these isolates show any connection with *Eupenicillium javanicum*?

FASSATIOVA: We have found four strains of *P. simplicissimum* with sclerotium-like structures, but have not seen ascospores.

REVISION OF *PENICILLIUM* SUBGENUS *FURCATUM* BASED ON SECONDARY METABOLITES AND CONVENTIONAL CHARACTERS

J.C. Frisvad and O. Filtenborg

Department of Biotechnology
Technical University of Denmark
2800 LYNGBY, Denmark

SUMMARY

Penicillium subgenus *Furcatum* contains species that are important in soil ecology and as contaminants of foods and feedstuffs. Taxa in this subgenus are difficult to identify and a high number of new species have been described since the monograph of Raper and Thom in 1949. We have examined all described taxa in that subgenus using morphological, physiological and chemical characters, with emphasis on profiles of secondary metabolites. A number of taxa, such as *P. megasporum* and *P. marneffei*, placed in subgenus *Furcatum* by one or more authors, are transferred to subgenus *Biverticillium*. Several species synonymised by Pitt in 1979 are reestablished on the basis of morphology and profiles of secondary metabolites. All species in subgen. *Furcatum* produced a high number of secondary metabolites and nearly all species produced known antibiotics or mycotoxins. The frequency of mycotoxin production in each species was high, often 100%. Some of the individual secondary metabolites produced by taxa in subgen. *Furcatum*, for example roquefortine C and oxaline, are also produced by taxa in subgen. *Penicillium*, while mycotoxins such as patulin, penicillic acid, griseofulvin and citrinin are quite widespread in all subgenera in *Penicillium* except subgen. *Biverticillium*. Proposed synonyms and secondary metabolite profiles of each species are listed.

INTRODUCTION

Penicillium subgen. *Furcatum* Pitt contains many different species of great importance in soil ecology, foods and feedstuffs and industrial production of enzymes. A number of these species have been reported to produce mycotoxins (Frisvad, 1986) and other secondary metabolites, but misidentified isolates (Frisvad, 1989) have obscured the connections between taxa in subgen. *Furcatum* and their mycotoxins.

Many species included in subgen. *Furcatum* by Pitt (1979) were placed in subgen. *Biverticillium* Dierckx (= sect. *Biverticillata-Symmetrica*; Raper and Thom, 1949) by Raper and Thom (1949) or Ramírez (1982). These included *P. atrovenetum* G. Smith, *P. novae-zeelandiae* van Beyma, *P. herquei* Bainier, *P. estinogenum* G. Smith, *P. paraherquei* Abe ex Udagawa, *P. coralligerum* Nicot and *P. asperosporum* G. Smith. Furthermore, Ramírez (1982) regarded as distinct a number of species synonymised by Pitt (1979). Abe *et al.* (1983 a,b,c) revised subgen. *Furcatum*, emphazising surface ornamentation of conidia; Ramírez (1985) evaluated several species described since 1979. Fassatiová (1965) and Stolk and Samson (1983) also evaluated several species placed in subgen. *Furcatum* by Pitt (1979).

We reexamined these important Penicillia with special emphasis on profiles of secondary metabolites. A summary of our results is reported below.

Modern Concepts in Penicillium and Aspergillus Classification
Edited by R. A. Samson and J. I. Pitt
Plenum Press, New York, 1990

MATERIAL AND METHODS

Ex type cultures, authentic cultures and several fresh cultures of each taxon allocated to subgen. *Furcatum* by different authors were grown on Czapek yeast extract agar (CYA), Malt extract agar (MEA), yeast extract sucrose agar (with Difco yeast extract) (YES), 25% glycerol nitrate agar (G25N) (Pitt, 1979) at 25° and on CYA at 37°. Profiles of secondary metabolites were examined by thin layer chromatography (TLC) and high performance liquid chromatography (HPLC) with diode array detection (DAD) as described by Frisvad *et al.* (1989a) and Frisvad and Thrane (1987).

RESULTS AND DISCUSSION

Subgen. *Furcatum* Pitt is not easy to define and other subgeneric and sectional arrangements have been proposed (Stolk and Samson, 1985). Morphologically members of subgen. *Furcatum* have a strong resemblance to members of subgen. *Aspergilloides* and *Eupenicillium*. All species in these supraspecific taxa are predominantly soil borne fungi.

Table 1. Species placed in *Penicillium* subgen. *Furcatum* (Pitt (1979), Abe *et al.* (1983a) or Stolk and Samson (1983) but reclassified here in subgen. *Biverticillium* on the basis of physiology and profiles of secondary metabolites.

P. aculeatum Raper et Fennell 1948	*P. mirabile* Belyakova et Mil'ko 1972

P. allahabadense Mehrotra et Kumar 1962
   P. korosum Rai *et al.* 1969
   P. zacinthae Ramírez et Martínez 1981
	P. sabulosum Pitt et Hocking 1985

	P. tardum Thom 1930 *P. elongatum* Bain. 1907 (nom. illegit.)

P. asperosporum G. Smith 1965
   = *P. echinosporum* G. Smith 1962
   P. resinae Qi et Kong 1983
	P. rugulosum var. *atricolum* Thom 1930
	   P. echinosporum Nehira 1933
	   P. scorteum Takedo *et al.* 1934
	   P. phialosporum Udagawa 1959

P. inflatum Stolk et Malla 1971	*P. verruculosum* Peyronel 1913
P. marneffei Segretain *et al.* 1960	*P. aculeatum* var. *apiculatum* Abe 1956

P. minioluteum Dierckx 1901
   P. gaditanum Ramírez et Martínez 1981
   P. samsonii Quintanilla 1983	

These morphological resemblances are supported by biochemical characters. Secondary metabolites such as citreoviridin, verruculogen, gliotoxin, curvularin, agroclavine, citromycetin, kojic acid, brefeldin A and spinulosin have been reported from *Eupenicillium* (Horie *et al.*, 1985), and subgens. *Aspergilloides* and *Furcatum* (Frisvad, 1986). Other metabolites such as patulin, griseofulvin, penicillic acid, penitrem A, xanthomegnin, citrinin, mycophenolic acid, roquefortine C, oxaline, isochromantoxin, viridicatumtoxin and carolic acid are also produced by members of subgen. *Penicillium* (Frisvad and Filtenborg, 1989), while secondary metabolites such as mitorubrinic acid, rugulosin, rubratoxin, vermicellin, glauconic acid, wortmannin, cyclochlorotine, luteoskyrin and talaron are exclusively produced by a high number of species in subgen. *Biverticillium* and

Table 2. Accepted species, synonyms and isolates examined in *Penicillium* subgen. *Furcatum*.

1.   *P. atrovenetum* G. Smith 1956
       P. coeruloviride G. Smith 1965

CBS 241.56(T), CBS 243.56, IMI 61836ii, IMI 161964, IMI 103148, CBS 240.65

2.   *P. brasilianum* Batista 1957
       P. paraherquei Abe ex G. Smith 1963
       P. skrjabinii Schmotina & Golovleva 1974

CBS 253.55(T), NRRL A-14996, 79S9FC9, Leistner sp. 863, NRRL 5881, FRR 1859,
IMI 297548, CBS 442.65, CBS 349.77, IMI 177905

3.   *P. canescens* Sopp 1912
       P. swiecickii Zaleski 1927

NRRL 910(T), NRRL 918

4.   *P. castellonense* Ramírez et Martínez 1981

IMI 253791 (T)

5.   *P. chalybeum* Pitt et Hocking 1985

FRR 2660(T), FRR 2659, FRR 2658

6.   *P. citrinum* Thom 1910
       Citromyces subtilis Bain. et Sartory 1912
       P. subtile (Bain. et Sartory) Biourge 1923
       P. aurifluum Biourge 1923
       P. fellutanum Biourge 1923
       P. implicatum Biourge 1923
       P. phaeo-janthinellum Biourge 1923
       P. sartoryi Thom 1930
       P. botryosum Batista et Maia 1957

NRRL 1841(T), NRRL 1843, NRRL 783, NRRL3463, IMI 92267, CBS 232.38, IMI 92229ii, CCM F-
391, IMI 276863, NRRL 2148

7.   *P. coralligerum* Nicot et Pionnat

NRRL 3465(T), CBS 1963, CBS 193.72, CBS 290.73

8.   *P. corylophilum* Dierckx 1901
       P. candidofulvum Dierckx 1901
       P. obscurum Biourge 1923
       P. citreovirens Abe ex Ramírez 1982

NRRL 802(T), CBS 320.59, CBS 330.79, NRRL A-23347, CBS 254.37

9.   *P. cremeogriseum* Chalabuda 1950
       P. glaucolanosum Chalabuda 1950
       ? *P. yarmokense* Baghdadi 1968
       ? *P. iriense* Boretti *et al.* 1973
       P. dodgei Pitt 1979
       P. klebahnii Pitt 1979
       ? *P. sajarovii* Quintanilla 1982

NRRL 1022, NRRL 3389, CBS 410.69, IMI 154289

10. *P. daleae* Zaleski 1927
       P. krzemieniewskii Zaleski 1927

IMI 89338 (T), NRRL 922, FRR 1914, IMI 228583

(cont.)

Table 2 (cont.)

11. *P. dierckxii* Biourge 1923
      ?*P. cinerascens* Biourge 1923 (a)
      P. charlesii G. Smith 1933
      P. fellutanum var. *nigrocastaneum* Abe 1956
      P. eben-bitarianum Baghdadi 1968
      P. gerundense Ramírez et Martínez 1980

CBS 185.81(T), NRRL 746, NRRL 1887, CBS 415.69, IMI 68224, NRRL 778, CBS 304.48

12. *P. echinulonalgiovense* Abe 1956 (to be validated)

IMI 68213(T)

13. *P. estinogenum* Komatsu et Abe ex G. Smith 1963
      P. estinogenum Komatsu et Abe 1956

CBS 319.59(T), IMI 304278, IMI 304279

14. *P. flavidostipitatum* Ramírez et González 1984

IJFM 7824(T)

15. *P. griseopurpureum* G. Smith 1965

CBS 406.65

16. *P. herquei* Bain. et Sartory 1912
      P. luteocoeruleum Saito 1949

NRRL 1040(T), NRRL 3450, CCM F-289, ATCC 46327

17. *P. humuli* van Beyma 1939

NRRL 872(T)

18. *P. janczewskii* Zaleski 1927
      ? *P. albidum* Sopp 1912
      P. kapuscinskii Zaleski 1927
      P. nigricans Bain. ex Thom 1930
      P. nigricans var. *sulphuratum* Abe1956
      P. radiatolobatum Lorinczi 1972*
      P. granatense Ramírez et al. 1980
      P. murcianum Ramírez et Martínez1981*

NRRL 919(T), FRR 538, NRRL 2043, NRRL A-23333, CBS 218.28, FRR 74, IMI 253795,
IMI 253800, IMI 228664, IJFM 7845, FRR 97, ATCC 32023, IMI 149218, IMI 121617, FRR 80, CBS
341.79

19. *P. jensenii* Zaleski 1927
      P. rivolii Zaleski 1927
      P. griseoazureum C. et M. Moreau 1941
      ? *P. corylophiloides* Abe 1956

NRRL 909(T), FRR 531, NRRL 906, CBS 325.59, CBS 162.42

20. *P. lanosum* Westling 1911
      P. kojigenum G. Smith 1961

IMI 40224, IMI 90463, NRRL 3442

21. *P. maclennaniae* Yip 1981

IJFM 7852(T)

22. **P. manginii** Duché et Heim 1931
    P. pedemontanum Mosca et Fontana 1963
    P. syriacum Baghdadi 1968
    P. atrosanguineum Dong 1973

NRRL 2134(T), IMI 99085, IJFM 7870, CBS 343.52, CSIR 1405, CSIR 1406, FRR 2004, CBS 380.75, CBS 418.69

23. **P. mariaecrucis** Quintanilla

CBS 270.83(T), ATCC 48476, Quintanilla 1118, 1022, 1049

24. **P. matriti** G. Smith 1961

NRRL 3452(T), IMI 96506, CBS 347.61, CBS 188.89

25. **P. megasporum** Orput et Fennel 1955
    P. giganteum Roy et Singh 1968

NRRL 2232, CBS 121.65, CBS 529.65

26. **P. melinii** Thom 1930
    P. radulatum G. Smith 1957
    ? *P. echinatum* Dale 1923

CBS 218.30(T), NRRL 848, IMI 85494, NRRL 3672, IMI 158663, FRR 263

27. **P. miczynskii** Zaleski 1927

NRRL 1077(T), FRR 1672

28. **P. moldavicum** Mil'ko et Beljakova 1967
    P. kabunicum Baghdadi 1968

CBS 627.67, CBS 409.69

29. **P. nalgiovense** Laxa 1932

NRRL 911 (ATCC 46455, CBS 138.70, FRR 1647)

30. **P. novae-zeelandiae** van Beyma 1940

NRRL 2128(T), IMI 38496, CBS 109.66, FRR 1905

31. **P. ochrochloron** Biourge 1923
    P. biforme var. *vitriolum* Sato 1939
    P. cuprophilum Sato 1939

NRRL 926(T), CCM F-158, FRR 2141, NRRL 924, NRRL 925

32. **P. onobense** Ramírez et Martínez 1981

CBS 174.81

33. **P. oxalicum** Currie et Thom 1915
    P. aragonense Ramírez et Martínez 1981
    P. asturianum Ramírez et Martínez 1981

IMI 192232(T), IMI 253788, ATCC 32024, CCM F-338, CBS 358.48

34. **P. paxilli** Bain. 1907

NRRL 2008(T), ATCC, 26601, Leistner Sp. 1263

(cont.)

Table 2 (cont.)

35. *P. piscarium* Westling 1911
     P. zonatum Hodges et Perry 1973

NRRL 1075(T)

36. *P. pulvillorum* Turfitt 1939
     P. novae-caledoniae G. Smith 1965
     P. novae-caledoniae var. *album* Ramírez et Martínez 1981
     P. ciegleri Quintanilla 1982

NRRL 2026(T), IMI 140441, IJFM 7184, IJFM 7673

37. *P. raciborskii* Zaleski 1927
     P. fagi Martínez et Ramírez 1978
     P. caerulescens Quintanilla 1983

NRRL 2150(T), CBS 689.77, FRR 1708, Quintanilla 1162, 1155, 1300, 1152, 1147,
IMI 166199, ATCC 32789, CBS 683.77, CBS 685.77, FRR 1695

38. *P. raistrickii* G. Smith 1933
     P. castellae Quintanilla 1983

NRRL 1044(T), FRR 1821, CBS 215.71, FRR 1576, IMI 137808, Quintanilla 1036, 1349, 1012, 1024

39. *P. rolfsii* Thom 1930

NRRL 1078(T), CCM F-337

40. *P. rubefaciens* Quintanilla 1982

Quintanilla 1133(T)

41. *P. sclerotigenum* Yamamoto 1955

IMI 68616, IMI 267703

42. *P. simplicissimum* (Oudem.) Thom 1930
     = *Spicaria simplicissima* Oudem. 1903
     P. janthinellum Biourge 1923
     P. raperi G. Smith 1957
     P. indonesiae Pitt 1979

NRRL 2016(T), NRRL 2674, CBS 191.67, CBS 346.68, IMI 108033, NRRL 904, ATCC 13154, ATCC 42743

43. *P. sizovae* Baghdadi 1968

IMI 140344(T)

44. *P. smithii* Quintanilla 1982
     P. corynephorum Pitt et Hocking

Quintanilla 1097(T), CBS 276.83, FRR 2663, FRR 2676, CBS 349.78

45. *P. soppii* Zaleski 1927
     ? *P. matris-meae* Zaleski 1927
     P. severskii Schechovtsov 1981
     P. michaelis Quintanilla 1982

CBS 226.28(T), NRRL 912, CBS 698.70, CBS 271.73, IJFM 19000, Quintanilla 1150, NRRL A-23325

46. *P. steckii* Zaleski 1927
     P. baradicum Baghdadi 1968

NRRL 2140(T), NRRL 6336, CBS 416.69, CBS 789.70, CBS 993.73, CBS 222.73

47. *P. vasconiae* Ramírez et Martínez 1980

CBS 339.79(T), CBS 175.81, NRRL 721

48. *P. velutinum* van Beyma 1935

NRRL 2069, FRR 270

49. *P. waksmanii* Zaleski 1927
    P. sumatrense von Szilvinyi 1936
    ? *P. decumbens* var. *atrovirens* Abe 1956
    ? *P. atrovirens* G. Smith 1963

NRRL 777(T), IMI 68223, NRRL 779

50. *P. westlingii* Zaleski 1927
    P. chrzaszczii Zaleski 1927
    ? *P. godlewskii* Zaleski 1927
    ? *P. charlesii* var. *rapidum* Abe 1956
    P. gorlenkoanum Baghdadi 1968
    P. damascenum Baghdadi 1968
    P. turolense Ramírez et Martínez 1981

IMI 92272(T), IMI 140337, NRRL 903, NRRL 2111, CBS 318.59, CBS 176.81, CBS 408.69, CBS 200.86, CBS 552.86

(a) Species with a question mark have not been tested for secondary metabolites
*   These species are morphologically closer to *P. canescens* than to *P. janczewskii*.

Talaromyces (Frisvad, 1986 and unpublished results). As an example, *P. phialosporum* Udagawa, considered to be a synonym of *P. oxalicum* Currie & Thom by Abe *et al.* (1983b), produced large amounts of mitorubrin and rugulosin, similar to the ex type culture of *P. tardum* (Table 1), supporting the opinion of Pitt (1979) that *P. phialosporum* and *P. tardum* are the same species.

Because of the form and arrangement of the phialides and metulae, and production of secondary metabolites typical of subgen. *Biverticillium*, a number of species have been excluded from subgen. *Furcatum* in this study and placed in subgen. *Biverticillium* (Table 1). Furthermore, the ex type culture of *P. griseoroseum* Dierckx, placed in subgen. *Furcatum* by Pitt (1979), has been shown to be identical with *P. chrysogenum* in subgen. *Penicillium* (Cruickshank and Pitt, 1987; Frisvad and Filtenborg, 1989). The remaining species, listed in table 2, are confidently placed in subgen. *Furcatum*. Three taxa, *P. victoriae* von Szilvinyi, *P. melagrinum* var. *viridiflavum* Abe and *P. kabunicum* Baghdadi, placed in *P. janthinellum* by Pitt (1979), were found not to belong to that species, but could not be placed in other species listed in Table 2. They will be examined further by HPLC-DAD.

Most species in subgen. *Furcatum* produced known mycotoxins with additional antibiotic activity, such as griseofulvin, patulin, penicillic acid, brefeldin A, citrinin, citreoviridin, kojic acid, mycophenolic acid or asperentin (Table 3). It is interesting to note that the most abundantly occurring *Penicillium* species in rhizosphere soil, *P. janczewskii* Zaleski and *P. westlingii* Biourge, consistently produced large amounts of griseofulvin and citrinin, respectively.

A number of common species produced tremorgens such as paxilline (*P. paxilli* Bainier and *Eupenicillium sheari* Stolk et Scott), penitrem A (*P. janczewskii* and an undescribed species), verruculogen (*P. brasilianum* Batista and another undescribed species previously placed in *P. paxilli*) or janthitrems (*P. piscarium* Westling). HPLC-DAD results showed that *P. janthinellum* FRR 1893 and *P. solitum* CBS 288.36 produced verruculogen, but these isolates did not fit readily in any of the species listed in Table 2.

Table 3. Production of mycotoxins and other secondary metabolites by taxa in *Penicillium* subgen. *Furcatum*.

Taxon	Mycotoxins and secondary metabolites
P. atrovenetum	1. 3-nitropropionic acid (a,b,c) 2. Atrovenetin (c), naphthalic anhydride (a,b,c)
P. brasilianum	1. Penicillic acid (a,b) 2. Viridicatumtoxin (a,b) 3. Verruculotoxin (a,b,c) 4. Verruculogen (a,b,c), fumitremorgin B & C (a,b), acetoxyverruculogen (c) 5. Paraherquamide (c) 6. Paraherquonine (c)
P. canescens	1. Griseofulvin (a,b,c)
P. castellonense	1. Penicillin
P. chalybeum	—
P. citrinum	1. Citrinin (a,b,c)
P. coralligerum	—
P. corylophilum	—
P. cremeogriseum	1. Brefeldin A (a,b,c) 2. Palitantin (a,b,c) 3. Penicillic acid (a,b,c)
P. daleae	1. Asperentin (a,b), 5'-hydroxyasperentin (a,b)
P. dierckxii	1. Carolic acid (a,b,c), carlosic acid (a,b,c), carlic acid (c), dehydrocarolic acid (a,b,c), carolinic acid (c), viridicatic acid (c)
P. echinulonalgiovense	—
P. estinogenum	1. Estin (c) 2. Geodin, dihydrogeodin, erdin (c)
P. flavidostipitatum	—
P. griseopurpureum	—
P. herquei	1. Atrovenetin, herqueinone, deoxy- herqueinone, norherqueinone,herqueichrysin, duclauxin (c), naphthalic anhydride (a,b,c) 2. Physicon (c), flavoskyrin (c) 3. Herquline (c)
P. humuli	—

P. janczewskii	1. Griseofulvin (a,b,c) 2. Penitrem A (a,b,c) 3. Amauromine (=nigrifortine) (c) 4. Penicillic acid (a)
P. jensenii	1. Griseofulvin (a,b)
P. lanosum	1. Griseofulvin (a,b) 2. Kojic acid (a,b,c) 3. Compactin (a,b)
P. maclennaniae	1. Spinulosin (b)
P. manginii	1. Citreoviridin (a,b,c), citreomontanin (c)
P. mariaecrucis	1. Xanthomegnin, viomellein (a,b)
P. matriti	1. Penicillic acid , orsellinic acid (a,b,c)
P. megasprum	1. Penicillic acid (c) 2. Physicon (c) 3. Asperphenamate (c) 4. Megasporizine (c) 5. Phyllostine (c) 6. 7-hydroxy-4,6-dimethylphtalide (c)
P. melinii	1. Patulin (a) (only in ex type culture)
P. miczynskii	1. Citreoviridin (a,b,c) 2. Citrinin (a)
P. moldavicum	—
P. novae-zeelandiae	1. Patulin (a,b,c)
P. ochrochloron	—
P. onobense	1. Brefeldin A
P. oxalicum	1. Roquefortine C (a,b,c) 2. Oxaline (a,b,c) 3. Secalonic acid D (a,b,c)
P. paxilli	1. Paxillin (a,b,c), dehydropaxillin (a,b), 1'-O-acetylpaxillin(a,b)
P. piscarium	1. Brefeldin A (a,b,c) 2. Janthitrem B (a,b,c)
P. pulvillorum	1. Penicillic acid (a,b)
P. raciborski	1. Mycophenolic acid (a,b,c)
P. raistricki	1. Penicillic acid (a,b) 2. Griseofulvin (a,b,c)
P. rolfsii	—
P. rubefaciens	—
P. sclerotigenum	1. Griseofulvin (a,b) 2. Roquefortine C (a,b)
P. simplicissimum = *P. janthinellum*	1. Xanthomegnin (a,b,c), viomellein, 3,4-dehydroxan- thomegnin, semivioxanthin, 7-De-O-methyl- semivioxanthin (c)
P. sizovae	1. Agroclavine I, epoxyagroclavine I (c)

(cont.)

Table 3 (cont.)

P. smithii	1. Citreoviridin (a,b)
	2. Canescin (a,b)
P. soppii	1. Terrein (a,b)
	2. Mycochromenic acid (c)
P. steckii	1. Curvularin (a,b,c), dehydroxy-curvularin (a,b,c)
	2. 3,7-dimethyl-8-hydroxy-6-methoxy- isochroman
	(= isochromantoxin) (a,b,c)
P. vasconiae	—
P. velutinum	—
P. waksmanii	—
P. westlingii	1. Citrinin (a,b,c)

a. Confirmed by TLC in two eluents (TEF & CAP) and two chemical treatments (anisaldehyde spray and heating, cerium sulphate spray followed by aluminium chloride spray and heating), compared with standards.
b. Confirmed by reversed phase HPLC and diode array detection, compared with standards.
c. The identity of the original producer has been confirmed, but a standard was not available.

Three important problems in the taxonomy of subgen. *Furcatum* have not been solved during this work. *P. albidum* Sopp was revived by Fassatiová (1965) as an earlier name for the very common species *P. janczewskii* (= *P. nigricans* Bainier ex Thom) because isolates of these species "showed numerous series of different transitional forms varying from velvety grey-green and strongly sporing through olive-green and light green to white colonies with a floccose surface and light sporulation" and "conidia varied from rough to spiny" (Fassatiová, 1965). We have seen the same variation in *P. janczewskii*. Furthermore isolates of *P. janczewskii*, *P. albidum*, *P. canescens* Sopp and *P. jensenii* Zaleski, together with their synonyms (Table 2) appear to form something approaching a continuum. All have a bright yellow or a dark orange to brown reverse on CYA. All these taxa produce griseofulvin. These taxa probably have a commmon ancestor, but they are sufficiently different, both morphologically and chemically, to be recognised as three or four species. Even though the conidia of *P. janczewskii* (including *P. kapuscinskii* and the intermediate forms between *P. canescens* and *P. janczewskii* mentioned by Pitt, 1979) may vary from roughened to spinose, they look quite different from those of *P. albidum* NRRL 2043 (Ramírez, 1982, pp. 824, 825), so the status of *P. albidum* remains unresolved. However, as the name *P. albidum* remains untypified and of uncertain application, it should probably remain as a *nomen dubium*. *P. murcianum* and *P. radiatolobatum* (Table 2) are morphologically most closely related to *P. canescens*, but biochemically to *P. janczewskii*, as they produce the same profile of secondary metabolites. Unlike *P. canescens*, *P. griseopurpureum* produce smooth stipes, but the conidia and colony morphology of the two species are identical, as are the profiles of secondary metabolites. Profiles of secondary metabolites suggest a classification of these difficult species somewhat at variance with the major distinguishing features used by Pitt (1979), i.e. stipe and conidium roughening. Reexamination of this cluster of allied species using a wider range of techniques is recommended.

Producers of brefeldin A, including *Eupenicillium ehrlichii* Klebahn and possible synonyms (Frisvad *et al.*, 1990), *P. piscarium*, *P. onobense* Ramírez & Martinez, *P.*

indicated in Table 2. However the conidia of *P. piscarium* (spinose), *E. ehrlichii* (finely roughened) and *P. onobense* (spirally roughened) appear to be quite different. Again other taxonomic methods may help in solving this problem. Some morphological variation seem to be common in abundantly occurring taxa in subgen. *Furcatum*. The species mentioned above appear to be related also to *E. javanicum* (van Beyma) Stolk & Scott and perhaps more distantly to *P. pulvillorum* Turfitt, *P. mariaecrucis* Quintanilla and *P. brasilianum* Batista.

Unlike all other isolates of this species and *P. radulatum* G. Smith, the ex type isolate of *P. melinii* produce patulin, so perhaps the name *P. radulatum* should be revived. Some other species, including *P. atrovenetum* G. Smith, *P. estinogenum* G. Smith and *P. westlingii*, the first two included in *P. melinii* by Pitt (1979), are regarded by us as distinct both morphologically and biochemically. Of these colourful species we have isolated *P. westlingii* and *P. estinogenum* most often, and always from soil.

P. manginii Duché & Heim and *P. soppii* Zaleski were included in *P. miczynskii* Sopp by Pitt (1979). While we believe that *P. manginii*, *P. miczynskii* and *P. citreonigrum* are certainly related, differences in morphology and profiles of secondary metabolites indicate to us that they are separate taxa. *P. soppii*, on the other hand, is quite different from these three species, all of which produce citreoviridin (Table 3). All four species are common in soil. *P. smithii* Quintanilla, another abundant producer of citreoviridin, is common in European and Canadian soils (Frisvad, unpublished), but has also been isolated from Indonesian dried fish (as *P. corynephorum*; Pitt and Hocking, 1985).

P. coralligerum, *P. echinulonalgiovense* Abe, *P. lanosum* Westling, *P. raciborskii* Zaleski, *P. sizovae* Baghdadi, *P. steckii* Zaleski and *P. westlingii* were not accepted as separate species by Pitt (1979), but are all regarded by us as distinct species. The last five species named are common in soil.

Species in subgen. *Furcatum* are not usually the most common taxa isolated from food or feed samples. Exceptions are *P. citrinum* Thom, which is very commmon in spices and on vegetables, *P. sclerotigenum* Yamamoto, which causes a rot of sweet potatoes, and *P. oxalicum*, which invades corn. Other species which have been found in foods and feeds include *P. brasilianum*, *P. corylophilum* Dierckx, *P. dierckxii* Biourge, *P. janczewskii*, *P. lanosum*, *P. manginii*, *P. miczynskii*, *P. novae-zeelandiae*, *P. paxilli*, *P. piscarium*, *P. pulvillorum*, *P. raistrickii* G. Smith, *P. simplicissimum* (Oudem.) Thom, *P. smithii*, *P. soppii* and *P. steckii*. These species are usually indicators of soil contamination. Apart from *P. citrinum* and *P. sclerotigenum*, they seem to be poor competitors on foods compared to species in subgen. *Penicillium*. *P. sclerotigenum* may be related to the latter subgenus, because its conidiophores may be terverticillate and it produces roquefortine C. *P. oxalicum* produces oxaline and roquefortine C, and may also have an affinity with species in subgen. *Penicillium*.

This study has shown that species in subgenus *Furcatum* are very efficient producers of secondary metabolites, often with antibiotic or mycotoxic activity, and that all the taxa are well defined by their profiles of secondary metabolites. Some of the more common species may vary in the degree of roughening of the conidiophores or conidia, even though they have nearly the same profile of secondary metabolites. It is recommended that these taxa should be examined using other biochemical techniques to fix their taxonomic position and determine their morphological variation.

ACKNOWLEDGEMENTS

We sincerely thank Amelia C. Stolk and Robert A. Samson for advice and many discussions on the taxa treated in this paper.

REFERENCES

ABE, S., IWAI, M. and TANAKA. H. 1983a. Taxonomic studies of *Penicillium*. III. Species in the section *Divaricatum*. *Transactions of the Mycological Society of Japan* 24: 95-108.

ABE, S., IWAI, M. and AWANO, M. 1983b. Taxonomic studies of *Penicillium*. IV. Species in the sections *Divaricatum* (continued) and *Furcatum*. *Transactions of the Mycological Society of Japan* 24: 109-120.

ABE, S., IWAI, M. and ISHIKAWA, T. 1983c. Taxonomic studies of *Penicillium*. IV. Species in the section *Furcatum* (continued). *Transactions of the Mycological Society of Japan* 24: 409-418.

CRUICKSHANK, R.H. and PITT, J.I. 1987. Identification of species in *Penicillium* subgenus *Penicillium* by enzyme electrophoresis. *Mycologia* 79: 614-620.

FASSATIOVA, O. 1965. Studies on the variability of *Penicillium albidum* Sopp emend. and the development of the conidia. *Ceskà Mykologie* 19: 104-110.

FRISVAD, J.C. 1986. Taxonomic approaches to mycotoxin identification. *In* Modern Methods in the Analysis and Structural Elucidation of Mycotoxins, ed. R.J. Cole, pp. 415-457. London: Academic Press.

—— 1989. The connection between the Penicillia and Aspergilli and mycotoxins with special emphasis on misidentified isolates. *Archives of Environmental Contamination and Toxicology* 18: 452-467.

FRISVAD, J.C. and FILTENBORG, O. 1989. Terverticillate Penicillia: chemotaxonomy and mycotoxin production. *Mycologia* 81 (in press).

FRISVAD, J.C. and THRANE, U. 1987. Standardized high-performance liquid chromatography of 182 mycotoxins and other fungal metabolites based on alkylphenone retention indices and UV-VIS spectra (diode array detection). *Journal of Chromatography* 404: 195-214.

FRISVAD, J.C., FILTENBORG, O. and THRANE, U. 1989a. Analysis and screening for mycotoxins and other secondary metabolites in fungal cultures by thin layer chromatography and high-performance liquid chromatography. *Archives of Environmental Contamination and Toxicology* 18: 331-335.

FRISVAD, J.C., SAMSON, R.A. and STOLK, A.C. 1990. Chemotaxonomic evidence for teleomorph-anamorph connections in *Eupenicillium javanicum* and related species. *In* Modern Concepts in *Penicillium* and *Aspergillus* Classification, eds. R.A. Samson and J.I. Pitt, pp. 445-454. New York and London: Plenum Press.

HORIE, Y., MAEBAYASHI, Y. and YAMAZAKI, M. 1985. Survey of productivity of tremorgenic mycotoxin, verruculogen by *Eupenicillium* spp. *Proceedings of the Japanese Association of Mycotoxicology* 22: 35-37.

PITT, J.I. 1979. The Genus *Penicillium* and its Teleomorphic States *Eupenicillium* and *Talaromyces*. London: Academic Press.

PITT, J.I. and HOCKING, A.D. 1985. New species of fungi from Indonesian dried fish. *Mycotaxon* 22: 197-208.

RAMIREZ, C. 1982. Manual and Atlas of the Penicillia. Amsterdam: Elsevier Biomedical.

—— 1985. Revision of recently described *Penicillium* taxa. *In* Advances in *Penicillium* and *Aspergillus* Systematics, eds. R.A. Samson and J.I. Pitt, pp. 135-142. New York and London: Plenum Press.

RAPER, K.B. and THOM, C. 1949. Manual of the Penicillia. Baltimore: Williams and Wilkins.

STOLK, A.C. and SAMSON, R.A. 1983. The ascomycete genus *Eupenicillium* and related *Penicillium* anamorphs. *Studies in Mycology, Baarn* 23: 1-149.

—— 1985. A new taxonomic scheme for *Penicillium* anamorphs. *In* Advances in *Penicillium* and *Aspergillus* Systematics, eds. R.A. Samson and J.I. Pitt, pp. 163-192. New York: Plenum Press.

DIALOGUE FOLLOWING DR. FRISVAD'S PRESENTATION

PITT: There are a few species, such as *P. westlingii*, that I placed in synonymy because I had seen only a single isolate. If you have seen more, then I am happy to have the species reinstated. Again, separation between *P. soppii* and *P. miczinskii* is quite welcome now that good evidence is available. The relationship between *P. corylophilum* with the monoverticillates: I have always considered that there is a continuum between the monoverticillates and species in section *Divaricatum*. Separation in The Genus *Penicillium* was done for practical reason. The possible conflict with secondary metabolite data does not concern me.

FRISVAD: I agree with you that there is a continuum. For example, *Eupenicillium javanicum* normally has monoverticillate penicilli, but we believe that there is a connection with *P. janthinellum*, which often has metulae.

CHRISTENSEN: It seems logical to speculate that these mycotoxins evolved as defense mechanisms against other species in the soil. Is there any correlation between mycotoxins and habitats?

FRISVAD: A student of Dr. Gams, Michael Schlag, made a study of microfungi associated with roots of *Pseudotsuga menziesii*. He found many isolates of *P. westlingii* that all made enormous amounts of citrinin. Other species were also present. Citrinin has very strong antibiotic activity and is certain to play a role in soil ecology. But isolates of *P. westlingii* from other soils also show the same profile of secondary metabolites. So, this seems to be related more to the actual species than the particular ecological habitat. There may be mycotoxins that are produced only in certain habitats, such as those produced by some of the kangaroo rat isolates, but these were mostly terverticillate Penicillia.

CHRISTENSEN: I'm pleased to see the restoration of several species, such as *P. coralligerum* and *P. pedemontanum*, because we have found these in soil in the US. We have just finished a survey of Penicillia in *Pseudotsuga menziesii* forests in the US. It would be quite interesting to compare this with Mr. Schlag's results.

FRISVAD: In Danish soil samples, we found a continuum between *P. canescens* and *P. jensenii*. Morphologically, these fit *P. canescens*, but the secondary metabolites were more like those of *P. jensenii* than *P. janczewskii*.

CHRISTENSEN: *P. jensenii* is a dominant species in sage brush soils in Wyoming. It would be interesting to compare these isolates with those from other habitats.

PATERSON: In some of the strains you mention only two metabolites. Am I right in assuming that other metabolites are also produced.

FRISVAD: Well, with HPLC you might see 96 different metabolites. What we see with TLC are the largest peaks from the HPLC. So, we might see as many as 15 with two dimensional TLC. HPLC is not used so much for identification, but for more fundamental understanding of species relationships.

PATERSON: I'm quite interested in comparing the total profiles, and seeing the percentage similarity on that basis.

FRISVAD: In some isolates of *P. jensenii*, *P. canescens* and *P. griseo-purpureum* we saw a red compound, based on its UV spectrum. We don't know what it is and it has quite a complicated structure. It's a big help if you can find out what such unknown compounds are, but for now we can only call them compound A, B and C and so on.

PATERSON: Well, that's alright. How do you chose the metabolites to list for each species?

FRISVAD: We first look for mycotoxins and antibiotic substances that we have in our stock of standards so I can compare the spectra. Then we look at all the others, which provide with broader information. Sometimes, you can obtain identifications by searching through ultraviolet spectra: that's how we identified compactin.

PITT: In my original morphological studies, I saw *P. kapusczinskii* between *P. canescens* and *P. janczewskii*. I decided that a morphologist would be unable to distinguish these species. It's interesting that you also saw *P. kapusczinskii* as a bump in this continuum of three species. Clearly, they are very difficult to separate using your criteria as well. Again, with *P. manginii* and *P. miczynskii*, I kept them together because a morphologist would have difficulty separating them, even though there were some differences. I'm still concerned about the typification of *P. simplicissimum*. I asked the University of Leiden for a type, and finding none, I neotypified *P. simplicissimum* in the sense that Raper had used. Stolk and Samson subsequently published a drawing of Oudemans which they accepted as the type in a very different sense from my neotype. In my opinion, this drawing could just as easily represent a *Paecilomyces*: after all, this was described as a *Spicaria*. Taking up this type destroyed a concept that had been in use for forty years. I think this was regrettable.

SAMSON: We were reluctant to take up this type. But Oudemans' drawings are very good, and we have isolated *P. simplicissimum* from the same area that Koning collected it. The colour on the drawing also indicated *P. simplicissimum*. I think that Oudemans would have recognized a *Paecilomyces*. He described it as green.

FRISVAD: The name *P. simplicissimum* is very descriptive of the type of penicillus that Oudemans drew. I think it is also quite regrettable that we have to sacrifice this name.

THE *PENICILLIUM FUNICULOSUM* COMPLEX – WELL DEFINED
SPECIES AND PROBLEMATIC TAXA

E.S. Van Reenen-Hoekstra[1], J.C. Frisvad[2], R.A. Samson[1] and A.C. Stolk[1]

[1]*Centraalbureau voor Schimmelcultures*
3740 AG Baarn, The Netherlands

[2]*Department of Biotechnology*
The Technical University of Denmark
2800 Lyngby, Denmark.

SUMMARY

The morphology and production of secondary metabolites of 96 isolates of *Penicillium funiculosum* and related taxa have been examined. On basis of both morphology and chemistry *P. pinophilum* and *P. minioluteum* are kept separate. *P. varians* placed in synonymy with *P. funiculosum* by Pitt (1979), proved to be morphologically distinct from *P. funiculosum* by its strongly pigmented stipes, cylindrical conidia and by its secondary metabolite production. Isolates FRR 1714 and CBS 642.68 both ex type of *P. minioluteum* proved to be different. Both isolates were believed to represent the original Biourge isolate no. 60. CBS 642.68 fully fits the original description of the species, but FRR 1714 resembles *Penicillium rubrum* sensu Raper and Thom. *P. allahabadense* is kept as a separate species and not included in *P. pinophilum*, because it differs from *P. pinophilum* in growth rate at 37°C and the shape of conidia. *P. diversum* is characterized by its poor growth on Czapek agars. The species accepted are shortly described and illustrated, while isolates of *P. funiculosum* from major culture collections were reidentified. A list of mycotoxins and other secondary metabolites specific for each species is given.

INTRODUCTION

Together with *Penicillium chrysogenum* Thom, *P. funiculosum* Thom is the most widely used *Penicillium* species for biotechnological purposes (Hamlyn et al., 1987). *P. funiculosum* in the broad sense of Raper and Thom (1949) is widely used for the production of biochemicals, such as cellulases and dextranases, but it is not known whether these products are produced by *P. funiculosum*, or similar species such as *P. pinophilum* Hedgcock or *P. minioluteum* Dierckx. Pitt (1979) divided *P. funiculosum* into three taxa, *P. funiculosum*, *P. pinophilum* and *P. minioluteum*, on the basis of characters such as colony diameters, colours, texture, stipe length, and penicillus shape. However, species concepts in *P. funiculosum* and related taxa remain unclear. To assist in clarifying this problem, cultures and fresh isolates deposited at the Centraalbureau voor Schimmelcultures were examined. In addition to this material many cultures from other collections were investigated together with some recently described taxa belonging to *Penicillium* section *Simplicium* Pitt (1979).

In the following paper we present our observations on 96 strains of *P. funiculosum* and related taxa primarily on the basis of morphological structures supplemented with data on the production of secondary metabolites.

Modern Concepts in Penicillium and Aspergillus Classification
Edited by R. A. Samson and J. I. Pitt
Plenum Press, New York, 1990

Table 1. Identity of examined cultures of P. funiculosum sensu Raper and Thom (1949) and related taxa.

Received as	Collection nr.	Actual identity
P. allahabadense	CBS 304.63 (=NRRL 3397)	*P. allahabadense*
P. allahabadense	CBS 441.89	*P. allahabadense*
P. aurantiacum	CBS 314.59 (=NRRL 3398)	*P. aurantiacum*
P. diversum	CBS 320.48 (=NRRL 2121)	*P. diversum* ch. I
P. diversum	NRRL 2122	*P. diversum* ch. I
P. diversum	CBS 242.73	*P. diversum* ch. II
P. diversum	CBS 276.58	*P. diversum* ch. II
P. diversum	IMI 303611	*P.* cf. *erythromellis*
P. diversum	FRR 977 (=IFO 6580 as *P. rubrum*)	*P.* cf. *erythromellis*
P. funiculosum	CBS 330.48 (=NRRL 1768)	*P.* cf. *pinophilum*
P. funiculosum	CBS 433.62	*P. rubrum*
P. funiculosum	CBS 631.66 (=IMI 114933)	*P. pinophilum*
P. funiculosum	CBS 104.71	*P. minioluteum*
P. funiculosum	CBS 884.72	"*P. rubrisclerotium*"*
P. funiculosum	CBS 996.72	*P. minioluteum*
P. funiculosum	CBS 272.86 (=FRR 1630)	*P. funiculosum*
P. funiculosum	CBS 195.88 (=NRRL 1159)	*P. rubrum*
P. funiculosum	IMI 61383	"*P. rubrisclerotium*"*
P. funiculosum	IMI 61385	"*P. rubrisclerotium*"*
P. funiculosum	IMI 63903	*P. pinophilum*
P. funiculosum	IMI 104624	"*P. rubrisclerotium*"*
P. funiculosum	IMI 132677	*P.* cf. *funiculosum*
P. funiculosum	IMI 134755	"*P. rubrisclerotium*"*
P. funiculosum	IMI 134756	"*P. rubrisclerotium*"*
P. funiculosum	IMI 142752	*P. funiculosum*
P. funiculosum	IMI 163167	*P.* cf. *erythromellis*
P. funiculosum	IMI 170614	*P. funiculosum*
P. funiculosum	IMI 211742	*P.* cf. *pinophilum*
P. funiculosum	FRR 883	*P. funiculosum*
P. funiculosum	NRRL 1033	*P. funiculosum*
P. funiculosum	NRRL 1035	*P. funiculosum*
P. funiculosum	NRRL 2119	*P.* sp. 1
P. funiculosum	NRRL 3114	*P.* cf. *resedanum*
P. funiculosum	NRRL 3363	*P. proteolyticum*
P. funiculosum	NRRL 6014	?
P. funiculosum	CBS 433.89	*P. funiculosum*
P. funiculosum	CBS 434.89	*P. funiculosum*
P. funiculosum	CBS 435.89	*P. funiculosum*
P. funiculosum	CBS 436.89	*P. funiculosum*
P. funiculosum	CBS 437.89	*P. funiculosum*
P. funiculosum	CBS 438.89	*P. funiculosum*
P. gaditanum	CBS 169.81	*P. minioluteum*
P. korosum	CBS 762.68	*P. allahabadense*
P. luteum	CBS 170.60	*P. pinophilum*
P. luteoviride	NRRL 2127	*Paecilomyces* sp.
*P. minioluteum***	CBS 252.31	*P. rubrum*
P. minioluteum	CBS 642.68 (=IMI 89377)	*P. minioluteum*
P. minioluteum	CBS 196.88 (=FRR 1714)	*P. rubrum*
P. minioluteum	IMI 68219	*P. rubrum*
P. minioluteum	IMI 113729	*P. rubrum*
P. minioluteum	IMI 139462	*P. rubrum*
P. minioluteum	IMI 147406	"*P. rubrisclerotium*"*

P. minioluteum	IMI 178519	*P. rubrum*
P. minioluteum	CBS 442.89	*P. minioluteum*
P. minioluteum	CBS 443.89	*P. minioluteum*
P. minioluteum	CBS 444.89	*P. minioluteum*
*P. pinophilum***	CBS 329.48	*P. rubrum*
*P. pinophilum***	CBS 631.66	*P. pinophilum*
P. pinophilum	FRR 1397	*P. pinophilum*
P. pinophilum	FRR 1487	*P. pinophilum*
P. pinophilum	CBS 439.89	*P. pinophilum*
P. pinophilum	CBS 440.89	*P. pinophilum*
*P. proteolyticum***	CBS 303.67 (=NRRL 3378)	*P. proteolyticum*
P. purpurogenum var. *rubrisclerotium***	CBS 270.35 (=NRRL 1064 =NRRL 1142)	*P. pinophilum*
P. purpurogenum var. *rubrisclerotium***	CBS 365.48	"*P. rubrisclerotium*"
P. purpurogenum var. *rubrisclerotium***	CBS 436.62	"*P. rubrisclerotium*"
P. rademiricii	CBS 140.84	?*P. diversum*
P. rubicundum	CBS 342.59 (=IMI 99723 =NRRL 3400)	*P. rubicundum*
P. samsonii	CBS 137.84 (=Q1032)	*P. minioluteum*
P. samsonii	CBS 138.84	*P. minioluteum*
P. varians	CBS 386.48 (=IMI 40586 =NRRL 2096)	*P. varians*
P. varians	CBS 884.71	*P. funiculosum*
P. varians	CBS 885.71	*P. funiculosum*
P. varians	CBS 272.73	*P.* cf. *funiculosum*
P. varians	ATCC 20150	*P.* cf. *minioluteum*
P. varians	ATCC 28070	*P. citrinum* ch. II
P. zacinthae	CBS 178.81 (=IMI 285805)	*P. allahabadense*

*   "*P. rubrisclerotium*" is based on *P. purpurogenum* var. *rubrisclerotium* CBS 365.48. This strain still produces dark reddish brown sclerotia (Stolk, 1973). The "culture ex type" NRRL 1064 = CBS 270.35 may have been a contaminant because it resembles *P. pinophilum*, which never produces sclerotia.

**   These strains are listed under *P. funiculosum* in the CBS catalogue (1987)

MATERIAL AND METHODS

Isolates of *P. funiculosum* and related species examined are shown in Table 1. Isolates were grown on CZ, CYA, 2% MEA, and YES agars (Samson and Van Reenen-Hoekstra, 1988) at 25°C and on CYA and MEA at 37°C. They were examined after one week for morphological, physiological and chemical characters. Colony colours are described according to Ridgway (1912).

For the analysis of secondary metabolites, the isolates were examined using the agar plug method (Filtenborg et al., 1983), while representative strains, including ex type cultures, of all species were examined by high-performance liquid chromatography (HPLC) with diode array detection (Frisvad and Thrane, 1987).

RESULTS AND DISCUSSION

Descriptions of the recognized taxa with short discussions about their taxonomic placement are given. The isolates examined and their identity are summarized in Table 1. The isolates of *P. funiculosum* and allied species produced a large number of secondary

Table 2. Production of mycotoxins and other secondary metabolites by species related to *P. funiculosum*.

Species	Secondary metabolites
P. allahabadense	Mitorubrinic acid, mitorubrin
P. aurantiacum	–
P. diversum	Alternariol monomethylether* Austinol* and isoaustin Diversonols Lichexanthone
P. funiculosum	Secalonic acid D* (traces)
P. minioluteum	Spiculisporic acid* Minioluteic acid Secalonic acid D* Mitorubrinic acid, mitorubrin, Mitorubrinol, mitorubrinol-acetat
P. pinophilum	Vermiculin Vermicellin Mitorubrinic acid, mitorubrin, Mitorubrinol, mitorubrinol acetate
P. rubicundum	Skyrin
P. varians	–

*   Recognized as mycotoxins.

metabolites. The profiles of secondary metabolites of the following species were specific (Table 2). *P. pinophilum, P. minioluteum* and *P. rubicundum* have secondary metabolites in common, while *P. rubicundum* and *P. allahabadense* also had secondary metabolites in common with *P. islandicum, P. variabile* and related species (Frisvad, unpubl. data).

DESCRIPTIONS

Penicillium funiculosum Thom – Fig. 1-2.
   Penicillium funiculosum Thom (sensu stricto)-Bull. Bur. Anim. Ind. US Dept. Agric. 118: 69, 1910

On MEA colony attaining a diameter of 30-52 mm in one week at 25°C; floccose, cottony, funiculose, often with well developed funicles. Mycelium white, salmon to buff, pinkish, or pale yellowish. Sporulation grey green, yellow green (Tea Green, Celandine Green; Ridgway Pl. XLVII). Reverse pale, yellow-orange, ochraceous, pinkish, cinnamon, salmon or dark brown-red. Exudate absent or a few hyaline drops present, soluble pigment lacking.
   On CZ 15-30 mm diam after one week; on CYA 25-45 mm diam; on YES 30-45 mm diam.
   At 37°C on MEA good growth, colony reaching a diameter of 20-45 mm, sporulating; on CYA 21-38 mm diam.

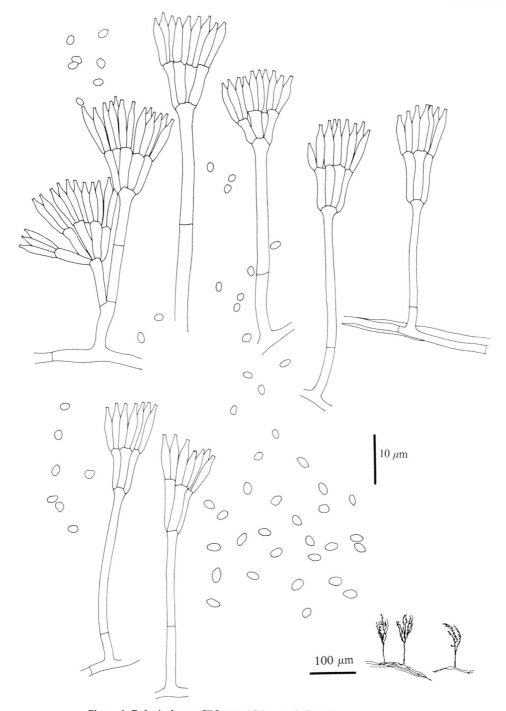

Figure 1. *P. funiculosum* CBS 272.86 (Neotype). Conidiophores and conidia.

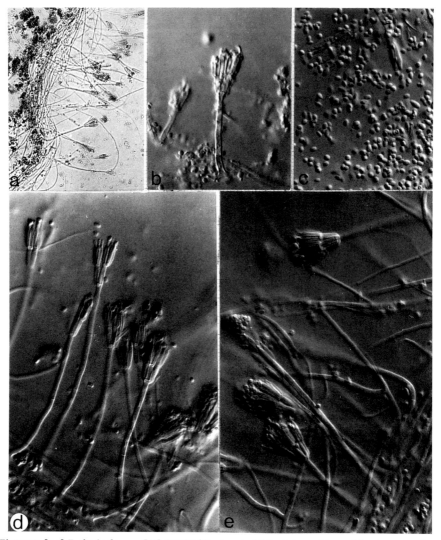

Figure 2. [a-c] *P. funiculosum* CBS 272.86 (Neotype); [a] funicles with conidiophores and conidia. (x350); [b-c] conidiophores and conidia (x1200); [d-e] CBS 884.71 (x1200)

Conidiophores on MEA arising both from the funicles and the substrate. Stipe length ranging from 25-150 μm, if borne from the funicles the shorter ones are dominant if borne from the substrate the longer ones are dominant, not or very slightly pigmented, smooth walled or sometimes very finely rough. Typically biverticillate, sometimes branches below the metulae may be present. Metulae 3-5, adpressed, measuring 8-10(12) x 2-3 μm. Phialides acerose, 10-12 x 2 μm. Conidia varying from subglobose to ellipsoidal, up to cylindrical (1,5)2-3 μm long, hyaline, mostly smooth-walled.

Isolates examined:
Culture ex neotype: CBS 272.86 = IMI 193019 = FRR 1630 from *Lagenaria vulgaris*, India 1975, J.S. Chohan-rather short stipes, typical of the species.

NRRL 1033, unrecorded source, Miss Bottomley, Pretoria, S. Africa-larger stipes compared to the neotype and more or less pigmented. Conidia ellipsoidal to fusiform with rather thick walls.

NRRL 1035, from G. Smith, London-morphology comparable with the neotype.

FRR 833, isolated from pasture grass, Qld., Australia, 1971, M.D. Connole-typical of the species.

CBS 884.71 and CBS 885.71, from air Indonesia, 1971, originally identified as *P. varians*-intermediates between *P. funiculosum* and *P. varians*, pigmentation much less than in *P. varians*, chemically the same as *P. funiculosum*.

IMI 142752, from air, Egypt, 1969-representative isolate of *P. funiculosum*

IMI 170614, from soil in rhizosphere of *Arachis hypogaea*, India 1972-representative strain, somewhat restricted growth, pigmentation not as strong as mentioned by Pitt (1979).

Singh 65a (=CBS 433.89), 69 (=CBS 434.89), 70b (=CBS 435.89), 80a (=CBS 436.89), 90 (=CBS 437.89), and 92 (=CBS 438.89) ex *Zea mays*, India 1987, K. Singh-representative isolates.

IMI 99723 = CBS 342.59 = NRRL 3400, ex type of *Penicillium rubicundum* Miller et al. from soil, Georgia, USA, 1956.

CBS 314.59 = IMI 99722 = ATCC 13216 ex type *P. aurantiacum* Miller et al. from cultivated soil, Georgia, USA 1955, A.A. Foster.

In *Penicillium funiculosum* sensu lato as mentioned by Raper and Thom (1949) also *P. minioluteum* and *P. pinophilum* are included. Thom (1910, 1930) characterised *P. funiculosum* having broadly spreading floccose, tufted, greyish-green colonies and conidiophores with appressed metulae and short stipes, 20-80(100) μm. Pitt (1979) neotypified *P. funiculosum* which is accepted here. However, we disagree with Pitt (1979) in placing *P. varians*, *P. aurantiacum* and *P. rubicundum* in synonymy with *P. funiculosum*. *P. varians* differs by its dark pigmented stipes and small cylindrical conidia. *P. aurantiaceum* differs in having larger and narrow ellipsoidal conidia, similar to those of *P. variabile*. The isolates of this species, all derived from the type, are now in a relatively poor condition and the production of secondary metabolites is also poor. Besides the morphological differences also the secondary metabolite profiles were distinct from *P. funiculosum*. *P. rubicundum* Miller et al. was considered by Pitt (1979) as a synonym of *P. funiculosum*. The CMI strain of this species is in a good condition and produces several characteristic secondary metabolites, different from those of *P. funiculosum*.

P. minioluteum* and *P. pinophilum* are considered separate species here. *P. minioluteum* has more or less restricted colonies with striking yellow mycelium, fails to grow at 37°C or only shows micro-colonies and has longer, often coloured stipes. *P. pinophilum* has a greater number of diverged metulae and also larger stipes compared to *P. funiculosum*. Besides the morphological characters also differences in secondary metabolite patterns justifies to keep both *P. minioluteum* and *P. pinophilum* separate from *P. funiculosum*.

Some isolates of *P. funiculosum* have been reported to produce mycotoxins such as wortmannin (Haefliger and Hauser, 1973) and spiculisporic acid (Mantle, 1987), but these isolates probably have been misidentified. The isolate of *P. funiculosum* NRRL 3363 which was claimed to produce wortmannin is a representative of *P. proteolyticum* Kamyschko.

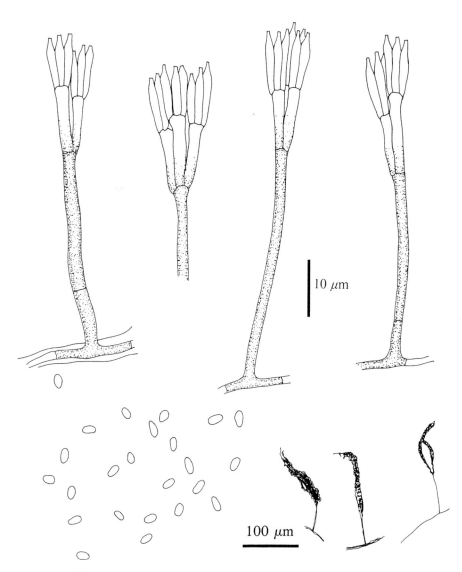

Figure 3. *P. varians* CBS 386.48 (Type). Conidiophores and conidia.

Penicillium varians G. Smith – Fig. 3-4

Penicillium varians G. Smith in Trans. Br. mycol. Soc. 18: 89-90, 1933.

On MEA colony attaining a diameter of 28-34 mm in one week at 25°C; cottony and funiculose, especially in the central part. Mycelium whitish to pinkish. Sporulation pale grey-green (Tea Green, Ridgway Pl.XLVII). Reverse dark green, with or without salmon or buff coloured areas. Exudate and soluble pigment lacking.

On CZ reaching a diameter of 14-18 mm, with a thinner colony than on MEA; on CYA 15-23 mm diam, strongly funiculose with salmon-pinkish mycelial ropes; On YES 25-37 mm diam.

At 37°C on MEA and CYA good growth, colony diameter of 25-30 mm.

Conidiophores on MEA arising in right angles from the mycelium. Stipes short 25-100 μm x 2-2,5(3) μm, typically darkly pigmented, smooth to finely rough-walled, septate. Penicilli typically biverticillate, but monoverticillate conidiophores also present. Metulae dominantly 2-3, strongly adpressed, 10-15 x 2-2.5 μm, darkly pigmented. Phialides acerose, 10-15 x 1.5-2 μm. Conidia cylindrical, later becoming more or less ellipsoidal, 2-3 μm long, hyaline, smooth walled.

Isolates examined:
Culture ex type: CBS 386.48 = IMI 40586 = FRR 2096 = NRRL 2096 = ATCC 10509 = IFO 6112, ex cotton yarn, Great Harwood, GB, 1927, G. Smith-all strains are similar and show all characteristics of the species.

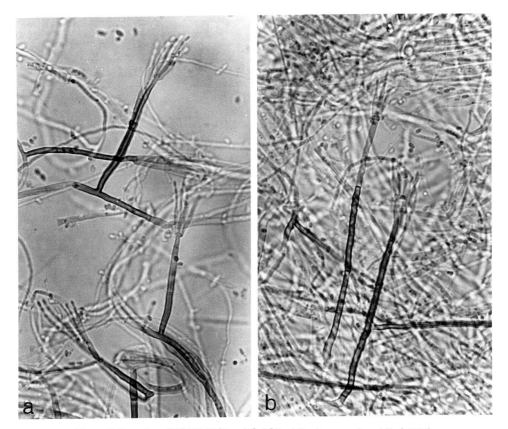

Figure 4. *P. varians* CBS 386.48 (Type). [a-b]Conidiophores and conidia (x1200).

P. varians is characterised by its strongly pigmented conidiophores and small smooth-walled cylindrical conidia. These pronounced features could not been observed in representative isolates of *P. funiculosum*. The illustrations by Smith (1933) show the type strain in its original well-developed form. It may have deteriorated during years of subculturing but most characters are still preserved.

When examined by HPLC-DAD, *P. varians* produces some unique secondary metabolites not shared with other species of subgen. *Biverticillium*.

Penicillium pinophilum Hedgcock – Fig. 5-6.

Penicillium pinophilum Hedgcock apud Thom, Bull. Bur. Anim. Ind. US Dept. Agric. 118: 37, 1910.
? *P. purpurogenum* var. *rubrisclerotium* Thom 1915

On MEA colony attaining a diameter of 35-40(45) mm in one week at 25°C; cottony, more or less floccose with small funicles. Mycelium pale yellow, often dominating colony appearance. Sporulation in blue green colours (Artemisia-Celandine, Ridgway Pl.XLVII). Reverse yellowish to brown (Vinaceous Buff, Ridgway Pl. XL). Exudate hyaline to yellowish, soluble pigment absent.

On CZ reaching a diameter of 15-25 mm. On CYA 25-35 mm diam. At 37°C on MEA and CYA good growth reaching 35 mm diameter with good sporulation.

Conidiophores on MEA arising from aerial mycelium on the small funicles. Stipes vary from 50-240 x 2.5-3 µm, smooth-walled. Metulae (3)5 to 8(10), often vesiculate with more than 5 metulae, and divergent, 10-14 µm x 2.5-3 µm. Phialides acerose, (8)10-12(15) x 2-2.5 µm. Conidia subglobose to ellipsoidal 2.5-3 µm long also larger ones present, smooth or finely roughened, more or less thick walled.

Representative isolates examined:
Culture ex neotype: CBS 631.66 from PVC, France, DSM, 1944
CBS 330.48 = IMI 40235 = ATCC 10446 = NRRL 1768, from soil, USA 1941-chemically atypical and good producer of duclauxin.

CBS 170.60 = IMI 87160 = ATCC 9644, unknown source, Harvard USA 1944, originally identified as *P. luteum*.

NRRL 1064 = NRRL 1142 = CBS 270.35, from *Zea mays*, USA L.B. Lockwood, 1935-all three strains similar.

IMI 63903, from soil Mexico, 1956, J. Nicot, originally identified as *P. funiculosum*-representative of species.

IMI 211742, from unknown source, Mass. USA, E.G. Simmons-originally identified as *P. funiculosum*, but its long stipes, the shape of the conidia and the yellow mycelium are entirely characteristic of *P. pinophilum*.

Singh 92a (=CBS 439.89) and 93 (=CBS 440.89), from *Zea mays*, India 1987, K. Singh.

Pitt (1979) designated IMI 114933 = CBS 631.66 as neotype of *P. pinophilum*, because the original isolate had lost most of the important characters. His neotypification is accepted here. It is doubtful whether *P. purpurogenum* var. *rubrisclerotium* is a synonym of *P. pinophilum*: Raper and Thom (1949) mentioned that there was little to suggest that NRRL 1064 (= CBS 270.35) was representative of the original concept of *P. purpurogenum* var. *rubrisclerotium*, and indeed we have never observed sclerotia in any strain of *P. pinophilum*. Another problem is that NRRL 1064 has more ellipsoidal conidia, pink aerial mycelium and a more floccose colony than *P. pinophilum*, but these may just be insignificant variations. All other isolates of *P. purpurogenum* var. *rubrisclerotium* (CBS 365.48, NRRL 1132, CBS 884.72, FRR 1095, IMI 61363, IMI 61385, IMI 147406, IMI 104624) are unique morphologically and chemically and should be described as a new species.

Reported producers of dextranase could be included in *P. pinophilum* (CBS 170.60), *P. pinophilum* (CBS 330.48) and *P. purpurogenum* var. *rubrisclerotium* (NRRL 1132). Only CBS

330.48 was found to produce large amounts of the mycotoxin duclauxin, so other isolates of *P. pinophilum* may be the most suited producers of dextranase.

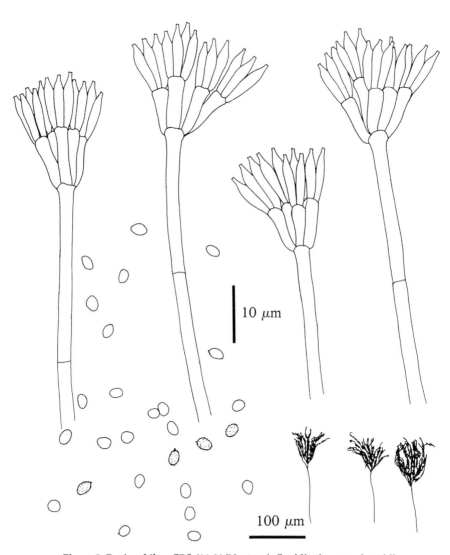

Figure 5. *P. pinophilum* CBS 631.66 (Neotype). Conidiophores and conidia

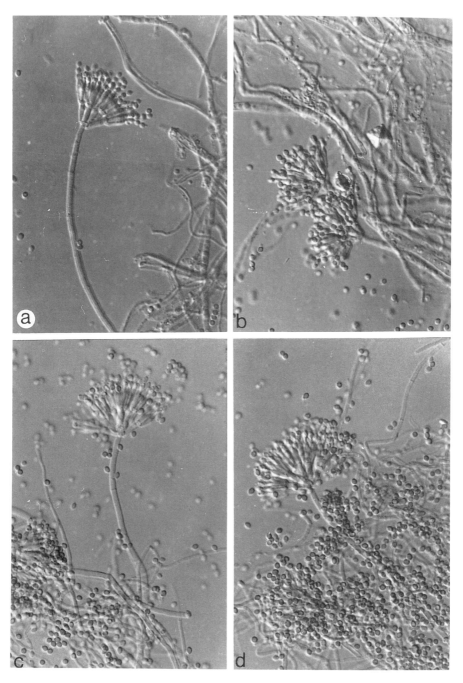

Figure 6. *P. pinophilum* CBS 631.66 (Neotype). Conidiophores with diverging metulae. [a] CBS 170.60; [b] CBS 631.66 (Neotype); [c-d] CBS 440.89 (all x1200).

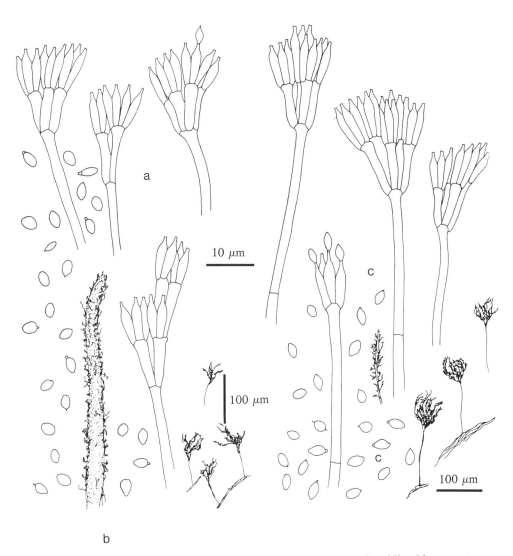

Figure 7 [a-b]. P. *allahabadense* CBS 304.63 (Type). Conidiophores and conidia with remnants of the connectives. After prolonged incubation synnema development can be observed (b). [c]. P. *zacinthae* CBS 178.81 (Type). Conidiophores and conidia.

Penicillium allahabadense Mehrotra et Kumar – Fig. 7.

Penicillium allahabadense Mehrotra et Kumar, Can. J. Bot. 40: 1399, 1962.
P. korosum Rai et al., Antonie van Leeuwenhoek 35: 430, 1969
P. zacinthae Ramírez et Martinez, Mycopathologia 74: 167, 1981

On MEA colony attaining a diameter of 25 mm in one week at 25 C; strongly funiculose, in central part also floccose, and after a prolonged incubation period becoming synnematous. Mycelium white to yellowish. Sporulation olive to blue green (Slate Olive to Artemisia Green, Ridgway Pl. XLVII). Reverse yellowish to cream. On Czapek agars somewhat smaller colony diameter, 20-23 mm. Colony morphology more or less the same as on MEA. At 37 C on MEA very restricted growth, 3-10 mm diam. funiculose and rich sporulation. Conidiophores biverticillate sometimes irregularly branched. Stipes up to 250 x 3 μm, smooth walled. Metulae 2-5, appressed. Phialides with a narrow tapering neck, 8-11 x 2-2.5 μm. Conidia ellipsoidal to somewhat fusiform, often apiculate at one side or with a remnant of the connective still visible, 3-4 μm long, smooth-walled (occasionally more or less finely roughened).

Representative isolates examined:
Culture ex type *P. allahabadense*: CBS 304.63 = NRRL 3397 = FRR 3397 = ATCC 15067, isolated from soil, Allahabad, India, B.S. Mehrotra, 1962-NRRL 3397 is in a good condition.

CBS 178.81 = IMI 253805 = ATCC 48474, ex type of *Penicillium zacinthae*, isolated from leaves of *Zacintha verrucosa*, Alicante Spain, 1979, C. Ramírez.
IBT TKO6 (=CBS 441.89) , isolated from ground coriander seeds, 1989, J.C. Frisvad.
CBS 762.68 ex type of *P. korosum*: from rhizosphere *Brassica campestics* var. *toria*, J.N. Rai-in poor condition, showing little growth at 37°C.

P. korosum and *P. allahabadense* were included in the synonomy of *P. pinophilum* by Pitt (1979). However, both these species grow slowly at 37°C and the conidia are more ellipsoidal than in *P. pinophilum*. The isolate NRRL 3397 of *P. allahabadense* is in a good condition, and descriptions and drawings made when accessioned at CBS showed that it is a separate species. *P. korosum* grows poorly at 37°C and has degenerated, but biverticillate penicilli can still be observed. *P. zacinthae* shows all the characteristics of *P. allahabadense* and should be regarded as a synonym. These conclusions are strongly supported by secondary metabolite production.

Penicillum minioluteum Dierckx – Fig. 8-9.

Penicillum minioluteum Dierckx, Ann. Soc. Scient. Brux. 25: 87, 1901
Penicillium gaditanum, Ramírez & Martinez, Mycopathologia 74: 163-171, 1981
Penicillium samsonii Quintanilla, Mycopathologia 91: 68-78, 1985.

On MEA colony attaining a diameter of 15-20 mm in one week at 25°C; more or less velvety with a raised central area at first, becoming lanose to funiculose after two weeks. Mycelium white and yellow. Sporulation olive to dark green (Leaf Green to Pois Green, Ridgway Pl. XLI). Reverse orange to pinkish red (Japan Rose, Ridgway Pl. XXVIII).
On CZ 5-8 mm in one week (about 10 mm after two weeks), with white to bright yellow mycelium and dark green sporulation. Reverse orange to dark red brown.
On CYA reaching up to 15 mm in one week. Mycelium white, pinkish or orange-yellow. Soluble pigment red, diffusing into agar. Exudate present as hyaline to pinkish drops. Reverse in orange shades to dark red-brown (Indian Red, Ridgway Pl. XXVII).
On Yes up to 25 mm, mycelium pinkish to yellow. Reverse red to dark red. (Dragon's blood Red to Pompeion Red, Ridgway Pl. XIII). At 37°C no growth on MEA and CYA.
Conidiophore stipe 100-250 x 3 μm, hyaline sometimes more or less pigmented, thick-walled, smooth. Metulae 3-5(7) appressed, measuring 10-12 x 2.5-3 μm. Phialides acerose, 10-12 x 2-3 μm. Conidia ellipsoidal to subglobose, sometimes more or less pointed or 2.5-3 μm, often thick-walled, smooth to finely rough, brownish.

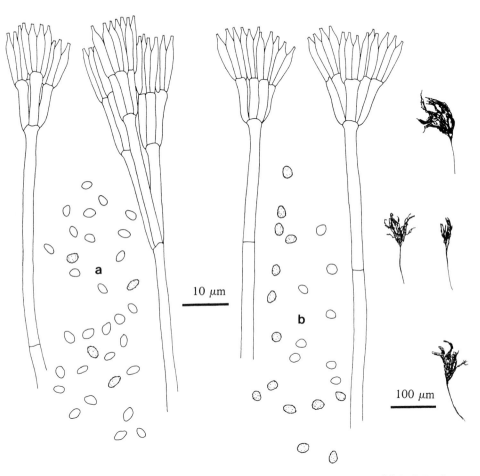

Figure 8. [a-b]. *P. minioluteum* (Neotype). Conidiophores and conidia; [a].CBS 642.68; [b] IMI 89377

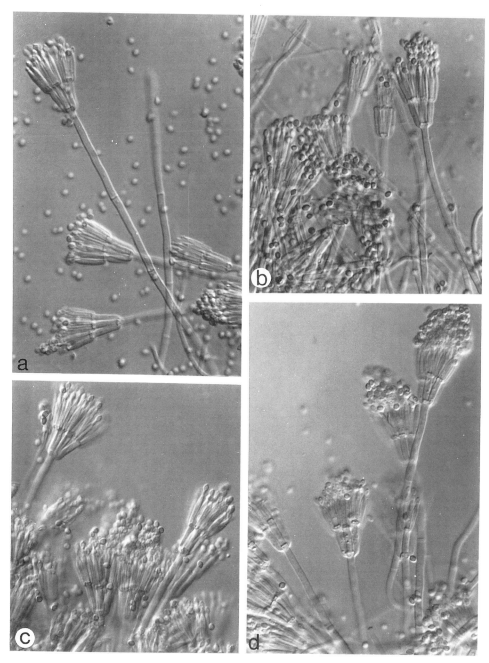

Figure 9 [a-d]. *P. minioluteum* . Conidiophores and conidia; [a-b].CBS 642.68 (Neotype); [c-d] CBS 996.72 (all x1200).

Representative isolates examined:
Culture ex neotype: CBS 642.68 = IMI 89377, unknown source, P. Biourge (no. 60).

CBS 169.81 = IMI 253792 = ATCC 42230. ex type *P. gaditanum* Ramírez & Martinez, from air, Madrid, Spain, 1978.

CBS 137.84 = Q1032 = IMI 327872 ex type *P. samsonii* Quintanilla, from apples damaged by insects Valladolid, Spain. The typical yellow mycelium is lost in CBS 137.84, but it is still very striking in Q1032 and CBS 138.84.

CBS 138.84 (= Q1283), received as *P. samsonii* Quintanilla, the same source as CBS 137.84

CBS 104.71, from *Tulipa* sp., the Netherlands

CBS 996.72, from jute sugar bags, the Netherlands

P. minioluteum differs from *P. funiculosum* by its more compact and rather restricted colonies, the bright yellow to orange mycelium and the dark green to olive green sporulation. At 37°C on MEA and CYA no growth or only microcolonies up to 2 mm diameter were observed. *P. funiculosum* generally shows more spreading colonies with whitish to pinkish mycelium and grey green sporulation. At 37°C on MEA and CYA *P. funiculosum* grows well. Chemically *P. minioluteum* and *P. funiculosum* are different (Table 2).

The culture FRR 1714, which Pitt (1979) used for the description of *P. minioluteum* differs markedly from CBS 642.68, although both strains were believed to represent the original Biourge isolate no. 60. CBS 642.68 showed a more floccose and compact colony with yellow to orange mycelium and dark green sporulation and on most media an yellow or brownish red reverse. FRR 1714 grows more rapidly and thinner (especially on MEA) often with red transparant areas, showing ropes in the central part after a prolonged incubation and with a greenish transparant reverse with red. The conidia are more fusiform than in CBS 642.68. The correct desposition of *P. minioluteum* sensu Pitt must be determined, but the isolate FRR 1714 morphologically and chemically resemble NRRL 1062, which Raper and Thom (1949) used to define *P. rubrum*. However, this isolate is not the orginal type culture and therefore a more detailed study of isolates belonging to *P. rubrum* sensu Raper and Thom will be carried out in a future study.

The type cultures of *P. gaditanum* and *P. samsonii* morphologically and chemically resemble *P. minioluteum* and are here considered synonyms of *P. minioluteum*.

Penicillium diversum Raper & Fennell – Fig. 10

Penicillium diversum Raper & Fennell – Mycologia 40: 539, 1948.
? *Penicillium rademirici* Quintanilla – Mycopathologia 91: 69-78, 1985.

On MEA colony attaining a diameter of 20-30 mm in one week at 25°C; velvety, plane in concentric zones, sporulating area alternating with submerged mycelium. Mycelium whitish to yellow. Sporulation olive to dark green (Slate Olive to Vetiver Green, Ridgway Pl. XLVII), Grayish-Olive (RidgwayPl. XLVI) or Storm Gray (Ridgway Pl. LII). Reverse uncoloured or yellowish (Naples Yellow Ridgway Pl. XVI). Exudate absent or sometimes present as hyaline drops.
On CZ restricted and poor growth, 2-5 mm in diam. Sporulation sparse, grey-green.
On CYA restricted growth attaining a diameter of 5 mm. Sporulation poor, grey green. Reverse buff to brownish.
On YES agar, a diameter of 5-9 mm. Sporulation grey or light mouse gray (Mouse Gray Ridgway Pl. LI). Reverse brownish. At 37°C on MEA 20-25 mm in diam., on CYA 2-5 mm diam.

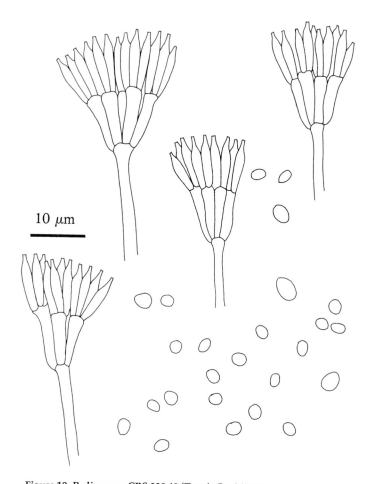

Figure 10. *P. diversum* CBS 320.48 (Type). Conidiophores and conidia.

Conidiophores on MEA borne from submerged or from basal mycelium. Stipes long, 150-300 (350) x 2.5-3 µm. Phialides with a long tapering neck, 8-10 x 1.5-2.5 µm. Conidia subglobose to ellipsoidal 2-3 µm, hyaline, smooth to finely rough walled.

Isolates examined:
Culture ex type: CBS 320.48 = NRRL 2121 = ATCC 10437 = IMI 40579 = IFO 7759 = QM 1921, from mouldy leather, USA-at present CBS 320.48 often shows many monoverticillate conidiophores, but typical biverticillate conidiophores were observed when the isolate was deposited, designated here as chemotype I.

NRRL 2122, from soil, Sweden, restricted growth at 37 C on MEA, chemotype I.

CBS 242.73, from lung *Rana macrodon*, 1972 – rough striate conidia, designated here as chemotype II.
CBS 276.58 from *Beta vulgaris*, 1958, chemotype II.
CBS 140.84 ex type of *P. rademirici* Quintanilla = Q 1248, from air under willow trees in the bank of Duero river, Herrea (Valladolid), Spain.

P. diversum is characterized by its poor growth on Czapek agar. Cultural and micromorphological characters are similar to those of *P. rademirici* Quintanilla, which also grows weakly at best on Czapek. In CBS 140.84, the type of *P. rademirici*, we have observed only biverticillate structures, not branches as drawn and described by Quintanilla (1985). Also the striking yellow to orange yellow mycelium is no longer present. Quintanilla (1985) placed the species between *P. primulinum* Pitt and *P. islandicum* Sopp. The type strain of *P. primulinum* at the CBS has lost the original yellow mycelium and looks now identical with *P. diversum*, but the two taxa differ significantly in their production of secondary metabolites. *P. rademirici* also differs from *P. diversum* in secondary metabolites produced and perhaps should be considered as a distinct species.

ACKNOWLEDGEMENTS

The authors thank the NATO Scientific Affairs Division (Brussel, Belgium) for the research grant (0216/86) for international collaboration.

REFERENCES

FILTENBORG, O., FRISVAD, J.C. and THRANE, U. 1983. Simple screening method for molds producing intracellular mycotoxins in pure culture. *Applied and Environmental Microbiology* 45: 581-585.

FRISVAD, J.C. and THRANE, U. 1987. Standardized high-performance liquid chromatography of 182 mycotoxins and other fungal secondary metabolites based on alkylphenone retention indices and UV-VIS spectra (diode array detection). *Journal of Chromatography* 404: 195-214.

HAEFLIGER, W. and HAUSER, D. 1973. Isolierung und Strukturaufklarung von 11-Desacetoxy-wortmannin. *Helvetica Chimica Acta* 56: 2901-2904.

HAMLYN, P.F., WALES, D.S. and SAGAR, B.F. 1987. Extracellular enzymes of *Penicillium*. In *Penicillium* and *Acremonium*. Biotechnology Handbooks 1, ed. J.F. PEBERDY, pp. 245-284. New York and London: Plenum Press.

MANTLE, P.G. 1987. Secondary metabolites of *Penicillium* and *Acremonium*. In *Penicillium* and *Acremonium*. Biotechnology Handbooks 1. ed. J.F. PEBERDY, pp. 161-243. New York and London: Plenum Press.

PITT, J.I. 1979. The genus *Penicillium* and its teleomorphic states *Eupenicillium* and *Talaromyces*. London and New York: Academic Press.

QUINTANILLA, J.A. 1985. Three new species of *Penicillium* belonging to subgenus *Biverticillium* Dierckx, isolated from different substrates. *Mycopathologia* 91: 69-78

RAMIREZ, C. and MARTINEZ, A.T. 1981. Four new species of *Penicillium* isolated from different substrates. *Mycopathologia* 74: 163-171.

RIDGWAY, R. 1912. Color standards and color nomenclature. Washington, DC.

RAPER, K.B. and C. THOM. 1949. Manual of the Penicillia. Williams and Wilkins, Baltimore.

SAMSON, R.A. and REENEN-HOEKSTRA, van E.S. 1988. An introduction to food-borne fungi. Third Edition. Baarn: Centraalbureau voor Schimmelcultures.

SMITH, G. 1933. Some new species of *Penicillium*. *Transactions of the British Mycological Society* 18: 88-91.

STOLK, A.C. 1973. *Penicillium donkii* sp. nov. and some observations on sclerotial strains of *Penicillium funiculosum*. *Persoonia* 7: 333-337.

THOM, C. 1910. Cultural studies of species of *Penicillium*. Bulletin of the Bureau of Animal Industry United States Department of Agriculture 118: 1-109.

THOM, C. 1930. The Penicillia. Williams and Wilkins, Baltimore.

DIALOGUE FOLLOWING THE PRESENTATION OF THE PAPER BY ELLEN VAN REENEN-HOEKSTRA

CHRISTENSEN: Did I notice an isolate called *P. diversum* var. *aureum* or *P. primulinum*?

SAMSON: Yes, I believe this is a good species. Dr. Pitt also included an isolate under that name that has been called *P. marneffei*. The type isolate of *P. marneffei* is in poor condition, but recently we have obtained some twenty new isolates from bamboo rats and AIDS patients and it seems to be a very good species. It is similar to *P. primulinum*.

PITT: I am pleased to see that you might resurrect *P. allahabadense*. The extype isolate was badly deteriorated, and I had little choice but to place it in synonymy. I still am unprepared to regard *P. aurantiacium* and *P. rubicundum* as separate species. They are quite closely related to *P. funiculosum*. Regarding *P. minioluteum*, I have only just realized the error that I made. The neotype isolate on which I based my concept of *P. minioluteum* came from the CMI collection and was clearly quite different from the CBS strain of the same isolate. This is regrettable but can easily happen. I tried to obtain the best isolates I could in the mid 1970's, but I failed in this case. *P. minioluteum* sensu Pitt will have to be given a different name.

SAMSON: This is a real problem. We also discovered that the CBS strain of the extype of *P. solitum* is quite different from the real type. It's very important to make sure that the cultures we work with match the original descriptions.

HENNEBERT: This is worth emphasizing. Many of the strains of Biourge's isolates kept in other culture collections show a lot of sectoring. When I discovered this, I started making single conidium isolates from *Penicillium* cultures. I often find three or four different colony morphologies can be derived from a single strain. Single spore cultures from extype strains should be carefully studied by traditional morphological methods and new techniques, such as secondary metabolites, to establish their identities. The other problem is that the descriptions of older authors, such as Dierckx, were made from totally different culture media than we now use. They used gelatin, not agar. Biourge added some agar to solidify the gelatin. In my lab, we are now studying the type strains of Biourge, the neotypes for Dierckx's species, using the original culture media. We eventually hope to provide a comparative morphology to allow a better comparison with the original descriptions.

IDENTIFICATION OF *PENICILLIUM* SPECIES ISOLATED FROM AN AGRICULTURAL LOESS SOIL IN GERMANY

H.I. Nirenberg and B. Metzler

Institut für Microbiologie
Biologische Bundesanstalt für Land- und Forstwirtschaft
1000 Berlin 33 Dahlem, FRG

SUMMARY

Of more than 5000 fungal isolates which were recovered from an agricultural loess soil in Germany, 20% were *Penicillium* and were identified as 15 species. One species close to *P.janthinellum* of unclear identity is described and discussed in detail.

INTRODUCTION

Since April 1987 a soil protection program has been carried out at the Biological Research Center for Agriculture and Forestry (BBA) in the Federal Republic of Germany. Five institutes have investigated the long-term impact of an intensive use of pesticides and fertilizers on chemical residues and soil biology of a loess soil in the vicinity of Braunschweig. Winter wheat, winter barley, and sugar beets have been grown in rotation. For seven years the production intensity and cropping system for each plot has been constant.

The Institute of Microbiology at BBA has analysed the fungal and algal soil flora in the cropping systems of two different production intensities. This paper is mainly concerned with the method of qualitative and quantitative assessment of Penicillia and especially with the taxonomy of a *Penicillium* species, which proved to be related to *P. janthinellum* and difficult to identify.

MATERIAL AND METHODS

Soil samples from the upper 5 cm of loess soils were taken every two months during a period when winter barley was grown (5.10.87-8.8.88). The recommended soil washing technique (Gams and Domsch, 1967) imposes serious problems if applied to loess soil because of its tiny particle sizes. More than 95% of particles are less than 200 μm in diameter (Feyk and Pretsch, 1988). The remaining bigger particles are not considered to be representative for the soil. Therefore we used the following method: subsamples were homogenized by shaking and sieving, then soil particles of 0.5 to 0.63 mm diameter are taken and placed individually on SNA (Synthetischer Nähragar, Nirenberg, 1981) in plastic Petri dishes (120 per sampling date). Bacterial growth was suppressed by three antibiotics (chlortetracy cline 10 mg/l, dihydrostreptomycin-sulphate 50 mg/l, penicillin G 100 mg/l). Incubation was at 20°C, the first five days in darkness, then the following week under continuous near ultraviolet light (to induce sporulation of certain fungi). Petri dishes were then stored at 20 to 25°C under natural day/night lighting cycles for an additional four weeks. Plates were examined under the microscope after one, three and six

Modern Concepts in Penicillium and Aspergillus Classification
Edited by R. A. Samson and J. I. Pitt
Plenum Press, New York, 1990

weeks. Fungi were identified during this time, if necessary after isolation. Species were counted only once per soil particle.

As *Penicillium* species usually sporulate rapidly, they can be transferred to other SNA plates after the first five days. On this medium the following species were determined: *P. canescens* Sopp, *P. melinii* Thom, *P. janczewskii* Zaleski, *P. expansum* Link, *P. griseofulvum* Dierckx, *P. brevicompactum* Dierckx, *P. glabrum* (Wehmer) Westling (*P. frequentans*), and *P. restrictum* Gilman & Abbott. *P. hordei* Stolk and an unknown *Penicillium* species under discussion below, could often be identified on the original plate. The other Penicillia were cultured for species determination as recommended by Raper and Thom (1949) and Pitt (1979).

RESULTS

From 720 soil particles, 5734 fungi were counted belonging to 62 genera and 148 species. *Penicillium* accounted for 20% of the isolates, representing 15 species (Table 1). The most frequent were *P. janczewskii, P. melinii, P. hordei, P. canescens*, and an unidentified species. A clear influence of the seasons could not be determined.

Table 1. Penicillia found in the soil samples (treatments summarized)

Sampling date Numbersof particles examined	5-10-87 120	1-12-87 120	1-2-88 120	18-4-88 120	13-6-88 120	8-8-88 120	Exi 720
P. aurantiogriseum	4	0	2	0	0	6	12
P. brevicompactum	0	0	0	3	2	0	5
P. canescens	6	25	17	14	21	19	102
P. expansum	2	9	3	2	7	11	34
P. glabrum	0	0	1	0	2	0	3
P. griseofulvum	0	0	1	0	1	10	12
P. hordei	32	12	8	19	37	32	140
P. janczewskii	81	80	63	58	81	52	415
P. janthinellum	5	3	0	0	0	0	8
P. jensenii	0	0	0	0	1	0	1
P. meliniip	66	61	57	32	42	41	299
P. restrictum	3	1	2	2	3	3	14
P. rugulosum	1	0	0	0	0	0	1
P. simplicissimum	2	2	1	0	0	2	7
P. spec. "trimorphum"	12	15	14	13	15	7	74

On SNA, most of the *Penicillium* species produced more or less velutinous growth; mycelial and reverse colours as well as soluble pigments were usually not pronounced. Those Penicillia which could be recognized after 7 days on this medium are listed below in the order of decreasing growth rate. Characterisation on SNA included features visible under the low power microscope.

P. expansum Link: Fast and loosely growing colonies, forming concentric rings, penicilli tervertcillate, conidia densely packed in columns, bright green.

P. griseofulvum Dierckx: Conidiophores more or less terverticillate, varying from very simple to quite complex, crowded at the inoculation point, conidial masses greyish green.

P. hordei Stolk: Conidiophores soon forming typical fascicles, often with a drop of exudate, often crowded at the inoculation point. Penicilli terverticillate, conidial masses brownish green.

P. melinii Thom: Spreading broadly, penicilli irregular (divaricate), conidia in columns, brownish green.

P. janczewskii Zaleski: Penicilli divaricate, conidial chains disordered, colonies appearing dark grey.

P. canescens Sopp: Growing more restrictedly, penicilli divaricate, delicate ramification and conidial columns, blue green.

P. restrictum Gilman & Abbott: Growing very restrictedly, very short monoverticillate conidiophores.

Penicillium species: Conidiophores long and more or less monoverticillate or medium long with one or two metulae; conidial chains divergent, coloured Isabelline; on short conidiophores especially after prolonged cultivation brownish conidia in false heads (Figs. 1 a-c). On the original plate usually identifiable after three weeks.

Since this fungus is hard to classify, a detailed description according to the model of Pitt (1979) is given:

CYA, 25°C, 7 days: Colonies 30-40 mm diam, radially sulcate, surface texture floccose, low; mycelium white; margins entire, low, rarely deep; conidiogenesis light, inconspicious; no exudate; no soluble pigment; reverse pale.
MEA, 25°C, 7 days: Colonies 30-50 mm diam, surface floccose, plane, margins entire, low or subsurface; mycelium white, sometimes with yellowish sectors; conidiogenesis inconspicous to greyish green; no exudate, no soluble pigment, reverse pale.
G25N, 25°C, 7 days: Colonies 0-6 mm diam; other characters as on CYA.
CYA, 5°C, 7 days: Growth varying from germination to colonies of 1-2 mm. 37°C, CYA, 7 days: Usually no growth, sometimes germination and formation of a microcolony of large vesicles (Fig. 1d).

Conidiophores borne from surface or aerial hyphae, stipes variable in length 15-300(-500) x 2-3 μm, flexible to stiff, more or less thin walled, smooth; penicilli mostly monoverticillate or with one, rarely 2-3 metulae, sometimes irregular; metulae more or less divergent, usually 10-15 x 2-2.5 μm; phialides in verticils of (1-4-)5-8, cylindroidal to flask shaped, 8-12(-15) x 2.5-3 μm (Fig.1e) ; conidia broadly ellipsoidal to apiculate 3.5-4 x 2.5-3 μm, smooth walled, sometimes faintly rugulose, born in long disordered chains, or in false heads when formed on short conidiophores with divergent phialides near the surface of SNA (Fig. 1c), in the latter case with brown pigment.

DISCUSSION

The unidentified *Penicillium* species, which we provisially will name: "*P. trimorphum*" is clearly a member of *Penicillium* series *Janthinella* because of its long collula and divaricate penicilli tending to be monoverticillate. In Table 2 a synoptic key is given for representative strains and their respective species descriptions.
The new species resembles most closely *P. ochrochloron* Biourge *sensu* Pitt in its smooth, broadly ellipsoidal apiculate conidia of 3.5-4 μm length; in the phialides with long cylindrical collula, the penicilli with only few metulae, and the quite rapid growth.

Neither species grows at 37°C, both show sparse conidiogenesis and colouration. However, *"P. trimorphum"* grows at 5°C on CYA better than *P. ochrochloron*, and much slower on G25N. Moreover, we have never found a strain with rugose conidiophores, which can be present in *P. ochrochloron*. The major problem is that the original description of Biourge (1923) and the description by Pitt (1979) differ in the evidently pronounced

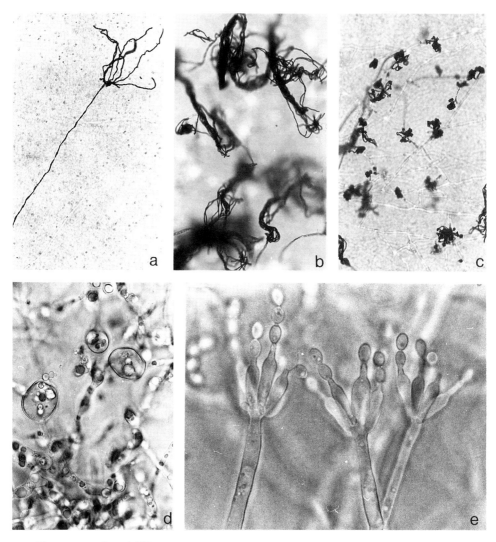

Figure 1 a-e. *"Penicillium trimorphum"*, a. Long and slender type of conidiophore, monoverticillate with long disordered conidial chains; grown on SNA; 125 x, b. Medium long type of conidiophores, divaricate, with disordered conidial chains; grown on SNA; 125 x, c. Short type of conidiophores with conidia forming false heads; after one month on SNA; 125 x, d. Extraordinary vesicles among the substrate mycelium on Czapek agar; 750 x, e. Nearly monoverticillate penicilli, phialides with long collula, conidia smooth, ellipsoidal; grown on SNA; 1500 x roughness of the conidia and the conidiophores in Biourge's description. Therefore, it is hard to follow Pitt's neotypification base dof an isolate from C. Thom. An isolate, which more closely fits Biourge's description exists in the collection of the BBA (no. 64184).

Another similar species proposed by Samson (pers. comm.) is *P. cremeogriseum* Chalabuda (1950). The identity is here rejected because the type culture (CBS 223.66) of *P. cremeogriseum* produces smaller conidiophores and smaller, rather globose conidia which are definitely rugulose. In addition it has a different growth on CYA.

P. janthinellum Biourge seems less closely related to "*P. trimorphum*" because of its globose rough conidia, more branched conidiophores and the vivid colouration of the cultures. As discussed by Pitt (1979), neotypification of this species is not unequivocal.

After examination of Pitt's neotypification of *P. ochrochloron*, we propose that it be rejected on the grounds that it is inconsistent with the protologue of Biourge (1923) and therefore a new species "*Penicillium trimorphum*" with *P. ochrochloron* sensu Pitt non Biourge as a synonym. Another neotype for *P. ochrochloron* Biourge should be selected; strain BBA 64184 represents a possible choice.

Table 2. Synoptic key with critical characteristics of the discussed Penicillia

Species													
"P. trimorphum" (BBA 65203)	+	0	0	±	±	0	+	±	±	+	+	±	0
P. ochrochloron (Biourge, 1923)	−	−	−	+	0	+	±	+	0	−	+	+	0
P. ochrochloron (Pitt, 1979)	0	0	+	0	0	±	±	+	0	+	+	0	0
P. ochrochloron (Raper & Thom, 1949)	−	−	−	±	±	+	±	+	0	+	+	±	±
P. ochrochloron (BBA 64184)	+	0	+	0	±	+	+	+	0	+	+	+	0
P. janthinellum (Biourge, 1923)	−	−	−	+	0	0	0	+	0	0	0	0?	+
P. janthinellum (Pitt, 1979)	0	+	+	0	0	0	±	+	0	0	±	±	+
P. janthinellum (Raper & Thom, 1949)	−	−	−	±	±	±	±	+	0	±	+	+	0
P. cremeogriseum (Chlabuda, 1950)	−	−	−	+	0	+	+	±	0	0	0	+	+
P. cremeogriseum (ex-type CBS 223.66)	0	±	+	±	0	±	±	±	±	±	0	+	±

Explanation of abbreviations and signs: +: with the property indicated, 0: without this property, ±: property less pronounced, −: not specified

ACKNOWLEDGEMENT

This study was partly supported by a grant of the German "Bundesminister für Forschung und Technologie" to Prof. Dr. W. Sauthoff. We are also grateful to Mrs. Marlene Katz for her skillfull technical assistance.

REFERENCES

BIOURGE, P. 1923. Les moisisssures du groupe *Penicillium* Link. La Cellule 33: 7-331.

CHALABUDA, T.V. 1950. Species novae e genere *Penicillium* Link. *Botaniceskie Materialy Otdela Sporovych Rastenij/Akademija Nauk SSSR* 6: 168.

FEYK, M. and PRETSCH, K. 1988. Bodenkundliche Spezialkartierung des BBA-Versuchsstandortes Ahlum bei Wolfenbüttel. Hannover: Niedersächsisches Landesamt für Bodenforschung.

GAMS, W. and DOMSCH, K.H. 1967. Beiträge zur Anwendung der Bodenwaschtechnik für die Isolierung von Bodenpilzen. *Archiv für Mikrobiologie* 58: 134-144.

NIRENBERG, H.I. 1981. A simplified method for identifying *Fusarium* spp. occurring on wheat. *Canadian Journal of Botany* 59: 1599-1609.

PITT, J.I. 1979. The genus *Penicillium* and its teleomorphic states *Eupenicillium* and *Talaromyces*. London: Academic Press.

RAPER, K.B. and THOM, C. 1949: A manual of the *Penicillia*. Blatimore: Williams and Wilkins

DIALOGUE FOLLOWING DR. NIRENBERG'S PRESENTATION

PITT: Dr. Nirenberg, I would place your unknown species, without too much hesitation, in *Penicillium janthinellum*. This species, in my work and in all previous works, is a variable species. It's quite possible to have a pale reverse and with many monoverticillate penicilli. Someday, someone may be able to split this species, but I can't do it on morphological grounds.

NIRENBERG: We have at the same time found *P. janthinellum* from the same sand that was treated with sulphuric acid and it is a completely different species. If you compare the microscopic features, such as the metulae, and see the differences in colony differences at 37°C and 5°C, it is really not the same fungus. We are less sure of the distinction from *P. ochrochloron* or *P. simplicissimum*. Our isolate has certain similarities with your concept of *P. ochrochloron*. We have a *P. ochrochloron* in our culture collection that matches the original description perfectly. It has the rugose conidia and stipes and therefore, I really question whether one should have neotype strains that are morphologically degenerated and no longer showing the original morphology. The *P. ochrochloron* that we received from your culture collection has conidia that are up to 5 μm long, whereas the original description has them 4 um long.

CHRISTENSEN: As a community ecologist, I would ask if you have compared this with what has been called *P. simplicissimum* in the literature?

NIRENBERG: Yes. We have found *P. simplicissimum* in the same soil, and it has compact metulae, rugose stipes and the conidia are much more rounded. In the isolates of our unidentified species, conidia are not globose or subglobose but strongly apiculate. In 14 day old cultures, the conidia are really dark brown.

CHRISTENSEN: I was quite interested in your total list, because these are species that I associate with grassland soils, *P. restrictum*, *P. canescens* and others. Did you also find *Paecilomyces lilacinus* or *Fusarium* species?

NIRENBEG: Yes. Of course, we found lots of *Fusarium* species. *Paecilomyces lilacinus* and *P. marquandii* were both common, particularly the latter. The soil was once a grassland soil, but initially and later mustard were grown there. The species spectrum has changed somewhat, but the Penicillia have not.

5

TAXONOMIC SCHEMES OF ASPERGILLUS

CHEMOTAXONOMY AND MORPHOLOGY OF *ASPERGILLUS FUMIGATUS* AND RELATED TAXA

J.C. Frisvad[1] and R.A. Samson[2]

[1]*Department of Biotechnology*
Technical University of Denmark
2800 Lyngby, Denmark

[2]*Centraalbureau voor Schimmelcultures*
3740 AG Baarn, The Netherlands

SUMMARY

Many isolates of *Aspergillus fumigatus* and related taxa in *Aspergillus* sect. *Fumigati* were examined morphologically and for profiles of secondary metabolites. The isolates of secondary metabolite production by *A. fumigatus* was very homogeneous: tryptiquivalins, fumigaclavine A, verruculogen, other fumitremorgins and fumagillin were produced by all isolates tested, Fumigatin and sulochrin like metabolites were produced, less frequently, while gliotoxin and helvolic acid were not detected. *A. brevipes*, *A. viridinutans*, *A. duricaulis* and A. *unilateralis*, produced unique profiles of secondary metabolites different from those above. *A. brevipes* and *A. viridinutans* produced viriditoxin, while *A. duricaulis* produced cyclopaldic acid and chromanols 1, 2 and 3. It is concluded that species of *Aspergillus* sect. *Fumigati* are well defined, both morphologically and by their profiles of secondary metabolites.

INTRODUCTION

Aspergillus fumigatus Fres. is a widespread thermophilic species. Its importance as a pathogenic fungus has been stressed repeatedly (Schonheyder, 1987). It is also allergenic (Cole and Samson, 1984; Wilken-Jensen and Gravesen, 1984) and produces several mycotoxins (Cole and Cox, 1981; Turner and Aldridge, 1983) (Table 1).

A. *fumigatus*, *A. brevipes* G. Smith, *A. duricaulis* Raper & Fennell, *A. unilateralis* Thrower and *A. viridinutans* Ducker & Thrower are classified in *Aspergillus* sect. *Fumigati* (Gams *et al.*, 1985) (= the *Aspergillus fumigatus* group; Raper and Fennell, 1965) together with anamorphs of *Neosartorya* species. *A. fumigatus* and *N. fischeri* (Wehmer) Malloch & Cain) var. *fischeri* produces fumitremorgin A, B and C and verruculogen (Nielsen *et al.*, 1988), while *A. duricaulis* and *N. quadricincta*, have been reported to produce cyclopolic acid or similar metabolites (Turner & Aldridge, 1983). These data suggest that *Neosartorya* and *Aspergillus* sect. *Fumigati* are related both morphologically and biochemically.

The identity of several varieties of *A. fumigatus* (Samson, 1979) is still uncertain and their mycotoxin production has been little investigated. In addition, little data exists on secondary metabolite production A. *fumigatus* isolates from different habitats. Clinical isolates often differ quite markedly from food-borne or soil isolates. This paper compares morphology and secondary metabolite production by isolates of *A. fumigatus* from different sources, and by other taxa in sect. *Fumigati*.

Modern Concepts in Penicillium and Aspergillus Classification
Edited by R. A. Samson and J. I. Pitt
Plenum Press, New York, 1990

Table 1. Secondary metabolites reported from *Aspergillus fumigatus* and related species

A. fumigatus
Gliotoxin (Waksman and Geiger, 1944)
Helvolic acid (Chain *et al.*, 1943)
Agroclavine, chanoclavine, elymoclavine and festuclavine (Spilsbury and Wilkinson, 1961)
Fumagillin (Tarbell *et al.*, 1960)
Fumitoxin A, B, C and D (Debeaupuis and Lafont, 1978)
Fumigaclavine A, B and C (Cole *et al.*, 1977)
Fumigatin and spinulosin (Anslow and Raistrick, 1938a, b)
Small phenols (Turner, 1971; Turner and Aldridge, 1983)
Fumitremorgin A and B (Yamazaki *et al.*, 1971), verruculogen and TR-2 (Cole *et al.*, 1977)
Tryptiquivaline C-H, I, and J, L, M (Yamazaki *et al.*, 1976, 1977, 1978, 1979)
2-chloro-1,3,8-trihydroxy-6-methylanthrone, AO-1 and 2-chloro-1,3,8-trihydroxy-6-
hydroxymethylanthrone (Yamamoto *et al.*, 1968),
Monomethylsulochrin (Turner, 1965)
Trypacidin (Balan *et al.*, 1965)
2,6-dihydroxyphenylacetic acid (Turner and Aldridge, 1983)
4-carboxy-5,5'-dihydroxy-3,3'-dimethyldiphenyl ether (Yamamoto *et al.*, 1972).

A. viridi-nutans
Viriditoxin (Weisleder and Lillehoj, 1971)
4-acetyl-6,8-dihydroxy-5-methylisocoumarin (Aldridge *et al.*, 1966),
4-acetyl-6,8-dihydroxy-5-methyl-3,4-dihydroxyisocoumarin and 4-acetyl-3,8-dihydroxy-6-
methoxy-5-methyl-3,4-dihydroisocoumarin (Turner and Aldridge, 1983).

A. duricaulis
Cyclopaldic acid, 3-O-methyl-cyclopolic acid and the related Chromanol 1, 2 and 3 (Brillinger *et al.*, 1978; Achenbach et al., 1982 a & b).

MATERIAL AND METHODS

For morphological examination isolates were grown on 2% maltextract agar, oatmeal agar and Czapek agar and incubated for 7 days at 25 and 37 C. For secondary metabolite analysis, isolates were grown on Czapek yeast extract agar (CYA), Difco yeast extract sucrose agar (YES), Sigma (Y-4000) yeast extract sucrose agar (SYES), and Blakeslee malt extract agar (MEA) (Samson and Pitt, 1985) for two weeks at 25° C. Three Petri dishes, one of CYA, YES and SYES were placed in a Stomacher bag and extracted with chloroform/methanol (2:1) for 3 min on a Colworth Stomacher. Further procedures were as described by Frisvad and Thrane (1987). The resulting extracts were analyzed by high performance liquid chromatography (HPLC) with diode array detection DAD as described by Frisvad and Thrane (1987) and the production of particular secondary metabolites was confirmed by thin layer chromatography (TLC) viewed under UV light before and after 50% sulphuric acid treatment (Frisvad *et al.*, 1989).

RESULTS AND DISCUSSION

Aspergillus fumigatus is morphologically a quite variable species, as documented by the varieties described since 1965 (De Vries and Cormane, 1969; Samson, 1979). Varsney and Sarbhoy (1981) concurred with Samson (1979) in placing nearly all of these varieties in synonymy with A. fumigatus, but upgraded *A. fumigatus* var. *acolumnaris* as *A. acolumnaris* (Rai *et al.*) Varsney & Sarbhoy to species status. HPLC analysis of extracts of isolates from

Table 2. Production of mycotoxins and other secondary metabolites by members of *Aspergillus* sect. *Fumigati*.

		1	2	3	4	5	6	7	8	9	10	11	12	13	14	15	16
A. brevipes	CBS 118.53	–	–	–	–	–	–	–	–	–	–	–	–	–	–	–	–
A. duricaulis	CBS 481.65	–	+	–	–	–	–	–	–	–	–	–	–	–	–	–	–
? *A. duricaulis*	IMI 217288	–	–	+	+	–	–	–	–	–	–	–	–	–	–	–	–
A. fumigatus	CBS 113.26	–	–	–	–	–	+	+	+	+	+	+	+	–	+	–	–
	CBS 132.54	–	–	–	–	–	+	+	+	–	–	+	+	–	+	–	–
	CBS 192.65	–	–	–	–	–	+	+	+	+	+	+	+	+	+	–	–
	CBS 143.89	–	–	–	–	+	+	+	+	–	–	+	+	–	+	–	–
	CBS 144.89	–	–	–	–	+	+	+	+	–	–	+	+	–	+	–	–
	CBS 145.89	–	–	–	–	+	+	+	+	+	+	+	+	–	+	–	–
	CBS 146.89	–	–	–	–	+	+	+	+	+	–	+	+	–	+	–	–
	CBS 147.89	–	–	–	–	+	+	+	+	+	+	+	+	–	+	–	–
	CBS 148.89	–	–	–	–	+	+	+	+	+	+	+	+	–	+	–	–
	CBS 149.89	–	–	–	–	+	+	+	+	–	–	+	+	–	+	–	–
	CBS 150.89	–	–	–	–	+	+	+	+	+	+	+	+	–	+	–	–
	CBS 151.89	–	–	–	–	–	+	+	+	+	+	+	+	+	+	–	–
	CBS 152.89	–	–	–	–	+	+	+	+	+	+	+	+	+	+	–	–
	CBS 153.89	–	–	–	–	–	–	–	+	–	–	+	+	–	+	+	–
	CBS 154.89	–	–	–	–	+	+	–	+	–	–	+	–	+	–	–	–
A. fumigatus var. *acolumnaris*	CBS 457.75	–	–	–	–	+	+	+	+	–	–	+	+	–	+	–	–
A. fumigatus var. *phialiseptus*	CBS 542.75	–	–	–	–	+	+	+	+	–	–	+	+	–	+	–	–
A. fumigatus var. *ellipticus*	CBS 407.65	–	–	–	–	+	+	–	+	–	–	+	+	–	+	–	–
A. fumigatus var. *sclerotiorum*	CBS 458.75	–	–	–	–	+	+	–	–	–	–	–	–	+	–	–	–
A. unilateralis	CBS 283.66	–	–	–	–	–	–	–	–	–	–	–	–	–	–	–	–
	CBS 126.56	–	–	–	–	–	–	–	–	–	–	–	–	–	–	–	+
A. viridinutans	CBS 127.56	+	–	–	–	–	–	–	–	–	–	–	–	–	–	–	–
(? mixed)	IMI 240496	–	–	–	–	–	–	–	–	–	–	–	–	+	–	–	–

1. viriditoxin
2. cyclopaldic acid, 3–O–methyl– cyclopolic acid, chromanol 1, 2 & 3
3. rugulovasine
4. canesin–like metabolite
5. fumitoxins
6. fumagillin
7. verrucologen, fumitremorgin C
8. tryptiquivalins
9. trypacidin
10. monomethylsulochrin
11. FUA
12. FUB
13. fumigatin
14. fumigaclavine A
15. terrein
16. mycophenolic acid

different habitats showed that *A. fumigatus* is a very homogeneous species with respect to profiles of secondary metabolites. Nearly all isolates produced fumigaclavine A, fumagillin, fumitoxins, trypacidin, monomethylsulochrin, several tryptoquivalins, verruculogen, fumitremorgin C and metabolites of two chromophore families, named here FUA and FUB. *A. acolumnaris* produced the same secondary metabolites as other isolates of *A. fumigatus*. Synonymy with *A. fumigatus*, as concluded by Samson (1979) is confirmed. The only chemically different isolate of those identified as *A. fumigatus*, CBS 153.89, produces many metabolites in common with *Neosartorya fischeri* var. *fischeri*.

A. fumigatus var. *sclerotiorum* is also different. It produces several metabolites in common with *A. viridi–nutans* IMI 240496, which may be a mixed culture.

Table 2 lists metabolites produced by the isolates examined. Metabolites from clinical, soil and food isolates of *A. fumigatus* are very similar, indicating that at least biochemically this is a homogeneous species.

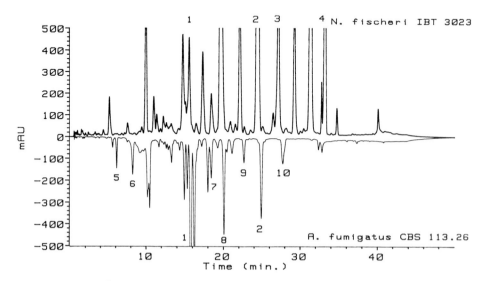

Fig. 1. Comparison of HPLC traces of *Neosartorya fischeri* var. *fischeri* and *Aspergillus fumigatus*. Note that retention times are about 0.75 min delayed for *N. fischeri*. The alkylphenone indices were the same. Metabolite 1 is fumitremorgin C, 2 is verruculogen, 3 is fumitremorgin B, 4 is fumitremorgin A, 5 is fumigaclavine A, 6 is FUA, 7 is trypacidin, 8 is monomethylsulochrin, 9 is FUB and 10 is fumagillin.

The other species in *Aspergillus* sect. *Fumigati*, *A. brevipes, A. duricaulis, A. unilateralis* and *A. viridinutans* produced distinct secondary metabolites, with few in common with *A. fumigatus*. Considerable variability was observed. Metabolites from the two isolates identified as *A. duricaulis* differed (Table 1). IMI 217288 produced rugulovasine A and a canescin–like metabolite. Canescin has been reported to be produced by *Neosartorya fischeri* var. *glabra* (Samson *et al.*, 1989). Growth rates and morphology also indicated that IMI 217288 was correctly classified as an anamorph of *N. fischeri* var. *glabra*. Apart from viriditoxin, as reported by Weisleder and Lillehoj (1971), *A. brevipes* and *A. viridinutans* did not have any secondary metabolites in common.

Possible connections between *Aspergillus* sect. *Fumigati* and *Neosartorya* are also of interest. Fig. 1 shows a comparison of HPLC traces of *N. fischeri* IBT 3023 and *A. fumigatus*. Most isolates of both species produce fumitremorgins, verruculogen and some tryptoquivalins, but differ in the capacity to produce several other known and unknown secondary metabolites. Tryptoquivalins, trypacidin and cyclopaldic acid, metabolites from three separate biochemical pathways, were found in species both with or without a teleomorph state (Table 3). Frisvad (1989) considered that species with more than three

Table 3. Production of mycotoxins and antibiotics in *Neosartorya* and *Aspergillus* sect. *Fumigati*.

Species	Mycotoxins and antibiotics
A. brevipes	Viriditoxin
A. duricaulis	Cyclopaldic acid, chromanol 1, 2, and 3
A. fumigatus	Fumagillin, fumitoxins, fumigaclavine A, fumigatin, fumitremorgin A, B, & C, gliotoxin, monomethylsulochrin, trypacidin, tryptoquivalins, FUA,FUB, helvolic acid
A. unilateralis	Mycophenolic acid
A. viridinutans	Viriditoxin
N. aurata	Helvolic acid
N. aureola	FUA, trypacidin, tryptoquivalins
N. fennelliae	FUA, trypacidin, viridicatumtoxin
N. fischeri	FUB, fumitremorgin A, B, C,verruculogen, tryptoquivalins
N. glabra ch. I	FUB
N. glabra ch. II	Mevinolins
N. quadricincta	FUA, cyclopaldic acid
N. spinosa	FUA, terrein, tryptoquivalins
N. stramenia	Canadensolide

secondary metabolite families in common are identical, so these results suggest that *Neosartorya* and *Aspergillus* sect. *Fumigati* are a natural group. These fungi share secondary metabolites only with species in *Aspergillus* subgen. *Clavati* (tryptoquivalins) and subgen. *Versicolores* sect. *Terrei* (terrein). The discovery that *Neosartorya fennelliae* synthesises the tetracycline–like mycotoxin viridicatumtoxin was unexpected, as it has only been found to date in *Penicillium aethiopicum* and *P. brasilianum*. However, several other mycotoxins are produced by species in both these genera (Frisvad, 1986).

Brief descriptions follow of the morphology and secondary metabolites of the species studied.

Aspergillus fumigatus Fresenius – Beitrage zur Mykologie: 81, 1863
A. fumigatus var. *ellipticus* Raper et Fennell 1965
? A. anomalus Pidoplichko et Kirilenko 1969
A. fumigatus var. *acolumnaris* Rai et al., 1971
? A. fumigatus var. *albus* Rai, Tewari et Agarwal 1974
? A. fumigatus var. *fulvoruber* Rai, Tewari et Agarwal 1974
? A. fumigatus var. *griseobrunneus* Rai et Singh 1974
A. fumigatus var. *sclerotiorum* Rai, Agarwal et Tewari 1971
A. phialiseptus Kwon–Chung 1975
See also Raper and Fennell (1965)

A. fumigatus is morphologically somewhat more variable than described by Raper and Fennell (1965). Conidia may be nearly smooth–walled occasionally, phialides may be septate (*A. phialiseptus*) and heads may be somewhat radiate (*A. fumigatus* var. *acolumnaris*). The ex–type isolate of *A. fumigatus* var. *ellipticus* seems distinct because of its smooth walled, ellipsoidal conidia, but chemically it is identical with *A. fumigatus* var. *fumigatus*. Kozakiewicz (1989) raised the var. *ellipticus* to species level as *A. neoellipticus* on the basis of differences of conidial ornamentation. This taxonomy is not followed, because the great variation seen in numerous *A. fumigatus* isolates. Kozakiewicz (1989) based her conclusions only on one isolate. All recently described varieties of *A. fumigatus* produce similar secondary metabolites to the species and should be synonymised. Clinical isolates were often more floccose, with less conidia, but could not be separated from the ex type isolate or other isolates from soil or feedstuffs, by techniques used here.

Aspergillus brevipes G. Smith – Trans. Brit. Mycol. Soc. 35: 241, 1952

Short and heavy walled stipes and finely spinulose conidia make *A. brevipes* morphologically distinct. It is also characterized by its vesicles borne at an angle to the stipe, as in *A. viridinutans* and *A. duricaulis*, a reddish brown reverse on CYA, coloured vesicles and phialides and dark blue green conidia. It produced three families of secondary metabolites (as determined by HPLC DAD). Weisleder and Lillehoj (1971) and Cole and Cox (1981) reported that *A. brevipes* produces viriditoxin. The observations were based on NRRL 576 (= NRRL 4365) which is *A. viridinutans* and *A. brevipes* (S. Peterson, pers. comm.)
Isolates examined: CBS 118.53 (ex type) from soil, Australia, G. Smith

Aspergillus duricaulis Raper & Fennell – Gen. *Aspergillus*: 249, 1965

A. duricaulis differs from *A. brevipes* by conspicously echinulate conidia and weakly coloured reverse on CYA. No secondary metabolites are synthesised in common with other species in *Aspergillus* sect. *Fumigati*, except cyclopaldic acid, also produced by *Neosartorya quadricincta*. Cyclopaldic acid, chromanol 1, 2, and 3 and other unknown secondary metabolites, are produced. This taxon differs in many respects from *A. brevipes* and therefore we do not follow Kozakiewicz (1989) who synonymize these species.
Isolates examined: CBS 481.65 (ex type) from soil, Buenos Aires , Argentina, J. Winitzky.

Aspergillus unilateralis Thrower – Aust. J. Bot. 2: 356, 1954

A. unilateralis is distinguished by phialides clustered on one side of the vesicle, echinulate conidia, a slow growth rate and a dark reverse on CYA. The two strains investigated produced several unique secondary metabolites. One of them produced small amounts of mycophenolic acid.
Isolates examined: CBS 126.56 from soil, Australia and CBS 283.66 from a similar source.

Aspergillus viridinutans Ducker & Thrower – Aust. J. Bot. 2: 357, 1954

A. viridinutans also produces vesicles borne at an angle to the stipe. The species is distinguished by thin–walled, sinuous stipes. Vesicles are uncoloured, conidia are pale blue green and only finely roughened, and conidium production is quite weak compared to the other species in the section. This species produces viriditoxin.
Isolates examined: CBS 127.56 from rabbit dung, Frankston, Australia and IMI 280490 from Zambia, J.N. Zulu, perhaps mixed with *A. fumigatus*.

ACKNOWLEDGEMENTS

The authors thank the NATO Scientific Affairs Division (Brussel, Belgium) for the research grant (0216/86) for international collaboration.

REFERENCES

ACHENBACH, H., MÜHLENFELD, A., WEBER, B., KOHL, W. and BRILLINGER, G.U. 1982a. Stoffwechselprodukte von Mikroorganismen, XXVII. Cyclopaldsaüre und 3–O–Methyl–cyclopolsaüre, zwei antibiotisch wirksame Substanzen aus *Aspergillus duricaulis*. *Zeitschrift für Naturforschung* 37b: 1091–1097.

ACHENBACH, H., MÜHLENFELD, A., WEBER, B. and BRILLINGER, G.U. 1982b. Stoffwechselprodukte von Mikroorganismen. XXVIII. Highly substituted chromanols from *Aspergillus duricaulis*. *Tetrahedron Letters* 23: 4659–4660.

ALDRIDGE, D.C., GROVE, J.F. and TURNER, W.B. 1966. 4–acetyl–6,8–dihydroxy–5–methyl–2–benzopyran–1 –one, a metabolite of *Aspergillus viridi–nutans*. *Journal of the Chemical Society*, C 1966: 126–129.

ANSLOW, W.K. and RAISTRICK, H. 1938a. The biochemistry of microorganisms. LVII. Fumigatin (3–hydroxy–4–methoxy–2,5–toluquinone) and spinulosin, metabolic products respectively of *Aspergillus fumigatus* Fresenius and *Penicillium spinulosum* Thom. *Biochemical Journal* 32: 687–696.

ANSLOW, W.K. and RAISTRICK, H. 1938b. The biochemistry of microorganisms. LIX. Spinulosin (3,6–dihydroxy–4–methoxy–2,5–toluquinone) a metabolic product of a strain of *Aspergillus fumigatus* Fresenius. *Biochemical Journal* 32: 2288–2289.

BALAN, J., KJAER, A., KOVACS, S. and SHAPIRO, R.H. 1965. The structure of trypacidin. *Acta Chemica Scandinavica* 19: 528–530.

BRILLINGER, G.–U., HEBERLE, W., WEBER, B. and ACHENBACH, H. 1978. Metabolic products of microorganisms. 167. Cyclopaldic acid from *Aspergillus duricaulis*. 1. Production, isolation and biological properties. *Archives of Microbiology* 116: 245–252.

CHAIN, E., FLOREY, H.W. JENNINGS, M.A. and WILLIAMS, T.I. 1943. Helvolic acid, an antibiotic produced by *Aspergillus fumigatus* mut. helvola Yuill. *British Journal of Experimental Pathology* 24: 108–118.

COLE, R.J. and COX, R.H. 1981. Handbook of Toxic Fungal Metabolites. New York: Academis Press.

COLE, R.J., KIRKSEY, J.W., DORNER, J.W., WILSON, D.M., JOHNSON, J.C., JOHNSON, A.N., BEDELL, D.M., J.P. SPRINGER, K.K. CHEXAL, J.C. CLARDY and R.J. COX. 1977. Mycotoxins produced by *Aspergillus fumigatus* species isolated from molded silage. *Journal of Agricultural and Food Chemistry* 25: 826–830.

COLE, G.R. and SAMSON, R.A. 1984. The conidia. *In* Mould allergy, eds. Y.Al–Doory and J.F. Domson, pp. 66–103. Philadelphia: Lea and Febiger

DEBEAUPUIS, J.P. and LAFONT, P. 1978. Toxinogenese in vitro d'*Aspergillus* du groupe *fumigatus*. Production de fumitoxins. *Mycopathologia* 66: 10–16.

DE VRIES, G.A. and CORMANE, R.H. 1969. A study on the possible relationships between certain mophological and physiological properties of *Aspergillus fumigatus* Fres. and its presence in or on human and animal (pulmonary) tissues. *Mycologia Mycopathologiga applicata* 39: 241–253.

FRISVAD, J.C. 1986. Taxonomic approaches to mycotoxin identification. *In* Modern Methods in the Analysis and Structural Elucidation of Mycotoxins, ed. R.J. Cole, pp. 415–457. New York: Academic Press.

—— 1989. The use of high–performance liquid chromatography and diode array detection in fungal chemotaxonomy based on profiles of secondary metabolites. *Botanical Journal of the Linnean Society* 99: 81–95.

FRISVAD, J.C. and THRANE, U. 1987. Standardized high–performance liquid chromatography of 182 mycotoxins and other fungal metabolites based on alkylphenone retention indices and UV–VIS spectra (diode array detection). *Journal of Chromatography* 404: 195–214.

FRISVAD, J.C., FILTENBORG, O. and THRANE, U. 1989. Analysis and screening for mycotoxins and other secondary metabolites in fungal cultures by thin–layer chromatography and high performance liquid chromatography. *Archives of Environmental Contamination and Toxicology* 18: 331–335.

GAMS, W., CHRISTENSEN, M., ONIONS, A.H.S., PITT, J.I. and SAMSON, R.A. 1985. Infrageric taxa of *Aspergillus*. *In* Advances in *Penicillium* and *Aspergillus* Systematics, eds. R.A. Samson and J.I. Pitt pp. 55–62. New York and London: Plenum Press.

KOZAKIEWICZ, Z. 1989. *Aspergillus* species on stored products. *Mycological Papers* 161: 1–188.

NIELSEN, P.V., BEUCHAT, L.R. and FRISVAD J.C. 1988. Growth of and fumitremorgin production by *Neosartorya fischeri* as affected by temperature, light and water activity. *Applied and Environmental Microbiology* 54: 1504–1510.

RAPER, K.B. and D.I. FENNELL. 1965. The Genus Aspergillus. Baltimore: Williams and Wilkins.

SAMSON, R.A. 1979. A compilation of the Aspergilli described since 1965. *Studies in Mycology, Baarn* 18: 1–38.

SAMSON, R.A. and PITT, J.I. eds. 1985 Advances in *Penicillium* and *Aspergillus* Systematics. New York and London: Plenum Press.

SAMSON, R.A., NIELSEN, P. and FRISVAD, J.C. 1989. The genus *Neosartorya*: differentiation on the basis of ascospore morphology and secondary metabolites. *In* Modern Concepts in *Penicillium* and *Aspergillus* Classification, eds. R.A. Samson and J.I. Pitt, pp. 455-467. New York and London: Plenum Press.

SCHONHEYDER, H. 1987. Pathogenic and serological aspects of pulmonary aspergillosis. *Scandinavian Journal of Infectious Diseases Supplementum* 51: 1–62.

SPILSBURY, J.F. and WILKINSON, S. 1961. Isolation of festuclavine and two new clavine alkaloids from *Aspergillus fumigatus*. *Journal of the Chemical Society* 1961: 2085–2091.

TARBELL, D.S., CARMAN, R.M., CHAPMAN, D.D., HUFFMAN, K.R. and MCCORKINDALE, N.J. 1960. The structure of fumagillin. *Journal of the American Chemical Society* 82: 1005–1007.

TURNER, W.B. 1965. The production of trypacidin and monomethylsulochrin by *Aspergillus fumigatus*. *Journal of the Chemical Society* 1965: 6658–6659.

——— 1971. Fungal Metabolites. London: Academic Press.

TURNER, W.B. and ALDRIDGE, D.C. 1983. Fungal metabolites. II. London: Academic Press.

VARSNEY, J.L. and SARBHOY, A.K. 1981. A new species of *Aspergillus fumigatus* group and comments upon its synonomy. *Mycopathologia* 73: 89–92.

WAKSMAN, S.A. and GEIGER, W.B. 1944. The nature of the antibiotic substances produced by *Aspergillus fumigatus*. *Journal of Bacteriology* 47: 391–395.

WEISLEDER, D. and LILLEHOJ, E.B. 1971. Structure of viriditoxin, a toxic metabolite of *Aspergillus viridi-nutans*. *Tetrahedron Letters* 48: 4705–4706.

WILKEN–JENSEN, K. and GRAVESEN, S. 1984. Atlas of moulds in Europe causing respiratory allergy. ASK Publishing, Copenhagen.

YAMAMOTO, Y., KINYAMA, N. and ARAHATA, S. 1968. Studies on the metabolic products of *Aspergillus fumigatus* (J–4). Chemical structure of metabolic products. *Chemical and Pharmaceutical Bulletin* 16: 304–310.

YAMAMOTO, Y., NITTAI, K., OOHATA, Y. and FURUKAWA, T. 1972. Studies of the metabolic products of *Aspergillus fumigatus* DH 413. V. A new metabolite produced by ethionine inhibition. *Chemical and Pharmaceutical Bulletin* 20: 931–935.

YAMAZAKI, M., SUZUKI, S. and MIYAKI, K. 1971. Tremorgenic mycotoxins from *Aspergillus fumigatus*. *Chemical and Phamaceutical Bulletin* 19: 1739–1740.

YAMAZAKI, M., FUJIMOTO, H. and OKUYAMA, E. 1976. Structure determination of six tryptoquivaline related metabolites from *Aspergillus fumigatus*. *Tetrahedron Letters* 33: 2861–2864.

——— 1977. Structure of tryptoquivaline C and D. Novel fungal metabolites from *Aspergillus fumigatus*. *Chemical and Pharmaceutical Bulletin* 25: 2554–2560.

——— 1978. Structure determination of six fungal metabolites, tryptoquivaline E, F, G, H, I, J from *Aspergillus fumigatus*. *Chemical and Pharmaceutical Bulletin* 26: 111–117.

YAMAZAKI, M., OKUYAMA, E. and MAEBAYASHI, Y. 1979. Isolation of some new tryptoquivaline–related metabolites from *Aspergillus fumigatus*. *Chemical and Pharmaceutical Bulletin* 27: 1611–1617.

EXOCELLULAR POLYSACCHARIDES FROM *ASPERGILLUS FUMIGATUS* AND RELATED TAXA

J.P. Debeaupuis[1], J. Sarfati[1], A. Goris[2], D. Stynen[2], M. Diaquin[1] and J.P. Latgé[1]

[1]*Unite de Mycologie*
Institut Pasteur
75724 Paris Cedex 15, France

[2]*Eco-Bio/Diagnostics Pasteur*
3600 Genk, Belgium

SUMMARY

Exocellular antigenic (EP) material excreted into the culture medium and composed of proteins and polysaccharides was studied in order to characterize isolates of various species of the *Aspergillus fumigatus* group (*Aspergillus* sect. *Fumigati*).
The EP produced by forty five strains of the *A. fumigatus* group (including teleomorphs of the genus *Neosartorya*) were analysed. The polysaccharide moiety of the extracellular fraction varies from 10 to 65% of the total lyophilized product. Galactose, mannose, and glucose are the main constituents of the EP of all strains studied. Their proportions vary considerably from one strain to another (galactose 4 to 58%, mannose 11 to 61%, glucose 4 to 68%). In spite of these variations, pathogenic and non-pathogenic strains *A. fumigatus sensu stricto* or teleomorphs and anamorphs cannot be separated on the basis of monosaccharidic composition. All isolates tested reacted positively with one monoclonal antibody directed against *A. fumigatus* galactomannan indicating the presence of galactofuran in all EP studied. The ELISA results with the rabbit antiserum indicated that *A. brevipes*, *A. duricaulis*, *A. unilateralis* are different from *A. fumigatus* and suggested the presence of specific sugar linkages in the EP of these species.

INTRODUCTION

Polysaccharides produced by zoopathogenic fungi possess antigenic properties and are important in the immune response of the host during fungal infection (Huppert, 1983). The galactomannans represent one of the most widely distributed classes of polysaccharides among fungi and have been described as integral components of the mycelial cell wall (Johnston, 1965) as well as exocellular extracts (Gorin and Spencer, 1968). The majority of studies on the structure and antigenic properties of galactomannans have involved polysaccharides which have been either chemically extracted from cell wall preparations of mycelium or spores (Barreto-Bergter and Travassos, 1980; Barreto-Bergter et al., 1981; Webster and McGinley, 1980), or from cytoplasmic material after total disruption of the mycelium (Bennett et al., 1985). Galactomannans, with molecular weights greater than 10,000 daltons, can be also released during growth (Preston et al., 1969; Gander et al., 1974; Notermans et al., 1987).

During growth, several *Aspergillus* species have been shown to produce exocellular polysaccharides containing galactomannan (Azuma et al., 1971; Bardalaye and Nordin, 1977; Reiss and Lehmann, 1979; Barreto-Bergter and Travassos 1980; Gorin and Spencer, 1968). Recently, we have isolated an exocellular antigen which is actively released during

Modern Concepts in Penicillium and Aspergillus Classification
Edited by R. A. Samson and J. I. Pitt
Plenum Press, New York, 1990

vegetative growth. The polysaccharide portion of this antigen is a galactomannan linked to protein and polyhexosamines (Latgé *et al.*, 1988).

Suzuki and Takeda (1975) have demonstrated an immunological cross-reactivity of galactomannans extracted from the cell wall of *Aspergillus fumigatus, A. niger, Trichophyton rubrum* and *Hormodendrum pedrosoi* against *H. pedrosoi* serum. It was postulated that galactosyl groups on these antigens may be immunodominant as removal of these residues, which are acid-labile, decreased the reactivity with rabbit antiserum. More recently, Bennett *et al.* (1985) confirmed that galactofuranosyl groups of a galactomannan isolated from *A. fumigatus* mycelium were immunodominant.

In contrast to the results of Suzuki and Takeda (1975), Notermans and Heuvelman (1985) and Notermans and Soentoro (1986) reported that the exocellular polysaccharides produced by various mould species may be genus specific. Cross reactivity occurred only between the closely related genera *Penicillium* and *Aspergillus*.

In the following study, we evaluated the variability of the chemical and antigenic composition of the exocellular material released by different isolates of *A. fumigatus* and taxonomically related species with emphasis on the polysaccharide components of the exoantigen.

MATERIAL and METHODS

Fungal isolates and culture methods.

Forty five isolates of *A. fumigatus* and related taxa from *Aspergillus* sect. *Fumigati* and *Neosartorya* were used for this study (Table I). Isolates were maintained on malt extract agar slants. Conidia were harvested and inoculated into 150 ml flasks containing 50 ml of sterilized, liquid glucose (2%) peptone (1%) medium. Cultures were incubated at 25°C on a rotary shaker at 120 rpm. The duration of cultivation, due to different growth rates, varied among the isolates (Table I).

Purification of the exoantigen.

Culture filtrates (two flasks per isolate) were precipitated with 4 volumes of ethanol overnight at 2-4°C. After centrifugation (10 min at 3000 g at 4°C) precipitates were washed (2x) with ethanol and freeze dried.

Chemical analysis.

Hexoses were analysed by the phenolsulfuric acid method (Dubois *et al.*, 1956) and proteins by using the Bio-Rad technique based on the method of Bradford (1976). Hexosamines were evaluated after acid hydrolysis (8 N HCl, 2 hours at 100°C), using the reaction of Elson-Morgan with p-dimethylaminobenzaldehyde (Ashwell, 1966).

Monosaccharide analysis.

Ethanol precipitates (EP) were hydrolysed and methylated in 0.5 M methylchloride overnight at 80°C. Resulting methylglycoside residues were trifluoroacetylated with a mixture of trifluoroacetic anhydride and dichloromethane (1:1) twice, for 5 min at 150°C in a sand bath. Chromatographic analysis was performed on a DI200 (Delsi Instruments, Suresnes, France) gas liquid chromatograph (GLC), at an oven temperature gradient between 110°C and 240°C (2°C per min), using a column (3 m x 3 mm) of OV 210 on Chromosorb WHP (Zanetta *et al.* 1972). Meso-inositol was used as an internal standard.

Table 1. Strains of *Aspergillus fumigatus* and related taxa used in this study.

Nomenclatures CBS *	Others	Species	Duration of culture (h)	Nx for figures
118.53		*Aspergillus brevipes*	164	AB
481.65		*A. duricaulis*	164	AD1
	2172.88	*A. duricaulis*	114	AD2
148.89	B617B	*A. fumigatus*	89	AF1
153.89	I FAS.19	*A. fumigatus*	89	AF2
154.89	I FAS.20	*A. fumigatus*	120	AF3
2458.56		*A. fumigatus*	114	AF4
113.26		*A. fumigatus*	89	AF5
143.89	1028	~ *A. fumigatus*	89	AF6
152.89	152. E1	*A. fumigatus*	89	AF7
146.89	2109	~ *A. fumigatus*	66	AF8
132.54		*A. fumigatus*	89	AF9
192.65		*A. fumigatu*	89	AF10
542.75		*A. fumigatu*	140	AF11
145.89	2140	~ *A. fumigatu*	89	AF12
147.89	DUV.IP	~ *A. fumigatus*	66	AF13
150.89	P8	*A. fumigatus*	89	AF14
151.89	152.W3	*A. fumigatus*	96	AF15
149.89	B614A	*A. fumigatus*	89	AF16
144.89	DAL.IP	~ *A. fumigatus*	66	AF17
457.75		*A. fumigatus* var. *acolumnaris*	164	AFA
487.65		*A. fumigatus* var. *ellipticus*	140	AFE
458.75		*A. fumigatus* var. *sclerotiorum*	114	AFS
283.65		*A. unilateralis*	114	AU1
126.56		*A. unilateralis*	140	AU2
	2404.90	*A. viridi-nutans*	114	AV1
127.56		*A. viridi-nutans*	94	AV2
466.65		*Neosartoria aurata*	140	NAU
105.55		*N. aureola*	140	NA1
106.55		*N. aureola*	94	NA2
598.74		*N. fennelliae*	137	NFE1
599.74		*N. fennelliae*	137	NFE2
	NHL 2951	*N. fennelliae*	137	NFE3
	NHL 2952	*N. fennelliae*	137	NFE4
544.65		*N. fischeri*	92	NFI1
404.67		*N. fischeri*	92	NFI2
111.55		*N. fischeri* var. *glabra*	164	NFG1
112.55		*N. fischeri* var. *glabra*	140	NFG2
483.65		*N. fischeri* var. *spinosa*	164	NFS1
297.67		*N. fischeri* var. *spinosa*	168	NFS2
135.52		*N. quadricincta*	160	NQ1
941.73		*N. quadricincta*	164	NQ2
	NHL 2948	*N. spathulata*	137	NS1
	NHL 2949	*N. spathulata*	137	NS2
498.65		*N. stramenia*	236	NST

* CBS: Centraalbureau voor schimmelcultures, Baarn, The Netherlands
~ Strains isolated from human pathogenic lesions

Binding of lectins.

EP were diluted (1 µg ml of total hexose) in 0.05 M carbonate buffer pH 9.6 and coated on wells of microtitration plates (Greiner, Bischwiller, France). Plates were incubated for 1 hour at 37x then overnight at 4°C. Wells were washed 5 times with phosphate buffer saline containing 0.1% Tween 20 (PBST). The washed wells coated with exoantigen were treated with peroxidase labelled lectines, diluted in PBST containing 1% bovine serum albumin. Concanavaline A (ConA-perox type VISigma, St Louis, MO, USA) was diluted at 1.10^{-3} and Wheat Germ Agglutinin (WGA-perox, Sigma) was diluted at 1.10^{-2}. After incubation 1 hour at 37°C, the plates were washed as before. Peroxidase was revealed by H_2O_2 and orthophenylene diamine hydrochloride (OPD) in 0.05 M citrate buffer at pH 5.2. The reaction was stopped by adding 50µl of 3.6 N sulfuric acid.

Monoclonal antibodies.

The monoclonal antibodies were produced according to the procedure described by Stynen *et al.* (1989). Basically Lou/c rats were immunized with the supernatant of homogenized *A. fumigatus* mycelium. Spleen cells of the immunized animal were fused with IR 983 F rat plasmacytoma cells as described by Bazin (1987). Interesting hybridoma cultures were cloned three times by limiting dilution. Antibody EB-A2, an IgM, was selected because of its specificity for galactomannan. It was produced *in vivo* in AMORAT rats and purified on an affinity column with an allotype specific murine monoclonal anti-rat Ig (Bazin, 1987). EB-Y8, a rat monoclonal IgM, specific for lipopolysaccharides of *Yersinia enterocolitica* O:8 (Stynen *et al.*, 1989) was used as a control antibody.

Rabbit antiserum.

Antiserum was raised in New Zealand White rabbits using EP from *A. fumigatus* isolate CBS 143.89 injected subcutaneously. In the two first injections the EP was mixed with Freund's complete adjuvant and following injections with Freund's incomplete adjuvant. The working dilutions of the antiserum in ELISA were 1.10^{-3} to 5.10^{-4}.

Enzyme-linked Immunosorbent Assay (ELISA).

Precipitates (EP) were diluted and coated on microtiter plates, incubated and washed as described previously for lectin binding studies. Samples (sera from immunized or control rabbits, or rat monoclonal antibodies) diluted in PBS-Tween-BSA were applied to washed microtiter wells. After incubation 1 hour at 37°C and washing 5 times with PBST, the peroxidase conjugates (Ab anti IgG Rat-peroxidase, Sigma or anti IgG H+L-peroxidase, Bio-Sys) were added, followed by incubation and staining by OPD as described above.

Inhibition.

Inhibition studies were carried out using the ELISA method. Inhibitors were pools of EP from either *A. fumigatus* isolates or *N. fischeri*, *N. fischeri* var. *glabrata* or *N. fischeri* var. *spinosa* diluted in PBS (20 µg/ml). Inhibitors were incubated overnight at 37°C with rabbit antisera (immunized and control) (1:1) and added to washed EP coated wells prepared as described previously. After incubation for 1 hour at 37°C and washing 5 times with PBST, the peroxidase conjuguate (Ab anti IgG H+L-peroxidase, Bio-Sys) was added, followed by incubation and staining by OPD as described above.

Table 2. Composition of exoantigen from isolates of *Aspergillus fumigatus* and related taxa

			Percentages of	
Nomenclatures	Genus and Species	Hexosamines	Proteins	Hexoses
CBS 118.53	Aspergillus brevipes	4	25	71
CBS 481.65	A. duricaulis	5	16	79
2172.88	A. duricaulis	6	25	69
CBS 148.89	A. fumigatus	10	31	59
CBS 153.89	A. fumigatus	7	30	63
CBS 154.89	A. fumigatus	20	34	45
2458.56	A. fumigatus	37	54	9
CBS 113.26	A. fumigatus	10	61	29
CBS 143.89	A. fumigatus	10	45	45
CBS 152.89	A. fumigatus	23	56	22
CBS 146.89	A. fumigatus	14	51	36
CBS 132.54	A. fumigatus	36	42	23
CBS 192.65	A. fumigatus	20	57	23
CBS 542.75	A. fumigatus	25	40	35
CBS 145.89	A. fumigatus	35	48	17
CBS 147.89	A. fumigatus	33	51	15
CBS 150.89	A. fumigatus	30	521	8
CBS 151.89	A. fumigatus	31	52	17
CBS 149.89	A. fumigatus	24	54	22
CBS 144.89	A. fumigatus	23	51	25
CBS 457.75	A. fumigatus var. acolumnaris	6	38	56
CBS 487.65	A. fumigatus var. ellipticus	14	24	62
CBS 458.75	A. fumigatus var. sclerotiorum	11	47	42
CBS 283.65	A. unilateralis	9	40	51
CBS 126.56	A. unilateralis	7	20	73
2404.90	A. viridi-nutans	28	57	15
CBS 127.56	A. viridi-nutans	4	40	56
CBS 466.65	Neosartorya aurata	7	23	69
CBS 105.55	N. aureola	33	43	24
CBS 106.55	N. aureola	40	37	23
CBS 598.74	N. fennelliae	3	38	60
CBS 599.74	N. fennelliae1	7	31	52
NHL 2951	N. fennelliae1	5	34	51
NHL 2952	N. fennelliae1	43	5	51
CBS 544.65	N. fischeri	18	30	52
CBS 404.67	N. fischeri	9	29	62
CBS 111.55	N. fischeri var. glabra	8	37	55
CBS 112.55	N. fischeri var. glabra	17	33	50
CBS 483.65	N. fischeri var. spinosa	29	42	29
CBS 297.67	N. fischeri var. spinosa	15	54	31
CBS 135.52	N. quadricincta	4	25	72
CBS 941.73	N. quadricincta	25	34	41
NHL 2948	N. spathulata	0	0	n
NHL 2949	N. spathulata	0	0	n
CBS 498.65	N. stramenia	4	21	75

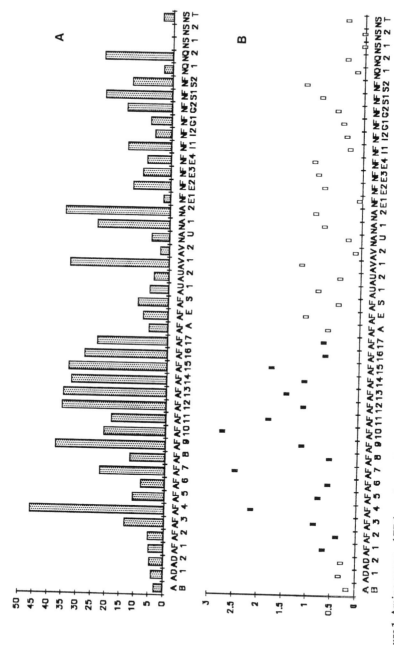

Figure 1. Amino sugars of EP from *Aspergillus fumigatus* and related taxa. A. Total hexosamines according to the Elson-Morgan reaction (μg/mg EP; B. Relation between hexosamine levels (μg) and WGA-peroxidase binding (o.d.) for 1mg EP. Filled box: *A. fumigatus* isolate

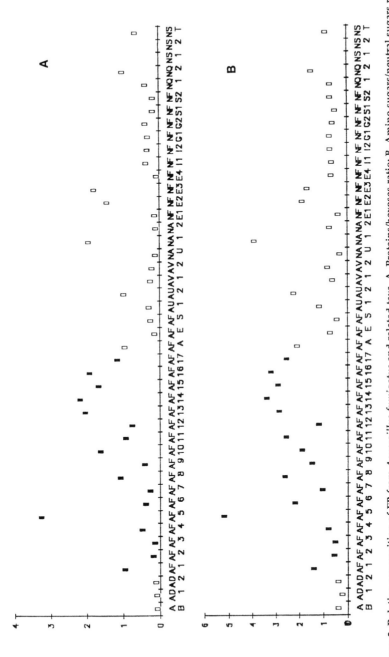

Figure 2. Relative composition of EP from *Aspergillus fumigatus* and related taxa. A. Proteins/hexoses ratio; B. Amino sugars/neutral sugars ratio. Filled box: *A. fumigatus* isolates.

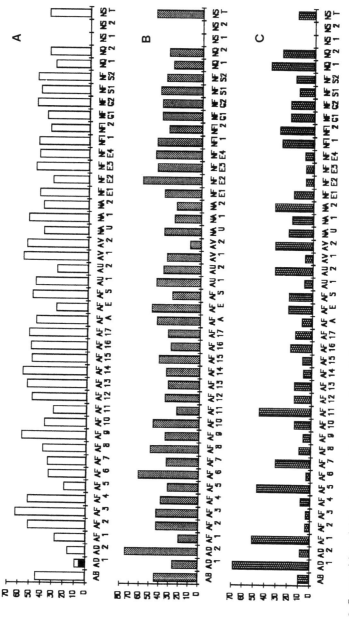

Figure 3. Composition of the polysaccharide moiety of the EP from *A. fumigatus* and related taxa. Percentages of galactose (A), fucose (A, in AD1), mannose (B) and glucose (C).

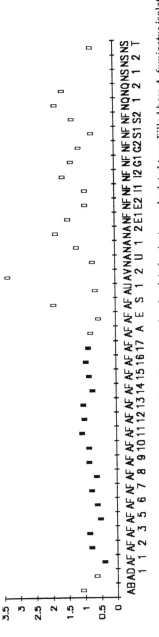

Figure 4. Relation between ConA-peroxidase affinity to EP and mannose levels of *A. fumigatus* and related taxa. Filled box: *A. fumigatus* isolates.

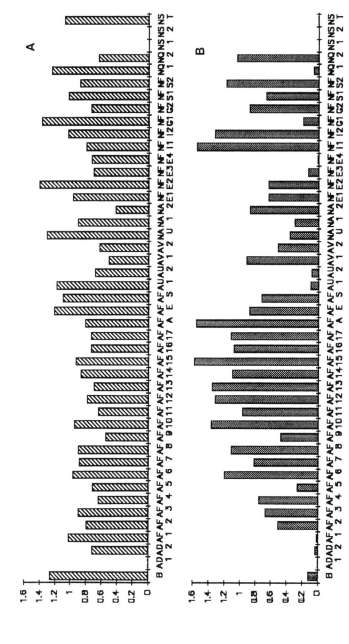

Figure 5. Binding of monoclonal Rat antibody specific of galactofuran (A) and polyclonal Rabbit antiserum specific of *A. fumigatus* (B) to EP from different isolates of *A. fumigatus* and related taxa. A. Monoclonal Rat antibody EB-A2; B. Polyclonal Rabbit anti EP (CBS 143.89) antiserum. Results are expressed as the difference between optical densities obtained with the positive and the control antisera.

RESULTS

Production of exoantigens.
All the isolates studied produced exoantigens when grown in the peptone glucose media, except *Neosartorya spathulata* (2 isolates tested). Relative amounts of material released during growth varied among isolates. Most of the *A. fumigatus* isolates produced the highest levels of exocellular material after 2 to 3 days growth (Table 1). The majority of species had to be grown for 5 to 7 days to obtain a precipitate, however, the growth of *N. stramenia* was very slow and ten days growth was required to produce a precipitate sufficient for chemical analysis.

General composition of culture filtrates.
The main constituents of all EP from broth filtrates were amino acids, hexosamines and neutral sugars (Table 2). *A. fumigatus* isolates produced significatively higher amounts of proteins and hexosamines than most other species (Table 2). The exception was *N. aureola*, which produced exocellular material containing 30-40% hexosamines (Fig. 1A). Hexosamine production was more irregular in *A. fumigatus* than protein production. In Fig. 1B the comparison between total hexosamine produced by the various moulds according to both the Elson-Morgan reaction and the binding of WGA, specific to 2-acetamide, 2-deoxy oligosaccharides to exocellular filtrates are summarised. Results suggest that the hexosamine moiety in the filtrates is a heteropolymer of hexosamines: N-acetyl-hexosamines. The level of WGA binding compunds are lower in most *A.fumigatus* than in the other taxa. Additionally, *A. fumigatus* isolates produced the largest quantities of hexosamines, but levels varied from 5 and 50% of the total dry weight of the precipitate. *A. fumigatus* produced higher levels of proteins but a lower percentage of neutral sugars (15 to 25% of the total precipitate) than other fungal species which produced EP containing 35 to 75% hexoses (Fig. 2).

Sugar composition of the EP.
Figure 3 shows the results of the GLC analysis of monosaccharide composition of the exopolysaccharide moiety in the EP. All isolates studied produced galactose, mannose and glucose, and one isolate, *A. duricaulis* (CBS 481.65), produced fucose. Galactomannan was the major component of the exopolysaccharide, especially for *A. fumigatus*. Levels of galactose were significant in all species studied, with the exception of *A. duricaulis*, one isolate of which produced 5% fucose while producing only 4% galactose. Fucose was not found in other species tested isolates. Hydrolysis of exopolysaccharides with 0.01N HC1 caused a significant reduction (76 to 97%) in the galactose content of the EP suggesting that galactose moiety was mainly composed of galactofuranosyl units. Glucose present in the EP could result also from nonspecific hydrogen binding of the monosaccharides to the EP. Such an hypothesis is strenghtened by the significant decrease of the glucose content observed in EP from *A. fumigatus* (isolates CBS 148.89, CBS 113.26, 152.89) after 0.01 N HCl hydrolysis. In these isolates the concentration of glucose, accounting for up to 30% of the polysaccharidic moiety of the EP became to 10% after HCl 0.01 N hydrolysis. A similar decrease in the glucose content was observed after 0.01 HCl hydrolysis of EP from *Neosartorya aurata*, *N. aureola*, *N. quadricincta* which were characterized by a high glucose concentration (Fig. 3).

With the *A. fumigatus* isolates, binding of exocellular precipitates with ConA were relatively low but similar for all the isolates of this species (figure 4). ConA binding was 2 to 3 time more important in some of the other taxa.

Table 3. Binding of the different EP to 4 different rabbit antisera (expressed as optical density obtained by ELISA)

Nomenclature (CBS)	Species	Optical Density I	II	III	IV
118.53	*Aspergillus brevipes*	0.372	0.242	0.255	0.224
481.65	*A. duricaulis*	0.260	0.261	0.259	0.244
148.89 ~	*A. fumigatus*	0.612	0.149	0.133	0.138
153.89	*A. fumigatus*	0.534	0.127	0.890	0.176
154.89	*A. fumigatus*	0.626	0.084	0.107	0.152
113.26 ~	*A. fumigatus*	0.932	0.152	0.359	0.252
143.89 ~	*A. fumigatus*	1.470	0.093	0.463	0.387
152.89	*A. fumigatus*	1.044	0.129	0.235	0.272
146.89	*A. fumigatus*	1.303	0.099	0.318	0.416
132.54 ~	*A.fumigatus*	1.390	0.240	0.309	0.347
192.65 ~	*A. fumigatus*	1.426	0.113	0.569	0.667
542.75 ~	*A. fumigatus*	0.843	0.113	0.133	0.219
145.89 ~	*A. fumigatus*	1.403	0.125	0.406	0.406
147.89	*A. fumigatus*	1.611	0.167	0.626	0.406
150.89	*A fumigatus*	1.540	0.114	0.322	0.505
151.89	*A fumigatus*	1.598	0.112	0.441	0.609
149.89 ~	*A fumigatus*	1.548	0.120	0.280	0.474
144.89	*A fumigatus*	1.451	0.147	0.461	0.589
457.75	*A fumigatus* var. *acolumnaris*	0.817	0.216	0.398	0.573
487.65	*A fumigatus* var. *ellipticus*	0.939	0.086	0.212	0.267
458.75	*A fumigatus* var. *sclerotiorum*	0.672	0.207	0.290	0.366
283.65	*A unilateralis*	0.351	0.256	0.249	0.224
127.56	*A viridi-nutans*	0.814	0.318	0.411	0.580
466.65	*Neosartorya aurata*	0.334	0.192	0.203	0.216
105.55	*N. aureola*	0.985	0.339	0.380	0.531
106.55	*N. aureola*	0.743	0.382	0.386	0.524
598.74	*N. fennelliae*	0.844	0.203	0.258	0.287
599.74	*N. fennelliae*	0.628	0.216	0.415	0.467
544.65 ~~	*N. fischeri*	1.119	0.242	0.321	0.448
404.67 ~~	*N. fischeri*	1.100	0.152	0.509	0.412
111.55 ~~	*N. fischeri* var. *glabra*	0.605	0.248	0.306	0.351
112.55 ~~	*N. fischeri* var. *glabra*	0.810	0.195	0.211	0.248
483.65 ~~	*N. fischeri* var. *spinosa*	1.168	0.222	0.298	0.527
297.67 ~~	*N. fischeri* var. *spinosa*	1.230	0.173	0.291	0.315
135.52	*N. quadricincta*	0.425	0.312	0.264	0.276
941.73	*N. quadricincta*	0.962	0.232	0.256	0.346
498.65	*N. stramenia*	0.394	0.262	0.255	0.250

I   : Anti 1028 (CBS 143. 89) EP antiserum
II   : Negative antiserum (control)
III : Anti 1028 EP antiserum adsorbed with EP from isolates of *A fumigatus* (~)
IV   : Anti 1028 EP antiserum adsorbed with EP from isolates of *N fischeri* and related varieties (~~).

Antigenic activities of EP.

As shown in Fig. 5A, monoclonal antibody EB-A2, which is specific of galactofuran, reacted with EP from all species studied except *A. unilateralis* CBS 481.65, which produced only 4% galactose.

The polyclonal antiserum from rabbit immunized with *A. fumigatus* EP (CBS 143.89) was more discriminatory than the monoclonal antibody, reacting strongly with all *A. fumigatus* isolates. Variable responses against other species were obtained. *A. brevipes, A. unilateralis, A. duricaulis, N. aureola,* and *N. stremenia* gave a lower reaction. *N. fischeri* and *A. viridinutans* gave a potent response, similar to *A. fumigatus* (Fig. 5B).

Inhibition. The results of the ELISA inhibition are summarized in the Table 3. Polyclonal rabbit serum was inhibited by pools of EP from isolates of *A. fumigatus, N. fischeri* and related varieties. Antigenic community between these two species was confirmed by the high inhibition (approximately 100%) obtained with the two types of adsorbed antisera.

DISCUSSION

Identification of *A. fumigatus* is important because it is one of the most important mycopathogen associated with immunocompromised patients. This comparative study aimied to determine if chemical and/or antigenic properties of the exocellular material produced by *A. fumigatus* could be used to delineate this species from non-pathogenic *Aspergillus* and other related taxa. The results indicate extensive homology among the species examined. Galactofuranosyl groups were detected in all EP examined and cannot be considered to be species specific. In agreement with Notermans *et al.* (1987, 1988) who reported antigens common to various *Aspergillus* and *Penicillium* species, we have found here that the immunodominant exopolysaccharide present in *A. fumigatus* is also present in related species as well as in *A. flavus, A. nidulans* and *A. niger* (unpubl. data).

Study of the types of linkages in the polysaccharide fractions is now desirable, especially for several isolates of *A. fumigatus* which showed unusual features (isolates CBS 148.89, CBS 153.89, CBS 154.89, 2458.56) and for other taxa which are most different (*A. duricaulis, A. unilateralis, N. aurata, N. stramenia*).

None of the chemical or antigenic criteria investigated in this study are species specific. However, the conjunction of several parameters can be helpful in defining species in the *A. fumigatus* group. As we have seen *A. fumigatus* is characterized by 1) an abundant EP with high proportions of proteins, 2) a heteropolymer of amino sugars, mostly hexosamines, 3) a moderate reactivity to ConA, and 4) a strong antigenic response to a polyclonal antiserum. These conclusions are based on a study of sixteen *A. fumigatus* isolates but only one or two from other taxa. More are needed.

From our results, sugars in the EP provide less effective separation than the protein moiety (Hearn *et al.*, 1989). However, the identification of specific sugar linkages of the exocellular polysaccharides may lead to the selection of monoclonal antibodies more specific that the one we used, which was produced from a very crude antigen.

Intracytoplasmic, carbohydrate containing molecules, should be also studied and compared to exocellular polysaccharides, as common antigenic reactions have been demonstrated between these two antigenic sources (Hearn, 1988).

ACKNOWLEDGEMENTS

The project at Eco-Bio was partially funded by I.W.O.N.L.

REFERENCES

ASHWELL, G. 1966. New colorimetric methods of sugar analysis. *Methods in Enzymology* 8: 85-95.

AZUMA, I., KIMURA, H., HIRAO, F., TSUBURA, E., YAMAMURA, Y. and MISAKI, A. 1971. Biochemical and immunological studies on *Aspergillus* III.Chemical and immunological properties of glycopeptide obtained from *Aspergillus fumigatus*. *Japanese Journal of Microbiology* 15: 237-246.

BARDALAYE, P.C and NORDIN J.H. 1977. Chemical structure of the galactomannan from the cell wall of *Aspergillus niger*. *Journal of Biological Chemistry* 252: 2584-2591.

BARRETO-BERGTER, E.M., and TRAVASSOS, L.R. 1980. Chemical structure of the galacto-D-mannan component from hyphae of *Aspergillus* spp. *Carbohydrate Research* 86: 273-285.

BARRETO-BERGTER, E.M., GORIN, P.A.J. and TRAVASSOS, L.R. 1981. Cell constituents of mycelia and conidia of *Aspergillus fumigatus.Carbohydrate Research* 95: 205-218.

BAZIN, H. 1987. Rat hybridoma formation and rat monoclonal methods. *In* Methods of Hybridoma Formation, eds. A.H. Bartal and Y. Hirsault, pp. 337-348. Clifton: Humana Press,

BENNETT, J.E., BHATTACHARJEE, A.K. and GLAUDEMANS, C.P.J. 1985. Galactofuranosyl groups are immunodominant in *Aspergillus fumigatus* galactomannan. *Molecular Immunology* 22: 251-254.

BRADFORD, M.M. 1976. Rapid sensitive method for the quantitation of protein utilizing the principles of protein dye-binding. *Analytical Biochemistry* 72: 248-254.

DUBOIS, M., GILLES, K.A., HAMILTON, J.K., REBERS, P.A. and SMITH, F. 1956. Colorimetric method for determination of sugars and related substances. *Analytical Chememistry* 28: 350-356.

GANDER J.E., JENTOFT N.H., DREWS L.R. and RICK D.P. 1974. The 5-O-a-D-galactofuranosyl-containing exocellular glycopeptide of *Penicillium charlesii*. *Journal of Biological Chemistry* 249: 2063-2072.

GORIN, P.A.J. AND SPENSER. J.F.T. 1968. Structural chemistry of fungal polysaccharides. *Advances in Carbohydrate Chemistry* 23: 367-414.

HEARN, V. 1988. Serodiagnosis of Aspergillosis. *In Aspergillus* and Aspergillosis, eds. H. Vanden Bossche, D.W.R. Mackenzie and G. Cauwenbergh, pp. 43-71. New York and London: Plenum Press.

HEARN, V., MOUTAOUAKIL, M. and LATGE, J.P. 1990. Analysis of components of *Aspergillus* and *Neosartorya* mycelial preparations by gelelectophoresis and Western blotting techniques. *In* Modern Concepts In *Penicillium* and *Aspergillus* classification, eds. R.A.Samson and J.I. Pitt, pp. 235-245, New York and London: Plenum Press.

HUPPERT, M., 1983. Antigens for measuring immunological reactivity, *In* Fungi pathogenic for human and animal. Part B. Pathogenicity and detection", ed. D. Howard, p. 219-301. New York & Basel: Marcel Dekker.

JOHNSTON, I.R. 1965. The composition of the cell wall of *Aspergillus niger*. *Biochemical Journal* 96: 651-658.

LATGE, J.P., SARFATI, J., FOURNET, B., DIAQUIN, M., DEBEAUPUIS, J.P., PREVOST, M.C. and DROUHET, E. 1988. The wall of *Aspergillus fumigatus* and its related antigenic slime. *Revista Iberica de Micologia* 5 (Supplement 1): 28.

NOTERMANS, S. and HEUVELMAN, C.J. 1985. Immunological detection of moulds in food by using the enzyme-linked immunosorbent assay (ELISA); Preparation of antigens. *International Journal of Food Microbiology* 2: 247-258.

NOTERMANS, S. and SOENTORO, P.S.S. 1986. Immunological relationship of extracellular polysaccharide antigens produced by different mould species. *Antonie van Leeuwenhoek* 52: 393-401.

NOTERMANS, S., VEENEMAN, G.H., VAN ZUYLEN, C.W.E.M., HOOGERHOUT, P. and VAN BOOM, J.H. 1988. (1,5)-linked A-D-galactofuranosides are immunodominant in extracellular polysaccharides of *Penicillium* and *Aspergillus* species. *Molecular Immunology* 25: 975-979.

NOTERMANS, S., WIETEN, G., ENGEL, H.W.B., ROMBOUTS, F.M., HOOGERHOOT, P. and VAN BOOM, J.H. 1987. Purification and properties of extracellular polysaccharide antigens produced by different mould species. *Journal of Applied Bacteriology* 62: 157-166.

PRESTON, J.F., LAPIS, E. and GANDER, J.E. 1969. Isolation and partial characterization of the extracellular polysaccharides of *Penicillium charlesii*. III. Heterogeneity in size and composition of high molecular weight exocellular polysaccharides. *Archives of Biochemistry and Biophysic* 134: 324-334.

REISS, E. and LEHMANN, P.F. 1979. Galactomannan antigenemia in invasive aspergillosis. *Infectious and Immunity* 25: 357-365.

STYNEN, D., GORIS, A., MEULEMANS, L. and BRIERS, E. 1989. A specific rat monoclonal antibody against *Yersinia enterocolitica* Serogroup 0: 8. *In* Rat hybridomas and rat monoclonal antibodies eds. H.Bazin Boca Raton, Fl.: CRC Press Inc. (in press).

SUZUKI, S. and TAKEDA, N. 1975. Serologic cross-reactivy of the D-galacto-D-mannans isolated from several pathogenic fungi against anti-*Hormodendrum pedrosoi* serum. *Carbohydrate Research* 40: 193-197.

WEBSTER, S.F. and McGINLEY, K.J. 1980. Serobiological analysis of an extractable carbohydrate antigen of *Pityrosporum ovale. Microbios* 28: 41-45.

ZANETTA, J.P., BRECKENRIDGE, W.C and VINCENDON, G. 1972. GLC of trifluoroacetate of o-methyl glycosides. *Journal of Chromatography* 69: 291-304.

BIOTYPING OF *ASPERGILLUS FUMIGATUS* AND RELATED TAXA BY THE YEAST KILLER SYSTEM

L. Polonelli, S. Conti, L. Campani and F. Fanti

Istituto di Microbiologia
Università degli Studi di Parma
Parma, Italy

SUMMARY

The yeast killer system has proven to be an effective procedure for intraspecific differentiation of filamentous fungi. This biotyping method may have an useful application as an epidemiological marker in outbreaks of mycotic infections, especially when other approaches are not readily available. The opportunistic pathogen *Aspergillus fumigatus* Fresenius is the most common etiologic agent of the different clinical types of human aspergillosis. Strain differentiation of either morphologically identified *A. fumigatus* or related species as well as related teleomorphs by the yeast killer system is described.

INTRODUCTION

Aspergillus fumigatus is the most common etiologic agent of human aspergillosis. The species is an ubiquitous saprophyte in nature. In man, the respiratory tract is the normal portal of entry as the small diameter of the conidia allows them to reach and germinate in the pulmonary alveolar spaces. Complications may occur in immunocompromised hosts, ranging from allergic bronchopulmonary aspergillosis to aspergilloma and, by hematogenous spreading, invasive aspergillosis.

Hospitals provide a striking concentration of patients receiving immunosuppressive therapy (Aisner et al., 1976; Allo et al., 1987; Arnow et al., 1978; Attah and Cerruti 1979; Brandt et al., 1985; Fanti et al., 1989; Glotzbach, 1982; Kyriakides et al., 1976; Mawk et al., 1983; Meyer et al., 1973; Opal et al., 1986; Petheram and Seal 1976; Ross et al., 1968; Simpson et al., 1977; Tack et al., 1982; Viollier et al., 1986; Weems et al., 1987; Wellens et al., 1982). *Aspergillus* nosocomial infections have emerged as a major cause of morbidity and mortality in hospitalized patients during the last two decades. A six year report (1970-1975) from the Centers for Disease Control (CDC) revealed an increase of 158% in the incidence of nosocomial aspergillosis (Fraser et al., 1979). The estimate should be considered quite conservative since invasive aspergillosis often is diagnosed or misdiagnosed only at postmortem examinations.

Nosocomial mycoses provide an intolerable discrepancy between the extraordinary progress in the management of seriously ill patients and the very slow advancements made in elucidating the pathogenesis and dynamics of fungal infections. More efforts need to be directed toward clinical research in medical mycology for diagnosis, prevention and infection control. Differentiation of biotypes of fungi responsible for human mycoses may represent a powerful tool in the investigation on the epidemiology of fungal outbreaks as well as in correlating associations among different pathology characters such as virulence, morphology and serology.

Modern Concepts in Penicillium and Aspergillus Classification
Edited by R. A. Samson and J. I. Pitt
Plenum Press, New York, 1990

Table 1. *Aspergillus* sect. *Fumigati* isolates tested for sensitivity to killer yeasts.

Study number	Species	Collection number
1	Aspergillus fumigatus	CBS 113.26 [1]
2	A. fumigatus	CBS 132.54
3	A. fumigatus	CBS 143.89
4	A. fumigatus	CBS 144.89
5	A. fumigatus	CBS 145.89
6	A. fumigatus	CBS 146.89
7	A. fumigatus	CBS 147.89
8	A. fumigatus	CBS 148.89
9	A. fumigatus	CBS 149.89
10	A. fumigatus	CBS 150.89
11	A. fumigatus	CBS 151.89
12	A. fumigatus	CBS 152.89
13	A. fumigatus	CBS 153.89
14	A. fumigatus	CBS 154.89
15	A. fumigatus	CBS 192.65
16	A. brevipes	CBS 118.53
17	A. fumigatus var. acolumnaris	CBS 457.75
18	A. fumigatus var. ellipticus	CBS 487.65
19	A. fumigatus	CBS 542.75
20	A. fumigatus var. sclerotiorum	CBS 458.75
21	A. unilateralis	CBS 283.66
22	A. viridinutans	CBS 127.56
23	Neosartorya aurata	CBS 466.65
24	N. aureola	CBS 105.55
25	N. aureola	CBS 106.55
26	N. fennelliae	CBS 598.74
27	N. fennelliae	CBS 599.74
28	N. fennelliae	CBS 410.89
29	N. fennelliae	CBS 411.89
30	N. fischeri var. fischeri	CBS 544.65
31	N. fischeri var. fischeri	CBS 404.67
32	N. fischeri var. glabra	CBS 111.55
33	N. fischeri var. glabra	CBS 112.55
34	N. fischeri var. spinosa	CBS 483.65
35	N. fischeri var. spinosa	CBS 297.67
36	N. quadricincta	CBS 135.52
37	N. spathulata	NHL 2948
38	N. spathulata	NHL 2949

[1] Centraalbureau voor Schimmelcultures, Baarn, The Netherlands

In this study we report the application of a biotyping system to *A. fumigatus* isolates based on their differential sensitivity to the action of selected killer yeasts. The system, which proved to be very effective in the differentiation of cultures of filamentous fungi (Polonelli *et al.*, 1987) has been extended to comparisons with other species of *Aspergillus* sect. *Fumigati* as well as related *Neosartorya* species.

Table 2. Yeast killer strains used for studying the sensitivity of Aspergillus sect. Fumigati isolates to killer toxins.

Code	Species	Collection	Number
K 1	*Pichia sp.*	Stumm (1)	1034
K 2	*Pichia sp.*	Stumm	1035
K 3	*P. anomala*	UM (2)	3
K 4	*P. anomala*	CBS (3)	5739
K 5	*P. anomala*	Ahearn (4)	Um866
K 6	*P. californica*	Ahearn	WC40
K 7	*P. canadensis*	Ahearn	WC41
K 8	*P. dimennae*	Ahearn	WC44
K 9	*P. mrakii*	Ahearn	WC51
K 10	*P. kluyveri*	Stumm	1002
K 12	*P. bimundalis*	Ahearn	WC38
K 17	*P. bimundalis*	CBS	5642
K 18	*P. fabianii*	Ahearn	WC45
K 19	*P. holstii*	CBS	4140
K 20	*P. subpelliculosa*	CBS	5767
K 22	*Candida guilliermondii*	UCSC (5)	0
K 23	*C. maltosa*	UCSC	0
K 24	*P. spartiniae*	UCSC	0
K 25	*P. nonfermentans*	UM	200
K 26	*P. carsonii*	CBS	810
K 27	*P. farinosa*	CBS	185
K 28	*P. guilliermondii*	CBS	2031
K 29	*C. pseudotropicalis*	UP (6)	241
K 30	*C. pseudotropicalis*	UP	254
K 31	*C. pseudotropicalis*	UP	330
K 32	*P. kluyveri*	UP	5F
K 33	*P. kluyveri*	UP	6F
K 34	*P. membranaefaciens*	UP	10F
K 35	*P. kluyveri*	UP	11F
K 36	*P. anomala*	UP	25F
K 37	*P. mrakii*	UCSC	255
K 44	*Kluyveromyces lactis*	Gunge (7)	IFO 1267
K 48	*K. lactis*	CBS	2359
K 49	*K. lactis*	CBS	2360
K 50	*K. fragilis*	UP	0
K 51	*K. fragilis*	UP	1

(1) C. Stumm, University of Njimegen, Njimegen, The Netherlands, (2) Istituto di Igiene, Universita' di Milano, Milan, Italy, (3) Centraalbureau voor Schimmelcultures, Baarn, The Netherlands, (4) D.G. Ahearn, Georgia State University, Atlanta, Georgia, USA, (5) Istituto di Microbiologia, Universita' Cattolica del Sacro, Cuore, Rome, Italy, (6) Istituto di Microbiologia, Universita' degli Studi di Parma, Parma, Italy, (7) N. Gunge, Kumamoto Institute of Technology, Kumamoto, Japan

MATERIAL AND METHODS

Isolates examined. The 38 isolates of the *Aspergillus* sect. *Fumigati* investigated in this study were provided by the Centraalbureau voor Schimmelcultures and are listed in Table 1. All of them were selected for comparative evaluation by other investigative procedures, being reported elsewhere in these proceedings (Samson and Pitt, 1990).

The yeast isolates used in this study have previously been shown to have killer activity towards a number of potentially sensitive isolates (Polonelli and Morace, 1986). The 36 recognized killer yeasts were grouped into triplets to make the recording of their activity easier by using a conventional code. All of them are maintained in our culture collection. Many were obtained from different international sources (Table 2).

Media.

Yeast Extract Peptone Dextrose Agar (yeast extract 1%, peptone 2%, glucose 2%, agar 2%) (Difco Laboratories, Detroit, Michigan, U.S.A.) buffered at pH 4.5 and added with methylene blue (0.003%) was used in the studies on the killer system. Sabouraud Dextrose Agar (Difco) was used as the growth and maintenance medium for the the killer yeast and sensitive fungal isolates.

Figure 1 Differential activity of the yeast isolates *Pichia anomala* UM 3 (K3), *Pichia anomala* CBS 5739 (K4) (killer activity), *Pichia* sp. Stumm 1034 (K1) and *Pichia* sp. Stumm 1035 (K2) (no killer activity) on *Aspergillus fumigatus* isolate CBS 146.89. Left: view of the front plate. Right: view of the bottom plate.

Biotyping by the yeast killer system.

The differentiation of the *Aspergillus* sect. *Fumigati* isolates into biotypes by the yeast killer system was carried out according to the procedure of Polonelli *et al.* (1987). Briefly, seven day old mycelial cultures grown on Sabouraud Dextrose Agar slants, were scraped to obtain a suspension of approximately 20,000 conidia/ml in distilled sterile water. One ml of the suspension was mixed with 20 ml of the buffered medium maintained at 45oC and poured into a Petri dish. After cooling a heavy loopfool of each killer yeast grown on

Sabouraud Dextrose Agar for 48 hours at 25oC was streaked on the surface of the agar. The plates were then incubated at 25oC.

Reading and interpretation of the results.
The culture plates were read after two weeks. The killer effect was considered positive when a clear zone of inhibition surrounded the streak of killer yeast (Figure 1). A code was used to record the combined effect of the killer yeasts grouped into triplets for distinguishing purposes. The triplets were then further coded, for ease of use, by numbers (Table 3).

Table 3. Coding modalities of the activity of each triplet of selected killer yeasts used for the differentiation of *Aspergillus* sect. *Fumigati* isolates.

1st killer	2nd killer	3rd killer	Code
+	+	+	7
−	+	+	6
+	−	+	5
−	−	+	4
+	+	−	3
−	+	−	2
+	−	−	1
−	−	−	0

RESULTS

Most of the *Aspergillus* sect. *Fumigati* isolates tested proved to be sensitive to the action of at least one killer yeast (Table 4). Only, four of them (*A. fumigatus* var. *ellipticus* CBS 487.65, *Neosartorya fischeri* var. *fischeri* CBS 404.67, *N. fischeri* var. *glabra* CBS 112.55, *N. fischeri* var. *spinosa* CBS 483.65) appeared to be resistant to all of the killer yeasts used in this study. The pattern of sensitivity was recorded according to the adopted conventional code (Table 5). Among the 15 species and varieties investigated the yeast killer system was able to differentiate 15 different biotypes (Table 6). As expected, the largest number of different biotypes (seven) was detected among the *A. fumigatus* isolates which were the most numerous of these studied.

The different biotypes varied significantly in killer yeast sensitivity. Some biotypes proved to be sensitive to many killer yeasts (i.e. biotypes G, J, K, L, O,) while another (biotype E) was sensitive only to *Pichia guilliermondii* CBS 2031 (killer yeast 28). Different species and varieties grouped into the same killer biotype with no apparent relatedness to conventional identification. Also different biotypes occurred within the same species or variety.

A broad spectrum of killer biotypes was also detected among the *Neosartorya* isolates tested. With the possible exception of *N. spathulata*, all the species with more than one tested isolate, were differentiated into biotypes.

Table 4. Sensitivity of *Aspergillus* sect. *Fumigati* isolates to killer yeasts.

Killer Code	Aspergillus isolates study number (see table 1)
K 1	
K 2	
K 3	
K 4	
K 5	
K 6	
K 7	
K 8	
K 9	
K 10	
K 12	
K 17	
K 18	
K 19	
K 20	
K 22	
K 23	
K 24	
K 25	
K 26	
K 27	
K 28	
K 29	
K 30	
K 31	
K 32	
K 33	
K 34	
K 35	
K 36	
K 37	
K 44	
K 48	
K 49	
K 50	
K 51	

Table 5. Biotyping of *Aspergillus* sect. *Fumigati* isolates according to their sensitivity to yeast killer toxins

Study code	Numerical killer system biotype				Alphabetical killer system biotype
1	4 7 3	3 0 0	4 1 0	0 0 0	A
2	4 7 3	3 0 0	0 1 0	0 0 6	B
3	4 7 3	3 0 0	4 1 0	0 0 0	A
4	4 7 3	3 0 0	0 1 0	0 0 0	C
5	4 7 3	3 0 0	0 1 0	0 0 0	C
6	4 7 3	3 0 0	0 1 0	0 0 6	B
7	4 7 3	3 0 0	0 1 0	0 0 0	C
8	4 7 3	3 0 0	0 1 0	0 0 0	C
9	4 7 3	3 0 0	0 1 0	0 0 0	C
10	4 7 3	3 0 0	0 1 0	0 0 6	B
11	4 7 3	3 0 0	0 3 0	0 6 7	D
12	4 7 3	3 0 0	0 1 0	0 0 6	B
13	0 0 0	0 0 0	0 1 0	0 0 0	E
14	4 7 3	3 0 0	0 1 0	0 6 7	F
15	4 7 3	3 0 0	0 1 0	0 0 0	C
16	7 7 3	3 0 0	0 3 0	0 2 6	G
17	4 7 3	3 0 0	0 3 0	0 0 0	H
18	0 0 0	0 0 0	0 0 0	0 0 0	I
19	4 7 2	3 0 0	4 7 4	0 2 6	J
20	0 0 0	0 0 0	0 1 0	0 0 0	E
21	4 7 3	3 0 0	0 1 0	0 0 0	C
22	0 0 0	0 0 0	0 1 0	0 0 0	E
23	4 7 3	3 2 0	6 3 0	0 0 0	K
24	4 7 3	3 2 0	6 3 0	0 0 0	K
25	7 7 3	3 2 0	6 3 0	0 0 0	L
26	4 7 3	3 0 0	0 3 0	0 0 0	H
27	4 7 3	3 0 0	0 3 0	0 0 0	H
28	4 7 3	3 0 0	0 3 0	0 0 6	M
29	4 7 3	3 0 0	0 3 0	0 0 6	M
30	7 7 3	3 2 0	6 3 0	0 0 0	L
31	0 0 0	0 0 0	0 0 0	0 0 0	I
32	0 0 0	0 0 0	0 1 0	0 0 0	E
33	0 0 0	0 0 0	0 0 0	0 0 0	I
34	0 0 0	0 0 0	0 0 0	0 0 0	I
35	0 0 0	0 0 0	0 1 0	0 0 0	E
36	0 0 1	0 0 0	0 1 0	0 0 0	N
37	4 7 3	3 0 0	4 3 0	0 0 6	O
38	4 7 3	3 0 0	4 3 0	0 0 6	O

DISCUSSION

The definition and terminology of intraspecific variation in fungi is poor and often confusing. As Odds (1985) pointed out, the categories of subspecies, variety, subvariety, form, and biotype are inextricably bound to concepts of taxonomy. However, his difinitions do not exactly coincide with those of epidemiology. "Strain" is the most common term used by medical mycologists for intraspecific variation. "Biotype" is primarily used by epidemiologists interested in the correlation between individual strains, their habitats and mode of transmission.

A condition for an effective biotyping system is that it is independent from other unrelated identification tests so that a cumulative information may be acquired by evaluating all of the different characters. Moreover, a biotyping system should be a reasonable compromise between test simplicity and effectiveness in discriminating among the largest possible number of isolates to be processed.

The yeast killer system has proved able to match these expectations, by differentiating the numerous isolates of sect. *Fumigati* studied into biotypes regardless of morphological characteristics and, therefore, of the species. Biotypes were randomly distributed among the different isolates: that the same biotype may refer to different species, and as well different biotypes may occur within the same species. There was no obvious correlation with morphological criteria of identification.

In our hands, the yeast killer system had already proven to be a flexible and reliable method for investigating cases of presumptive nosocomial infections caused by *A. fumigatus* isolates (Fanti *et al.*, 1989). This study confirms that the system can be easily applied to related species and varieties of their anamorphic or telemomorphic characters.

The multidisciplinary approach to the characterization of the same isolates of sect. *Fumigati* used in this study (i.e. study of secondary metabolites including mycotoxins, extracellular polysaccharides, water soluble components by SDS PAGE and Western Blotting other than morphological studies) being reported elsewhere in these proceedings (Samson and Pitt, 1989) could throw more light upon the intra and inter specific linking of *Aspergillus* sect. *Fumigati*.

Table 6. Grouping of killer biotypes among teleomorph and anamorph of *Aspergillus* sect. *Fumigati* isolates.

species and variety	Number of tested isolates	A	B	C	D	E	F	G	H	I	J	K	L	M	N	O
Aspergillus fumigatus	16	2	4	6	1	1	1				1					
A. brevipes	1							1								
A. fumigatus var. *acolumnaris*	1								1							
A. fumigatus var. *ellipticus*	1									1						
A. fumigatus var. *sclerotiorum*	1					1										
A. unilateralis	1			1												
A. viridinutans	1					1										
Neosartorya aurata	1											1				
N. aureola	2											1	1			
N. fennelliae	4								2					2		
N. fischeri var. *fischeri*	2									1			1			
N. fischeri var. *glabra*	2					1				1						
N. fischeri var. *spinosa*	2					1				1						
N. quadricincta	1														1	
N. spathulata	2															2
Total tested isolates	38	2	4	7	1	5	1	1	3	4	1	2	2	2	1	2

REFERENCES

AISNER, J., SCHIMPFF, S.C, BENNETT, J.E., YOUNG, V.M and WIERNIK,P.H. 1976. *Aspergillus* infections in cancer patients: association with fireproofing materials in a new hospital. *Journal of American Medical Association* 235: 411-412.

ALLO, M.D., MILLER, J., TOWSEND, T. and TAN, C. 1987. Primary cutaneous aspergillosis associated with Hickman intravenous catheters. *New England Journal of Medicine* 317: 1105-1108.

ARNOW, P.M., ANDERSON, R.L. MAINOUS, P.D. and SMITH, E.J. 1987. Pulmonary aspergillosis during hospital renovation. *American Review of Respiratory Diseases* 118: 49-53.

ATTAH, C.A. and CERRUTI, M.M. 1979. *Aspergillus* osteomyelitis of sternum after cardiac surgery. *New York State Journal of Medicine* 79: 1420-1421.

BRANDT, S.J., THOMPSON, R.L. and WENZEL, R.P. 1985. Mycotic pseudoaneurysm of an aortic bypass graft and contiguous vertebral osteomyelitis due to *Aspergillus*. *American Journal of Medicine* 79: 259-262.

FANTI F., CONTI, S., CAMPANI, L., MORACE, G., DETTORI, G. and POLONELLI, L. 1989. Studies on the epidemiology of *Aspergillus fumigatus* infections in a university hospital. *European Journal of Epidemiology* 5: 8-14. FRASER, D.W., WARD, J.I., AJELLO, L, and PLIKATYS, B.D. 1979. Aspergillosis and other systemic mycoses: the growing problem. *Journal of the American Medical Association* 242: 1631-1635.

GLOTZBACH R.E. 1982. *Aspergillus terreus* infection of pseudoaneurysm of aortofemoral vascular graft with contiguous vertebral osteomyelitis. *American Journal of Clinical Pathology* 77: 224-227.

KYRIAKIDES, G.K., ZINNEMAN, H.H., HALL, W.H., ARORA, V.K., LIFTON J.. DE WOLF, W.E. and MILLER, J. 1976. Immunologic monitoring and aspergillosis in renal transplant patients. *American Journal of Surgery* 131: 246-252.

MAWK, J.R., ERICKSON, D.L., CHOU, S.N. and SELJESKOG E.L. 1983. *Aspergillus* infections of the lumbar disc spaces. *Journal of Neurosurgery* 58: 270-274.

MEYER, R.D.., YOUNG, L.S. and ARMSTRONG, D. 1973. Aspergillosis complicating neoplastic disease. *American Journal of Medicine* 54: 6-15.

ODDS, F.C. 1985. Biotyping of medically important fungi. *In* Current Topics in Medical Mycology, eds. M.R. McGinnes, p. 155-171. New York: Springer Verlag.

OPAL, S.M., ASP, A.A., CANNADY, P.B., MORSE, P.L., BURTON, L.J. and HAMMER, D.A. 1986. Efficacy of infection control measures during a nosocomial outbreak of disseminated aspergillosis associated with hospital construction. *Journal of Infectious Diseases* 153: 634-637.

PETHERAM, I.S. and SEAL, R.M.E. 1976. *Aspergillus* prosthetic valve endocarditis. *Thorax* 31: 380-390.

POLONELLI, L and MORACE, G. 1986. Reevaluation of the yeast killer phenomenon. *Journal of Clinical Microbiology* 24: 866-869.

POLONELLI, L., DETTORI, G., CATTEL, C. and MORACE, G. 1987. Biotyping of mycelial fungus cultures by the killer system. *European Journal of Epidemiology* 3: 237-242.

ROSS, D.A., ANDERSON, D.C. MACNAUGHTON, M.C. and STEWART, W.K. 1968. Fulminating disseminated aspergillosis complicating peritoneal dialysis in eclampsia. *Archives of Internal Medicine* 121: 183-188.

SAMSON, R.A. and PITT, J.I. eds. 1990. Modern Concepts in *Penicillium* and *Aspergillus* Classification. New York and London: Plenum Press.

SIMPSON, M.B., MERZ, W.G., KURLINSKI, S.P. and SOLOMON, M.H. 1977. Opportunistic mycotic osteomyelitis. Bone infections due to *Aspergillus* and *Candida* species. *Medicine Baltimore* 56: 475-481.

TACK, K.L., RHANE, F.S., BROWN, B. and THOMPSON, R.C. 1982. *Aspergillus* osteomyelitis: report of four cases and review of the literature. *American Journal of Medicine* 73: 295-300.

VIOLLIER, A.F., PETERSON, D.E. DE JONGH, C.A., NEWMAN, K.A., GRAY, W.C., SUTHERLAND, J.C., MOODY, H.A. and SCHIMPFF, S.C. 1986. *Aspergillus* sinusitis in cancer patients. *Cancer Philadelphia* 58: 366-371.

WEEMS, J.J. jr. ANDERMONT, A., DAVIS, B.J. TANCREDE, C.H., GUIGET, M., PADHYE, A.A., SQUINAZI, F. and HARTONE, W.J. 1987. Pseudoepidemic of aspergillosis after development of pulmonary infiltrates in a group of bone marrow transplant patients. *Journal of Clinical Microbiology* 25: 1459-1462.

WELLENS, F., POTULIEGE, C., DEUVART, F.E. and PRIMO, G. 1982. *Aspergillus* osteochondritis after median sternotomy. Combined operative treatment and drug therapy with amphotericin B. *Thoracic and Cardiovascolar Surgery* 30: 322-324.

ANALYSIS OF COMPONENTS OF *ASPERGILLUS* AND *NEOSARTORYA* MYCELIAL PREPARATIONS BY GEL ELECTROPHORESIS AND WESTERN BLOTTING PROCEDURES

V.M. Hearn[1], M. Moutaouakil[2] and J.-P. Latgé[2]

[1]*Mycological Reference Laboratory*
Central Public Health Laboratory
London NW9 5HT, UK

[2]*Unité de Mycologie*
Institut Pasteur
75724 PARIS, Cédex 15, France

SUMMARY

Twenty three isolates of *Aspergillus* spp. including 16 of *A.fumigatus* and 14 of *Neosartorya* spp. were analysed by SDS-polyacrylamide gel electrophoresis and immunoblotting. Qualitative and quantitative differences were observed among species and individual isolates especially in their reactivity towards human IgG. Grouping of many *A. fumigatus* isolates and their relation to *Neosartorya* isolates is indicated by the protein profiles seen with mycelial and culture filtrates on SDS-PAGE and on denaturing polyacrylamide gels.

INTRODUCTION

The taxonomic usefulness of protein profiles obtained by electrophoretic separations has often been demonstrated for fungal genera, including *Aspergillus* (Sorenson *et al.*, 1971). More recently, the frequency of occurrence of two cell sap components in a limited number of *Aspergillus* species, including *A. fumigatus*, *A. flavus*, *A. fischeri* and *A. fennelliae* has been investigated by two-dimensional electrophoresis (Piechura *et al.*, 1985). In the work reported here, we have compared molecular composition and associated immunogenicity of several *Aspergillus* and *Neosartorya* species in an attempt to further characterize these closely related fungi.

Proteins and glycoproteins of water soluble fractions of mycelial preparations, together with ethanol precipitates of culture filtrates, were separated by electrophoresis in polyacrylamide gels. The Coomassie Blue staining patterns of both intact molecules and molecular subunits were compared in nondenaturing gels and sodium dodecylsulphate containing gels, respectively. Possible variation among the different species in molecules possessing catalase activity was also examined. Finally, the antigenic reactivity of both *Aspergillus* and *Neosartorya* species was determined by combined SDS-PAGE and Western blotting procedures. Blots were probed with serum from patients with antibodies to *A. fumigatus* antigens, as measured by ELISA.

MATERIAL AND METHODS

Species examined.
The origin of the *Aspergillus* and *Neosartorya* species studied are shown in Table 1.

Modern Concepts in Penicillium and Aspergillus Classification
Edited by R. A. Samson and J. I. Pitt
Plenum Press, New York, 1990

Table 1. Isolates of *Aspergillus* and *Neosartorya* species

Reference		Collection	samples studied		
			a+b	*a*	*b*
1	CBS 111.55	*N. fischeri* var.*glabra*	+	+	+
2	CBS 599.74	*N. fennelliae*	+	+	+
3	CBS 127.56	*A. viridinutans*	+		+
4	CBS 404.67	*N. fischeri*	+	+	
5	CBS 487.65	*A. fumigatus* var. *ellipticus*	+		+
6	CBS 544.65	*N. fischeri*	+	+	+
7	CBS 598.74	*N. fennelliae*	+	+	+
8	CBS 118.53	*A. brevipes*	+		
9	CBS 941.73	*N. quadricincta*	+		+
10	CBS 457.75	*A. fumigatus* var. *acolumnaris*	+		+
11	CBS 483.65	*N. fischeri* var. *spinosa*	+	+	+
12	CBS 297.67	*N. fischeri* var. *spinosa*	+	+	+
13	CBS 112.55	*N. fischeri* var. *glabra*	+	+	+
14	CBS 135.52	*N. quadricincta*	+		+
15	CBS 481.65	*A. duricaulis*	+		+
16	CBS 466.65	*N. aurata*	+	+	+
17	CBS 192.65	*A. fumigatus*	+	+	
18	CBS 498.65	*N. stramenia*	+		+
19	CBS 105.55	*N. aureola*	+		+
20	CBS 283.65	*A. unilateralis*	+	+	+
21	1028	*A. fumigatus*	+	+	+
22	DUV.IP	*A. fumigatus*	+	+	
23	CBS 458.75	*A. fumigatus* var. *sclerotiorum*	+		
24	152.WS	*A.f umigatus*	+	+	
25	IFAS. 20	*A. fumigatus*	+	+	
26	B614B	*A. fumigatus*	+	+	
27	P8	*A. fumigatus*	+	+	+
28	CBS 481.65	*A. duricaulis*	+		
29	NCPF 2140	*A. fumigatus*	+	+	
30	B617B	*A. fumigatus*	+	+	
31	CBS 542.75	*A. fumigatus*	+	+	+
32	CBS 132.54	*A. fumigatus*	+	+	
33	NCPF 2109	*A. fumigatus*	+	+	
34	CBS 106.55	*N. aureola*	+		+
35	DAL.IP	*A. fumigatus*	+	+	
36	IFAS.l9	*A. fumigatus*	+	+	+
37	152.EI	*A. fumigatus*	+	+	
38	CBS 113.26	*A. fumigatus*	+	+	
39	CBS 126.56	*A. unilateralis*		+	
40	NHL 2951	*N. fennelliae*		+	
41	NHL 2952	*N. fennelliae*		+	
42	2172.88	*A. duricaulis*		+	+
43	2458.56	*A. fumigatus*		+	+

a. SDS-PAGE; b. Western blots.

Preparation of fractions.
The isolates were grown in 2% glucose, 1% peptone broth at 25°C. The time of growth varied with the isolate (Debeaupuis *et al.*, 1989). Mycelium was separated from the culture medium by paper filtration and washed extensively with water. Culture filtrates were treated with 4 volumes of ethanol overnight at 4°C. After centrifugation, precipitated material (EP) was washed with ethanol, resuspended in water and stored at -20°C. Water soluble fractions were prepared from the fungal mycelium by disruption in 50mM ammonium bicarbonate at 5°C in an MSK Braun cell homogenizer with 1 mm diam glass beads for approximately 4-5 min. Soluble constituents were separated from the insoluble residue by centrifugation at 100,000 g for 1 hr. Supernatants were concentrated approximately 5-fold by overnight dialysis at 4°C against polyethylene glycol 6000 to give the water-soluble (WS) fraction of each mycelial preparation (Hearn and Mackenzie, 1979).

Protein analysis.
The total protein of each fraction was determined using the Bio-Rad (Munich, G.F.R.) dye binding protein assay with bovine serum albumin as a standard. Measurement was done directly on WS fractions, whereas EP fractions were homogenized by ultrasonication and boiled 5 min in IN NaOH before protein measurement.

Polyacrylamide gel electrophoresis (SDS-PAGE).
Samples were subjected to electrophoresis in a Bio-Rad Mini Protean TM II apparatus (8.2 x 10 x 0.075cm) using a separating gel of 12.5% polyacrylamide, essentially according to Laemmli (1970). Samples (25 µg of protein) were diluted 1:1 in 'denaturing buffer' (2% SDS, 10% glycerol, 5% mercaptoethanol and a trace of bromophenol blue in 64 mMTris/HCl buffer, pH 6.8) and boiled for 3 min. A Tris-glycine tank buffer (pH 8.3) was used. Electrophoresis was carried out at a constant current of 30mA (Hearn *et al.*, 1989). Molecular weight (mol.wt.) protein standards (Gibco/BRL SARL, France) were run in parallel.

Protein and glycoprotein detection.
Separated components were stained for protein in 0.1% Coomassie Brilliant Blue R-250 in 25% methanol and 7.5% acetic acid for 1 hr at room temperature and destained in the same solvent without dye (Dzandu *et al.*, 1984).

Electroblotting.
The electrophoretically separated proteins and glycoproteins were electro-transferred to nitrocellulose membranes (Hybond C, Amersham Int., U.K.) in a transblotting chamber (LKB 2005 Transphor), using the method of Towbin *et al.* (1979).

Antigenic reactivity of WS preparations.
The free binding sites on the membrane were saturated with phosphate buffered saline, pH 7.2, containing 0.15% Tween 20 (PBS-T) for 1 hr at room temperature. Two separate pools of human serum samples (highly reactive in an ELISA for *A. fumigatus* antibodies) were incubated at dilutions of 1:600 or 1:800 in PBS-T for 2 hr at room temperature. Blots were then exposed to rabbit anti-human IgG conjugated to peroxidase (Dako, Denmark) at 1:1000 dilution in PBS-T. Antigen/antibody complexes were localised by staining for peroxidase activity with 3,3'-diaminobenzidine as substrate for a maximum incubation time of 10 min.

Electrophoresis in native gels.

PAGE was done with WS extracts in an LKB 2001 Vertical Electrophoresis Unit, initially in 7.5% but finally in gradient 5-15% polyacrylamide gels (16 x 18 x 0.1cm). Samples for analysis containing 30 μg protein were applied in 10% glycerol and a trace of bromophenol blue in 64 mM Tris/HCl buffer, pH 6.8; SDS and mercapto-ethanol were omitted from the system. The tank buffer was again Tris-glycine at pH 8.3. Electrophoresis was done at 30 mA per gel. High mol. wt. protein standards for use in native gels (Sigma Chemical Co.), were run in parallel. Following electrophoresis, gels were stained for the presence of protein and glycoprotein moieties as described above.

Detection of catalase activity.

The ferricyanide negative stain of Woodbury *et al.* (1971), was used to locate bands containing catalase following PAGE in native gels, as described by Wayne and Diaz (1986). The loading of each fraction was based on the amount necessary to optimize the separation of the multiple catalase bands present in many of the preparations.

RESULTS

SDS-PAGE analysis.

All WS preparations were subjected to SDS-PAGE; the separated components were stained for protein with Coomassie Blue (Fig. 1). Each sample contained an array of molecules which ranged in apparent mol. wt. from 10 kDa to approximately 100 kDa. Some extracts contained low concentrations of additional components with apparent mol. wts between 100-200 kDa. The profiles obtained showed qualitative and quantitative differences among samples.

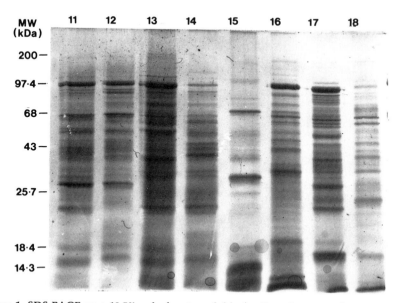

Figure 1. SDS-PAGE on a 12.5% gel of water soluble fractions from sample Nos. 11-18, stained for protein with Coomassie Blue. Mol. wts of protein standards are shown in the left margin.

Fewer protein bands were detected with Coomassie blue in the ethanol precipitate of the culture filtrates than in the WS fractions. Most of the protein bands had apparent mol. wts less than 67 kDa. With one exception (No. 43), all *A. fumigatus* isolates displayed similar protein profiles, with three major bands with apparent mol. wts between 50 and 30 kDa (Fig. 2). The profiles obtained with one isolate of each *A. unilateralis* and *A. duricaulis* (No. 39 and 42, respectively), were closely related to the typical *A. fumigatus* pattern, whereas the other isolate of *A. unilateralis* (No. 20) was quite different.

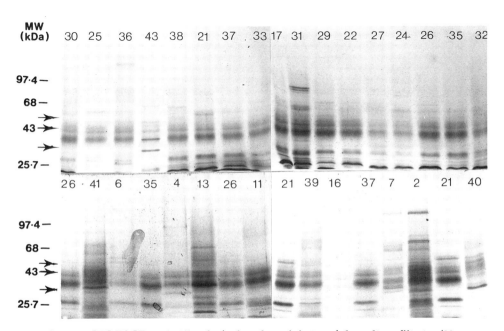

Figure 2. SDS-PAGE on 12.5% gel of ethanol precipitates of the culture filtrates (20 µg protein per well) stained for protein with Coomassie Blue. Mol. wts of standards are shown in the left margin. Arrows indicate recurring bands.

All the isolates of *Neosartorya fischeri* studied produced patterns very similar to those of *A. fumigatus*. In contrast, the patterns obtained with the *N. fennelliae* and *N. aurata* EP s differed from those shown by *A. fumigatus* and *N. fischeri*.

Reactivity of extracts towards Aspergillus-positive sera.
Two ELISA-positive human serum pools showed very different binding patterns when exposed to the WS fractions of the isolates examined. Thirty six from a total of thirty seven isolates tested gave a strong reaction with multiple bands in the 50-200 kDa range with human serum pool `a' (Fig. 3). The number of components of mol. wt. <50 kDa which showed antigenic reactivity towards *A. fumigatus* antibodies were relatively few and the

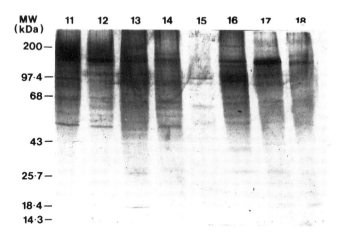

Figure 3. Water soluble fractions from sample Nos. 11-18 separated by SDS-PAGE and transferred to nitrocellulose for probing with anti-*Aspergillus* antiserum. Mol. wts of protein standards are shown in the left margin.

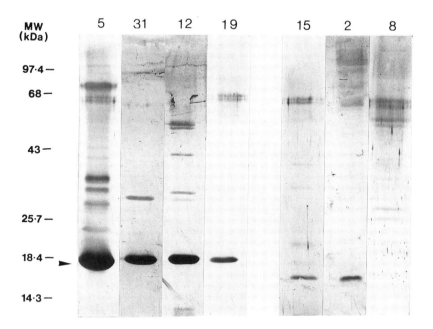

Figure 4. Ethanol precipitate separated by SDS-PAGE and transferred to nitrocellulose for probing with anti-*Aspergillus* human antiserum (10^{-3} dilution). Mol. wts of standards are shown in the left margin. The arrows indicates a dominant, recurring band at 17 kDa.

band staining was much less intense by comparison with high mol. wt. moieties. Only one *A. duricaulis* (No. 15) showed a very weak binding with *A. fumigatus* IgG. When another human serum pool (pool `b') was tested in the same system, 10 of 36 isolates were only weakly reactive, including five isolates of *A. fumigatus* (results not shown). Twenty four from a total of twenty six isolates tested gave a positive reaction when their EP fractions were separated by SDS-PAGE and probed with a pool of human antisera after transfer to nitrocellulose. The number of protein bands which showed antigenic reactivity toward *A. fumigatus* antibodies was relatively small and never exceeded 15 bands. Most of the bands were in the 10-90 kDa range, with a major antigen of mol. wt. of 16-18 kDa (Fig 4). Based on the presence or absence of this antigen band, two groups of species could be differentiated (Table 2). These results confirmed the grouping made on the basis of the Coomassie Blue protein staining of SDS-PAGE of EP: *A. fumigatus*, *N. fischeri* and *N. aureola* could be distinguished from *A. viridinutans*, *A. duricaulis*, *A. unilateralis*, *A. brevipes*, *N. fennelliae* and *N. quadricincta*.

Table 2. Presence or absence of the 16-18 kDa antigen band in *A. fumigatus* and related taxa.

Presence	*Absence*
A. fumigatus: isolates 36, 31, 43, 27, 26, 17, 10, 23	*A. viridinutans*: isolate 3
A. fisheri: isolates 6, 1, 13, 12, 11	*A. brevipes*: isolate 8
N. aureola: isolates 19, 34	*A. duricaulis*: isolates 15, 42
N. straemenia: isolate 18[1]	*A. unilateralis*: isolate 20
	N. fennelliae: isolates 7, 2[1]
	N. quadricincta: isolates 9, 14
	N. aurata: isolate 16[1]

[1] overall reactivity of the blot with human pool was very low

Analysis on native gel.
All WS preparations were examined following electrophoresis in native gels. In most samples Coomassie Blue stain revealed a number of bands, with a wide range of mol. wts (Fig.5). A few preparations showed a minimal number of Coomassie Blue stained bands (viz., *N. stramenia*, *A. unilateralis* and two *A. fumigatus*). Most *A. fumigatus* isolates (12 of 16, three only in trace amounts) showed the presence of a double band with mol. wts near 100 kDa. Exceptions were *A. fumigatus* Nos 32 and 36, where this double band appeared to be absent. Of the other species tested, only *A. fumigatus* var. *ellipticus* and *N. fischeri* var. *glabra* possessed double bands with mobilities indistinguishable from those of *A. fumigatus*. A smaller number of *A. fumigatus* isolates (No. 21, 26, 27, 30, 31, 35, 37 and 38) shared three additional bands, as did *A. fumigatus* var. *ellipticus* (Fig. 5).

Analysis of catalase band patterns on native gels.
Staining for catalase activity after electrophoresis on native gels showed the presence of at least one catalase component in all WS preparations. Most *A. fumigatus* isolates produced two major bands, one with approximate mol. wt above and one below 272 kDa called slow and fast bands, respectively. The slow band was consistently present and its mobility appeared constant in all *A. fumigatus* isolates; the fast component was absent from sample

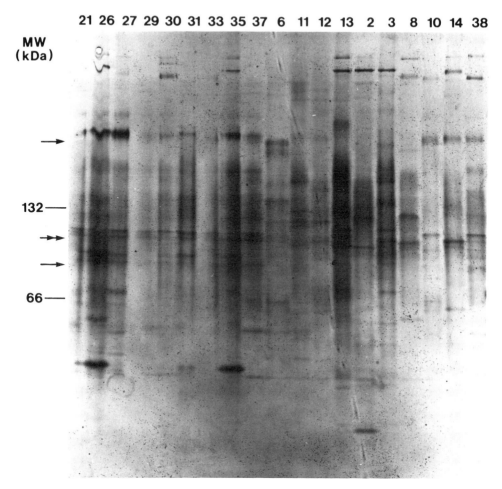

Figure 5. PAGE on a native gel (5-15% gradient) of water soluble fractions: from left to right, 9 isolates of *A. fumigatus*, 9 other species, 1 isolate of *A. fumigatus* and 1 other species. The gel was stained for protein with Coomassie Blue. Arrows indicate the most commonly shared bands (->->), and other shared bands (->) present in the different species. Mol. wts of bovine serum albumin (monomer and dimer) are shown in the left margin.

No. 36 and its mobility showed isolate to isolate variation (Fig. 6). One sample (No. 25) gave a catalase pattern which differed markedly from all other *A. fumigatus* isolates tested. Several *A. fumigatus* WS preparations contained additional, minor catalase bands, but these components showed no constant pattern. Some other species also produced two major catalase bands but they usually showed different mobilities from those seen in *A. fumigatus*. However, *A. fumigatus* var. *acolumnaris* (No. 10) and *N. fischeri* var. *glabra* (No. 13), produced two catalase bands of similar mobility to these of *A. fumigatus*. The two isolates of *N. fischeri* (Nos. 4 and 6), also contained two catalase bands with mobilities similar to *A. fumigatus*, but were only minor components. These results are summarized in Table 3 and Fig. 6.

Table 3. Catalase band patterns seen with *Aspergillus* and *Neosartorya* species

A. fumigatus isolates with 2 major catalase bands	*A. fumigatus isolates with >2 catalase bands*	*Other species with 2 major bands*
17	17	2 (*N. fennelliae*)
21	22	5 (*A. fumigatus* var. *ellipticus*)
22	24	
24	26	7 (*N. fennelliae*)
26	27	10 (*A. fumigatus* var. *acolumnaris*)
27	29	
29	30	13 (*N. fischeri* var. *glabra*)
30	35	
31	37	20 (*A. unilateralis*)
32		
33		23 (*A. fumigatus* var. *sclerotium*)
35		
37		
38		
TOTAL 14	9	7

DISCUSSION

Separation of WS preparations of *Aspergillus* and *Neosartorya* species by SDS-PAGE showed very complex protein patterns, when visualized with Coomassie Blue. Since differences among these preparations were readily apparent, the method offers the potential of determining protein homologies (i.e., the number of common protein bands) among isolates and species of *Aspergillus* and *Neosartorya*. However, the multiplicity of constituent molecules present in each fraction could make it difficult to assess the reproducibility of the gel electrophoresis method and the batch to batch variation of the samples. Because the number of protein bands detected is lower, EP profiles obtained by SDS-PAGE may prove more suitable than WS extracts to cluster isolates or species. In addition EP bands lie in the 10-80 kDa range where separation of proteins is better than in the high mol. wt. range.

The antigenic patterns of these WS preparations, examined by immunoblotting, varied considerably with the two pools of human serum tested. Some isolates (20 out of 36) bound strongly to antibodies present in each of the two pools; with others (9 of 36), the binding profile changed from strong (for serum pool `a') to weak (for serum pool `b'). Thus, heterogeneity among immunogens was readily detectable. More precise information may be obtained by using antibody specific to certain immunodominant antigens, as indicated by the profiles achieved with EP preparations.

Electrophoresis of the intact molecules gave a much simplified protein pattern, with some evidence of recurring components in *A. fumigatus* isolates. Analysis showed two closely associated bands with mol. wts. in the 100 kDa range. These were present in 12 out of 16 WS preparations, when monitored in both 7.5% and gradient polyacrylamide gels.

Enzymes, including catalase, have been shown by several workers to possess antigenic activity when tested with anti-*Aspergillus* antisera (Tran van Ky *et al.*, 1968;

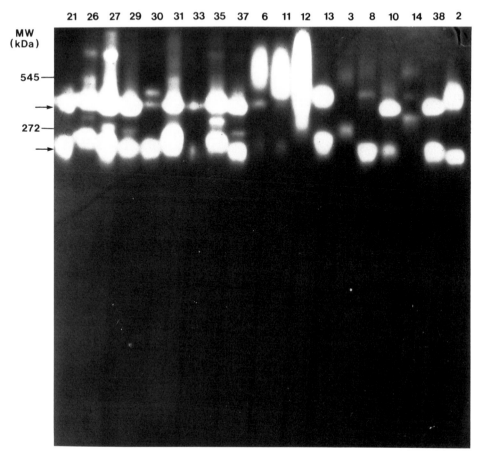

Figure 6. PAGE on a native gel (5-15% gradient) of water soluble fractions: from left to right, 9 isolates of *A. fumigatus*, 9 other species, 1 isolate of *A. fumigatus*, 1 other species. The gel was stained for catalase by the ferricyanide method. Arrows indicate the 2 major catalase components. Mol. wts of urease (trimer and hexamer) are shown in the left margin.

Schonheyder and Andersen, 1984). PAGE revealed 1-5 bands with catalase activity when *Aspergillus* and *Neosartorya* species were examined. Catalase band patterns of *A. fumigatus* isolates showed a measure of similarity, with one major exception (No. 25). No characteristic pattern was obvious among other species, but numbers tested were small. In this study, two batches of *A. fumigatus* No. 33 (NCPF 2109) were monitored for catalase activity; one batch contained significantly higher levels of both the fast and slow components, at comparable protein concentrations. In a separate study where ten batches of this isolate were analysed, 3 of 10 batches appeared to lack the fast catalase component (Hearn, unpublished). Thus batch variation can represent an important factor in such analyses and requires investigation in order to fully interpret the results obtained.

Four parameters have been used to analayse relatedness among the *Aspergillus* sect. *Fumigati* and *Neosartorya*. While *A. fumigatus* isolates showed a tendency to cluster, some isolate differentiation was evident. Tests which compare constituent molecules of WS on native gels plus associated catalase activity, in addition to protein and antigenic patterns, of EP tended to align *N. fischeri* and *N. aureola* with *Aspergillus* sect. *Fumigati*.

ACKNOWLEDGEMENT

The authors wish to express their thanks to Dr. C.K. Campbell, of the Mycological Reference Laboratory, for helpful discussion in the preparation of the manuscript.

REFERENCES

DEBEAUPUIS, J.P., SARFATI, J., GORIS, A., STYNEN, D., DIAQUIN, M. and LATGÉ, J.P. 1989. Exocellular polysaccharides from *Aspergillus fumigatus* and related taxa. *In* Modern Concepts in *Penicillium* and *Aspergillus* Classification, eds. R.A. Samson and J.I. Pitt, pp. 209-223. New York and London: Plenum Press.

DZANDU, J.K., DEH, M.E., BARRATT, D.L. and WISE, G.E. 1984. Detection of erythrocyte membrane proteins, sialoglycoproteins and lipids in the same polyacrylamide gel using a double-staining technique. *Proceedings of the National Academy of Sciences, U.S.A.* 81: 1733-1737.

HEARN, V.M. and MACKENZIE, D.W.R. 1979. The preparation and chemical composition of fractions from *Aspergillus fumigatus* wall and protoplasts possessing antigenic activity. *Journal of General Microbiology* 112: 35-44.

HEARN, V.M., GRIFFITHS, B.L. and GORIN, P.A.J. 1989. Structural analysis of water-soluble fractions obtained from *Aspergillus fumigatus* mycelium. *Glycoconjugate Journal* 6: 85-100.

LAEMMLI, U.K. 1970. Cleavage of structural proteins during the assembly of the head of bacteriophage T4. *Nature, London* 227: 680-685.

PIECHURA, J.E., KURUP, V.P., FINK, J.N. and CALVANICO, N.J. 1985. Antigens of *Aspergillus fumigatus*.III. Comparative immunochemical analyses of clinically relevant aspergilli and related fungal taxa. *Clinical and Experimental Immunology* 59: 716-724.

SCHONHEYDER, H. and ANDERSEN, P. 1984. IgG antibodies to purified *Aspergillus fumigatus* antigens determined by enzyme-linked immunosorbent assay. *International Archives of Allergy and Applied Immunology* 74: 262-269.

SORENSON, W.G., LARSH, H.W. and HAMP, S. 1971. Acrylamide gel electrophoresis of proteins from *Aspergillus* species. *American Journal of Botany* 58: 588-593.

TOWBIN, H., STACHELIN, T. and GORDON, J. 1979. Electrophoretic transfer of proteins from polyacrylamide gels to nitrocellulose sheets: procedure and some applications. *Proceedings of the National Academy of Sciences, U.S.A.* 76: 4350-4354.

TRAN van KY, P., BIGUET, J. and VAUCELLE, T. 1968. Etudè d'une fraction antigènique d'*Aspergillus fumigatus* support d'une activité catalasique. Conséquence sur le diagnostic immunologique de l'aspergillose. *Revue d'Immunologie* 32: 37-52.

WAYNE, L.G. and DIAZ, G.A. 1986. A double staining method for differentiating between two classes of mycobacterial catalase in polyacrylamide electrophoresis gels. *Analytical Biochemistry* 157: 89-92.

WOODBURY, W., SPENCER, A.K. and STAHMANN, M.A. 1971. An improved procedure using ferricyanide for detecting catalase isozymes. *Analytical Biochemistry* 44: 301-305.

DIALOGUE FOLLOWING THE FOUR PRESENTATIONS ON ASPERGILLUS FUMIGATUS

HEARN: We were unable to distinguish between pathogenic and nonpathogenic isolates of *Aspergillus fumigatus* using our antisera. In fact, all the isolates we tested reacted with the antiserum for *A. fumigatus*. We realize now that the antiserum may be reacting with sugars and that these might be common not only in Aspergilli, but in other fungi as well. Upon reflection, if we had taken a different approach to the problem, perhaps we would have had better results. If we had removed the carbohydrate moieties from the outer surfaces of the walls and raised the antibodies to the underlying protein parts of the molecule, perhaps the resulting serum would have been more specific.

LATGÉ: One problem we faced was that we had twenty isolates of *Aspergillus fumigatus*, and only one or two strains of the other species. Would it be possible to get more strains of the other species?

SAMSON: Unfortunately, many of the isolates of these other species in culture collections are misidentified, and they are very rare in any case.

CHRISTENSEN: An obvious idea would be to return to the type locality. For example, *A. brevipes* was originally isolated from tundra soil on Mt. Koskiusko in Australia. It would be fun to go back there and try to find it.

HEARN: If these species are so rare, do we really have to worry about them?

SAMSON: For our purposes, no.

CHRISTENSEN: It seems to me that they may not be rare. Perhaps we just haven't been looking in the habitat in which they are most abundant.

SAMSON: Ecologists or morphologists may be interested in seeing more isolates. But when you are considering medical importance, then *A. fumigatus* is sufficient.

CHRISTENSEN: *Aspergillus nutans*, *A. parvulus* and *A. kanagawaensis* are species that have pinkish conidia, but they are otherwise very similar to the species of sect. *Fumigati*. An Indian worker also reported a tan-coloured *A. fumigatus*. Have these been considered in these studies ?

FRISVAD: No, but it would be interesting to see them.

CHRISTENSEN: *A. nutans* also produces nodding conidiophores and so do *A. brevipes* and *A. unilateralis*.

SAMSON: I was unable to obtain that tan-coloured mutant from India.

PATERSON: What is the taxonomic significance of species of *Aspergillus* producing some of the same secondary metabolites as *Penicillium* species?

FRISVAD: If you have two species, one in *Penicillium*, the other in *Aspergillus*, with globose conidia, you would not say they are closely related only for that reason. I feel the same way about secondary metabolites. If you are working with smaller groups of taxa, then the secondary metabolites have taxonomic significance. For example, emodin is produced by *P. islandicum*, but it is also produced by rhubarb.

CHRISTENSEN: Would the collaborators in this project be prepared to subject their information to Principal Components Analysis (PCA)?

FRISVAD: There are some programs available to do that, of course.

BRIDGE: If you are going to do PCA, you have to be certain that the correlation between the organisms in Euclidian. Immunological data is ideal for this, because it is representative of a distance matrix.

CHRISTENSEN: Is there suitable immunological data that could be used?

POLONELLI: We have tried to correlate the killer yeast data with the immunological data we have from Western blot analysis from antibodies against *Candida albicans*. Apparently, there was a real correlation between the two different biotyping approaches. We don't know why. We have investigated cases of sexually transmitted infection and some cases of nosocomial infection from mother to newborne baby with both systems. Biotyping with the killer yeast system and the Western blot analyses using monoclonal antibodies corresponded.

TAXONOMY OF *ASPERGILLUS* SECTION *RESTRICTA*

J.I. Pitt[1] and R.A. Samson[2]

[1]*CSIRO*
Division of Food Processing
North Ryde, N.S.W. 2113, Australia

[2]*Centraalbureau voor Schimmelcultures*
Oosterstraat 1, 3742 SK Baarn, The Netherlands

SUMMARY

The taxonomy of *Aspergillus* Section *Restricti* (the *Aspergillus restrictus* Group) has been reviewed and revised on the basis of morphological taxonomy, a bibliographic review and a herbarium search. Three species are accepted: *Aspergillus caesiellus* Saito, *A. penicillioides* Spegazzini and *A. restrictus* G. Smith. *A. conicus* Blockwitz and *A. gracilis* Bainier are of uncertain application. *A. itaconicus*, also studied, is not related to species in this Section. A key and descriptions are included.

INTRODUCTION

Presumably because of their xerophilic character, *Aspergillus restrictus* and related species were regarded by Thom and Raper (1945) as a Series within the "*Aspergillus glaucus* Group", now more commonly known as the genus *Eurotium*. Raper and Fennell (1965) raised this Series to "Group" status as the "*A. restrictus* Group". Gams *et al.* (1985) renamed this set of related taxa as *Aspergillus* Section *Restricti* in accordance with the Botanical Code.

Taxonomy within this Section has remained unclear. The best known species, *A. restrictus* G. Smith 1931 has the lowest priority under the Code of any of the species included by Raper and Fennell (1965). Perhaps the most common species, and certainly the most xerophilic, is *A. penicillioides* Spegazzini 1894. It grows poorly if at all on commonly used (high water activity) media, is often overlooked and infrequently recognised.

Other species in Sect. *Restricti* are *A. caesiellus* Saito 1904, *A. conicus* Blockwitz 1914 and *A. gracilis* Bainier 1907. *A. itaconicus* Kinoshita 1931 was also examined, because Thom and Raper (1945) included it in their concept of the *A. glaucus* Group. Each of these is a rarely encountered and little known species.

A. conicus, *A. gracilis* and *A. penicillioides* were neotypified by Samson and Gams (1985). However, the picture is confused by the fact that isolates used as a basis of the descriptions of *A. gracilis* and *A. penicillioides* by Thom and Raper (1945) were both placed in *A. restrictus* by Raper and Fennell (1965). There is also doubt about the accuracy of the neotypifications of *A. conicus* and *A. gracilis*.

This paper is intended to clarify the taxonomy of these species.

Modern Concepts in Penicillium and Aspergillus Classification
Edited by R. A. Samson and J. I. Pitt
Plenum Press, New York, 1990

MATERIAL AND METHODS

Cultures.

Representative isolates were examined from *A. caesiellus, A. gracilis, A. conicus, A. penicillioides, A. restrictus* and *A. itaconicus.* Cultures were obtained from the CBS, FRR, NRRL and ATCC collections.

Media.

Isolates were examined on Czapek yeast extract agar (CYA), malt extract agar (MEA), 25% glycerol nitrate agar (G25N; Pitt, 1979); and also 20% sucrose Czapek yeast extract agar (CY20S) and malt yeast 50% glucose agar (MY50G; Pitt and Hocking, 1985). All cultures were examined after 7 days incubation at 25°C, and on CYA at 37°C as well.

Capitalised colour names and adjacent numbers in brackets are from the Methuen "Handbook of Colour" (Kornerup and Wanscher, 1978).
A bibliographic search and a check on all available herbarium material was also conducted.

RESULTS

Examination of two strains of the extype isolate of *A. itaconicus* (CBS 115.32 and FRR 161) showed that morphology and growth rates were inconsistent with placement in *Aspergillus* Sect. *Restricti.* Growth rates, especially on CY20S and G25N, were much faster than observed for other species considered here. Vesicles were large and spherical, clearly indicating that Raper and Fennell (1965) were correct in placing this species in a section distant from Sect. *Restricta.*

Other isolates examined formed a cohesive pattern of growth rates and microscopic morphology which indicated a close relationship, and a natural series. Two distinct clusters of isolates were observed; one was homogeneous, the other could be split into two species on the basis of a variety of minor features. Hence three species in all are recognised here.

Naming of these species was not so simple. The homogeneous cluster included a variety of isolates which had been identified as *A. penicillioides.* Examination of the packet labelled "*A. penicillioides*" from Spegazzini's herbarium by one of us (R.A.S.) revealed no useful information, other than that Spegazzini had clearly intended *A. penicillioides* to be spelled that way, rather than "*A. penicilloides*" (Fig. 1), a variant which has been in common use for many years (Raper and Thom, 1945; Raper and Fennell, 1965). Neotypification, based on concepts with which the authors were in close agreement (Pitt and Hocking, 1985; Samson and van Reenen-Hoekstra, 1988) was the obvious procedure for stabilising this name in the sense in which it is in common use. This was carried out by Samson and Gams (1985), who designated Herb. IMI 211342 (= CBS 540.65, WB 4548) as neotype.

It is notable that Raper and his coworkers' concept of *A. penicillioides* was not so clear as it has become in more recent times. NRRL 151, regarded by Raper and Thom (1945) as "fit satisfactorily" and similar to Thom 4197.3 "agreed to by Spegazzini" was not mentioned under this species by Raper and Fennell (1965), but placed under *A. restrictus.*

Isolates in the second cluster were regarded as representative of *A. caesiellus, A. conicus, A. gracilis* and *A. restrictus* by various previous authors and collections. Of these, only *A. caesiellus* and *A. restrictus* were represented by cultures derived from types. One taxon in this cluster was represented by the oldest name, *A. caesiellus*, which was therefore accepted as a distinct species.

Of the other three names, considerable doubt exists about the validity of *A. conicus* and *A. gracilis*, despite the neotypification of Samson and Gams (1985). Using a name suggested by Blockwitz, Dale (1914) published *A. conicus*, based on her description of "a *Penicillium*" in Dale (1912), but without further description. According to Thom and Raper (1945), Blockwitz (1929) expressed a lack of confidence in Dale's usage of his name. Nevertheless, Raper and Thom (1945) accepted it. In the absence of a type, and in view of the obvious confusion over what this species represented, we see no reason to continue its usage. In consequence we reject *A. conicus*, as of uncertain application.

A. gracilis was described and illustrated by Bainier (1907) in terms which are not clearly indicative of a species belonging in Sect. *Restricta*. Indeed, *A. gracilis* may equally well have been related to *A. fumigatus*. NRRL 145 (= Thom 4246), used as a basis of the description of *A. gracilis* in Thom and Raper (1945: 139), was later "believed to represent an exceptionally slow-growing *A. restrictus*". On the basis of these inconsistencies, the neotypification of *A. gracilis* by Samson and Gams (1985) is rejected. Like *A. conicus*, this species is also regarded as of uncertain application.

The valid name for the taxon under discussion at this point is therefore *A. restrictus*, a widely used name over a number of years, though almost certainly confused with *A. penicillioides* in much of the literature.

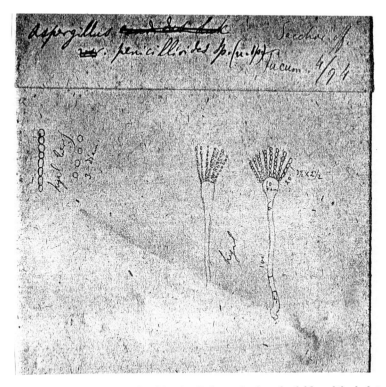

Figure 1. Original handdrawing and writing by C. Spegazinni on the folder of the holotype of *Aspergillus penicillioides* (herb. LP).

TAXONOMY

Key to the species in Aspergilli sect. Restricti

1. Conidia borne as cylinders; at maturity usually cylindrical to doliiform (barrel shaped) 2
   Conidia borne as ellipsoids; at maturity usually ellipsoidal to subspheroidal .... *A. penicillioides*

2. Colonies on CY20S at 7 days not exceeding 20 mm in diameter .... *A. restrictus*
   Colonies on CY20S at 7 days exceeding 20 mm in diameter .... *A. caesiellus*

Aspergillus caesiellus Saito - J. Fac. Sci. Coll. Imp. Univ. Tokyo 18: 49, 1904 – Fig. 2.
   Aspergillus gracilis var. *sartoryi* Batista *et al.* - Mycopath. Mycol. Appl. 8: 196, 1957.

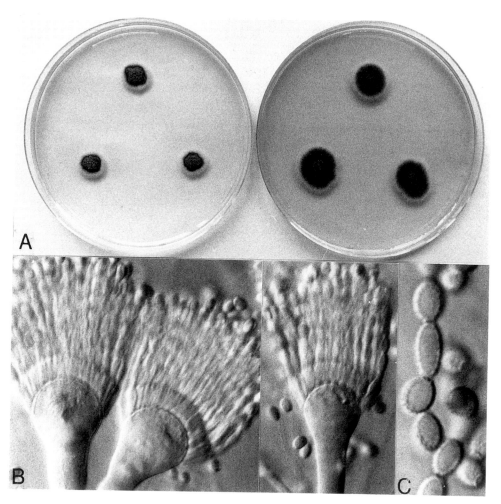

Figure 2. *Aspergillus caesiellus*. A. Colonies on CYA and MEA, 7 days. B. Conidial heads, x 750. C. Conidia, x 1875.

Colonies on CYA 6-12 mm in diameter, dense, usually umbonate, often irregularly wrinkled, of very close, velutinous texture; mycelium white; conidial heads sparse to numerous, Greenish Grey to Dark Green (26D-F2-3); reverse pale to dull green. Colonies on MEA 10-16 mm in diameter, low and plane, sparse to moderately dense, velutinous or lightly funiculose; mycelium inconspicuous, white; conidial production moderate to heavy, Dark Green (26F3-4); reverse pale to very deep green. Colonies on G25N variable, usually 4-12 mm in diameter, often irregular in shape, dense, surface velutinous to somewhat floccose or, in some isolates, mucoid; condia absent to abundant, in the latter case dark green; reverse colourless to grey green. Colonies on CY20S showing optimal development, 20 to 30 mm in diameter, plane or lightly wrinkled, dense, velutinous to floccose-funiculose; mycelium sometimes conspicuous in overgrowths, white; conidial production heavy, Dark Green (26-27F3-5); reverse pale to deep green. Colonies on MY50G 12-18 mm in diameter, plane, sometimes with irregular margins, often floccose; mycelium white; conidial production absent or moderate, in the latter case Greenish Grey (26-27D-E2), paler than on the other media; reverse pale to pale green. Sometimes slow growth on CYA at 37°C, with colonies up to 6 mm in diameter, of white mycelium.

Conidiophores borne from surface hyphae, developing optimally on CY20S, stipes varying widely with isolate, measuring (25-)50-150(-300) x 4-7(-9) μm, with thin, smooth, colourless walls, sometimes septate or of irregular diameter; vesicles pyriform to funnel-shaped, 10-16 μm in diameter, bearing phialides only, often confined to more or less flattened apices; phialides crowded, but seldom numerous, commonly 6-8 μm long; conidia borne as cylinders and often swelling in diameter only when separated from the phialide by 4-6 younger conidia, at maturity ellipsoidal to doliiform, (4-)5-6(-9) x 3-4 μm, with spinose walls.

Typification. Herb. IMI 172278, derived from Saito's type, was designated as lectotype by Samson and Gams (1985). Cultures derived from type include IMI 172278 and CBS 470.65.

Distinctive features. A. *caesiellus* and A. *restrictus* are readily distinguished from A. *penicillioides* by more vigorous growth on CYA and MEA, and more fundamentally by the method of formation of conidia: conidia in A. *caesiellus* and A. *restrictus* are borne as cylinders, and only slowly become more round, while those of A. *penicillioides* are borne as ellipsoids, and rapidly become ellipsoidal to subspheroidal.

A. *caesiellus* and A. *restricus* are closely related. A. *caesiellus* grows more rapidly on CY20S than A. *restrictus*. Colonies of A. *caesiellus* usually exceed 20 mm in diameter on CY20S, while those of A. *restrictus* rarely do. A. *caesiellus* produces somewhat larger conidia and, in the isolates examined here, shorter, less numerous phialides. The isolates of A. *caesiellus* examined grew at 37°C, though weakly. It is not known if this is a consistent character.

Occurrence. This has been a quite rarely recognised species, as it has probably been confused with A. *restrictus*. An understanding of its distribution must await better recognition.

Isolates examined. CBS 470.65 & NRRL 5061 (= FRR 3697), ex type, K. Saito; CBS 116.55 & NRRL 4745 & ATCC 11969 (= WB 4745, FRR 3696), ex type of A. *gracilis* var. *sartoryi* Batista *et al.*, laboratory contaminant, Brazil, A. C. Batista; NRRL 2652 (= FRR 3695), source unknown; FRR 2176, from dried chili, Papua New Guinea, A.D. Hocking.

Aspergillus restrictus G. Smith - J. Textile Inst. 22: 115, 1931 – Fig. 3.
> *Aspergillus restrictus* var. *B* G. Smith, J. Textile Inst. 22: 115, 1931.
> *Penicillium fusco-flavum* Abe, J. Gen. Appl. Microbiol. 2: 64, 1956.

Colonies on CYA 6-12 mm in diameter, sulcate or wrinkled, low, dense and velutinous; mycelium inconspicuous, white; conidial heads often poorly formed, sparse to numerous, in the latter case Dull Green (26-27C-E3); reverse pale to very dark green. Colonies on MEA 6-12 mm in diameter, occasionally smaller, similar to those on CYA or centrally raised, conidial production heavy, but heads poorly formed, coloured Dull Green to Dark Green (27C-F8); reverse usually pale. Colonies on G25N 10-14 mm in diameter, plane or umbonate, usually similar to those on MEA, but heads well formed, producing long columns of conidia when mature; reverse pale or sometimes dark green. Colonies on CY20S 16-20 mm in diameter, generally similar to those on G25N apart from slightly more rapid growth. Colonies on MY50G 12-16 mm in diameter, plane or umbonate, with aerial growth and conidial production usually sparse, coloured Greenish Grey to Dull Green (27C-D3); reverse pale. No growth on CYA at 37°C.

Conidiophores borne from surface hyphae, developing optimally on CY20S, stipes 75-200 µm long, sometimes sinuous, with thin, colourless, smooth walls, enlarging from the base gradually, then more abruptly to pyriform vesicles; vesicles 10-15 µm in diameter, fertile over the apical hemisphere, or less, bearing phialides only; phialides crowded, 8-10 µm long; conidia borne as cylinders, in long appressed columns, adhering in liquid mounts, when mature nearly cylindrical or doliiform, 4.0-5.5 µm long, with rough walls.

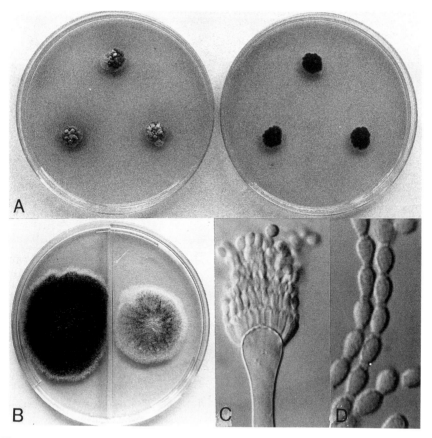

Figure 3. *Aspergillus restrictus.* A. Colonies on CYA and MEA, 7 days. B. Colonies on CY20S (left) and MY50G (right), 14 days. C. Conidial heads, x 750. D. Conidia, x 1875.

Typification. Samson and Gams (1985) designated Herb. IMI 16267 as lectotype of *A. restrictus.* Cultures ex type include IMI 16267, CBS 117.33, CBS 541.65, NRRL 154 and ATCC 16912.

Distinctive features. See *A. caesiellus* above.

Occurrence. *A. restrictus* is the most commonly recognised species in Sect. *Restricta*, but it seems likely that it has often been confused with *A. caesiellus* and *A. penicillioides.* It is widely distributed in cereals and other dried foods (Pitt and Hocking, 1985).

Isolates examined. CBS 541.65 & NRRL 154 (= CBS 117.33, FRR 3689, IMI 16267, ATCC 16912), ex type, from mouldy cloth, United Kingdom, G. Smith; CBS 118.33 & NRRL 148 (= FRR 3726, IMI 16268) ex type of *A. restrictus* var. *B*, from cotton fabric, United Kingdom, G. Smith; CBS 331.59 & FRR 1173 (= IMI 68226), ex type of *Penicillium fusco-flavum* Abe, from ?soil, Japan, S. Abe; NRRL 158 (= FRR 3692), received as *A. conicus*; NRRL 159 (= FRR 3693), received as *A. conicus*; NRRL 160 (= FRR 3694), received as *A. conicus*; NRRL 156 (= FRR 3690), received as *A. gracilis*; FRR 1992, from dried split peas, Australia, A.D. King; FRR 2967 and FRR 2973, both from Indonesian dried fish, K.A. Wheeler.

Aspergillus penicillioides **Spegazzini** - Revta La Plata Univ. Fac. Agron. Vet. 2: 246, 1896 – Fig. 4.
    Aspergillus vitricola Ohtsuki, Botan. Mag., Tokyo 75: 436, 1962.
    Aspergillus glaucus var. *tonophilus* Ohtsuki, Antiques and Art Crafts Japan 13: 22, 1956.

Colonies on CYA up to 5 mm in diameter, but sometimes only microcolonies, of white mycelium only. Growth on MEA usually limited to microcolonies, occasionally colonies up to 5 mm in diameter formed, similar to those on CYA. Colonies on G25N 8-14 mm in diameter, plane or centrally raised, sometimes sulcate or irregularly wrinkled, texture velutinous or lightly floccose; mycelium usually inconspicuous, white; conidial production moderate, heads typically radiate, uncommonly in loose columns also, coloured Dull Green to Dark Green (27C-F8); reverse pale to dark green. Colonies on CY20S varying from microcolonies up to 10 mm in diameter, similar to those on CYA, occasionally some dull green conidial production but conidiophores poorly formed; reverse pale. Colonies on MY50G 10-16 mm in diameter, plane or umbonate, relatively sparse, velutinous to floccose; conidial production moderate, Greyish Green to Dull Green (27C-D3); reverse pale. No growth on CYA at 37°C.
    Conidiophores borne from surface or aerial hyphae, showing optimal development on G25N or MY50G, stipes 150-300(-500) µm long, sometimes sinuous, with thin, smooth, colourless walls, enlarging gradually from the base, then rather abruptly to pyriform to spathulate vesicles; vesicles mostly 10-20 µm in diameter, usually fertile over two thirds of the area, bearing phialides only; phialides (7-)8-11 µm long; conidia borne as ellipsoids, at maturity ellipsoidal to subspheroidal, 4.0-5.0 µm in diameter or in length, with spinose walls.

Typification. In the absence of type material, Samson and Gams (1985) designated Herb. IMI 211342 as neotype of *A. penicillioides.* Cultures ex neotype include CBS 540.65, WB 4548 and FRR 3722.

Nomenclature. Although usually written "*A. penicilloides*", Spegazzini called this species "*A. penicillioides*". This is also more correct orthographically (Stearn, 1966).

Distinctive features. Apart from its more xerophilic character, resulting in smaller colonies on the more dilute media than the other species, *A. penicillioides* produces conidia as ellipsoids rather than cylinders.

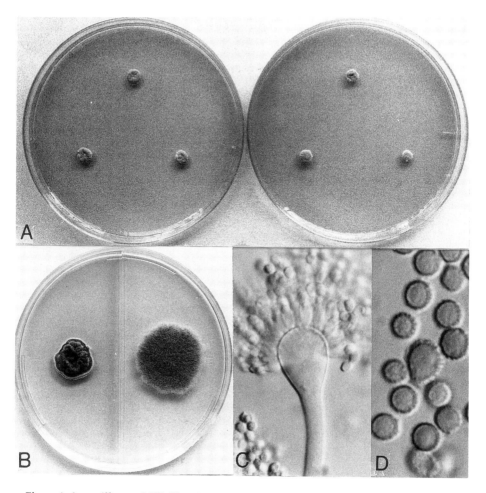

Figure 4. *Aspergillus penicillioides*. A. Colonies on CYA and MEA, 7 days. B. Colonies on CY20SC (left) and MY50G (right), 14 days. C. Conidial heads, x 750. C. Conidia, x 1875.

Occurrence. In our experience (Pitt and Hocking, 1985; Samson and van Reenen-Hoekstra, 1988), *A. penicillioides* is quite a common species in many kinds of dried foods and related habitats, but it has frequently been overlooked or misidentified. It is readily missed, because it requires media of reduced water activity for efficient enumeration or isolation. For example Hocking (1981) reported high counts of *A. penicillioides* from capsicums and other dried foods when dichloran 18% glycerol agar was used as the enumeration medium, but complete failure in detection using dichloran rose begal chloramphenicol agar, a medium of water activity much closer to those usually used in routine isolation work.

Christensen (1955) reported the frequent isolation of *A. restrictus* from grains stored just above the safe moisture limit, and described the damage inflicted on the grain. From details in his paper, it is clear that that he was referring to *A. penicillioides.*

Isolates examined. CBS 540.65 & ATCC 16910 (= FRR 3722, IMI 211392, WB 4548), ex neotype, from human skin, Brazil, D. Borelli; ATCC 16905 (= FRR 3734, IMI 108298), ex type of *Aspergillus vitricola*, from binocular lens, Japan, T. Ohtsuki; ATCC 14567 (= FRR 3735, IFO 6529), from binocular lens, Japan, T. Ohtsuki; CBS 116.26 & NRRL 151 (= FRR 151), from sugar cane product, Louisiana, U.S.A., Owen; CBS 539.65 & ATCC 16906 (= FRR 3725, WB 4962), from a gun firing mechanism, R. Emerson; ATCC 26634 (= FRR 3733), from spoiled rice, Japan, H. Ito; CBS 118.55 (= FRR 3720), from man, Netherlands; FRR 2766 and FRR 2855, both from Indonesian dried fish, Australia, K.A. Wheeler; FRR 2177 and FRR 2188, both from dried chili, Papua New Guinea, A.D. Hocking; numerous isolates from dry habitats including house dust (Samson and van der Lustgraaf, 1978), furniture, human skeletons etc.

ACKNOWLEDGEMENTS

The authors thank the curators of the culture collections and herbaria mentioned in the text for providing cultures and type specimens.

REFERENCES

BAINIER, G. 1907. Mycotheque de l'Ecole de Pharmacie. XII. *Bulletin trimestrielle de Societe de Mycologique de France* 23: 90-93.

BLOCHWITZ, A. 1929. Die Gattung *Aspergillus*. Neue Spezies, Diagnosen, Synonyme. *Annales Mycologici* 27: 205-240.

DALE, E. 1912. On the fungi of the soil. *Annales Mycologici* 10: 452-477.

—— 1914. On the fungi of the soil. II. *Annales Mycologici* 12: 33-62.

CHRISTENSEN, C.M. 1955. Grain storage studies. XVIII. Mold invasion of wheat stored for sixteen months at moisture contents below 15 percent. *Cereal Chemistry* 32: 107-116.

GAMS, W., CHRISTENSEN, M., ONIONS, A.H.S., PITT, J.I. and SAMSON, R.A. 1985. Infrageneric taxa of *Aspergillus*. In Advances in *Penicillium* and *Aspergillus* Systematics, eds. R.A. Samson and J.I. Pitt, pp. 55-62. New York and London: Plenum Press.

HOCKING, A.D. 1981. Improved media for isolation of fungi from foods. *CSIRO Food Research Quaterly* 41: 7-11.

KORNERUP, A. and WANSCHER, J.H. 1978. Methuen Handbook of Colour. London: Eyre Methuen.

PITT, J.I. 1979. The Genus *Penicillium* and its Teleomorphic States *Eupenicillium* and *Talaromyces*. London: Academic Press.

PITT, J.I. and HOCKING, A.D. 1985. Fungi and Food Spoilage. Sydney: Academic Press.

RAPER, K.B. and FENNELL, D.I. 1965. The Genus *Aspergillus*. Baltimore, Maryland: Williams and Wilkins.

SAMSON, R.A. 1985. Typification of the species of *Aspergillus* and associated teleomorphs. In Advances in *Penicillium* and *Aspergillus* Systematics, eds. R.A. Samson and J.I. Pitt, pp. 31-54. New York and London: Plenum Press.

SAMSON. R.A. and LUSTGRAAF, B. van der. 1978. *Aspergillus penicilloides* and *Eurotium halophilicum* in association with house-dust mites. *Mycopathologia* 64: 13-16.

SAMSON, R.A. and VAN REENEN-HOEKSTRA, E.S. 1988. An Introduction to Food-borne Fungi. Baarn, Netherlands: Centraalbureau voor Schimmelcultures.

STEARN, W.T. 1966. Botanical Latin. New York: Hafner Publishing Company.

THOM, C. and RAPER, K.B. 1945. A Manual of the Aspergilli. Baltimore, Maryland: Williams and Wilkins.

ISOENZYME PATTERNS IN *ASPERGILLUS FLAVUS* AND CLOSELY RELATED SPECIES

R.H. Cruickshank[1] and J. I. Pitt[2]

[1]*Department of Agricultural Science*
University of Tasmania
Hobart, Tas. 7000

[2]*CSIRO*
Division of Food Processing
North Ryde, N.S.W. 2113, Australia

SUMMARY

Polyacrylamide gel electrophoresis was used to examine several kinds of exoenzymes from a number of isolates of *Aspergillus flavus, A. parasiticus, A. oryzae, A. sojae, A. tamarii* and *A. nomius.* Enzymes studied were pectinases, ribonucleases, amylases and proteases. Methods used were similar to those previously published, with some additional media.
Very well defined patterns were obtained for each enzyme and substrate system. *A. flavus, A. parasiticus, A. tamarii* and *A. nomius* produced distinct patterns. However, *A. oryzae* produced patterns very similar to those of *A. flavus,* and *A. sojae* to those of *A. parasiticus.* Consequently, isoenzyme patterns could not be used to distinguish the domesticated species from their wild types.

INTRODUCTION

Patterns of isoenzymes produced by electrophoretic techniques (zymograms) have proved to be of considerable taxonomic value with species in *Penicillium* subgenus *Penicillium* (Cruickshank and Pitt 1987 a, b). Continuing interest in the taxonomy of the mycotoxigenic and fermentation species in *Aspergillus* sect. *Flavi* prompted examination of the same techniques with important species in that series.

The taxonomy of *Aspergillus* sect. *Flavi* has had some controversial aspects recently. Kurtzman *et al.* (1986) reduced the important aflatoxin producing species *A. parasiticus* to the status of a subspecies of *A. flavus.* The food fermentation species *A. oryzae* and *A. sojae* were reduced to the status of varieties. Klich and Pitt (1985, 1988) and others have argued that such action is both taxonomically unsound and unacceptable for practical reasons: taxonomic separation of potent toxin producers and fermentation cultures at species level is essential.

This paper reports a study on some species in *Aspergillus* sect. *Flavi* by studies of patterns of enzyme electrophoresis.

MATERIAL AND METHODS

Fungi.

Almost 130 isolates from the FRR collection at CSIRO Division of Food Processing, North Ryde, were examined. The isolates were selected from the large range studied by Klich and Pitt (1988) and included the species *A. flavus, A. parasiticus, A. oryzae, A. sojae,* and

Modern Concepts in Penicillium and Aspergillus Classification
Edited by R. A. Samson and J. I. Pitt
Plenum Press, New York, 1990

A. tamarii, plus the newly described species *A. nomius* (Kurtzman *et al.*, 1987). Conventional taxonomy was carried out by standard methods (Raper and Fennell, 1965) and newly developed criteria (Klich and Pitt, 1988). Isolates for enzyme electrophoretic studies were examined blind, i.e. without any indication of conventional taxonomic results.

Enzyme electrophoresis.
Methods for the production, electrophoresis and detection of enzymes were generally as described by Cruickshank and Pitt (1987), but with minor changes and additions. Pectic enzymes were produced and examined as described. To extend the range of pectic enzyme detection to include lyases active at high pH, separate gels were incubated for 1 hr in 0.05 M tris-HCl buffer, pH 7.5, containing 4 mM $CaCl_2$, before staining with ruthenium red.

Amylase and ribonuclease were produced in a decoction from 10% fresh weight of potato in distilled water, before detection as described. Protease was also produced in this potato decoction. It was examined in 10.25% acrylamide (2.5% bisacrylamide) gels containing 0.05% glycinin. Glycinin was extracted from defatted soybean flour using a method based on that of Smith and Circle (1938). After electrophoresis, gels were incubated for 1 hr in 0.1 M acetate buffer, pH 5.0, then stained for 2 hr in 0.1% crocein scarlet. Excess stain was removed by changes of distilled water.

RESULTS

The 129 isolates could be placed in four major groups by pectic zymogram evidence obtained under acidic conditions (Fig. 1). and under alkaline conditions in the presence of calcium ions (Fig. 2). This grouping was supported by characteristics of acid protease zymograms (Fig. 3). One group coresponded with *A. tamarii*, a second with *A. parasiticus*, and a third with *A. flavus*. The fourth group included isolates which were classifiable morphologically as *A. flavus* but were atypical of that species, in that both B and G aflatoxins were produced, a characteristic of *A. parasiticus* (Klich and Pitt, 1988). These isolates belonged to *A. nomius* (Kurtzman *et al.*, 1987).

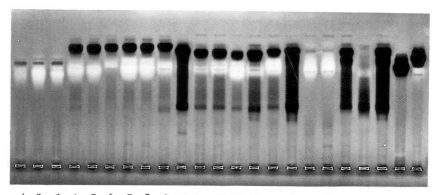

Figure 1. Pectic zymograms of gels obtained under acidic conditions (0.1 M malic acid). Numbers refer to isolates in Table 1.

Table 1. Isolates used in the production of figures

Lane	Isolate[a]	Species	Source
1	ATCC 269501	A. tamarii	Black pepper, USA
2	MRC 2080	A. tamarii	Dried bean, Mozambique
3	VDR 32	A. tamarii	Sunflower seed, Swaziland
4	ATCC 26804	A. parasiticus	Peanut, USA
5	MRC 3425	A. parasiticus	Oats, S. Africa
6	VDR 4	A. parasiticus	Sunflower seed, S. Africa
7	ATCC 26768	A. parasiticus	Dried sausage, Poland
8	MRC 200	A. parasiticus	Corn, Mozambique
9	VDR 3	A. parasiticus	Soil, S. Africa
10	ATCC 10196	A. oryzae	Pine board, USA
11	MRC 4243	A. flavus	Apricot kernel, S. Africa
12	VDR 18	A. flavus	Sunflower seed, Namibia
13	ATCC 24109	A. flavus	Black pepper, USA
14	MRC 1174	A. flavus	Sorghum, S. africa
15	VDR 8	A. flavus	Sunflower seed, Lesotho
16	ATCC 22546	A. flavus	Corn, USA
17	MRC 1318	A. flavus	Maize malt, S. Africa
18	VDR 23	A. flavus	Sunflower seed, S. Africa
19	ATCC 28540	A. flavus	Buckwheat, Japan
20	MRC 196	A. oryzae	Cassava, Mozambique
21	VDR 20	A. flavus	Fenugreek seed, S. Africa
22	ATCC 15546	A. nomius	Wheat, USA
23	IMI 190557	A. nomius	Turmeric, india

[a] Culture collections: ATCC, American Type Culture Collection, Rockville, MD, USA; IMI, CABI International Mycological Institute, Kew, Surrey, UK; MRC, National Research Institute for Nutritional Diseases, Tygerberg, South Africa; VDR, W. B. van der Reit, Pretoria, South Africa.

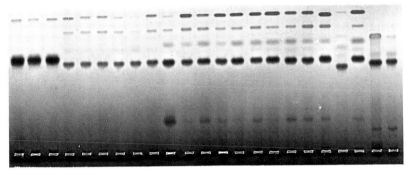

Figure 2. Pectic zymograms of gels incubated for 1 hr in 0.05 M tris-HCl buffer, pH 7.5, containing 4 mM $CaCl_2$. Numbers refer to isolates in Table 1.

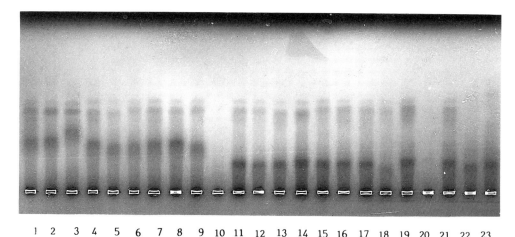

Figure 3. Acid protease zymograms. Numbers refer to isolates in Table 1.

The amylase zymograms of *A. flavus* (Fig. 4) and the ribonuclease zymograms of *A. parasiticus* and *A. flavus* (Fig. 5) provided evidence of subgrouping. For example, the *A. parasiticus* isolates could be placed in two subgroups by the Rf of their fast-migrating ribonuclease isoenzymes, as in lanes 4-6 and 7-9 in Fig. 5. These subgroups did not correspond with specific morphological properties, so far as could be determined, or, in *A. flavus*, with the presence or absence of aflatoxin production.

The isolates of *A. sojae* examined grouped with *A. parasiticus* and those of *A. oryzae* with *A. flavus*. One isolate of *A. oryzae*, MRC 196, gave aberrant results (lane 20 in each zymogram). The reason for this is not understood.

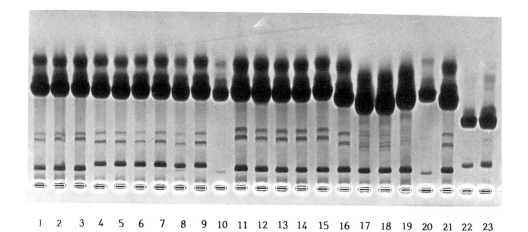

1 2 3 4 5 6 7 8 9 10 11 12 13 14 15 16 17 18 19 20 21 22 23

Figure 4. Amylase zymograms. Numbers refer to isolates in Table 1.

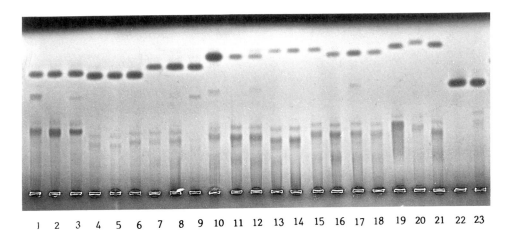

Figure 5. Ribonucelase zymograms. Numbers refer to isolates in Table 1.

DISCUSSION

This study has shown that the major species in *Aspergillus* sect. *Flavi*: *A. flavus*, *A. parasiticus*, *A. tamarii* and *A. nomius* can be effectively differentiated on the basis of enzyme electrophoretic patterns. The correlation with morphological taxonomy, as previously determined by Klich and Pitt (1988), was absolute for the 129 isolates studied. This provides independent evidence of the effective nature of the taxonomic criteria selected by Klich and Pitt (1988) for differentiating *A. flavus* from *A. parasiticus*, particularly spore surface texture. Moreover, the correlation between enzyme electrophoretic patterns and secondary metabolite production is also absolute for the 129 isolates: *A. flavus* produces aflatoxin B_1 and B_2 or cyclopiazonic acid, or both, while *A. parasiticus* produces aflatoxin G_1 and G_2 as well as B_1 and B_2, but never cyclopiazonic acid.

Correlations shown between electrophoretic patterns, secondary metabolite production and morphological taxonomy in *Penicillium* subgen. *Penicillium* are also evident in the *Aspergillus* section studied here.

The anomolous isolates reported by Klich and Pitt (1988) to have the morphology of *A. flavus* but to make G aflatoxins have been shown to belong to the newly described species *A. nomius*, which is clearly distinct. *A. nomius* isolates have the morphology of *A. flavus*, except that sclerotia, produced by some isolates, are smaller and vertically elongate, not spherical like those of *A. flavus*. This morphological distinction correlates with their distinctive isoenzyme patterns and the production of both B and G aflatoxins.

Enzyme electrophoretic patterns do not enable differentation of *A. flavus* and *A. oryzae*, or *A. parasiticus* and *A. sojae*. It has already been established (Kurtzman *et al.*, 1986; Klich and Pitt, 1988) that these two pairs of species are very closely related. However, as has been argued previously (Klich and Pitt, 1988), a very strong case exists for maintaining separate species for the food fermentation Aspergilli, regardless of the closeness of those relationships. Most isolates of *A. oryzae*, and all known of *A. sojae*, can accurately be described as domesticated fungi. Taxonomy and the world in general is best served by maintaining separate species names for such important fungi.

REFERENCES

CRUICKSHANK, R.H. and PITT, J.I. 1987a. The zymogram technique: isoenzyme patterns as an aid in *Penicillium* classification. *Microbiological Sciences* 4: 14-17.

—— 1987b. Identification of species in *Penicillium* subgenus *Penicillium* by enzyme electrophoresis. *Mycologia* 79: 614-620.

KLICH, M.A. and PITT, J.I. 1985. The theory and practice of distinguishing species of the *Aspergillus flavus* group. *In* Advances in *Penicillium* and *Aspergillus* Systematics, eds. R.A. Samson and J.I. Pitt, pp. 211-220. New York and London: Plenum Press.

—— 1988. Differentiation of *Aspergillus flavus* from *A. parasiticus* and other closely related species. *Transactions of the British Mycological Society* 91: 99-108.

KURTZMAN, C.P., SMILEY, M.J., ROBNETT, C.J. and WICKLOW, D.T. 1986. DNA relatedness among wild and domesticated species in the *Aspergillus flavus* group. *Mycologia* 78: 955-959.

KURTZMAN, C.P., HORN, B.W. and HESSELTINE, C.W. 1987. *Aspergillus nomius*, a new aflatoxin-producing species related to *Aspergillus flavus* and *Aspergillus tamarii*. *Antonie van Leeuwenhoek* 53: 147-158.

RAPER, K.B. and FENNELL, D.I. 1965. The Genus *Aspergillus*. Baltimore, Maryland: Williams and Wilkins.

SMITH, A.K. and CIRCLE, S.J. 1938. Peptization of soybean proteins - extraction of nitrogenous constituents from oil-free meal by acids and bases with and without added salts. *Industrial and Engineering Chemistry* 30: 1414-1418.

DIALOGUE FOLLOWING DR. PITT'S PRESENTATION

SAMSON: Doesn't your data prove, in fact, that Kurtzman et al. were right to lump these species together? If your data show that the species are so very closely related, doesn't this support their point of view?

PITT: Our data doesn't necessarily support their point of view, but it doesn't contradict it either. *Aspergillus flavus* and *A. parasiticus* are really quite distinct species. Klich and Pitt (1988) showed how to separate these species quickly and accurately with the microscope. There are very important differences between these species, and we must keep them separately. I have recently been carrying out a study on the physiology of toxigenic fungi and sorting out what we know about water relations, temperature relations and pH. Most of the papers in the literature that talk about *A. flavus* and the production of aflatoxin in relation to environmental conditions are really talking about *A. parasiticus*. The critical difference is that *A. flavus* only produces B aflatoxins. *A. parasiticus* produces both B and G aflatoxins. Most people who have reported on the physiology of *A. flavus* have in fact been working with *A. parasiticus*, and this is clear from their reports of which aflatoxins were produced. The literature, in fact, is very confused. Concerning *A. oryzae* and *A. sojae*, the question of whether they are the same as *A. flavus* and *A. parasiticus* is, in practical terms, irrelevant. Regulatory authorities and the food industry must have distinct names for the species that are important in food fermentations.

PATERSON: Could you explain why the two *A. oryzae* isolates looked so different in their pectinase patterns?

PITT: No. Dr. Cruikshank did not regard the differences as significant. I might add that the correlation between the isoenzyme patterns and types of mycotoxin produced was absolute.

CHRISTENSEN: What is the source of *A. nomius*?

SAMSON: It was isolated from mouldy wheat and diseased alkali bees.

PITT: It has sclerotia that are indeterminate and kind of ant hill shaped. Some isolates don't make these sclerotia, and they look just like *A. flavus*. However, they make B and G aflatoxins like *A. parasiticus*. It was described by Kurtzman *et al.* (1987) and is a good species.

CHRISTENSEN: I remember seeing such a culture in Raper's collection. How does *A. nomius* differ from *A. leporis*. It is characterized by columnar heads and also has indeterminate sclerotia.

FRISVAD: *A. leporis* produces lots of beautiful secondary metabolites and is a distinct species.

CHRISTENSEN: *A. leporis* is quite a characteristic species in sage brush grasslands in Wyoming, as are Artemesia and antelope.

PITT: I've never seen *A. leporis*. I'm interested in aflatoxin production, not in dung.

SUGIYAMA: *A. leporis* has a ubiquinone Q-10 system. In contrast, most of the other species in section *Flavi* have a Q-10(H_2) system. I think *A. leporis* should be excluded from the section *Flavi*.

6

COMPUTER-ASSISTED IDENTIFICATION OF PENICILLIA AND ASPERGILLIA

COMPUTER APPLICATIONS IN *PENICILLIUM* AND *ASPERGILLUS* SYSTEMATICS

M.A. Klich

United States Department of Agriculture
Southern Regional Research Center
New Orleans, Louisiana 70179, USA

SUMMARY

Computer technology is having an impact on several major areas of concern to *Penicillium* and *Aspergillus* systematists. Use of Database Management Systems and other computer programs is allowing curators of culture collections to readily obtain and distribute information on strains, species and other taxa. Numeric classification systems require computers to analyse the empirical data on which they are based. Computers are also being used in fungal identification systems. The need for a multiuser database system for *Penicillium* and *Aspergillus* systematics is discussed.

INTRODUCTION

At the *Penicillium* identification workshop held by Dr J. I. Pitt and myself in Athens, Georgia, USA in 1985, the need for an improved or alternate identification system for this genus became obvious. The idea of developing a computer-assisted key to the common species of *Penicillium* was conceived at that time. Since computer keys had not been developed for many filamentous fungi, and users were not familiar with them, we decided to simply computerize a key form that most mycologists are familiar with, a synoptic key. I wrote the program in FORTRAN, using data from Pitt's books (Pitt, 1979; Pitt, 1985) and files, information in other published descriptions (e.g. Raper and Thom, 1949; Samson *et al.*, 1976), data gathered by participants in the workshop, and data gathered in my laboratory. This provided information on several isolates from each species, grown at more than one location. The data clearly demonstrated that many characters, including colony diameter, colour and texture, varied when a single isolate was grown at different times and in different locations. Several microscopic characteristics, however, were very consistent. These characters were emphasised in PENKEY.

Most of the isolates used in the initial development of PENKEY had been held in collections for some years. To obtain information on more recent isolates, PENKEY was distributed on an experimental basis to researchers and university personnel in Australia and the USA. Users were asked to return completed data sheets both for isolates for which PENKEY provided logical identifications and for ones which it did not. Data sheets were then checked against cultures in one of our laboratories. Data were also compiled on fresh isolates sent directly to us for identification. Variability, especially in macroscopic characters, remained high. However, when PENKEY was used in the *Penicillium* and *Aspergillus* identification workshop taught by us in August, 1988 in Japan, about 80% of identifications were correct. By this time, the ultimate need for a databank/database with information from many isolates grown at many locations was obvious. From such a database, systems could be devised wherein probabilities of occurence for the various character states, such as a colony diameter, could be utilized. Such a database cannot be

Modern Concepts in Penicillium and Aspergillus Classification
Edited by R. A. Samson and J. I. Pitt
Plenum Press, New York, 1990

created in two or three years. So while we continue to offer identifications in return for completed data sheets for the various versions of PENKEY, time exists to consider methods for establishing the best possible *Penicillium* database system. The rest of this paper is dedicated to background, thoughts and ideas that may be useful in creating computer-based systems for *Penicillium* and *Aspergillus* systematics.

Computers and computer usage have evolved rapidly since the first commerical computers were developed in the 1950s (Gore and Stubbe, 1983). At that time the hardware contained vacuum tubes and software programming was difficult and very tedious. By the mid 1960s vacuum tubes were replaced by transistors and computer languages such as FORTRAN and COBOL were introduced. These mainframe computers were large and expensive, and for the most part, data were processed at central locations using punched cards. As integrated circuits were developed, smaller computers became feasible, leading to an explosion in the development of minicomputers and microcomputers since the late 1970s. Computers are now relatively inexpensive and accessible to virtually everyone in developed countries. With the introduction of 'user friendly' software packages, a computer user no longer needs to have any real understanding of computer programming. Of special interest to those of us needing to organize large volumes of data has been the development of data base systems in which data, entered once, may be retrieved for many purposes. Commercial software packages called Data Base Management Systems (DBMS) are now available and contain programs to create, access and update databases.

It should be noted that coevolution of computer technology and the use of computers in *Aspergillus* and *Penicillium* systematics or in other disciplines has not been straightforward. A strong human factor exists in the decision to use computers. The feeling that computers might take over peoples' jobs, and the inherent tendency to avoid the unfamiliar, have delayed the acceptance of computers by many potential users. Until recently, computer software was not 'user friendly' and potential users had to rely on the expertise of computer scientists.

An amazing parallel exists in peoples' perceptions of computer science and systematics. Both disciplines have a large public service component. Both computer scientists and systematicists must develop usable systems for non-experts. Both sciences have been accused of using jargon indecipherable to those outside the discipline and maintaining an aura of being an exclusive priesthood. In short, some of us have not been 'user friendly'. Both disciplines have an inherent subjectivity such that even experts in the field sometimes disagree. Both disciplines are dynamic. New computer systems seem to appear with great frequency which is confusing to casual computer users. Users of fungal systematics often become disgruntled at simple name changes let alone new taxonomic systems!

In spite of the problems in using computers, their potential utility has not been disregarded by systematists. There are three main areas of computer application in systematics: bookkeeping, e.g. cataloging curatorial, bibliographic and biogeograpic information; classification, e.g. numerical taxonomy; and identification. I will briefly review relevant work in each area and then discuss potential applications to an *Aspergillus* or *Penicillium* database system.

BOOKKEEPING

TAXIR, written for mainframe computers in the 1960s, was one of the earliest computer systems for organization of data on organisms. In the U.K. it became known as EXIR.

TAXIR has a flat file structure which means that the information must be in a form similar to a two-dimensional table. This is an efficient and rapid system, but not suited for data structures that cannot be construed as flat files. Various solutions to this problem have been devised. TAXIR has a limit of 112 characters per state and problems such as more than one authority, etc. are difficult to solve.

Adey *et al.* (1984) used EXIR to develop a database for the Vicieae (the tribe of legumes including vetches and peas), consisting of five database files (nomenclatorial, morphological, geographical, curatorial, and chemical). Programs associated with EXIR include EXIRPOST, used to tabulate retrieved material or to prepare it for input into other application programs, while CONFOR is used to create printed descriptions and SYNONYMS produces nomenclatorial lists.

Anderson *et al.* (1976) used a series of FORTRAN programs to assemble data on microscopic soil fungi on the computer so that these data could be used at a later date in published monographs. The information included reference lists, taxonomic, physiological, and ecological data. The information from these files was used to create the excellent reference book "Compendium of Soil Fungi" (Domsch *et al.*, 1980).

Many curatorial databases have been developed for individual herbaria and culture collections. Wetmore (1979) used a DBMS called SYSTEM 2000, and keypunched cards to computerize the herbarium at University of Minnesota. The commercial packages Datastar, Supersort, and Wordstar were used in a system developed at the University of Surrey for the culture collection there (Bryant, 1983). Wolff *et al.* (1982) used dBASE II to create a culture collection database at the University of Waterloo, Ontario, Canada . The New York Botanical Garden uses dBASE III PLUS for its record keeping (Thiers, 1988).

Larger microbial database systems have included the Canadian Fungal Collection Database (National Research Council of Canada, Halifax); the Microbial Culture Information Service (Department of Trade and Industry, UK) which includes physiological and biochemical data useful to industry; the Microbial Strain Data Network (UK); and the World Data Center on Microorganisms sponsored by the World Federation of Culture Collections (RIKEN, Saitama, Japan). These systems are coordinating several collections and making collection information more readily accessible to outside users.

Recently, the Microbial Information Network Europe (MINE) has been developed for microbial culture collections in Europe (Gams *et al.*, 1988). The main function of MINE is to provide a standardized access system for information in the major European culture collections. The curators of the institutions participating in MINE have agreed to a general format for computerized strain data which thereby can easily be exchanged. In this hierarchical system (using BASIS software), a strain record is created for each strain, and a species record to which strains may be linked contains more general data on the species. MINE also contains files for synonyms and for teleomorph/anamorph connections. Input into this system is strictly standardized and data is validated before being entered into the database. The system also includes a security system to prevent unauthorized use.

NUMERICAL TAXONOMY

It is perhaps unfortunate that the first introduction of many taxonomists to the use of computers in taxonomy was "Principles of Numerical Taxonomy" (Sokal and Sneath, 1963). Several negative comments at the beginning of the book on the training, background and intelligence of taxonomists could not have endeared the authors to many taxonomists at the time. Perhaps these comments helped engender a negative attitute toward the proposed methods. The 1963 book was also idealistic in what numerical taxonomy could accomplish. McNeill (1984) summarized the perceptions of early numerical taxonomy as a four step process: 1) select a large number of attributes; 2) measure the attributes on the population in question; 3) feed the data into the computer (black box); and 4) obtain THE objective classification. By the time when Sneath and Sokal published their second book (Sneath and Sokal, 1973), numerical taxonomists were much less idealistic.

In the ten years between 1963 and 1973, 13 papers were published which used numerical taxonomic techniques on filamentous fungi. Some studies supported existing taxonomic interpretations. For instance, a study of *Helminthosporium* and closely related taxa supported the separation of *Drechslera, Bipolaris* and *Stemphylium* as distinct genera (Ibrahim and Threlfall, 1966). However, the results of other studies were often unacceptable, even to the authors. Kendrick and Weresub (1966) used the methods suggested by Sokal and Sneath (1963) in a study of the resupinate Basidiomycetes. They noted in their abstract that "using haphazardly assembled, equally weighted characters results in a haphazard classification, at least in the Basidiomycetes." The discussion of this method in the paper itself was even less kind. Sneath and Sokal (1973, p. 420) responded to this criticism by stating again that phenetic techniques will not necessarily yield the same relationships as traditional methods.

Some taxonomists did not realize immediately that the purpose of numerical taxonomy was to construct classes rather than to help taxonomists assign indiviuals to previously established classes. There has been a continuing conflict between proponents of phylogenetic classifications and hierarchical (phenetic) classifications. Computer based systems are not intended to reflect evolution (Sneath and Sokal, 1973). However, as Gower (1975) states, "Evolutionists ask that, if the hierarchies are not estimates of phylogeny and usually cannot be used in diagnostic keys because they are polythetically based, what use are they?" Numerical classification has not lived up to its promise of providing unambiguous and respected methods for constructing classifications, because, according to Heywood (1984), there is no one 'true' or 'correct' classification. Different classifications will be produced when different methods are used.

A different kind of criticism of numerical classifications is that they are so frequently concordant with conventionally produced classifications that they are not worth the effort involved to produce them (Heywood, 1984). This argument could be interpreted as either indicating that the previous classifications were essentially correct or that it has not been possible to eliminate subjectivity in the numerical techniques.

The concept of unweighted characters in the Sokal and Sneath (1963) method was another heavily criticised aspect of numerical taxonomy (Hogeweg, 1976; Kendrick, 1965). Various methods have been developed to give more weight to some characters than to others. To Sokal and Sneath (1963), this process was seen as adding subjective factors into the numerical taxonomic system. Kendrick and Proctor (1964) found that in unweighted systems similarities or differences in secondary characters could outweigh similarities or differences in primary characters, leading to anomalies in the relationships between the taxa as measured by the similarity coefficients. In *Aspergillus* and *Penicillium*, for instance, presence of metulae would be considered a primary character, and size, shape, or number of metulae secondary characters. In an unweighted analysis, the similarity coefficient for species with metulae might be artificially high or low because of similarities or differences in the secondary characters. In their study, Kendrick and Proctor (1964) gave additional weight to primary characters, and were able to construct a satisfactory dendrogram for the Hyphomycetous species they were examining. They did not define 'satisfactory' but apparently meant a system close to that derived by standard taxonomic methods.

One benefit of numerical taxonomy has been that it has forced taxonomists to reappraise their methodology. For the most part, taxonomists have accepted the fact that there is an inherent subjective component in this discipline, but at the same time acknowledge that empirical analysis of data is a vital part of developing modern classification systems. Frank and Zwanziger (1988) summed up the situation well: "the initial enthusiasm for numerical methods has smoothed down; the main reasons are both misinterpretation of the aims and exaggerated expectations in the infallibility of the mathematical apparatus." They suggest that taxonomists remain aware that computers are only an aid in taxonomy: computers can only make suggestions, and should never dictate solutions.

IDENTIFICATION

Part of taxonomy is a service industry. Once a classification system has been established, it must be communicated in the most efficient way possible to users. Computers can help to define characters most appropriate for identification keys and even write identification keys based on information in a classification system. Payne and Preece (1980) reviewed the theory and practice of keys and key development as well as the statistical theory involved in creating efficient keys of all kinds.

Several systems are available for generating keys. Hall (1970) created a dichotomous key system in which the couplets were chosen by the computer from a data matrix. At the time this required a mainframe computer. Johnston (1980) used the computer to generate polyclaves (punch card systems) from data matrices. Rhoades (1988) has developed a taxonomic database which will generate synoptic keys and has constructed such keys to several groups of fungi.

Kendrick (1972) used computer graphics to write a key to the didymosporous Hyphomycetes. In this ingenious key, images appear on the screen and the user picks one. After a series of choices, the final image contains the name of the fungus, a diagram of its salient features and a brief written description. Margot *et al.* (1984) devised an on-line identification program for mushrooms. Written in FORTRAN, this key is based on a data matrix and is essentially synoptic.

One of the earliest large computer-assisted mycological identification keys was to the yeasts (Barnett and Pankhurst, 1974). In this key, 52 physiological tests were used to identify 434 species of yeasts. The species were divided into eight groups, and then a dichotomous key was generated to the species within each group. The six physiological tests used to divide the species into eight groups were selected based on their separation coefficient values. The character which best divided the species into two equal groups had the highest separation coefficient. The dichotomous key for each group was then generated by the computer, using the characters with the lowest levels of intraspecific variability (Barnett and Pankhurst, 1974).

The first major review of biological identification systems using computers was published in 1975 (Pankhurst, 1975). It was generally agreed that computer-stored dichotomous keys had little advantage over written dichotomous keys. However, several dichotomous keys had been generated by the computer using 'recursive algorithms', which repeatedly divided the taxa into two mutually exclusive subsets, with each application producing one couplet in the key. Discriminant analysis, which involves weighting characters by their ability to help distinguish between two taxa or clusters of taxa, was already being used to aid identification procedures involving taxa with many overlapping characters. The disadvantage of discriminant analysis is that it requires the observer to record a large number of characterisitics per specimen. Cluster analyses and similarity coefficients of various sorts were also being used at that time.

Several problems were recognized with the computer techniques (Pankhurst, 1975). Taxonomists were often not concerned with empirical analysis, so were often inconsistent in their species descriptions, leaving out characters which would be necessary for developing computer-based identification systems. There was apparently a feeling at the time that if computers could perform a task, they would do it better than humans. Also, many taxonomists were unfamiliar with the capabilities and limitations of computers in their field. Now, many taxonomists are familiar with these limitations. Species descriptions in recent works tend to be more detailed, and most people are much more realistic about use of computers because computers are more a part of daily life.

PREVIOUS USE OF NUMERICAL METHODOLOGIES IN STUDIES OF *ASPERGILLUS* AND *PENICILLIUM* SYSTEMATICS

A variety of approaches to *Penicillium* and *Aspergillus* taxonomy have made use of numerical methodology. Cluster analyses have been used to study relationships based on long chain fatty acids in both *Aspergillus* and *Penicillium*. Seventeen species of *Penicillium* fell into three major groups, but the groups bore no resemblance to any known taxonomic scheme (Dart *et al.*, 1976a). The same was true for cluster analyses of long chain fatty acids for 14 species of *Aspergillus*, although intraspecific variation in *Eurotium amstelodami* and *A. niger* was quite low (Stretton *et al.*, 1976; Dart *et al.*, 1976b). When the *Aspergillus* study was repeated at different temperatures, the clusters were quite different, with only two major clusters, one of which contained all four *Eurotium* species (Lee *et al.*, 1977). In a study of both *Aspergillus* and *Penicillium* it was concluded that the two genera could not be separated from one another by differences in their long chain fatty acid composition (Lee *et al.*, 1976).

In her revision of the black-spored Aspergilli, Al-Musallam (1980) conducted numerical analyses on 78 isolates related to *A. niger*. In unweighted and weighted analyses based on 28 morphological features, characters having the highest significance in forming the clusters were very similar. These included colony diameter on Czapek's medium, conidial head color, conidial ornamentation, stipe length and vesicle diameter. In each of the two analyses, two major clusters and several minor clusters (subclusters) were formed. The major clusters were recognized as species (*A. niger* and *A. foetidus*) and the subclusters as varieties and formae. A key to these taxa was written using the characters of importance from the numerical analyses.

To better understand the koji moulds used in Oriental fermentations, Murakami (1971) used both numerical and traditional methods to examined 406 isolates from *Aspergillus* sect. *Flavi*, including 306 strains of koji moulds. His numerical analysis was based on 20 features, each of which had three character states, some of which were qualitative (e.g. none, little, much) rather than empirical. Significant characters were established, including conidial diameter, conidial wall texture, pink conidial color on anisaldehyde medium and anisic acid medium, reverse coloration, production of aflatoxins, kojic acid and sclerotia. After these and other characters were tested in the laboratory, the most consistent characteristics were considered to be most suitable for use in identification keys. Many of the above characteristics, as well as presence or absence of metulae and wrinkling of the colony reverse, were used to write a key to the koji moulds.

Christensen (1981) generated similarity coefficients (C) from data in a synoptic key to members of *Aspergillus* sect. *Flavi* to assess species relationships. The most closely related species were *A. toxicarius* and *A. flavus* (C= 0.96). She suggested that *A. toxicarius* is an occasionally biseriate *A. parasiticus*. Others have also considered *A. parasiticus* and *A. toxicarius* to be synonymous (eg. Klich and Pitt, 1985). *A. parasiticus* was also found to be quite similar to *A. sojae* (C= 0.81). However, *A. oryzae* was 'comparatively distinct' from *A. flavus* (no value given).

Klich and Pitt (1985) also studied *Aspergillus* sect. *Flavi*. Based on discriminant analysis on morphological features of 62 isolates, five characters were judged most satisfactory for separating *A. flavus*, *A. oryzae*, *A. parasiticus*, *A. sojae*, and *A. tamarii*. These were colony colour; conidial size and surface texture; vesicle diameter; conidiophore length; and presence or absence of metulae. Several of these characters were similar to those used by Murajkami (1971) to distinguish koji moulds.

The Subcommission on *Penicillium* and *Aspergillus* Systematics (SPAS) of the International Commission on Taxonomy of Fungi is currently completing its first study. Isolates of the four closely related species *P. glabrum*, *P. spinulosum*, *P. purpurescens*, and *P. montanense* were examined for standard morphological features, secondary metabolites, isoenzymes, and DNA restriction fragment length polymorphisms. Species differentiation by the various approaches was in fairly good agreement. To determine the best discriminating characters, the morphological data were analyzed using discriminant analysis. In the preliminary analyses, conidial wall texture, maximum diameter on CYA and minumum on G25N (Pitt, 1979), minimum phialide width and maximum vesicle diameter were the most useful characters. Currently, a larger sample is being analyzed to test and refine the model.

A numerical taxonomic study of *Penicillium* subgen. *Penicillium* is currently being conducted at the CAB International Mycological Institute (CMI) (Bridge *et al.*, 1985). This large undertaking, involving the acquisition and analysis of 195 characters for each of 300 isolates, will be reported elsewhere in these Proceedings (Bridge *et al.*, 1990).

In the past few years, several computer assisted identification systems have been developed. PENKEY (Klich and Pitt, 1988) is a computer assisted synoptic key to common *Penicillium* species. PENNAME (Pitt, 1990) is a dBASE III Plus computer key to the same set of species. A key to *Penicillium* subgen. *Penicillium* has been developed in Apple BASIC for the food industry (Williams, 1990), and a further key to this subgenus (Bridge, 1990) is also presented in these proceedings.

TAXONOMIC DATABASE SYSTEMS FOR *ASPERGILLUS* AND *PENICILLIUM*

Development of a multiuser/multicontributor database systems for *Aspergillus* and *Penicillium* would faciltiate research efforts on these two genera. First they would allow us to attain a greater understanding of intra- and inter-specific variability so that future classification and identification systems may be based, as much as possible, on empirical data. Second, these database systems would serve as focal points for cooperative or coordinated efforts among researchers at different institutions. Crovello *et al.* (1984) pointed out that systematics has suffered from the lack of coordination among institutions. A shared system would prevent duplication of efforts, provide data for discussion, and provide taxonomic researchers with more information than any one researcher alone could obtain. As new biochemical and genetic methods are developed they will provide additional tools for taxonomy, and these can easily be integrated into database systems.

What should be the characteristics of the ideal *Aspergillus* or *Penicillium* database system? First it should be easy to create and update files as the systems expand. It should be easy to create and combine subsets of data from different files in the system. It should be amenable to statistical manipulation, key generation, etc. The information and current updates should be easy to transfer among users and capable of integration into other database systems such as those of culture collections. Files or subsets of files should be easy to print. Finally, the system should be inexpensive to run.

Several major decisions will be faced by the potential creators and users of such a database before commencing the work. One need only thumb through a recent computer magazine or a review such as "Databases in Systematics" (Allkin and Bisby, 1984) to realize that there are a number of potential Database Management Systems from which to choose. The strengths and weaknesses of these relative to the objectives and desirable characteristics of the *Aspergillus* and *Penicillium* systems should need to be carefully considered. Decisions would need to be made on the characters to be entered. Although it is not difficult to add characters to some systems, a well defined intial set would reduce complications later.

The technical difficulties are probably minor compared to the practical problems in setting up a taxonomic database system to be used by participants from many institutions. How would the work be funded? Who would have access to the system? How would use of the system be acknowledged in publications? How often would the system be updated, and how would the updates be distributed? Perhaps these and other practical considerations can be discussed during the final session of this meeting.

REFERENCES

ADEY, M. E., ALLKIN, R., BISBY, F. A., WHITE, R. J. and MACFARLANE, T. D. 1984. The Vicieae database: an experimental taxonomic monograph. *In* Databases in Systematics, eds. R. Allkin and F. A. Bisby. Orlando, Florida: Academic Press, pp. 175-188.

AL-MUSALLAM, A. 1980. Revision of the black *Aspergillus* species. Ph.D Thesis. Utrecht, Netherlands: University of Utrecht.

ALLKIN, R. and BISBY, F. A., eds. 1984. Databases in Systematics. Orlando, Florida: Academic Press.

ANDERSON, T. H., BODENSTEIN, J. and DOMSCH, K. H. 1976. A partially computerized documentation system for microscopic soil fungi. *Canadian Journal of Botany* 54: 1709-1713.

BARNETT, J. A. and PANKHURST, R. J. 1974. A New Key to the Yeasts. New York: Elsevier Press.

BRIDGE, P. D. 1990. Identification of terverticillate Penicillia from a matrix of percent positive test results. *In* Modern Concepts in *Penicillium* and *Aspergillus* Classification, eds. R. A. Samson and J. I. Pitt, pp. 283-287. New York and London: Plenum Press.

BRIDGE, P. D., HAWKSWORTH, D. L., KOZAKIEWICZ, Z., ONIONS, A. H. S., PATERSON, R. R. M. and SACKIN, M. J. 1985. An integrated approach to *Penicillium* systematics. *In* Advances in *Penicillium* and *Aspergillus* Systematics, eds. R. A. Samson and J. I. Pitt, pp. 281-309. New York and London: Plenum Press.

BRYANT, T. N. 1983. A microcomputer-based information storage and retrieval system for the maintenance of records for a culture collection. *Journal of Applied Bacteriology* 54: 101-107.

CHRISTENSEN, M. 1981. A synoptic key and evaluation of species in the *Aspergillus flavus* group. *Mycologia* 73: 1056-1084.

CROVELLO, T. J., HAUSER, L. A. and KELLER, C. A. 1984. BRASS BAND (The Brassicaceae databank at Notre Dame): an example of database concepts in systematics. *In* Databases in Systematics, eds. R. Allkin and F. A. Bisby, pp. 219-233. Orlando, Florida: Academic Press.

DART, R. K., STRETTON, R. J. and LEE, J. D. 1976a. Relationships of *Penicillium* species based on their long-chain fatty acids. *Transactions of the British Mycological Society* 66: 525-529.

—— 1976b. Strain variation in *Aspergillus niger*. *Microbios Letters* 3: 183-185.

DOMSCH, K. H., GAMS, W. and ANDERSON, T. H. 1980. Compendium of Soil Fungi. London: Academic Press.

FRANK, H. M. and ZWANZIGER, H. 1988. Macrochemical color reactions of macromycetes. VI. Cluster analysis of color data. *Mycotaxon* 33: 91-96.

GAMS, W., HENNEBERT, G. L., STALPERS, J. A., JANSSENS, D., SCHIPPER, M. M. A., SMITH, J., YARROW, D. and HAWKSWORTH, D. L. 1988. Structuring strain data for storage and retrieval of information on fungi and yeasts in MINE, the Microbial Information Network Europe. *Journal of General Microbiology* 134: 1667-1689.

GORE, M. and STUBBE, J. 1983. Elements of Systems Analysis. 3rd ed. Dubuque, Iowa: Wm. C. Brown.

GOWER, J. C. 1975. Relating classification to identification. *In* Biological Identification with Computers, ed. R. J. Pankhurst, pp. 251-263. New York: Academic Press.

HALL, A. V. 1970. A computer-based system for forming identification keys. *Taxon* 19: 12-18.

HEYWOOD, V. H. 1984. Electronic data processing in taxonomy and systematics. *In* Databases in Systematics, eds. R. Allkin and F. A. Bisby, pp. 1-16. Orlando, Florida: Academic Press.

HOGEWEG, P. 1976. Iterative character weighing in numerical taxonomy. *Computers in Biology and Medicine* 6: 199-211.

IBRAHIM, F. M. and THRELFALL, R. J. 1966. The application of numerical taxonomy to some graminicolous species of *Helminthosporium*. *Proceedings of the Royal Society of London, Series B*, 165: 362-388.

JOHNSTON, B. C. 1980. Computer programs for constructing polyclave keys from data matrices. *Taxon* 29: 47-51.

KENDRICK, W. B. 1965. Complexity and dependence in computer taxonomy. *Taxon* 14: 141-154.

—— 1972. Computer graphics in fungal identification. *Canadian Journal of Botany* 50: 2171-2175.

KENDRICK, B. W. and PROCTOR, J. R. 1964. Computer taxonomy in the Fungi Imperfecti. *Canadian Journal of Botany* 42: 65-88.

KENDRICK, B. W. and WERESUB, L. K. 1966. Attempting neo-Adansonian computer taxonomy at the ordinal level in the Basidiomycetes. *Systematic Zoology* 15: 307-329.

KLICH M. A. and PITT, J. I. 1985. The theory and practise of distinguishing species of the *Aspergillus flavus* group. *In* Advances in *Penicillium* and *Aspergillus* Systematics, eds. R. A. Samson and J. I. Pitt, pp. 211-220. New York and London: Plenum Press.

—— 1988. A computer-assisted synoptic key to common *Penicillium* species and their teleomorphs. New Orleans, Louisiana: privately published.

LEE, J. D., STRETTON, R. J. and DART, R. K. 1976. Classification of *Aspergillus* and *Penicillium* species. *Microbios Letters* 2: 163-167.

LEE, J. D., DART, R. K. and STRETTON, R. J. 1977. Computerized classification of *Aspergillus*. *Transactions of the British Mycological Society* 69: 137-141.

MARGOT, P., FARQUHAR, G. and WATLING, R. 1984. Identification of toxic mushrooms and toadstools (Agarics) - an on-line identification program. *In* Databases in Systematics, eds. R. Allkin and F. A. Bisby, pp. 249-261. Orlando, Florida: Academic Press.

McNEILL, J. 1984. Taximetrics To-day. *In* Current Concepts in Plant Taxonomy, eds. V. H. Heywood and D. M. Moore, pp. 281-299. Systematics Special Vol. 25. Orlando, Florida: Academic Press.

MURAKAMI, H. 1971. Classification of the koji mold. *Journal of General and Applied Microbiology* 17: 281-309.

PAYNE, R. W. and PREECE, D. A. 1980. Identification keys and diagnostic tables: a review. *Journal of the Royal Statistical Society, Ser. A*, 143: 253-292.

PANKHURST, R. J., Ed. 1975. Biological Identification with Computers. Systematics Association Special Vol. 7. London: Academic Press.

PITT, J. I. 1979. The Genus *Penicillium* and its Teleomorphic States *Eupenicillium* and *Talaromyces*. London: Academic Press.

—— 1988. A Laboratory Guide to Common *Penicillium* Species. North Ryde, N.S.W.: CSIRO Division of Food Research.

—— 1990. PENNAME, a new computer key to common *Penicillium* species. *In* Modern Concepts in *Penicillium* and *Aspergillus* Classification, eds. R. A. Samson and J. I. Pitt, pp. 279-281. New York and London: Plenum Press.

RAPER, K. B. and THOM, C. 1949. A Manual of the Penicillia. Baltimore, Maryland: Williams and Wilkins.

RHOADES, F. 1988. PC-TAXON: a Taxonomic Database. Wentworth, New Hampshire: COMPress.

SAMSON, R. A., STOLK, A. C. and HADLOK, R. 1976. Revision of the subsection *Fasciculata* of *Penicillium* and some allied species. *Studies in Mycology, Baarn* 11: 1-47.

SNEATH, P. H. A. and SOKAL, R. R. 1973. Numerical Taxonomy: the Principles and Practice of Numerical Classification. San Francisco: Freeman and Sons.

SOKAL, R. R. and SNEATH, P. H. A. 1963. Principles of Numerical Taxonomy. San Francisco: Freeman and Sons.

STRETTON, R. J., LEE, J. D. and DART, R. K. 1976. Relationship of *Aspergillus* species based on their long chain fatty acids. *Microbios Letters* 2: 89-93.

THIERS, B. M. 1988. Herbarium Manager User's Manual "C" Version 1.0. Privately published.

WETMORE, C. M. 1979. Herbarium computerization at the University of Minnesota. *Systematic Botany* 4: 339-350.

WILLIAMS, A.P. 1990. Identification of *Penicillium* and *Aspergillus* - computer assisted keying. *In* Modern Concepts in *Penicillium* and *Aspergillus* Classification, eds. R. A. Samson and J. I. Pitt, pp. 289-294. New York and London: Plenum Press.

WOLFF, B. R., GLICK, B. R. and PASTERNAK, J. J. 1986. A microcomputer program for establishing and maintaining a database for laboratory culture collections. *Letters in Applied Microbiology* 3: 45-48.

PENNAME, A NEW COMPUTER KEY TO COMMON *PENICILLIUM* SPECIES

J.I. Pitt

CSIRO Division of Food Processing
North Ryde, N.S.W. 2113, Australia

SUMMARY

Problems in identifying *Penicillium* species are well recognised. Fundamentally, these result from a lack of completely stable characters of taxonomic value. Conventional dichotomous keys use a relatively few characters, and hence are sensitive to variation in them. The synoptic key is inherently more effective for this difficult genus, but the large number of species makes synoptic keys unwieldy to use. Designed to overcome this problem, PENNAME is a new computer based synoptic key to 68 common *Penicillium, Eupenicillium* and *Talaromyces* species. It is designed to be used with "A Laboratory Guide to Common *Penicillium* Species" (Pitt, 1988), but can be of value with other modern taxonomies as well. PENNAME is "user friendly", accepting keyed in data on colony and microscopic characters directly on screen. Output in the current experimental version consists of a list of species, with the number of characters by which each differs from the test isolate. Later versions are intended to be "stand alone", with simple taxonomic treatments of common species included.

A fundamental problem in the identification of *Penicillium* species is that nearly all characters of taxonomic value lack complete stability. Conventional dichotomous keys use a relatively few characters, and hence are sensitive to variation in those characters. A second problem, common to most fungal genera, is that data available is often incomplete, or questionable, making dichotomous keys cumbersome and difficult to use. The synoptic key (Leenhouts, 1966) is inherently more effective for difficult genera, but the large number of species in *Penicillium* makes a synoptic key unwieldy to use.

Designed to help overcome this problem, PENNAME is a new computer key to *Penicillium* species. At present, it will key the 52 most common species of the genus, together with 10 *Eupenicillium* and 6 *Talaromyces* species, but in time it will be expanded to accommodate all currently accepted species. At present, it is designed primarily for use with the second edition of "Laboratory Guide to Common *Penicillium* Species" (Pitt, 1988), and it keys the species included in that work. However, it will be of value as an adjunct to any modern taxonomy, provided the media, general methods and plating regime of Pitt (1979) are used, i.e. CYA, MEA and G25N at 25°, plus CYA at 5° and 37°, and incubation for 7 days.

PENNAME is basically a synoptic key, and operates by comparing input data with standard information contained in the program's databases. In this respect PENNAME is similar to PENKEY (Klich and Pitt, 1988). Like PENKEY, PENNAME is a "stand alone" program, and it can be run on any IBM PC (registered) or compatible computer, using MS-DOS (registered) or similar operating system.

Unlike PENKEY, which is written in FORTRAN, PENNAME is written in dBase III Plus (registered) software, and compiled in Clipper (registered). This leads to a number of advantages. PENNAME also has several interesting features relating to the data input

Modern Concepts in Penicillium and Aspergillus Classification
Edited by R. A. Samson and J. I. Pitt
Plenum Press, New York, 1990

system, the primary data output, and various secondary choices. These are described in more detail below.

Data input.
Input of data into PENNAME is particularly simple. Raw data on colony diameters and pigmentation, together with microscopic features of the unknown isolate are input directly onto a screen entry form. The PENNAME program then interacts directly with this data, interpreting it in a form in which it can be compared with standard data. The program calculates certain other parameters, i.e. metulae to phialide length ratio, and conidial length to width ratio, and uses these data also. Depending on responses to questions in the data input, PENNAME may call for supplementary data on a second screen entry.

Primary output.
The primary output provided by PENNAME is a list of eight species names, which the program considers to be the most likely identifications for the unknown. Beside each species name is displayed a "count" figure, which represents the number of characters by which the unknown differs from the standard description of that species. PENNAME is quite a stringent key: a count of 0 indicates a high probability that the unknown has been correctly classified, while a count greater than 4 indicates a high probability of a mismatch.

Secondary output.
The user then has the choice of looking more closely at the characters by which the unknown differs from a particular species. Entry of a three letter code for the species at the appropriate prompt produces a display of entered characters or data ranges which do not match those of the chosen species, and a display of the corresponding characters or data ranges which the chosen species does exhibit. Up to four such sets of differing characters will be displayed. Although designed to be used with the most closely matching species shown on the primary output display, any species may be compared with the unknown by entering the appropriate three letter code. The codes are mostly the same as the first three letters of each species name, but have been modified where two species have the same first three letters. A complete list of the codes is provided with notes on use of the key.

For reasons of space, the secondary output is rather cryptic for characters which involve numerical ranges. However, an explanation is provided with the notes to PENNAME, and usage of the key will rapidly lead to familiarity.

Probability figure.
Accompanying the secondary output is a "probability figure" designed to indicate the likelihood of the unknown being the species selected for comparison in the secondary output. It is emphasised that this figure is based on the input data, and cannot for obvious reasons take into account inaccurate raw data. The figure is based on calculations which relate both to the number of species with a low "count" figure, and a comparison of data ranges input for the unknown with standard ones for the selected species. It is also emphasised that the probability figure can only be taken as a guide at present.

Species diagnosis.

PENNAME also can provide brief diagnoses of all the included species. The user can indicate a wish to see a diagnosis of a particular species, again by typing in the three letter code for the species at the appropriate prompt.

Advantages of PENNAME.

As with all synoptic key systems, data entered into PENNAME may be as much or as little as is available. This provides great flexibility in use, including applications other than determinative taxonomy. However, In the author's opinion, PENNAME possesses several additional advantages over similar keys now available:

1. PENNAME is up to date, being based on Pitt (1988).
2. The databases used may be readily modified by the program originator, to allow updating of "standard" information, changes in species names, or the addition of extra species, as required.
3. Data entry is simple and direct, with no user manipulation necessary. The program calculates further useful data from the input also.
4. Most of the data to be entered is simply acquired, and requires little or no knowledge of *Penicillium* taxonomy. In particular, lack of certainty about penicillus type will rarely produce a misleading output. For example, although the experimental version displayed at this Workshop asks the user to differentiate where possible between penicilli belonging to subgen. *Furcatum* and subgen. *Biverticillium*, later versions will not. PENNAME will permit quite effective determinations for many species when no penicillus type is specified at all. Later versions will take advantage of this by suggesting a first run where doubt on penicillus type exists, and a second more definitive run based on the output from the first.
5. PENNAME allows considerable interaction with the user. The secondary output enables the user to quantitatively compare data from an unknown isolate with any number of known species. Diagnoses may also be checked directly within the program. It is believed this user interaction will be of great value.

Detailed notes on the use of PENNAME will accompany the software, which may be obtained on a 5.25 inch disc from the author.

REFERENCES

KLICH M. A. and PITT, J. I. 1988. A computer-assisted synoptic key to common *Penicillium* species and their teleomorphs. New Orleans, Louisiana: privately published.

LEENHOUTS, P.W. 1966. Keys in Biology. A survey and a proposal of a new kind. Proceedings Koninklijke Nederlanse Akademie van Wetenschappen, Series C, 69: 571-596.

PITT, J.I. 1979. The Genus *Penicillium* and its Teleomorphic States *Eupenicillium* and *Talaromyces*. London: Academic Press.

—— 1988. A Laboratory Guide to Common *Penicillium* Species. 2nd ed. North Ryde, N.S.W.: CSIRO Division of Food Processing.

IDENTIFICATION OF TERVERTICILLATE PENICILLIA FROM A MATRIX OF PERCENT POSITIVE TEST RESULTS

P.D. Bridge

CAB International Mycological Institute
Ferry Lane, Kew, Surrey, TW9 3AF, UK

SUMMARY

The fifty-seven most discriminatory characters from an integrated multidisciplinary numerical taxonomy were used to produce a percent positive matrix for 37 species and species groups of terverticillate and similar penicillia. Identification of unknowns against the percent positive matrix was determined by two methods: the first gave a normalized probability score and the second calculated the Modal Likelihood Fraction (MLF), which was based on both the highest probability attained and the theoretical highest probability. The matrix and identification procedures were tested with the results from over 100 known and unknown cultures. In most cases a normalized probability score of 0.99 or better to the correct taxon was obtained. The MLF scores obtained were very low, reflecting the high level of variation in most taxa; 10^{-9} appeared to be a suitable cutoff point, all except one isolate giving values in excess of this to the correct taxon.

The merits of this quantitative identification scheme are discussed, and the selection of the most suitable identification score is considered.

INTRODUCTION

The use of percent positive tables and computer programs to produce computer assisted identification matrices is well established in microbiology (Sneath, 1978; Willcox *et al.*, 1980; D'Amato *et al.*, 1981). The detailed methodology of identification matrices is well described elsewhere, and methods also exist for assessing their performance and discrimination (Sneath and Sokal, 1973; Willcox *et al.*, 1973; Willcox *et al.*, 1980; Sneath, 1980 a, b, c). Unlike keys, identification from a percent positive matrix uses all of the available information and so the effect of a single erroneous result or an atypical isolate is minimised. Most computer assisted systems will operate on the data available for an unknown and so can provide identifications from incomplete data sets (e.g. Sneath, 1979a).

The results from an integrated multidisciplinary numerical taxonomy of terverticillate penicillia which defined 37 taxa with 100 characters were used to produce an identification matrix (Bridge *et al.*, 1989a, b). The full data set was examined for the most discriminatory characters with the programs CHARSEP and DIACHAR (Sneath, 1979b, 1980a) and a matrix containing 57 characters for the 37 taxa was produced. The identification scores obtainable for the most typical member of each group were determined with the program MOSTTYP and overlap between groups was estimated with the program OVERMAT (Sneath, 1980b, c). These last two procedures suggested that the groups within the matrix were distinct on the characters included and that typical isolates should emerge with high identification scores. The full matrix has been published elsewhere (Bridge *et al.*, 1989b) and this contribution considers some particular features of the data and the identification coefficients.

Modern Concepts in Penicillium and Aspergillus Classification
Edited by R. A. Samson and J. I. Pitt
Plenum Press, New York, 1990

IDENTIFICATION COEFFICIENTS

A large number of different identification coefficients have been proposed for use with identification matrices. Coefficients can be based on the similarity of an unknown to the taxa, the distances of unknowns from the taxa, or the probability of the unknown belonging to the taxa; for full details see Sneath and Sokal (1973) or Willcox *et al.* (1980). Probability based coefficients are generally used, and values in the percent positive table are considered to be the probabilities of properties being recorded for the taxa. An important consideration in the use of probability coefficients and matrices is that they can be used to calculate the likelihood of an unknown belonging to any one of the taxa included. They assume however that the different taxa are distinct and that the unknown belongs to a taxon included in the matrix.

A number of formulae are available for calculating identification scores from probabilities, the most generally used being the Willcox score, which is based on Bayes' theorem (see Willcox *et al.*, 1973):

$$P(U \mid t) = \frac{P(U)\, P(t \mid U)}{\Sigma u\, P(U)\, P(t \mid U)}$$

where $P(U \mid t)$ is the probability for taxon U on t characters; $P(U)$ is the prior probability for taxon U and $P(t \mid U)$ is the probability that a member of taxon U will show the characters t.

This theorem gives a relative or normalized probability by dividing the final best absolute likelihood by the sum of the scores to all taxa. This method acts as a correction if some taxa have extremely variable character states and also indicates if there is only little difference between the best and other identifications. A further feature of Bayes' theorem is that it includes values for prior probalities, so that final scores can be weighted by the likelihood of isolating a particular taxon (see Sneath and Sokal, 1973; Willcox *et al.*, 1980). This is particularly useful if the data matrix contains both commonly and rarely isolated taxa. In practice, however, it is difficult to accurately estimate prior probabilities in advance and so they are usually set equal for all taxa and cancel out.

Identification scores based on Bayes' theorem are to a certain extent dependent upon the other taxa in the matrix. As a result, if an unknown does not belong in any group in the matrix, it may still give a high identification score if it is significantly less unlikely that it belongs to one taxon as compared to the others. One way of controlling this is to take account of any mis-matches in characters, so that an unknown that gave a high identification score but only poorly matched to that taxon would be noticed.

Identification scores that do not consider the other taxa directly in their calculation can also be useful in these circumstances. One example is the Modal Likelihood Fraction (MLF; Dybowski and Franklin, 1968), which is the ratio of the absolute likelihood that an unknown belongs to a taxon and the absolute likelihood that would be obtained for the most typical isolate on the characters used.

DETERMINATION OF IDENTIFICATION

The identification matrix was used with the program MATRIX to identify a collection of 52 previously identified isolates and 51 unidentified isolates. One isolate remained unidentified from this set and one isolate was identified to the correct species but to an incorrect variety. The criteria used in the program MATRIX were the Willcox identification score and the MLF. Correct identifications were made when the Willcox score was greater than 0.9 and the MLF was greater than 10^{-9}. In six cases the identification score was less than 0.9 to the correct group, but the MLF score was satisfactory (Bridge *et al.*, 1989b).

The Willcox identification score has been used in bacteriology and the value used to determine a "good" identification has varied in practice from 0.85 to 0.999 (Willcox *et al.*, 1973; Feltham and Sneath, 1982; Williams *et al.*, 1983; Priest and Alexander, 1988). The performance of the *Penicillium* matrix is within this range, and is therefore considered satisfactory. However, several authors have pointed out the dangers of relying solely on a single Willcox score for identification (e.g. Feltham and Sneath, 1982; Priest and Alexander, 1988). The program MATRIX also gave an MLF score and listed any possibly incorrect test results.

The MLF score was introduced by Dybowski and Franklin (1968), but they did not suggest any cut off level for accurate identifications. The most typical member of a taxon would give a score of 1, while atypical members could give very reduced scores. The results from the *Penicillium* matrix showed that in this case there appeared to be a natural cutoff value at about 10^{-9}, although most clear cut identifications were made when the MLF was greater that 10^{-6} (see Bridge *et al.*, 1989b). However, a number of factors can affect the final value of an MLF score. MLF is dependent on the score obtained for the most typical member of the taxon: if this organism does not exist in practice, this would reduce the MLF. If a taxon consisted of a group of organisms that show significant variation in the characters used, then further isolates could be expected to be some distance from the most typical, and so again the MLF would be reduced. One significant advantage with the MLF score was that with the *Penicillium* data it acted to some extent as an indicator for excluded taxa. Strains belonging to some *Penicillium* species not included in the matrix were tested. The Willcox identification scores were often quite high and could have indicated an erroneous identification. However, the MLF scores were very low indeed in the order of 10^{-19} or less.

PERFORMANCE WITH REDUCED SETS

Unlike a traditional identification key, an identification from a matrix does not normally emphasize particular characters (Sneath, 1978). As a result an identification matrix procedure can be used with a reduced data set. This was tested with the *Penicillium* matrix by attempting to identify a number of strains chosen at random with a variety of reduced data sets. The data used in the construction of the *Penicillium* identification matrix were taken from a multidisciplinary study and included characters from physiological, biochemical and SEM techniques in addition to morphological characters (see Bridge *et al.*, 1989a; 1989b). Test reduction to produce reduced data sets was performed by excluding blocks of characters, such as all metabolite characters or all conidial characters, to give data sets that varied from 57 to 15 characters. Although the detailed performance of the matrix varied depending upon the isolate used, correct identification was possible with the MLF score and the Willcox identification score. As examples, some results from two isolates are given in Table 1. PP280 was a well characterized isolate used in the original numerical taxonomy, and strain IMI 154241 was a line of the ex-type culture of *P. granulatum* not included in the main study. These results showed that identification to a particular group was not necessarily dependent upon one type of the character, and that different character sets were in general complementary. This would be expected if the taxa were accurately described by all of the tests.

Table 1. Identification performance with reduced data sets.

(i) *Penicillium hirsutum* PP 280

set of characters used	number of tests	identification score to correct taxon	MLF score to correct taxon
All	57	0.9999998	4.9×10^{-5}
Spore characters excluded	41	0.9997911	4.7×10^{-3}
Spore and metabolite characters excluded	28	0.8587269	5.5×10^{-3}
Physiology and SEM	23	0.9398831	4.7×10^{-4}
Plate and slide morphology	21	0.9971092	4.4×10^{-4}
Plate morphology	15	0.9985357	0.105

(ii) *Penicillium granulatum* IMI 154241

set of characters used	number of tests	identification score to correct taxon	MLF score to correct taxon
All	57	0.9999982	1.4×10^{-4}
Spore characters excluded	41	0.9913316	5.9×10^{-4}
Spore and metabolite characters excluded	28	0.9942101	0.493
Physiology and SEM	23	0.7245676	0.493
Plate and slide morphology	21	0.9914898	0.119
Plate morphology	15	0.8101497	0.243

CONCLUSIONS

This is the first report of probability based identifications in the filamentous fungi, and the initial results are very encouraging. The identification matrix for the terverticillate penicillia and the associated computer programs can successfully identify unknown cultures at species level. Pitt (1985) suggested that 70-80% of isolates of *Penicillium* from common sources were readily identifiable using gross morphological and cultural features. This figure was substantially improved on in the current tests of the percent positive matrix (Bridge *et al.*, 1989b), although it is recognized that many more strains need to be studied for a complete evaluation. The performance of the identification procedure is ultimately a test of the classification embodied within it. Here the results supported the species concepts developed by Bridge *et al.* (1989a).

In this case, the Modal Likelihood Fraction proved to be a more reliable measure of identification than the usual Willcox identification score. However, the MLF has been used in few practical systems prior to this work, and so there is little additional experience for comparison. As the MLF is derived from the probabilities involving a single taxon and its most typical member, similar results may be obtained with one of the coefficients based on taxon radius (Sneath and Sokal, 1973).

This investigation into matrix performance with reduced data sets has been very encouraging. Although further investigation is needed, the results suggest that in most cases basic reliable identifications can be made with reduced data sets consisting of mainly gross morphological and cultural characters. However, the more critical characters will be needed in some cases.

ACKNOWLEDGEMENTS

This work was supported by Science and Engineering Research council contract SO/17/84 "Systematics of microfungi of biotechnical and industrial importance". I would like to thank Prof. P.H.A. Sneath and M.J. Sackin, Department of Microbiology, Leicester University, for their considerable help and discussions in developing the matrix.

REFERENCES

BRIDGE, P.D., HAWKSWORTH, D.L., KOZAKIEWICZ, Z., ONIONS, A.H.S., PATERSON, R.R.M., SACKIN, M.J. and SNEATH, P.H.A. 1989a. A reappraisal of the terverticillate penicillia using biochemical, physiological and morphological features I. Numerical taxonomy. *Journal of General Microbiology* (in press)

BRIDGE, P.D., HAWKSWORTH, D.L., KOZAKIEWICZ, Z., ONIONS, A.H.S., PATERSON, R.R.M. and SACKIN, M.J. 1989b. A reappraisal of the terverticillate penicillia using biochemical, physiological and morphological features II. Identification. *Journal of General Microbiology* (in press)

D'AMATO, R.F., HOLMES, B. and BOTTONE, E.J. 1981. The systems approach to diagnostic microbiology. *CRC Critical Reviews in Microbiology* 9: 1-44

DYBOWSKI, W. and FRANKLIN, D.A. 1968. Conditional probability and identification of bacteria: a pilot study. *Journal of General Microbiology* 54: 215-229

FELTHAM, R.K.A. and SNEATH, P.H.A. 1982. construction of matrices for computer-assisted identification of aerobic gram- positive cocci. *Journal of General Microbiology* 128: 713-720

PITT, J.I. 1985. A Laboratory Guide to Common *Penicillium* species. North Ryde, N.S.W.: CSIRO Division of Food Research

PRIEST, F.G. and ALEXANDER, B. 1988. A frequency matrix for probabilistic identification of some bacilli. *Journal of General Microbiology* 134: 3011-3018

SNEATH, P.H.A. 1978. Identification of microorganisms. In Essays in Microbiology, eds. J.R. Norris and M.H. Richmond, pp. 10/1-10/32. Chichester: John Wiley and Sons.

—— 1979a. BASIC program for identification of an unknown with presence-absence data against an identification matrix of percent positive characters. *Computers and Geosciences* 5: 195-213

—— 1979b. BASIC program for character separation indices from an identification matrix of percent positive characters. *Computers and Geosciences* 5: 349-357

—— 1980a. BASIC program for the most diagnostic properties of groups from an identification matrix of percent positive characters. *Computers and Geosciences* 6: 21-26

—— 1980b. BASIC program for determining the best identification scores possible from the most typical examples when compared with an identification matrix of percent positive characters. *Computers and Geosciences* 6: 27-34

—— 1980c. BASIC program for determining overlap between groups in an identification matrix of percent positive characters. *Computers and Geosciences* 6: 267-278

SNEATH, P.H.A. and SOKAL, R.R. 1973. Numerical Taxonomy. San Francisco: W.H. Freeman and Sons

WILLCOX, W.R., LAPAGE, S.P., BASCOMB, S. and CURTIS, M.A. 1973. Identification of bacteria by computer: theory and programming. *Journal of General Microbiology* 77: 317-330

WILLCOX, W.R., LAPAGE, S.P. and HOLMES, B. 1980. A review of numerical methods in bacterial identification. *Antonie van Leeuwenhoek* 46: 233-299

WILLIAMS, S.T., GOODFELLOW, M., WELLINGTON, E.M.H., VICKERS, J.C., ALDERSON, G., SNEATH, P.H.A., SACKIN, M.J. and MORTIMER, A.M. 1983. A probability matrix for identification of some Streptomycetes. *Journal of General Microbiology* 129: 1815-1830

IDENTIFICATION OF *PENICILLIUM* AND *ASPERGILLUS*: COMPUTER-ASSISTED KEYING

A.P. Williams

Leatherhead Food R.A.
Randalls Road, Leatherhead, Surrey, KT22 7RY, UK

SUMMARY

During the last few decades, practical identifications schemes for bacteria and yeasts have abandoned the use of dichotomous keys, in favour of the recognition of profiles of reactions typical of the organisms sought. Examples of such profile schemes include the A.P.I. bacterial and C.O.M.P.A.S.S. yeast identification systems. Identification by profile has several advantages over dichotomous keys, including the use of a constant range of criteria, the capacity to incorporate variable results and ready adaptation for speedy use on a microcomputer. Additionally, the principles of such identification schemes are readily understood by microbiologists who only occasionally need to identify moulds.

Now that more data on physiological characteristics of common moulds are available to supplement existing morphological information it is possible to construct identification keys that take into account various aspects of the whole organism. This presentation discusses ways of simplifying the identification of species of *Penicillium* subgenus *Penicillium* and *Aspergillus*.

INTRODUCTION

Although the following discussion is related primarily to the identification of species of *Penicillium* and *Aspergillus*, the intention is to encourage improved identification schemes for any suitable fungal genera. *Penicillium* subgenus *Penicillium* and *Aspergillus* have been selected because of their dominance of the spoilage mycoflora of many types of food, especially processed food. Additionally, subgenus *Penicillium* (species with terverticillate penicilli) includes species that are among the most difficult to distinguish by conventional methodology.

First, we must decide whether improved identification schemes are needed and, if so, who needs them. Certainly the professional mycologist is very likely to be familiar with the characteristics of moulds within a given area of interest and is more likely to be concerned with the taxonomic implications of the differences between those moulds. On the other hand the Quality Assurance or Control Microbiologist in industry, who rarely has had any formal mycological training, is occasionally faced with a mould spoilage problem. He then needs to know the significance of the species present, those properties of the mould that have given rise to the problem such as heat resistance, psychrotolerance, xerotolerance etc., the likely source of the contamination and the means to control it. Although accurate identification remains fundamental, the extent to which detailed speciation is necessary will be much more subjective. For example, where survival of heat-resistant spores is suspected, it is obviously desirable to recognise species in *Byssochlamys*, *Neosartorya* and *Talaromyces* that might be implicated. On the other hand, few would wish to speciate the occasional spoilage isolate of *Phoma* from a dairy product, because no further useful information would be gained.

Modern Concepts in Penicillium and Aspergillus Classification
Edited by R. A. Samson and J. I. Pitt
Plenum Press, New York, 1990

The busy microbiologist in industry has a number of requirements of any identification scheme, whether bacteriological or mycological. First and most important he must be told clearly what test to do and what to record. The criteria used must have reproducibility and should, as far as possible, be easy to observe and should not involve scarce or prohibitively expensive equipment. Second, the result, when obtained, should be easily placed into context. When a significant species has been recognised, it must be easy for the user to find examples of recorded habitats and known problems associated with that species. Equally, where a species is ubiquitous and unlikely to indicate a special problem, this also should be recognisable.

For those charged with the task of compiling practical identification schemes, the onus is on them to provide clear and unambiguous instructions for use and to supply a greater quantity of interpretive information than has hitherto been considered necessary or desirable.

THE MODERN APPROACH TO SELECTION OF IDENTIFICATION CRITERIA

Few, if any, mycologists would question that the taxonomic basis for the recognition of fungal species should be morphological. The problem in practical terms remains that identification by morphology alone requires a degree of specialist training that is not available to most microbiologists. On the other hand, in recent years much supplementary physiological and metabolic information has been acquired for moulds that are of economic significance, including *Penicillium* and *Aspergillus* (Bridge *et al.*, 1985; Frisvad, 1981; Pitt, 1979; Pitt and Hocking, 1985). In some cases such information has already been incorporated into practical identification keys (Pitt, 1979; Pitt and Hocking, 1985) although the basis of separation has remained either dichotomous or synoptic keying.

Most microbiologists are, however, more familiar with identification routines where a fixed range of tests is carried out on a bacterium or yeast that has already been presumptively identified to some extent (e.g. the A.P.I. bacterial identification schemes). Now that much more information is available about a variety of the characters possessed by many mould species, it is possible to simplify their identification by using a similar approach and building an overall identification strategy.

As in bacteriological analysis, identification can be broken into several stages, all of which are greatly enhanced by use of a well-programmed laboratory computer. Having obtained a culture that requires identification, the first stage, is to use a routine that will select the correct database to be used and indicate the test results that will be needed. This has an exact analogy with the first stage of the bacterial identification system used in Cowan and Steel's Manual for the Identification of Medical Bacteria (Cowan, 1974). for example, in *Penicillium* subgenus *Penicillium*, a relatively small range of conidial colours (white, grey, brown, blue and shades of green) needs to be recorded and teleomorphs are absent. On the other hand, in *Aspergillus*, a wider range of conidial colours is found and ascospore characteristics are used to distinguish several species. The second stage in identification is to carry out growth tests under standard conditions and to record the information requested by the instructions. The information recorded would include growth rate measurements, colony colours, substrate utilisations and microscopic characteristics. Having recorded the information and identified by comparison with a database prepared for all relevant taxa, the final stage is to present the user with information about the mould identified. This, although the most subjective stage of the operation, is also the most important in terms of the overall usefulness of the identification. Data, supplied at this stage or sources indicated, must include information about incidence, significance in the environment and relevant details on physiological and nutritional requirements.

The schemes described below are intended for use with isolates of *Penicillium* subgenus *Penicillium* and foodborne species of *Aspergillus*. The range of media used and characters recorded are the minimum needed to distinguish most common isolates. It is, however, accepted that better identification criteria may soon be available and that their incorporation into any identification scheme would be desirable, as would any accepted changes in the names of recognised taxa

CONSTRUCTION OF A COMPUTER KEY

Perhaps the major power of a computer key is its ability to analyse a range of characters simultaneously, and thus to reduce the errors that result from inaccurate or missing data, when dichotomous keys are used. Additionally, a standard amount of information about each species is used in each identification. The princile of building profiles is widely used in other microbiological disciplines. For example a *Pseudomonas* would not be primarily recognised by a dichotomous key, but would be defined by the profile "Gram negative, non-sporing, catalase positive, oxidase positive rod, attacking sugars oxidatively". Profiles are also the basis of the A.P.I. bacterial and C.O.M.P.A.S.S. yeast identification schemes.

Another powerful feature of the use of profile keys is the flexibility possible in the construction of the internal database. Many characteristics (either positive of negative) will be clear cut and easily incorporated. Others, however, may be less easy to read or subject to variability. Examples from our own culture collection include isolates of *Penicillium aurantiogriseum* with consistently smooth stipes; dairy isolates of *Penicillium roqueforti* with abnormally slow growth rates and isolates from several species, which show unusually roughened conidia. It is easy to adapt the database so that the less common character also scores as a positive and does not adversely affect the identification.

In the identification schemes described below, all information entered at the keyboard is taken into account in making the identification. For any characteristic, agreement between the unknown and each species in the database is awarded a score of one; a characteristic that is very rarely found scores 0.5 and a clear disagreement scores zero. It would be a simple matter to add a weighting factor to emphasise key characteristics, but this has not been found to be necessary in the present context.

IDENTIFICATION SCHEMES FOR *PENICILLIUM* SUBGENUS *PENICILLIUM* AND FOR COMMON SPECIES OF *ASPERGILLUS*

Cultural methods
All cultures are inoculated as triple points on appropriate media and incubated in air at 25°C for 7 days. Media used are listed in Table 1.

After incubation, a fixed number of characteristics is recorded for the isolate to be identified. These are listed in Table 2.

Table 1. Media Used in Identification Schemes

Penicillium	*Aspergillus*
Czapek yeast agar (CYA) (Pitt, 1979) Malt extract agar (MEA) (Pitt, 1979) Yeast extract sucrose agar (YES) (Frisvad, 1983) Creatine sucrose agar (CREA) (Frisvad, 1983) Nitrite sucrose agar (NSA) (Frisvad, 1983)	Czapek yeast agar (CYA) (Pitt, 1979) Malt extract agar (MEA) (Pitt, 1979) 25% glycerol nitrate (G25N) (Pitt, 1979)

Table 2. Cultural Characteristics Used for Identification

Penicillium	*Aspergillus*
Colony diameter on CYA Colony diameter on MEA Colony diameter on YES Strength of growth on CREA Acid on CREA Strength of growth on NSA Conidia on CYA Reverse colour on CYA Reverse colour on YES Texture of stipe surface Texture and shape of conidia Shape of penicillus Conidial crusts on MEA Colony texture on CYA	Colony diameter on CYA Colony diameter on MEA Colony diameter on G25N Conidial colour on CYA Presence of cleistothecia or sclerotia Vesicle shape Conidial shape Texture of conidial surface Phialides or metulae + phialides Ascospore ornamentation

The computer keys.
Both keys operate in the same manner. For each unknown isolate to be identified, the results are entered into the computer via a simple menu and are converted into a data string comprising fourteen subunits (*Penicillium*) or ten subunits (*Aspergillus*). The data string contains each positive result in a unique position that depends on the answer given to each question. This is compared sequentially with the master strings in the database (23 for subgenus *Penicillium* or 22 for *Aspergillus*), which have been compiled by examination of fresh isolates and published data (Pitt, 1979; Frisvad, 1981, 1985a,b; Pitt and Hocking, 1985). Tables 3 and 4 give the names of species included in the keys and sources in the literature that contain the nearest equivalent standard description. Agreement with each species in the database is calculated as a percentage, and results are sorted by the computer into descending numerical order.

Table 3. Species of Subgenus *Penicillium* in the Key

Current name	Original name (source)	
P. atramentosum	P. atramentosum	(1)
P. aurantiogriseum	P. puberulum	(1)
P. brevicompactum	P. brevicompactum	(1)
P. camemberti	P. camemberti	(1)
P. chrysogenum	P. chrysogenum	(1)
P. commune	P. commune	(2)
P. coprophilum	P. concentricum	(2)
P. crustosum	P. crustosum	(1)
P. cyclopium	P. aurantiogriseum	(1)
P. digitatum	P. digitatum	(1)
P. echinulatum	P. echinulatum	(1)
P. expansum	P. expansum	(1)
P. glandicola	P. granulatum	(1)
P. griseofulvum	P. griseofulvum	(1)
P. hirsutum	P. verrucosum var. corymbiferum	(2)
P. hordei	P. hordei	(2)
P. italicum	P. italicum	(1)
P. olsonii	P. olsonii	(1)
P. roqueforti	P. roqueforti	(1)
P. solitum	P. verrucosum var. melanochlorum	(2)
P. verrucosum	P. verrucosum	(1)
P. viridicatum	P. viridicatum	(1)
P. vulpinum	P. claviforme	(1)

(1) Pitt 1979
(2) Samson *et al.* (1976)

After analysis, the identification is displayed on the screen and the percentage agreement with that species is stated. There is an option to scan the percentage agreements for all other species in the database and to display details of any tests that failed to agree. A printed Result Form can be made, if required.

CONCLUSION

The two identification schemes that have been briefly described are intended to make the task of mould identification easier for the non specialist. The criteria used to make the identifications have been chosen, initially, because they work, but it is hoped that improvements may be made as better criteria become known.

Table 4. Species of *Aspergillus* in the Key

Neosartorya fischeri	(1)	*Aspergillus niger*	(1)
Eurotium amstelodami	(1)	*Aspergillus ochraceus*	(1)
Eurotium chevalieri	(1)	*Aspergillus parasiticus*	(1)
Eurotium herbariorum	(1)	*Aspergillus penicilloides*	(1)
Eurotium repens	(1)	*Aspergillus restrictus*	(1)
Eurotium rubrum	(1)	*Aspergillus sydowii*	(1)
Emericella nidulans	(1)	*Aspergillus tamarii*	(1)
Aspergillus candidus	(1)	*Aspergillus terreus*	(1)
Aspergillus clavatus	(1)	*Aspergillus ustus*	(1)
Aspergillus flavus	(1)	*Aspergillus versicolor*	(1)
Aspergillus fumigatus	(1)	*Aspergillus wentii*	(1)

(1) Pitt and Hocking (1985)

REFERENCES

BRIDGE, P.A., HAWKSWORTH, D.L., KOZAKIEWICZ, Z., ONIONS, A.H.S., PATERSON, R.R.M. and SACKIN, M.J. 1985. An integrated approach to *Penicillium* systematics. *In* Advances in *Penicillium* and *Aspergillus* Systematics, eds. R.A. Samson and J.I. Pitt, pp. 311-325. New York and London: Plenum Press.

COWAN, S.T. 1974. Cowan and Steel's Manual for the Identification of Medical Bacteria". Cambridge: Cambridge University Press.

FRISVAD, J.C. 1981. Physiological creiteria and mycotoxin production as aids in classificatiln of common asymmetric Penicillia. *Applied and environmental Microbiology* 41: 568- 579.

—— 1985a. Classification of asymmetri Penicillia using expressions of differentiation. *In* Advances in *Penicillium* and *Aspergillus* Systematics,eds. R.A. Samson and J.I. Pitt, pp. 327-335. New York and London: Plenum Press.

—— 1985b. Profiles of primary and secondary metabolites of value in classification of *Penicillium viridicatum*. *In* Advances in *Penicillium* and *Aspergillus* Systematics, eds. R.A. Samson and J.I. Pitt, pp. 311-325. New York and London: Plenum Press.

PITT, J.I. 1979. The genus *Penicillium* and its Teleomorphic States *Eupenicillium* and *Talaromyces*". London: Academic Press.

PITT, J.I. and HOCKING, A.D. 1985. Fungi and food spoilage. Sydney: Academic Press.

SAMSON R.A., STOLK, A.C. and HADLOK, R. 1976. Revision of subsection *Fasciculata* of *Penicillium* and some allied species. *Studies in Mycology, Baarn* 11: 1-47.

GENERAL DISCUSSION FOLLOWING AFTER THE SESSION ON COMPUTER-ASSISTED KEYS

GAMS: I'd like to ask Mr. Williams what language he used for his program.

WILLIAMS: I wrote the program on an Apple originally, and adapted it for use on an IBM using WilliamsBASIC.

GAMS: Dr. Bridge, you have told us about the software development, but not about the data you entered. Are the data used in this system the same as developed by CMI?

BRIDGE: The data is selected from our numerical taxonomy project.

GAMS: How easy would it be to extend the data set?

BRIDGE: It would be very easy. The data file is a standard ASCII file and can be handled by a word processor. The program itself uses any data file in the right format.

SAMSON: For whom are you writing your key?

Bridge: At the moment, I was interested to see what we could do with the data we had.

SAMSON: It looks like a key for ascomycetes, with words like spore, reticulations, lobate and so on. There are no characters in *Penicillium* like that.

BRIDGE: Certainly these terms are used for describing the ornamentation using the SEM.

SAMSON: So you need an SEM to use your key?

BRIDGE: At the moment, this is a demonstration of mixed data types running from a matrix. This may not be a practical identification scheme, but it is an example of what can be done. It is a methodology for an identification strategy.

GAMS: Dr. Klich has expressed some important thoughts about international coordination of data systems. Perhaps we can find some agreement on the basic data and the structures to be used. Most of the keys we have seen today start with Dr. Pitt's monograph. It would be workable, provided that the system is explanable and can grow steadily with each addition or correction being incorporated.

SEIFERT: Perhaps we should ask Dr. Pitt how he feels about everyone using his data as if it was public domain.

PITT: I don't have any problem with people using the data from my publications. The data in PENNAME is, in fact, derived from my private data using data from about 2600 isolates, correctly identified and filed in dBase III+. At the moment, this is not generally accessible but it could be made available to selected people for particular purposes.

GAMS: So your database is much more than just one record for each species?

PITT: Certainly. The database is one record for each isolate.

GAMS: How many species or records were sorted during your demonstration?

PITT: There are 68 species, and more than 100 characters. I plan to expand this to a larger data set.

GAMS: If we are to set up the cooperative database, we have to chose between two alternatives. Are we satisfied with a minimal amount of information, one record for each taxon, and distribute this as widely as possible or should we immediately try for a very sophisticated system incorporating all the doubtful isolates, synonyms and so on?

SEIFERT: I think that we have to decide what the database is going to be used for. Who needs records for 10,000 isolates? This might be valuable for taxonomists. But what we have seen today are keys designed by taxonomists for general use. Perhaps Dr. Klich can explain what purpose this international database should serve.

KLICH: I was thinking of a taxonomic database. I can see this will be much more complicated than I had imagined. I was thinking of separate files for morphology, one for biochemistry and so on. So often we publish a paper and the information disappears into our filing cabinets or personal computers, and this is the information that I'd like to be able to get at, information that we have developed and finished with. The problems are who should contribute and who should have access.

GAMS: There are only relative few who will use such an extensive database. General users may be satisfied with a limited data set, but for the working taxonomist, we need a centrally administered, up to date, database.

PITT: The data in the database would be meaningless if the identifications are inaccurate. The identifications would have to be checked.

GAMS: We have the same problem with *Fusarium*. Two or three authorities must check the identifications before the data is entered.

SAMSON: This is impractical. How can you ask three people to check two or three thousand isolates?

GAMS: It may not be a question of checking thousands of isolates. If we have deviating isolates, that may be incorrectly identified, these could be checked by the authorities.

KLICH: One of the fields in the database should be who made the identification. Then you can select data that you trust.

GAMS: Especially when we are talking about secondary metabolites.

KLICH: We have to use the information intelligently. I don't see this as a major problem. There could be different files for different kinds of information. We don't have to make all of it available.

GAMS: How about the data structure? Should we be using a database system, or should we consider using more sophisticated expert system software. These are user friendly programs, and they can be used for generating keys, or point to discrepancies in hand-made keys. For example, Super Expert or Prologue are already available.

SEIFERT: The problem with expert system software is that is all first generation, and the database systems available are all fourth generation or more. They have standard data formats and are often compatable with the dBase format. This is important for any international effort based on microcomputers.

GAMS: Expert system data can also be superimposed on dBase formats.

Additions to the dialogue at a later time

PITT: There are two aspects of such a database. One would be to have a strictly nomenclatural database when were the species described, who described them, are they typified, do the specimens exist, have they been neotypified? Most of this information is available in Pitt (1979). Perhaps this could be scanned into a word-processing file and then into a database file, if necessary. The second aspect would deal with the properties of isolates or strains that actually have been studied. My own database includes some 2500 isolates, and the data for many is very complete. This is a dBase III+ file with numeric fields on colony diameters and microscopic characters that allows me to assemble data from a single species for any character I choose.

HAWKSWORTH: It might be useful as a first step to have some standardized field names and definitions and so on. Perhaps SPAS could look after this?

CHRISTENSEN: I'd like to make an appeal for the inclusion of fields relating to ecological data such as geographical locations, substrate or habitat, and prior names for the isolate.

SEIFERT: It's easy to add these fields onto your own database.

ONIONS: Who would actually fund the development of a database on such a large scale? Who would do all this work?

SAMSON: Who would actually use such a database? You can already extract a lot of information on specific isolates from the databases of individual culture collections.

7

NEW APPROACHES FOR PENICILLIUM AND ASPERGILLUS SYSTEMATICS: MOLECULAR BIOLOGICAL TECHNIQUES

A REVIEW OF MOLECULAR BIOLOGICAL TECHNIQUES FOR SYSTEMATIC STUDIES OF *ASPERGILLUS* AND *PENICILLIUM*

E.J. Mullaney and M.A. Klich

United States Department of Agriculture
Southern Regional Research Center
New Orleans, Louisiana 70179, USA

SUMMARY

Over the last several years various molecular taxonomic techniques have been utilized for defining evolutionary relationships and for identifying species. A review of the different techniques, their limitations, and their impact to date on *Aspergillus* and *Penicillium* systematics are presented. Comparisons of results obtained from molecular techniques and conventional taxonomic methods, present trends in each technique's application, and possible application of recently developed molecular techniques are discussed.

INTRODUCTION

Several genetic engineering procedures have been developed in the last several years. Following closely behind these developments have been several efforts to adapt or develop some of these molecular biological techniques for taxonomic and phylogenetic research. The primary impetus for this move has been the perceived promise of this technology for greater precision and clarification of taxonomic questions not satisfactorily addressed by current methods. In fungal taxonomy, with little or no fossil record of many species, lack of a perfect stage in some species, and few available morphological characteristics on which to base taxonomic decisions in others, taxonomic decisions have been difficult to make. Molecular techniques were adopted by some scientists as a means to overcome these difficulties. A review of the previous fungal research reveals that most of the molecular research has been clustered in taxonomically uncertain groups and that no single technique has been uniformly employed. It is also significant that in a few cases when different techniques were utilized to study a single fungal group the findings were not always in agreement. Therefore, as in conventional taxonomic methods, selection of the appropriate molecular technique for the specific problem addressed is essential.

The scope of this paper is limited to a review of molecular biological techniques in *Aspergillus* and *Penicillium* taxonomy, genera not widely studied by these methods. However, *Aspergillus nidulellus* Eidam [more widely known as *Aspergillus nidulans*; telemorph = *Emericella nidulans* (Eidam) Vuillemin] has been extensively studied and this knowledge may enhance the potential development of molecular techniques in *Penicillium* and other *Aspergillus* species. Nucleic acid isolation techniques developed for *A. nidulellus* and other fungi have been used in both of these genera (Yelton *et al.*, 1984; Biel and Parrish, 1986; Kurtzman *et al.*, 1986; Klich and Mullaney, 1987; Lee *et al.*, 1988). Some of these techniques require a minimal amount of specialized equipment and training and should be readily adopted by nonspecialist laboratories. Additional purification of DNA and separation of nuclear, ribosomal and mitochondrial DNA by cesium chloride/bisbenzimide density gradient centrifugation (Garber and Yoder, 1983) or

Modern Concepts in Penicillium and Aspergillus Classification
Edited by R. A. Samson and J. I. Pitt
Plenum Press, New York, 1990

hydroxylapatite fractionation (Britten and Kohne, 1968) is also possible. To date most molecular taxonomic research on these genera has been on *Aspergillus* species.

Four molecular methods have been used in fungal systematics: G+C molar percentage, determining the relative frequency of the nucleotides guanine and cytosine in DNA: DNA complementarity, measuring the rate and extent of reassociation of single stranded DNA from two isolates; ribosomal RNA (rRNA) sequence comparison, comparing homology of nucleotide sequences from rRNAs of different species; and restriction fragment length polymorphism (RFLP), where DNA from different isolates is digested by a restriction enzyme, electrophoretically separated, and the resulting DNA fragment patterns compared. Each of these methods is discussed in more detail below.

G+C VALUES

Among the simplest molecular techiques used in taxonomy, the molar percentage of guanine plus cytosine (G+C) is a measure of DNA base composition. As G+C values among unrelated species may be similar, and G+C ratios provide no information on the arrangement of the bases, its taxonomic use is primarily exclusionary (Kurtzman, 1985). This technique is often used as a initial screening step in DNA-DNA complementarity studies.

G+C values in fungi are principally determined by thermal denaturation (Marmur and Doty, 1962; Mandel and Marmur, 1968) and cesium chloride density gradient equilibrium centrifugation (Schildkraut *et al.*, 1962; Mandel *et al.*, 1968). As the different types of DNA (nuclear, mitochondrial, ribosomal) will have different G+C values, they must be separated for precise measurements, preferably by cesium chloride/bisben-zimide density gradient centrifugation.

The G+C values for several species in *Aspergillus* sect. *Flavi* (Gams *et al.*, 1985) have been measured by Kurtzman *et al.* (1986, 1987), as follows: *A. flavus*, 49.1-49.9; *A. oryzae*, 49.1-49.6; *A. parasiticus*, 49.1-50.0; *A. sojae*, 49.1-49.4; *A. leporis*, 49.2; *A. nomius*, 49.7-50.4; *A. tamarii*, 48.6-48.6. These values, reported with DNA complementarity measurements, were not used independently for taxonomic purposes.

DNA COMPLEMENTARITY

DNA complementarity, the reassociation of DNA from two isolates, is used as a measure of relatedness among species. One of the first molecular techniques used in biological classification (Hoyer *et al.*, 1964), its utility has led to the development of several different methods. Some use radioisotopes to label DNA as a means to quantify reassociation, others monitor reassociation spectrophotometrically during temperature reduction (Seidler and Mandel, 1971; Ellis, 1985). Kurtzman (1985) has detailed these procedures and reviewed their use in fungi. Repeated DNA sequences can reassociate with each other, so their presence may obscure the true reassociation value. Therefore, the accuracy of these methods is enhanced by using highly purified nuclear DNA consisting exclusively of unique (non-repeated) sequences. Hydroxyapatite chromatography has been routinely used to remove repeated DNA sequences from samples to be tested. In most studies, the complementarity values are normalized by considering the level of reassociation of the control DNA to itself as 100%.

Measurements of DNA relatedness in *Aspergillus* and *Penicillium* have been limited to *Aspergillus* sect. *Flavi*. Based on reassociation at 42o, Kurtzman *et al.* (1987) proposed a new species, *Aspergillus nomius*, for several isolates formerly considered to be *A. flavus*. Based on data from a single isolate of each species, Kurtzman *et al.* (1986) suggested reduction of *A. oryzae*, *A. sojae*, and *A. parasiticus* to the ranks of subspecific taxa of *A. flavus*. The limitation on the number of isolates which can be studied due to the labour intensive requirements of these measurements has been one factor in limited adoption of this technique in taxonomic research in these genera.

RNA SEQUENCE COMPARISON

Ribosomes, which translate messenger RNA into protein, consist of protein and RNA subunits of various sizes. These subunits, referred to by their sedimentation rates (e.g. 5S, 18S, 25S), contain some of the most highly conserved nucleic acid sequences known. Ribosomal RNA sequences have been determined in several taxonomic or phylogenetic studies above the species level (Kurtzman, 1985; Hori and Osawa, 1987).

The small 5S rRNA has been sequenced in many filamenteous fungi. Limited success has been reported in taxonomic studies on Basidiomycetes. Recently Bartoszewki *et al.* (1987) have reported intraspecific heterogeneity in the 5S rRNAs of *A. nidulellus*. Hasegawa *et al.* (1985) suggested that the 5S rRNA molecule is too small (about 120 nucleotides) to be statistically reliable, and proposed that the larger subunit rRNAs are better suited for phylogenetic analysis. Preliminary sequence and secondary structure data from the 18S and 28S rRNA have supported this (Blanz and Unseld, 1987). The development of a new rapid rRNA sequencing technique using crude cellular lysates (Lane *et al.*, 1985), has the potential to provide much valuable information, but it is not suitable for use as a sole tool on which to base all taxonomic decisions.

RESTRICTION FRAGMENT LENGTH POLYMORPHISM

RFLP uses restriction endonucleases (restriction enzymes) to cleave double stranded DNA. This is followed by gel electrophoresis to separate the fragments by molecular weight and net charge. After staining the DNA with ethidium bromide, the resulting DNA fragment pattern may be observed under UV light. The same DNA will form different patterns when enzymes that cleave at different restriction sites are used.

Southern blot analysis (Southern, 1975), also known as RFLP with hybridization, (RLFPH) may be used to determine if isolates have regions with similar DNA sequences. After electrophoresis, the DNA fragments are denatured into single strands, and then transferred to a nitrocellulose membrane. A previously prepared sample of DNA (termed the probe) is then radiolabelled, and mixed with the membrane bound DNA under conditions which favor hybridization between fragments with identical or very similar DNA sequences. The membrane is then washed to remove excess probe and regions of hybridization visualized by autoradiography.

The proportion of common bands in an RFLP or RFLPH can be used as a measure of relatedness between isolates. This method has the advantage of being simpler than either DNA complementarity or sequencing and it correlates well with DNA sequence analysis (Nei, 1983). However, several factors can limit the accuracy of RFLP or RFLPH analyses (Taylor 1986; Natvig *et al.*, 1987; Förster & Kinscherf 1988). Mutations at a recognition site or of length (deletions or insertions), can yield a polymorphism which has little significance. In RFLPs, where a probe is not used, similarly sized non-homologous DNA fragments will be read as identical, giving a false impression of similarity. However, RFLP and RFLPH have served as useful tools in distinguishing closely related species and in phylogenetic studies.

RFLP both alone and with Southern blot analysis (RFLPH) have been used in *Aspergillus* and *Penicillium* systematics. Kozowski and Stepién (1982) used RFLPH analysis of mitochondrial DNA (mtDNA) to assess the morphological data which has formed the basis of *Aspergillus* taxonomy. They reported that two members of *Aspergillus* sect. *Flavi*, *A. oryzae* and *A. tamarii*, were almost identical. They questioned the apparent close relationship between *A. awamori* and *A. niger*, but acknowledged the limitations of RFLP analysis. In a review of *Aspergillus* genetic variation, Croft (1986) reported that no intraspecific sequence variation has been reported in *Aspergillus* mtDNA, unlike the case of *Neurospora*. However, distinct restriction site patterns have been reported to be associated with each different species in *Aspergillus* sect. *Nidulantes*. Length mutations, i.e. presence or absence of introns in certain regions of the mitochondrial genome, appear to be the principle reasons for these differences. Cloned random fragments from an *A. nidulellus* genomic library, used as probes in RFLPH analysis of different species in sect. *Nidulantes* also proved useful as a diagnostic tool (Croft, 1986). One interesting finding was that an *A. nidulellus* probe hybridized strongly to DNA from *A. terreus*, suggesting that these species may be more closely related than the present taxonomies suggest. Using probes from *A. nidulellus* developmental genes, Mullaney and Klich (1987) reported a quite high degree of hybridization between *A. nidulellus* probes and *A. terreus* DNA. The major morphological differences between *A. nidulellus* and *A. terreus* are colony colour, vesicle diameter, and stipe colouration, and the latter two characters are variable in sect. *Nidulantes*. These data suggest an overreliance on colony colour in *Aspergillus* classification (Raper and Fennell, 1965; Klich and Pitt, 1988).

Previous molecular research utilizing *A. nidulellus* suggests that the conidia and the complex, multicellular conidiophores that are the characteristic features of *Aspergillus* and *Penicillium* are coded for by several developmentally regulated genes expressed only during conidial development. Several of these genes have been cloned, including *tubC* (May *et al.*, 1985) *SpoC1* (Timberlake and Barnard, 1981) or cloned and sequenced such as *brlA* (Adams et al, 1988). Beta3-tubulin, the product of the *tubC* gene, functions in the development of microtubules during conidial mitosis. *SpoC1* is a 13.3 kb gene cluster that codes for several poly(A)+RNAs that are primarily expressed during conidiophore development. The product of the *brlA* gene is apparently a nucleic acid binding protein whose presence in vegetative cells results in the expression of several other developmentally regulated genes essential for the production of conidiophores. The mutant allele of the *brlA* locus blocks conidiation by producing abnormal sterile conidiophores with the phenotype known as "bristle".

This research to identify and define the specific function of genes coding for the *A. nidulellus* conidiophore has provided a new potential means of measuring phylogentic relationships in *Aspergillus* and *Penicillium*. Clones of these three *A. nidulellus* genes (*SpoC1, tubC, brlA*) have been used in RFLPH analysis of representative species of both *Aspergillus* and *Penicillium* (Mullaney and Klich, 1987; 1988). Under standard conditions, a considerable degree of hybridization was found between the *A. nidulellus SpoC1* gene cluster and the *tubC* gene and *A. fumigatus*, *A. terreus*, *A. niger*, *A. flavus*, *A. ochraceus*, *P. citreonigrum*, *P. montanense*, and *P. lividum* DNAs. The functions of the genes in the *SpoC1* gene cluster are unknown, but there is a widespread distribution of DNA sequence with homology to the *SpoC1* clone in representative Hyphomycetes (Mullaney and Klich, 1988). Less DNA sequence conservation appears to be found in the analogous brlA genes in other species of *Aspergillus* and *Penicillium*. When the *brlA* gene was used as a probe in the same study, this *A.nidulellus* gene only hybridized strongly to *A. terreus* DNA. This supports the observations of Croft (1986) noted above. As additional *A. nidulellus* developmental genes are characterized, both development of a model system to study conidiogenesis in other Hyphomycetes and a source of DNA probes coding for significant phenotypic characteristics of *Aspergillus* and *Penicillium* can be expected. Then the possibility of convergent evolution of conidiophores can be investigated on the molecular level.

In a practical application of RFLPs, a technique was developed to help distinguish the aflatoxigenic *A. flavus* from the koji mould *A. oryzae* (Klich and Mullaney, 1987). In this study, eleven *A. flavus* isolates and seven *A. oryzae* isolates were digested with a variety of restriction enzymes and subjected to gel electrophoresis. One enzyme, *Sma*I, yielded consistent species specific patterns. All of the *A. flavus* isolates showed a major 3.8 kb band, while all but one of the *A. oryzae* isolates showed major bands at 2.7 and 1.0 kb. The final *A. oryzae* isolate had slightly larger bands (3.1 and 1.2 kb). The observed differences in the banding patterns were shown to result from differences in nuclear rather than mitochondrial DNA. The combined size of the two unique *A. oryzae* bands (2.7 and 1.0) is approximately equal to the size of the unique *A. flavus* band (3.8). This suggests that *A. oryzae* contains one *Sma*I site not found in *A. flavus*. The *Sma*I restriction digest patterns of total DNA provided a useful adjunct to other taxonomic criteria distinguishing these two economically important secies. Apparently, the resolution of this method is greater than that of the reassociation method used by Kurtzman et al (1986), where the relative reassociation values for one isolate each of *A. flavus* and *A. oryzae* was 100%.

An RFLP analysis has recently been included in the first study by the Subcommission of *Penicillium* and *Aspergillus* Systematics of the ICTF which is nearing completion. In this study to clarify relationships among *P. glabrum*, *P. spinulosum* and other closely related species, RFLP results were compared to standard morphological methods, gross physiological methods, grouping by secondary metabolite, and classification based on isoenzymes. Agreement was reached by the majority of participants on 13 of the 15 isolates; RFLP data agreed with the majority for 10 of the 13 isolates. In this study total DNA was used and the major visible bands after electrophoresis were most likely mtDNA or rDNA. It is not surprising that RFLP results were in agreement with those of the other methods. If phenotype is indeed a reflection of genotype, one would expect results from different approaches to agree. The fact that such a simple method was effective is encouraging for future applications.

One factor limiting the increased use of molecular technology in many locations is the need to use radioactive material. Effective non-radioactive methods to label nucleic acids are being developed and have been used recently in fungal taxonomic studies (e.g. Taylor *et al.*, 1985). The technology for the use of non-radioactive oligonucleotide probes to selectively detect alleles differing in only a single nucleotide is now available (Bugawan *et al.*, 1988) and could potentially be used to distinguish morphologically similar fungal isolates. The commercial availability of universal primers for sequencing rRNA, the ability to rapidly sequence rRNA from crude cellular lysates (Lane *et al.*, 1985) and commercial development of automated sequencing technology will also have potential value. Another new technique that has a potential impact on fungal molecular genetics is the polymerase chain reaction (PCR) (Saiki et al, 1988), a procedure which allows the amplification of a few molecules of template DNA in a crude extract to several million copies in a few hours. This will generate enough DNA to allow sequencing without the need for cloning (Wong *et al.*, 1987; Stoflet *et al.*, 1988). A procedure to use PCR *in vitro* amplification of rRNA genes followed by direct sequencing is also now available (Medlin *et al.*, 1988).

REFERENCES

ADAMS, T.H., BOYLAN, M.T., and TIMBERLAKE, W.E. 1988. *brlA* is necessary and sufficient to direct conidiophore development in *Aspergillus nidulans*. *Cell* 54: 353-362.

BARTOSZEWSKI, S., BORSUK, P., KERM, I., BARTNIK, E. 1987. Microheterogeneity in *Aspergillus nidulans* 5S rRNA genes. *Current Genetics* 11: 571-573.

BIEL, S.W. and PARRISH, F.W. 1986. Isolation of DNA from fungal mycelia and sclerotia without use of density gradient ultracentrifugation. *Analytical Biochemistry* 154: 21-25.

BLANZ, P.A. and UNSELD, M. 1987. Ribosomal RNA as a taxonomic tool in mycology. *Studies in Mycology, Baarn* 30: 247-258.

BRITTEN, R.J. and KOHNE, D.E. 1968. Repeated sequences in DNA. *Science, N.Y.* 161: 529-540.

BUGAWAN, T.L., SAIKI, R.K., LEVENSON, C.H., WATSON, R.M. and ERLICH, H.A. 1988. The use of non-radioactive oligonucleotide probes to analyze enzymatically amplified DNA for prenatal diagnosis and forensic HLA typing. *Biotechnology* 6: 943-947.

CROFT, J.H. 1986. Genetic variation and evolution in *Aspergillus*. *In* Evolutionary Biology of the Fungi, eds. A.D.M. Rayner, C.M. Brasier, and D. Moore, pp. 311-323. New York: Cambridge University Press.

ELLIS, J.J. 1985. Species and varieties in the *Rhizopus arrhizus-Rhizopus oryzae* group as indicated by their DNA complementarity. *Mycologia* 77: 243-247.

FÖRSTER, H. and KINSCHERF, T.G. 1988. Estimation of relatedness between *Phytophthora* species by analysis of mitochondrial DNA. *Mycologia* 80: 466-478.

GAMS, W., CHRISTENSEN, M., ONIONS, A.H., PITT, J.I. and SAMSON, R.A. 1985. Infrageneric taxa of *Aspergillus*. *In* Advances in *Penicillium* and *Aspergillus* Systematics, eds. R. A. Samson and J. I. Pitt, pp. 55-62. New York and London: Plenum Press.

GARBER, R.C. and YODER, O.C. 1983. Isolation of DNA from filamentous fungi and separation into nuclear, mitochondrial, ribosomal, and plasmid components. *Analytical Biochemistry* 135: 416-422.

HASEGAWA, M., IIDA, Y., YANO, T., TAKAIWA, F. and IWABUCHI, M. 1985. Phylogenetic relationships among eukaryotic kingdoms inferred from ribosomal RNA sequences. *Journal of Molecular Evolution* 22: 32-38.

HORI, H. and OSAWA, S. 1987. Origin and evolution of organisms as deduced from 5S ribosomal RNA sequences. *Molecular Biology and Evolution* 4: 445-472.

HOYER, B.H., MCCARTHY, B.J. and BOLTON, E.T. 1964. A molecular approach in the systematics of higher organisms. *Science, N.Y.* 144: 959-967.

KLICH, M.A. and MULLANEY, E.J. 1987. DNA restriction enzyme fragment polymorphism as a tool for rapid differentiation of *Aspergillus flavus* from *Aspergillus oryzae*. *Experimental Mycology* 11: 170-175.

KLICH, M.A. and PITT, J.I. 1988. A Laboratory Guide to Common *Aspergillus* Species and their Teleomorphs. North Ryde, N.S.W.: CSIRO Division of Food Processing.

KOZOWSKI, M. and STEPIÉN, P.P. 1982. Restriction enzyme analysis of mitochondrial DNA of members of the genus *Aspergillus* as an aid in taxonomy. *Journal of General Microbiology* 128: 471-476.

KURTZMAN, C.P. 1985. Molecular taxonomy of the Fungi. *In* Gene Manipulations in Fungi, eds. J. W. Bennett and L. L. Lasure, pp. 35-63. Orlando, Florida: Academic Press.

KURTZMAN, C.P., SMILEY, M.J., ROBNETT, C.J. and WICKLOW, D.T. 1986. DNA relatedness among wild and domesticated species in the *Aspergillus flavus* group. *Mycologia* 78: 955-959.

KURTZMAN, C.P., HORN, B.W. and HESSELTINE, C.W. 1987. *Aspergillus nomius*, a new aflatoxin-producing species related to *Aspergillus flavus* and *Aspergillus tamarii*. *Antonie van Leeuwenhoek* 53: 147-158.

LANE, D.J., PACE, B., OLSEN, G.J., STAHL, D.A., SOGIN, M.L. and PACE, N.L. 1985. Rapid determination of 16S ribosomal RNA sequences for phylogenetic analyses. *Proceedings of the National Academy of Science USA*, 82: 6955-6959.

LEE, S.B., MILGROOM, M.G. and TAYLOR, J.W. 1988. A rapid, high yield mini-prep method for isolation of total genomic DNA from fungi. *Fungal Genetics Newsletter* 35: 23-24.

MANDEL, M. and MARMUR, J. 1968. Use of ultraviolet absorbance-temperature profile for determining the guanine plus cytosine content of DNA. *In* Methods in Enzymology. XII. Nucleic Acids Part B, eds. L. Grossman and K. Moldave, pp. 195-206. New York: Academic Press.

MANDEL, M., SCHILDKRAUT, C.L. and MARMUR, J. 1968. Use of CsCl density gradient analysis for determining the guanine plus cytosine content of DNA. *In* Methods in Enzymology. XII. Nucleic Acids Part B, eds. L. Grossman and K. Moldave, pp. 184-195. New York: Academic Press.

MARMUR, J. and DOTY, P. 1962. Determination of the base composition of deoxyribonucleic acid from its thermal denaturation temperature. *Journal of Molecular Biology* 5: 109-118.

MAY, G.S., GAMBINO, J., WEATHERBEE, J.A. and MORRIS, N.R. 1985. Identification and function analysis of Beta-Tubulin genes by site specific intergrative transformation in *Aspergillus nidulans*. *Journal of Cell Biology* 101: 712-719.

MEDLIN, L., ELWOOD, H.J., STICKEL, S. AND SOGIN, M.L. 1988. The characterization of enzymatically amplified eukaryotic 16S-like rRNA-coding regions. *Gene* 71: 491-499.

MULLANEY, E.J. and KLICH, M.A. 1987. Survey of representative species of *Aspergillus* for regions of DNA homology to *Aspergillus nidulans* developmental genes. *Applied Microbiology and Biotechnology* 25: 476-479.

—— 1988. Representative Hyphomycetes have regions of DNA homologous to *Aspergillus nidulans* developmental genes. *Mycologia* 80: 582-585.

NATVIG, D.O., JACKSON, D.A. and TAYLOR, J.W. 1987. Random-fragment hybridization analysis of evolution in the genus *Neurospora*: the status of four-spored strains. *Evolution* 41: 1003-1021.

NEI, M. 1983. Genetic polymorphism and the role of mutation in evolution. *In* Evolution of Genes and Proteins, eds. M. Nei and R.K. Koehn, pp. 165-190. Sunderland, Massachusetts, Sinauer Associates, Inc.

RAPER, K.B. and FENNELL, D.I. 1965. The genus *Aspergillus*. Baltimore: Williams and Wilkins.

SAIKI, R.K., GELFAND, D.H., STOFFELL, S., SCHARF, S.J., HIGUCHI, R., HORN, G.T., MULLIS, K.B. and ERLICH, H.A. 1988. Primer-directed enzymatic amplification of DNA with a thermostable DNA polymerase. *Science N.Y.* 239: 487-491.

SCHILDKRAUT, C.L., MARMUR, J. and DOTY, P. 1962. Determination of the base composition of deoxyribonucleic acid from its buoyant density in CsCl. *Journal of Molecular Biology* 4: 430-443.

SEIDLER, R.J. and MANDEL, M. 1971. Quantitative aspects of deoxyribonucleic acid renaturation: base composition, state of chromosome replication and polynucleotide homologies. *Journal of Bacteriology* 106: 608-614.

SOUTHERN, E.M. 1975. Detection of specific sequences among DNA fragments separated by gel electrophoresis. *Journal of Molecular Biology* 98: 503-517.

STOFLET, E.S., KOEBERL, D.D., SARKAR, G. and SOMMER, S.S. 1988. Genomic amplification with transcript sequencing. *Science, N.Y.* 239: 491-494.

TAYLOR, J.W. 1986. Fungal evolutionary biology and mitochondrial DNA. *Experimental Mycology* 10: 259-269.

TAYLOR, J.W., SMOLICH, B.D. and MAY, G. 1985. An evolutionary comparison of homologous mitochondrial plasmid DNAs from three *Neurospora* species. *Molecular and General Genetics* 201: 161-167.

TIMBERLAKE, W.E. and BARNARD, E.C. 1981. Organization of a gene cluster expressed specifically in the asexual spores of *A. nidulans*. *Cell* 26: 29-37.

WONG, C., DOWLING, C.E., SAIKI, R.K., HIGUCHI, R.G., ERLICH, H.A. and KAZAZIAN, H.H.,Jr. 1987. Characterization of ß-thalassaemia mutations using direct genomic sequencing of amplified single copy DNA. *Nature* 330: 384-386.

YELTON, M.M., HAMER, J.E. and TIMBERLAKE, W.E. 1984. Transformation of *Aspergillus nidulans* by using a trpC plasmid. *Proceedings of the National Academy of Science, USA* 81: 1470-1474.

RFLP ANALYSIS OF NUCLEAR AND MITOCHONDRIAL DNA AND ITS USE IN *ASPERGILLUS* SYSTEMATICS

J.H. Croft, V. Bhattacherjee and K.E. Chapman

School of Biological Sciences
University of Birmingham
Birmingham B15 2TT, UK

SUMMARY

The analysis of restriction fragment length polymorphisms (RFLP's) in both nuclear and mitochondrial DNA is proving useful in establishing phylogenetic relationships in the section *Nidulantes* of *Aspergillus*. Largely because of the presence of a highly diagnostic perfect stage, this group has a firm and well established taxonomy, and the well developed genetics (making use of sexual, parasexual and protoplast fusion methods) has enabled the concept of the biological species to be applied. An examination of RFLP's in the nuclear and mitochondrial DNA of these species has allowed a comparison to be made between the extent and nature of such variation in the DNA sequences with the group structure imposed by the various systematic studies and thus permits an interpretation of the RFLP's in terms of that systematic classification. The highly diagnostic nature of restriction fragment variation in the mitochondrial genomes and of certain probes for detecting nuclear RFLP's will be described.

INTRODUCTION

The classification of species in the genus *Aspergillus* has been based primarily on growth, morphological and physiological characters (e.g. Raper and Fennell, 1965). Specification is well defined in some sections, such as sect. *Nidulantes* where the presence of a sexual stage has resulted in a precise classification. However, other areas of the genus are less clearly defined, particularly those containing only anamorph species.

The genetic bases of the characters used for taxonomic purposes are generally unknown. Consequently little information is available concerning the phylogenetic and evolutionary relationships between the described taxa. A species should, perhaps, best be considered as a population which is reproductively isolated from other populations: that is, as an ideal, we should invoke the concept of the biological species in our schemes of classification. However, the practical necessity for simple and rapid identification of unknown isolates is recognised and accepted. Species concepts other than that of the biological species may be appropriate for many purposes, especially in anamorph genera where the identity of reproductively isolated populations is unknown.

Though many biochemical and physiological techniques have provided useful information for classification, it is possible to argue that variation in the composition, nucleotide sequence and organisation of DNA is the most likely to give a clear and sensitive discrimination between organisms and also to indicate clearly their evolutionary and phylogenetic relationships. Methods such as DNA base composition and overall DNA complementarity (Kurtzman, 1985) provide an idea of levels of relationship, but are relatively crude and difficult to interpret. For example, high complementarity is clearly an indicator of close relationship, but too much intrageneric variation probably exists in

Modern Concepts in Penicillium and Aspergillus Classification
Edited by R. A. Samson and J. I. Pitt
Plenum Press, New York, 1990

degree of complementarity to allow definition of species or precisely identify individual isolates.

An alternative method of assessing DNA sequence divergence is that of the analysis of restriction fragment length polymorphisms (RFLP 's). RFLP analysis of the relatively small fungal mitochondrial (mt) genome may be achieved by direct observation, following electrophoresis, of the number and size of the fragments produced by digestion with type II restriction endonucleases (Kozlowski and Stepien, 1982). Small regions of the much larger nuclear genome may be analysed by the use of cloned DNA probes in Southern blot hybridisation experiments (Southern, 1975). This technique has been used mainly to provide large numbers of genetic markers in eukaryotic genomes, particularly those of human origins in order to detect markers linked to genes known to cause inherited disorders (Botstein *et al.*, 1980; Donis-Keller *et al.*, 1987) and more recently in plants (Tanksley *et al.*, 1989) in order to provide markers linked to genes for economic traits such as resistance to diseases. The discovery of hypervariable probes based on highly polymorphic repeated sequences in human DNA led to the development of 'genetic fingerprinting' techniques in which each individual in the human population has a high probability of producing a unique pattern in a Southern blot experiment thus permitting the identification of individuals and accurate determination of their pedigrees (Jeffreys *et al.*, 1985). Direct observation of the bright bands produced by repeated sequences in electrophoretic smears of digested nuclear DNA have also been informative. Repeated sequences may also be highlighted by probing with labelled total nuclear DNA in Southern blot experiments.

The variation in all of these patterns is due to differences in the DNA sequence which alter the size and number of the fragments obtained following digestion with restriction endonucleases. The differences may result from single base pair substitutions, small insertions or deletions, or gross chromosomal structural changes. The generation and detection of RFLP's is illustrated in Fig. 1.

The amount of polymorphism observed will depend both on the amount of polymorphism actually present in the populations under study and experimental details, for example, in Southern blot experiments, on the nature of the probe being used. The amount of polymorphism present in a species may be expected to vary according to its breeding system. In outbreeding organisms, such as species of *Phanaerochaete*, each individual basidiospore progeny has been shown to differ when probed with random genomic sequences, reflecting a high level of polymorphism (Raeder and Broda, 1986). The situation in species where dispersal is probably mainly clonal such as *Aspergillus* or *Penicillium* may be quite different. Other factors may also be important, such as the time elapsed since the species differentiated and the history of natural selection of the species.

The types of probe used range from known characterised cloned genes of greater or lesser expected sequence conservation to those which reveal hypervariable sequences. Thus a probe consisting of a highly conserved gene such as that coding for ribosomal DNA, or for tubulin, will hybridise to DNA taken from several genera and will probably show relatively little polymorphism within a species or even within a genus. Other cloned genes, such as the pectin lyase D gene of *A. niger* may provide sufficient polymorphism to distinguish between clearly defined taxa, (Kusters-van Someren *et al.*, 1990). At the other extreme, some repeated DNA sequences may be located at several positions in the genome and be highly polymorphic, thus giving a hypervariable probe which may reveal a great deal of intraspecific variation between quite closely related isolates.

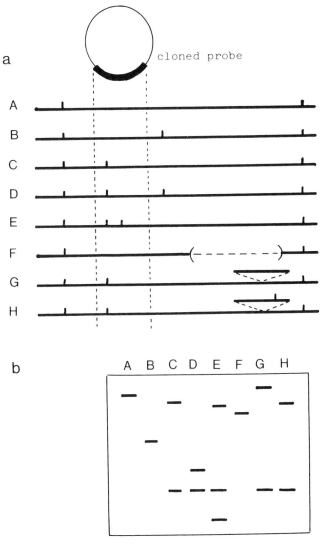

Figure 1. Generation and detection of RFLPs. A cloned genomic sequence from strain A was used as a probe in a Southern blot experiment in which DNA from strains A to H was isolated and digested to completion by a particular restriction enzyme. These digested DNA samples were then electrophoresed and transferred to a DNA binding membrane and then hybridised to the radiolabelled probe. The bands where the probe hybridised to the homologous fragments in the digested DNA were then revealed by autoradiography.

a] The cloned fragment used as a probe was thet sequence limited by the dotted lines. Strains A to H were polymorphic for the sites for the particular restriction enzyme used as indicated by the vertical bars. In addition strain F contained a deletion and strains G and H an insertion.

b] The fragments revealed for each of the eight strains by this probe are illustrated.

Thus, for purposes of taxonomy, it is important to choose a suitable probe which reveals variability at a level which is of relevance to the study being undertaken. At the present time the selection of probes is an empirical process in fungal genera, as information to enable a prediction of the nature of appropriate probes remains limited.

This paper will attempt to demonstrate that RFLP analysis is a useful and perhaps definitive approach to the determination of a phylogenetic classification of genera such as *Aspergillus* or *Penicillium*. A very precise distinction between closely related biological species should also be possible by this technique.

GENETIC ANALYSIS OF THE SECTION *NIDULANTES*

Aspergillus nidulans is one of the genetically best studied of fungal species. However, the vast majority of work on *A. nidulans* has been carried out on mutant derivatives of one isolate, NRRL 194, the so-called Glasgow strains (Pontecorvo *et al.*, 1953; Clutterbuck, 1974). The background of information available from these now classical studies has enabled a considerable amount of genetics to be carried out on other isolates of *A. nidulans* and also on other species or varieties belonging to *Aspergillus* sect. *Nidulantes*, thus providing a framework for the interpretation of RFLP studies.

Isolates which have been examined from species or varieties in the sect. *Nidulantes* fall into a number of heterokaryon compatibility groups (Grindle, 1963a, b; Croft, 1985). Such groups also exist in other sections of *Aspergillus*, including both teleomorph and anamorph species (Caten, 1971). Heterokaryon incompatibility is under heterogenic control: an allelic difference at any one of a series of specific *het* genes is sufficient for two isolates to be incompatible. In a sample of six heterokaryon compatibility groups, eight *het* loci have been demonstrated, with multiple alleles at two (Dales and Croft, 1977; Dales *et al.*, 1983; Dales and Croft, 1983; Croft, 1985). Most pairs of isolates from the natural environment are likely to belong to different heterokaryon compatibility groups and to differ at a number of *het* loci. Considerable genetic variation exists in a species such as *A. nidulans*, but detailed examination has shown that it is mainly confined to differences between heterokaryon compatibility groups and not between isolates within a group (Butcher *et al.*, 1972; Croft and Jinks, 1977). Members of the same compatibility group are clearly clonally related. It is tempting to suggest that this division of a species such as *A. nidulans* into sexually interfertile but genetically distinct heterokaryon incompatible groups which are probably efficiently dispersed by conidia could lead to their further divergence into genetically isolated sibling species. In reality, this is probably a simplistic model.

Nevertheless, what appear to be sibling species do occur in sect. *Nidulantes*. *A. quadrilineatus*, *A. nidulans* var. *echinulatus* and *A. rugulosus* are very closely related to each other and to *A. nidulans*, but are clearly distinct biologically, because sexual crosses with *A. nidulans* are not fully fertile. Hybrid cleistothecia have not been produced between *A. nidulans* and *A. nidulans* var. *echinulatus* despite many attempts. Crosses of *A. quadrilineatus* and *A. rugulosus* with *A. nidulans* have both given rare hybrid cleistothecia, but these contain only a very small number of recognisable ascospores and give rise to aneuploid and allodiploid progenies. However, when protoplast fusion is used to overcome heterokaryon incompatibility parasexual crosses are possible between these species and allodiploids can be isolated (Dales and Croft, 1977; Croft and Dales, 1983) Such crosses may then be haploidised and the progeny analysed. Such analysis has shown the free segregation of all eight linkage groups among the progeny of an *A. nidulans* plus *A. quadrilineatus* fusion but the presence of complex translocations in *A. nidulans* var.

echinulatus involving the chromosomes equivalent to the *A. nidulans* linkage groups III, V and VIII. Other genetic differences between the latter two taxa can be revealed by this analysis, including the presence of allelic differences at one *het* gene on each of the chromosome homeologues equivalent to the *A. nidulans* linkage groups II, VI and VII and at least one on the III-V-VIII complex. One further result from protoplast fusion experiments between these closely related species is that selectable mitochondrial markers, such as oligomycin resistance, can be transferred readily from one species to another. This has been shown to involve either the transfer of the whole mitochondrial genome or a very high frequency of mitochondrial recombination (Earl *et al.*, 1981).

At the other extreme, sect. *Nidulantes* includes species, such as *A. unguis, A. heterothallicus* and *A. stellatus,* which are clearly distantly related to *A. nidulans.* No genetic interaction at all is possible between *A. nidulans* and these species (Kevei and Peberdy, 1984). No fusion products have ever been recovered from protoplast fusion experiments and mitochondrial transfer experiments have never been successful.

RFLP ANALYSIS OF THE SECTION *NIDULANTES*

Analysis of RFLPs in the mitochondrial genome is becoming widely used for detection of variation in natural populations of fungal species, particularly of plant and animal pathogens. It has also been used successfully in phylogenetic and evolutionary studies of *Neurospora* (Taylor and Natvig, 1989). The mitochondrial genome of *A. nidulans* has been almost completely sequenced, maps of restriction sites are available and functions have been assigned to most regions of the molecule (Brown *et al.*, 1985). As a consequence, detailed interpretation of the restiction patterns obtained for other species of the sect. *Nidulantes* has become possible. The most surprising feature of the RFLPs of the mtDNA of those species so far examined is that no restriction site polymorphism has been detected within a species. This remains to be confirmed for a larger sample of isolates, species and restriction enzymes, but the results to date are quite striking. For example, some fifteen isolates of *A. nidulans* from the U.K., continental Europe, North and South America and Southern Africa all have identical mitochondrial restriction patterns for six restriction enzymes, and each species tested has a distinct, diagnostic restriction map.

The differences in mtDNA between the closely related species have been shown to be due almost entirely to the presence or absence of inserted sequences. These sequences are optional introns which tend to be located in certain regions of the genome (Fig. 2). They may make up a considerable proportion of the molecule, but may be absent in some species. The difference between the 30kb mitochondrial genome of *A. quadrilineatus* and the almost 40kb genome of *A. nidulans* var. *echinulatus* is due almost entirely to seven extra introns in the latter species. The exonic sequences of these species are almost identical, there being only about 0.2% base substitution (Jadayel, 1986).

This result contrasts with comparisons of the mtDNA of *A. nidulans* with sect. *Nidulantes* species which have no genetic interaction with *A. nidulans*, namely *A. unguis, A. heterothallicus* and *A. stellatus*. While it is possible to align their restriction maps with that of *A. nidulans* (Fig. 3) and sequence conservation in ribosomal RNA regions is evident, more polymorphism exists among these species. For example, comparison of a short (124bp) sequence located in the fourth exon of the *oxi*A gene of *A. nidulans* and of *A. heterothallicus* shows fifteen base substitutions (12%). Moreover, in *A. heterothallicus*, part of the mitochondrial genome has been rearranged (Fig. 4) ((Jadayel, 1986).

A. nidulans var. echinulatus

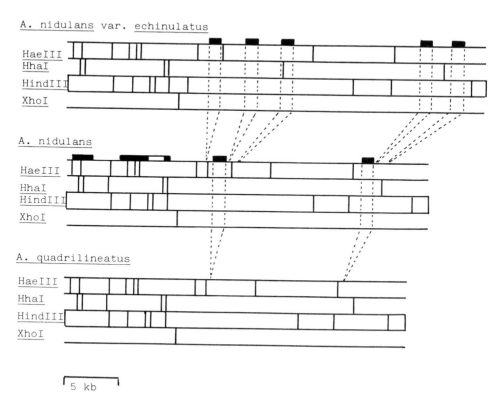

```
5 kb
```

Figure 2. Linearised restriction site maps of the circular mitochondrial genomes of *Aspergillus nidulans* var. *echinulatus*, *A. nidulans* and *A. quadrilineatus* aligned around a unique XhoI site found in a tRNA gene. The position of the small and large rRNA genes is indicated for *A. nidulans*. The three species are closely related with few restriction site differences. The differences in size and restriction fragment patterns is due to the presence or absence of optional introns, the positions of which are indicated by solid bars conected by vertical dotted lines.

The diagnostic nature of RFLPs in nuclear DNA was revealed by the use of random genomic probes chosen from a library of one isolate of *A. nidulans* cloned in the vector Lambda EMBL3. The inserts cloned in this vector are about 17 to 23 kb in size. Most of these clones, when used as probes in Southern blot experiments against digests of total DNA of the strain from which they were isolated, hybridised with from three to six bands. When these probes were hybridised with digested DNA from a range of isolates of *A. nidulans* belonging to different heterokaryon compatibility groups and from different geographical locations, very little or no polymorphism was revealed. The polymorphism that was present was limited to differences between heterokaryon compatibility groups, and was limited to simple single restriction site differences (Fig. 5). No polymorphism was detected within a compatibility group.

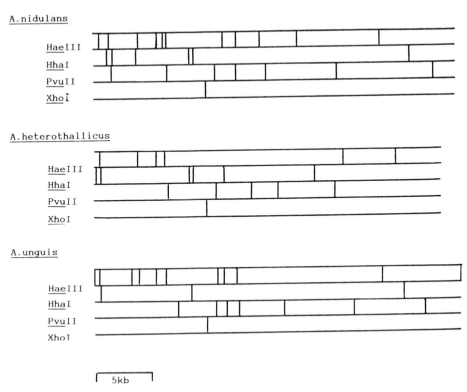

Figure 3. Linearised restriction maps of the circular mitochondrial genomes of *Aspergillus nidulans*, *A. heterothallicus* and *A. unguis* aligned around the unique XhoI site. These three species are less closely related than those in figure 2 and the site polymorphism is apparent. The intron status of *A. heterothallicus* and *A. unguis* is unknown (From Jadayel, 1986).

When these probes were used with digested DNA from other sect. *Nidulantes* species, *A. quadrilineatus, A. rugulosus* and *A. nidulans* var. *echinulatus*, hybridisation was detected, usually to a similar number of bands, but the patterns revealed considerable polymorphism. A more complex genetic basis for the polymorphism was indicated than that found between heterokaryon compatibility groups within a species (Fig. 6). These probes appear to be highly diagnostic for each of these closely related species.

When these random genomic probes were used in Southern blot experiments with digested DNA from *A. unguis*, no hybridisation could be detected at all. This result was also produced with DNA from *A. niger*. However, when DNA from *A. terreus* was probed, hybridisation was detected (Fig. 6), suggesting that this species has some relationship with *A. nidulans*.

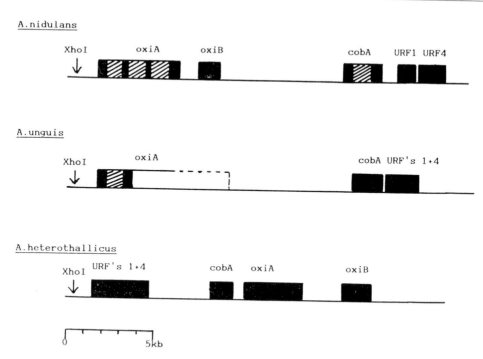

Figure 4. A comparison of the organisation of some genes in the mitochondrial genome of *Aspergillus nidulans, A. unguis* and *A. heterothallicus* showing structural rearrangements present in *A. heterothallicus*. The maps are aligned around the unique *XhoI* site. Known introns are cross-hatched, but the intron content of *A. unguis* and *A. heterothallicus* is not known. The maximum possible extent of the oxiA gene in *A. unguis* is indicated (From Jadayel, 1986).

DISCUSSION

The results, briefly summarised here, demonstrate the highly diagnostic nature of RFLP analysis in *Aspergillus* sect. *Nidulantes*, both in nuclear DNA, when revealed by the use of random genomic probes, and in mtDNA. The patterns revealed by the restriction cleavage of mtDNA are characteristic for each species as are the patterns of hybridisation revealed by most random genomic probes used in Southern blot experiments with nuclear DNA. The extreme conservation of the mitochondrial genome within each species is unexpected and perhaps still requires confirmation. Nevertheless, this intraspecific conservation, together with the presence or absence of optional introns in the otherwise similar mitochondrial genomes of many taxa in sect. *Nidulantes* allows a clear distinction between these taxa on the basis of their restriction fragment patterns, as the presence of introns introduces additional restriction sites. The evolution of these mitochondrial genomes is clearly of considerable interest. More taxa from sect. *Nidulantes* need to be examined to

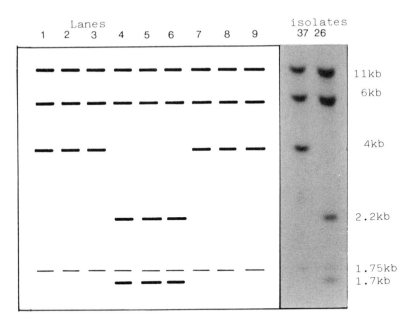

Figure 5. Diagrammatic representation of the autoradiograph obtained by the hybridisation of a ^{32}P labelled random genomic clone from *Aspergillus nidulans* to *Hind* III digests of total nuclear DNA of nine isolates of *A. nidulans* from three heterokaryon compatibility (h-c) groups The strains used were Birmingham isolates 3, 28 and 38 (h-c group A) in lanes 1 to 3, isolates 1, 26 and 36 (h-c group B) in lanes 4 to 6, and isolates 33, 37 and 74 (h-c group C) in lanes 7 to 9. The photograph shows this single site polymorphism for isolates 37 and 26 (Chapman and Croft, unpublished).

increase the sample size, but to date each taxon examined has shown a different restriction fragment pattern. If the variation seen so far is a general feature of the mitochondrial genomes of other sections in *Aspergillus* and also in *Penicillium*, then mitochondrial restriction fragment patterns will be vey useful for the assignment of isolates to genetically related groups. The relative simplicity of the procedure for the extraction, digestion and electrophoresis of mtDNA is an added attraction for this approach.

The nature of the random genomic probes remains unknown, but most appear to be single copy sequences. Some random genomic probes appear to be conserved in the species closely related to *A. nidulans*. Other probes appear to be based on highly repeated sequences and show hypervariability (O'Dell *et al.*, 1989) but so far these have not been isolated from *Aspergillus*.

Very little screening of clones for use in differentiation between species has been necessary. The conservation of these sequences within a single species is striking, the small amount of variation present being explicable by single site variation. The polymorphism between the closely related species is more complex, but the genetics of these differences has not been analysed yet. The hybridisation of random genomic probes from *A. nidulans* to sequences from closely related species but not to more distantly related taxa such as *A. unguis* or *A. niger* provides a dot blot method for the rapid grouping of related taxa. The hybridisation of *A. nidulans* probes to *A. terreus* suggest a closer relationship than conventional taxonomy indicates.

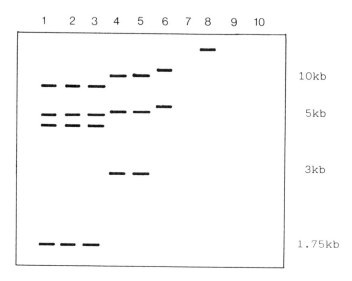

Figure 6. Diagrammatic representation of the autoradiograph obtained by the hybridisation of a random genomic probe from *Aspergillus nidulans* to EcoRI digests of DNA from different species of *Aspergillus*. The strains used were: lanes 1 to 3, *A. nidulans* isolates 26, 74 and 258; lanes 4 and 5, *A. nidulans* var. *echinulatus* isolates 25 and 209; lane 6, *A. quadrilineatus* isolate 12; lane 7, *A. unguis*; lane 8, *A. terreus*; lanes 9 and 10, *A. niger*, (Chapman and Croft, unpublished).

Despite the intraspecific conservation of both nuclear and mtDNA there is considerable genetic variation present within a species such as *A. nidulans*. This variation can be demonstrated at the molecular level by the use of two classes of probe. The first is a transposon-like sequence (supplied by Dr. M. Hynes and Dr. A. Upshall) which was isolated from *A. nidulans*. Preliminary experiments showed that hybridisation patterns produced by this probe in Southern blot experiments with nuclear DNA from other *A. nidulans* isolates are heterokaryon compatibility group specific, that is genotype specific. A probe of labelled wild-type M13 bacteriophage has produced a similar result. This virus has been shown to contain a short repeated sequence which hybridises to DNA of all organisms tested and possibly provides a universal hypervariable probe (Vassart *et al.*, 1987; Ryskov *et al.*, 1988). Repeated sequences, showing up as bright bands in electrophoretic digests of nuclear DNA, have been used to distinguish between strains of *A. flavus* and *A. oryzae* (Klich and Mullaney, 1987). Although the patterns of bands produced by repeated sequences within a species are generally consistent variation is sometimes observed. For example, additional or different bands of repeated DNA have been found in some isolates of *A. quadrilineatus*. Thus, although the random genomic probes have revealed strong intraspecific conservation of DNA sequences, the study of polymorphism in various classes of repeated sequence DNA suggest that they may vary considerably between isolates within a species and thus be of interest in terms of evolution within the genus.

The analysis of RFLPs found in *Aspergillus* sect. *Nidulantes* has been carried out against a good background of genetic information about the taxa used, which has enabled interpretation of the polymorphisms found in terms of genetic differences. This information will be of importance when attempting to interpret the DNA polymorphism found in anamorphtaxa such as those in *Aspergillus* sect. *Nigri* or in *Penicillium*. The methods will provide information concerning the phylogenetic relationships of the various taxa studied and thus give the taxonomist a genetic and evolutionary interpretation of the classification produced by systematic studies.

REFERENCES

BOTSTEIN, D., WHITE, R.L., SKOLNICK, M. and DAVIS, R.W. 1980. Construction of a genetic linkage map in man using restriction fragment length polymorphisms. *American Journal of Human Genetics* 32: 314-331.

BROWN, T.A., WARING, R.B., SCAZZOCCHIO, C. and DAVIS, R.W. 1985. The *Aspergillus nidulans* mitochondrial genome. *Current Genetics* 9: 113-117.

BUTCHER, A.C., CROFT, J.H. and GRINDLE, M. 1972. Use of genotype-environmental interaction analysis in the study of natural populations of *Aspergillus nidulans*. *Heredity* 29: 263-283.

CATEN, C.E. 1971. Heterokaryon incompatibility in imperfect species of *Aspergillus*. *Heredity* 26: 299-312.

CLUTTERBUCK, A.J. 1974. *Aspergillus nidulans*. In Handbook of Genetics, Vol. 1, ed. R.C. King, pp. 447-510. New York: Plenum Press.

CROFT, J.H. 1985. Protoplast fusion and incompatibility in *Aspergillus*. In Fungal protoplasts, applications in biochemistry and genetics, eds. J.F. Peberdy and L. Ferenczy, pp. 225-240. New York: Marcel Dekker.

CROFT, J.H. and JINKS, J.L. 1977. Aspects of the population genetics of *Aspergillus nidulans*. In Genetics and physiology of *Aspergillus*, J.E. Smith and J.A. Pateman pp. 339-360. London: Academic Press.

CROFT, J.H. and DALES, R.B.G. 1983. Interspecific somatic hybridisation in *Aspergillus*. In Protoplasts 1983, Proceedings of the 6th International Protoplast Symposium, eds. I. Potrykus, C.T. Harms, A. Hinnen, R. Hütter, P.J. King and R.D. Shillito, pp. 179-186. Basel: Birkhäuser Verlag.

DALES, R.B.G. and CROFT, J.H. 1977. Protoplast fusion and the isolation of heterokaryons and diploids from vegetatively incompatible strains of *Aspergillus nidulans*. *FEMS Microbiology Letters* 1: 201-203.

—— 1983. A chromosome assay method for the detection of heterokaryon incompatibility (het) genes operating between members of different heterokaryon compatibility (h-c) groups in *Aspergillus nidulans*. *Journal of General Microbiology* 129: 3643-3649.

DALES, R.B.G.., MOORHOUSE, J. and CROFT, J.H. 1983. The location and analysis of two heterokaryon incompatibility (*het*) loci in strains of *Aspergillus nidulans*. *Journal of General Microbiology* 129: 3637-3642.

DONIS-KELLER, H., GREEN, P., HELMS, C., CARTINHOUR, S., WEIFFENBACH, B., STEPHENS, K., KEITH, T.P., BOWDEN, D.W., SMITH, D.R., LANDER, E.S., BOTSTEIN, D., AKOTS, G., REDIKER, K.S., GRAVIUS, T., BROWN, V.A., RISING, M.B., PARKERS, C., POWERS, J.A., WATT, D.E., KAUFFMAN, E.R., BRICKER., A., PHIPPS, R., MÜLLER-KAHLE, H., FULTON, T.R., NG, S., SCHUMM, J.W., BRAMAN, J.C., KNOWLTON, R.G., BARKER, D.F., CROOKS, S.M., LINCOLN, S.E., DALY, M.J. and ABRAHAMSON, J. 1987. A genetic linkage map of the human genome. *Cell* 51: 319-337.

EARL, A.J., TURNER, G., CROFT, J.H., DALES, R.B.G., LAZARUS, C.M., LÜNSDORF, H. and KÜNTZEL, H. 1981. High frequency transfer of species specific mitochondrial DNA sequences between members of the Aspergillaceae. *Current Genetics* 3: 221-228.

GRINDLE, M. 1963a. Heterokaryon compatibility of unrelated strains in the *Aspergillus nidulans* group. *Heredity* 18: 191-204.

—— 1963b. Heterokaryon compatibility of closely related wild isolates of *Aspergillus nidulans*. *Heredity* 18: 397-405.

JADAYEL, D.M. 1986. Variation in the organisation and structure of the mitochondrial DNA of species of *Aspergillus*. Ph.D. thesis, University of Birmingham.

JEFFREYS, A.J., WILSON, V. and THEIN, S.L. 1985. Hypervariable 'minisatellite' regions in human DNA. *Nature* 314: 67-73.

KEVEI, F. and PEBERDY, J.F. 1984. Further studies on protoplast fusion and interspecific hybridisation within the *Aspergillus nidulans* group. *Journal of General Microbiology* 130: 2229-2236.

KLICH, M.A. and MULLANEY, E.J. 1987. DNA restriction enzyme fragment polymorphism as a tool for rapid differentiation of *Aspergillus flavus* from *Aspergillus oryzae*. *Experimental Mycology* 11: 170-175.

KOZLOWSKI, M. and STEPIEN, P.P. 1982. Restriction enzyme analysis of mitochondrial DNA of members of the genus *Aspergillus* as an aid in taxonomy. *Journal of General Microbiology* 128: 471-476.

KURTZMAN, C.P. 1985. Classification of fungi through nucleic acid relatedness. In Advances in *Aspergillus* and *Penicillium* systematics, eds. R.A. Samson and J.I. Pitt, pp. 233-254. New York and London: Plenum Press.

KUSTERS-VAN SOMEREN, M.A., KESTER, H.C.M., SAMSON, R.A. and VISSER, J. 1990. Variation in pectinolytic enzymes of the black Aspergilli: a biochemical and genetic approach. In Modern Concepts in *Penicillium* and *Aspergillus* classification, eds. R.A. Samson and J.I. Pitt, pp. 321-334. New York: Plenum Press.

O'DELL, M., WOLF, M.S., FLAVELL, R.B., SIMPSON, C. and SUMMERS, R.W. 1989. Molecular variation in populations of *Erysiphe graminis* on barley, oats and rye. *Plant Pathology* (in press).

PONTECORVO, G., ROPER, J.A., HEMMONS, L.M., MACDONALD, K.D. and BUFTON, A.W.J. 1953. The genetics of *Aspergillus nidulans*. *Advances in Genetics* 5: 141-238.

RAEDER, U. and BRODA, P. 1986. Meiotic segregation analysis of restriction site polymorphisms allows rapid genetic mapping. *EMBO Journal* 5: 1125-1128.

RAPER, K.B. and FENNELL, D.I. 1965. The genus *Aspergillus*. Baltimore: Williams and Wilkins.

RYSKOV, A.P., JINCHARADZE, A.G., PROSNYAK, M.I., IVANOV, P.L. and LIMBORSKA, S.A. 1988. M13 phage DNA as a universal marker for DNA fingerprinting of animals, plants and microorganisms. *FEBS Letters* 233: 388-392.

SOUTHERN, E.M. 1975. Detection of specific sequences among DNA fragments separated by gel electrophoresis. *Journal of Molecular Biology* 98: 503-517.

TANKSLEY, S.D., YOUNG, N.D., PATERSON, A.H. and BONIERBALE, M.W. 1989. RFLP mapping in plant breeding: new tools for an old science. *Biotechnology* 7: 257-264.

TAYLOR, J.W. and NATVIG, D.O. (1989). Mitochondrial DNA and evolution of heterothallic and pseudohomothallic *Neurospora* species. *Mycological Research* (in press).

VASSART, G., GEORGES, M., MONSIEUR, R., BROCAS, H., LEQUARRE, A.S. and CHRISTOPHE, D. 1987. A sequence in M13 phage detects hypervariable minisatellites in human and animal DNA. *Science* 235: 683-684.

VARIATION IN PECTINOLYTIC ENZYMES OF THE BLACK ASPERGILLI: A BIOCHEMICAL AND GENETIC APPROACH

M.A. Kusters - van Someren[1], H.C.M. Kester[1], R.A. Samson[2] and J. Visser[1]

[1]Department of Genetics, section Molecular Genetics
Agricultural University Wageningen, The Netherlands

[2]Centraalbureau voor Schimmelcultures
Oosterstraat 1, 3740 AG Baarn, The Netherlands

SUMMARY

Several black *Aspergillus* isolates have been analyzed by biochemical and genetic tools to investigate whether such techniques are useful for rapid, objective identification. Western blots were used to screen for pectin lyase production and Southern blots to look for homology in these isolates with respect to their pectin lyase gene(s). All *A. niger* isolates can easily be identified by these methods. They produce a similar pectin lyase, although differences in molecular weigth and reactivity with a monoclonal antibody exist. Also all *A. niger* isolates contain at least one conserved gene: the *pel*D gene. There are other hybridizing bands visible, probably resulting from heterologous hybridization with another pectin lyase gene. Classification on the basis of presence of these bands is different from classification on the basis of morphological features. The other species of sect. *Nigri* differ in one or more of these aspects with the *A. niger* aggregate, except *A. foetidus*.

INTRODUCTION

The black Aspergilli (*Aspergillus* sect. *Nigri*, Gams *et al.*, 1985) have been frequently studied, because of their industrial importance and significance as a plant or human pathogen. Species in this section appear to be closely related. Identification of individual species, however, is difficult and the several attempts made at classifying the section are debatable. For example, Mosseray (1934 a and b) recognized 35 species, while Thom and Raper (1945), and Raper and Fennell (1965) accepted 15 and 12 species, respectively. Al-Musallam (1980) reinvestigated the *Aspergillus* sect. *Nigri* and accepted 7 species, of which *A. niger* represented an aggregate of 7 varieties and 2 formae (see Table 1). Although a number of species can be readily distinguished on the basis of the conidiophore structure, shape and ornamentation of the conidia, several taxa are still difficult to identify.

The black *Aspergillus* species are well known for their ability to produce and secrete a large variety of pectinolytic enzymes which attack pectin either by hydrolysis (the polygalacturonases) or by transelimination (the pectin lyases). Pectin esterase saponifies the substrate. The variation in extracellular enzyme patterns between two different *A. niger* isolates (NW756 and CBS 120.49) has prompted us to investigate the use of variation in pectin lyase as an aid in the classification within *Aspergillus* sect. *Nigri*.

Two different experimental approaches were used. First, pectin lyase patterns were analyzed by Western blotting using antibodies raised against purified enzymes. Second, a previously cloned gene, pectin lyase D, was used as a probe to screen restriction enzyme digests of isolated DNA after Southern blotting. A variety of *Aspergillus* isolates was studied, including (neo)type cultures and typical representatives.

Modern Concepts in Penicillium and Aspergillus Classification
Edited by R. A. Samson and J. I. Pitt
Plenum Press, New York, 1990

Table 1. Species concepts of the black Aspergilli according to Al-Musallam (1980) and Raper and Thom (1965)

Al-Musallam (1980)	*Raper & Thom (1965)*
A. carbonarius	A. carbonarius
A. heteromorphus	A. heteromorphus
A. ellipticus A. helicothrix	A. ellipticus
A. japonicus var. japonicus A. japonicus var. aculeatus	A. japonicus A. aculeatus
A. niger var. niger	A. niger A. ficuum A. tubingensis
A. niger var. niger f. hennebergii	
A. niger var. phoenicisp A. niger var. phoenicis f. pulverulentus	A. phoenicis A. pulverulentus
A. niger var. awamori A. niger var. nanus A. niger var. usamii A. niger var. intermedius	A. awamori
A. foetidus	A. foetidus

MATERIAL AND METHODS

Isolates.
The cultures used were obtained from the Centraalbureau voor Schimmelcultures (Baarn, The Netherlands) and are shown in Table 2.

Media and growth conditions.
The minimal medium (MM) used had a composition as described by Pontecorvo *et al.* (1953). It consists of 70 mM NaNO$_3$, 11mM KH$_2$PO$_4$, 2 mM MgCl$_2$, 6 mM KCl and 0.9 mg ZnSO$_4$. 7H$_2$O, 0.2 mg MnCl$_2$, 0.06 mg CoCl$_2$, 0.06 mg CuSO$_4$. 5H$_2$O, 0.04 mg (NH$_4$)$_6$Mo$_7$O$_{24}$ 4H$_2$O, 0.29 mg CaCl$_2$. 2H$_2$O, 0.2 mg FeSO$_4$. 7H$_2$O per litre. Complete medium (CM) had the same basal composition as MM, but was supplemented with 2 g neopeptone, 1 g vitamin assay casamino acids, 1 g yeast extract, 0.3 g sodium ribonucleate and 2 ml vitamin stock solution (1 mg/ml thiamine, 1 mg/ml riboflavine, 0.1 mg/ml p-aminobenzoic acid, 1 mg/ml nicotinamide, 0.5 mg/ml pyridoxine HCl, 0.1 mg/ml panthothenic acid, and 2 μg/ml biotin) per litre. All media were adjusted to pH 6.0 with NaOH before autoclaving for 20 min at 120°C. For propagation of conidia, Petri dishes containing CM solidified with 1.2% agar supplemented with 20 mM sucrose as carbon source were inoculated by streaking out a conidial suspension. After growth for 4-6 days at 30°C, conidia were harvested by adding 5-10 ml saline/Tween [0.8% (v/v) NaCl, 0.005% (v/v) Tween 20] per Petri dish. The conidial suspension obtained was thoroughly shaken to break conidial chains and concentrations determined with a haemocytometer. To study the expression of pectolytic enzymes, MM with 1% (w/v) sugar beet slices was inoculated at 10^6 conidia per ml. Incubation was for two days at 28°C in a New Brunswick

Table 2. Isolates examined

A. carbonarius	CBS 111.26 (NT), CBS 112.80
A. ellipticus	CBS 707.79 (T)
A. helicothrix	CBS 677.79 (T)
A. heteromorphus	CBS 117.55 (T)
A. japonicus	CBS 114.51 (T), CBS 621.78
A. aculeatus	CBS 172.66, CBS 115.80
A. niger	CBS 554.65 (NT)
A. niger	CBS 120.49
A. awamori	CBS 557.65 (NT), CBS 563.65
A. phoenicis	CBS 126.49, CBS 135.48
A. nanus	CBS 136.52, CBS 131.52
A. usami	CBS 139.52 (T), CBS 553.65
A. intermedius	CBS 559.65, CBS 117.32
A. hennebergii	CBS 118.35 (T), CBS 125.52
A. pulverulentus	CBS 558.65, CBS 425.65
A. foetidus	CBS 121.28, CBS 618.78
A. niger	NW756

(T) ex type; (NT) ex neotype culture

orbital shaker at 250 rpm. The medium was separated from the mycelium by filtration over cheese cloth, dialyzed against 20 mM sodium acetate buffer (pH 5.5) and stored at –20°C.

To obtain mycelial biomass for the isolation of DNA, CM supplemented with 20 mM sucrose was inoculated with 10^6 conidia per ml. Incubation was as described above. The mycelia were harvested by filtration over cheese cloth, washed with cold saline, squeezed to remove excess liquid and stored at –80°C.

Isolation of chromosomal DNA.
DNA was isolated as described by de Graaff et al. (1988). Frozen mycelium (0.5 g) was disrupted with a microdismembrator (Braun) for 1 min and 4 ml freshly prepared, prewarmed (55°C) extraction buffer [0.2 M Tris, 0.26 M NaCl, 50 mM EGTA, 20 mg/ml tri-isopropyl-naphthalene sulphonic acid, 120 mg/ml p-aminosalicylic acid, pH 8.5 and 37.5% (v/v) phenol] was added. The suspension was mixed thoroughly on a vortex mixer for 1 min, 1 ml chloroform was added and the suspension was mixed again (1 min). After centrifugation (10^4 g, 10 min) in a Sorvall high speed centrifuge the aqueous phase was extracted once with phenol/chloroform (1:1) and once with chloroform. DNA was precipitated from the aqueous phase with 2 vol. ethanol at room temperature by centrifugation (10^4 g, 10 min) in a Sorvall high speed centrifuge. The pellet was dried under vacuum and dissolved in 500 µl sterile distilled water with 20 µg/ml RNase A and stored at –20°C.

Restriction enzyme analysis.
Approximately 2 µg chromosomal DNA was digested for 2 h in a 200 µl volume using 20 units of the appropriate restriction enzymes according to the manufacturer's conditions (BRL). The DNA was then precipitated by adding 0.1 vol. 3M NaAc (pH 5.6) and two vol. ethanol and centrifugation in a small table centrifuge (10 min, 10^4 g). The pellet was dried under vacuum and dissolved in 12 µl of sterile distilled water. After the addition of 5 µl

sample buffer [0.25% (w/v) bromophenol blue, 15% (w/v) Ficoll type 400], the samples were loaded on an agarose gel [0.6% in TBE (0.89 M Tris, 0.89 M Boric acid, 27 mM EDTA) containing 0.5 µg/ml ethidium bromide; the electrophoresis buffer also consisted of TBE/ethidium bromide]. Electrophoresis was carried out for 16 h at 30 V (gel dimensions 22 × 17 cm), or until the bromophenol blue dye marker had nearly run off the gel.

Southern blotting.
After electrophoresis the DNA was denatured by gently agitating the gel in a 0.5 M NaOH, 1.5 M NaCl solution for 30 min (solution changed after 15 min). The gel was neutralized by agitating for 30 min (solution changed after 15 min) in 0.5 M Tris-HCl (pH 7.5), 1.5 M NaCl. The DNA was blotted overnight on nitrocellulose (0.45 µm, Schleicher & Schuell) using 10 ∗ SSC (1.5 M NaCl, 0.15 M tri-sodium citrate dihydrate) as transfer buffer. The DNA was fixed on the nitrocellulose membrane by baking for 1 h in an oven at 80°C (Maniatis et al., 1982).

Isolation of the probe.
The plasmid which contains the pelD gene (pGW840) was propagated in E. coli MH1 (Goddard et al., 1983) grown in LB medium (10 g trypticase peptone, 5 g yeast extract, 10 g NaCl, pH 7.5 per litre) with 50 µg/ml ampicillin. Plasmid DNA was isolated as described by Maniatis et al. (1982). 1 µg DNA was digested with BamHI and PstI in a final volume of 20 µl under the conditions described by the manufacturer (BRL). The restriction fragments were separated on a 0.8% (w/v) agarose gel in TBE/ethidium bromide. After electrophoresis, the desired band was cut out and the DNA was isolated by electro-elution using sample cups from Isco Inc. (Lincoln, USA). The DNA was precipitated by adding 0.1 vol. 3M NaAc (pH 5.6) and 2 vol. ethanol and by subsequent centrifugation in a small table centrifuge at 10^4 g for 10 min. The DNA pellet was dried under vacuum, suspended in sterile distilled water and the concentration was determined by electrophoresis using lambda DNA as standardized concentration marker.

Preparation of the probe.
50 ng of the BamHI/PstI fragment on which the pelD gene is situated was labeled using the random priming method: 50 ng hexamers (Pharmacia) were added as well as 1 µl 10 ∗ priming buffer (0.1 M Tris-HCl pH 7.5, 50 mM MgCl$_2$) in a final volume of 10 µl. The mixture was heated at 100°C for 3 min and cooled on ice. dGTP, dCTP and dTTP were added (0.15 mM each), as well as 50 µCi α32P-dATP and 1 U Klenow enzyme (BRL), in a final volume of 20 µl. After incubation at 37°C for 15 min, the non-incorporated label was removed by centrifugation in a spun column through Sephadex G50 in TE (10 mM Tris-HCl pH 8.0, 1mM EDTA pH 8.0) (Maniatis et al., 1982).

Hybridization.
Prehybridization of the Southern blot was done for 1 h at 65°C in hybridization buffer. This buffer consists of 50 mM Tris (pH 7.5), 10 mM EDTA (pH 7.5), 1 M NaCl, 0.5% (w/v) SDS, 0.1% (w/v) tetra-sodium diphosphate-10-hydrate, 10 ∗ Denhardt [1% (w/v) Ficoll, 1% (w/v) polyvinylpyrolidone, 1% (w/v) BSA] and, in addition, 100 µg/ml heat-denatured, sheared herring sperm DNA (Boehringer, Mannheim) to minimize aspecific hybridization. The probe was heat-denatured (5 min, 100°C) and then added to the prehybridization mix. After 18 h incubation, the membrane was washed twice for 30 min with 2 ∗ SSC, 0.5% (w/v) SDS at 65°C and once with 0.2 ∗ SSC, 0.5% (w/v) SDS (30 min,

65°C), air dried, and exposed to a Konica X-ray film using Kodak intensifying screens for at least 16 h at -60°C.

Immunological screening for pectin lyases.

Small samples (30 µl) of the dialyzed culture fluids were mixed with 10 µl sample buffer [0.25 M Tris-HCl pH 6.8, 8% (w/v) SDS, 40% (v/v) glycerol, 20% (v/v) 2-mercaptoethanol, 0.05 mg/ml bromophenol blue] and heated (3 min, 100°C). SDS-polyacrylamide gel electrophoresis was performed according to Laemmli (1970) in a 10% slab gel. The electrophoresis was stopped when the internal bromophenol blue marker had reached the bottom of the gel. Proteins were transferred to nitrocellulose (0.45 µm, Schleicher & Schuell) by electroblotting (0.8 mA/cm² for 1 h) using the LKB Novablot system and the continuous buffer system [39 mM glycin, 48 mM Tris, 0.0375% (w/v) SDS and 20% (v/v) methanol] as described by the manufacturer. To minimize aspecific binding of the antibody to the nitrocellulose filter, the filter was incubated with 3% (w/v) gelatin in Tris buffered saline (TBS: 20 mM Tris-HCl pH 7.5, 0.5 M NaCl) for 1 h. After washing the blot (2 times 5 min) in TTBS [0.05% (v/v) Tween 20 in TBS] it was incubated overnight in 75 ml TTBS containing 1% (w/v) gelatin and 10 µl monoclonal antiserum or 0.1% (v/v) polyclonal antiserum. Excess antiserum was removed by washing the blot (2 times 5 min) in TTBS. Finally, the blot was incubated for 2 h in 75 ml TTBS containing 1% (w/v) gelatin and 10 µl goat-anti-mouse γ-globulin alkaline phosphatase conjugate (Biolabs) when monoclonal antibodies were used or 10 µl goat-anti-rabbit γ-globulin alkaline phosphatase conjugate when polyclonal antibodies were used. After washing in TTBS (2 times 5 min) to remove excess conjugate, the blot was washed in TBS (5 min) to remove Tween. Detection of alkaline phosphatase activity was accomplished by immersing the blot in 0.1 M NaHCO$_3$ buffer (pH 9.8), 1 mM MgCl$_2$, 0.3 mg/ml nitro blue tetrazolium (NBT) and 0.15 mg/ml 5-bromo-4-chloro-3-indolyl phosphate (BCIP). The staining was stopped by washing the blot in water.

The antisera used were polyclonal antibodies raised in rabbits against PLI and PLII, two *A. niger* pectin lyases isolated from the commercially available pectinolytic preparation Ultrazym^R (van Houdenhoven, 1975), and monoclonal antibodies raised in mice against PLI, which recognize both PLI and PLII (M. Flipphi, A. Schots, E. Egberts, H.C.M. Kester and J. Visser, in preparation).

RESULTS

Western blot analysis of pectin lyase.

Proteins from the culture fluids were separated on SDS polyacrylamide gels, blotted on nitrocellulose and incubated with a monoclonal antibody raised against PLI, but which reacts also with PLII (Fig. 1). From Fig. 1B it is clear, that *A. niger* CBS 120.49 produces one pectin lyase: PLII. Under the growth condition used the *pel*D gene product PLI cannot be detected. From the *A. niger* isolates only *A. intermedius* CBS 117.32 and *A. hennebergi* CBS 125.52 do not produce a pectin lyase reactive with the monoclonal antobody. However, when we use polyclonal antibodies raised against PLI and PLII,these isolates do show a pectin lyase of approximately the same molecular weight as PLII (not shown). Therefore, the epitope recognized by the monoclonal antibody is not present in the pectin lyase produced by these isolates. Further, there seem to be differences in the band intensities and in the apparent molecular weigths of the pectin lyases of *A. niger*. Since quantitative

M.A. Kusters-van Someren *et al.*

1. *A. carbonarius* CBS 111.26, 2. *A. carbonarius* CBS 112.80, 3. *A. ellipticus* CBS 707.79, 4. *A. helicothrix* CBS 677.79, 5. *A. heteromorphus* CBS 117.55, 6. *A. japonicus* CBS 114.51, 7. *A. japonicus* CBS 621.78, 8. *A. aculeatus* CBS 172.66, 9. *A. aculeatus* CBS 115.80, 10. *A. niger* CBS 554.65, 11. *A. awamori* CBS 557.65, 12. *A. awamori* CBS 563.65, 13. *A. phoenicis* CBS 126.49, 14. *A. phoenicis* CBS 135.48

1. *A. phoenicis* CBS 135.48, 2. *A. nanus* CBS 136.52, 3. *A. nanus* CBS 131.52, 4. *A. usami* CBS 139.52, 5. *A. usami* CBS 553.65, 6. *A. intermedius* CBS 559.65, 7. *A. intermedius* CBS 117.32, 8. *A. hennebergii* CBS 118.35, 9. *A. hennebergii* CBS 125.52, 10. *A. pulverulentus* CBS 558.65, 11. *A. pulverulentus* CBS 425.65, 12. *A. foetidus* CBS 121.28, 13. *A. foetidus* CBS 618.78, 14. *A. niger* CBS 120.49, 15. *A. niger* NW756

Figure 1. Western analysis of pectin lyases of various isolates of *Aspergillus* sect. *Nigri*. Extracellular proteins were separated by SDS-PAGE, blotted on nitrocellulose and incubated with monoclonal antibodies raised against PLI. Markers are PLI (46.3 kD) and PLII (45.5 kD).

Table 3. Division of the *A. niger* strains in groups on the basis of results of Western blotting using monoclonal antibodies reacting with PLI and PLII, and Southern blotting using the *A. niger* CBS 120.49 *pel*D gene as a probe.

strain	CBS number	Western				Southern			
		O	i	ii	iii	D	I	II	III
A. niger	CBS 554.65		x			x	x		
	CBS 120.49		x			x		x	
A. awamori	CBS 557.65		x			x		x	
	CBS 563.65		x			x	x		
A. phoenicis	CBS 126.49		x			x	x		
	CBS 135.48		x			x			x
A. nanus	CBS 136.52		x			x			x
	CBS 131.52			x		x		x	
A. usami	CBS 139.52			x		x	x		
	CBS 553.65			x		x	x		
A. intermedius	CBS 559.65				x	x			x
	CBS 117.32	x			x¹			x	
A. hennebergii	CBS 118.35			x		x		x	
	CBS 125.52	x		x¹				x	
A. pulverulentus	CBS 558.65		x			n.d.			
	CBS 425.65		x			n.d.			
A. niger NW756	—				x	x			x

Groups i, ii and iii are made on the basis of different apparent molecular weights as mentioned in the text. O: no pectin lyase detected when using monoclonal antibodies; 1: on the basis of Western blot patterns using polyclonal antibodies.

Groups I, II and III are made on the basis of absence/presence of fragments a, b and c as mentioned in the text. D: containing the 0.8 kb *pel*D fragment; n.d. not determined

analysis by Western blotting is not very accurate, we can not draw any firm conclusions on whether the slight differences in band intensity reflect true differences in expression level. Also, the faint band which is sometimes present just above the PLII band seems to correlate in intensity with the intensity of the PLII band itself. Strain differences have therefore been correlated with the position of the major band, and not with the occurrence of the minor band.

The *A. niger* isolates can thus be divided into three groups (see Table 3):
(i)   those with a PL of approximately the same molecular weight as that from *A. niger* CBS 120.49;
(ii)   those with a PL with a slightly lower apparent molecular weight, similar to the NW756 pectin lyase and PL II from Ultrazym^R;
(iii) isolates with a PL with a distinctly lower molecular weight.

Group (i) is the largest and contains besides *A. niger* CBS 120.49 *A. niger* CBS 554.65, both *A. niger (awamori)* strains, both *A. niger (phoenicis)* strains, *A. niger (nanus)* CBS 136.52, and both *A. niger (pulverulentus)* strains. Group (ii) contains *A. niger (nanus)* CBS 131.52, both *A. niger (usami)* strains, both *A. niger (hennebergii)* strains (CBS 125.52 included on the basis of the pectin lyase reactive with the polyclonal antibody), and *A. niger* NW756, and group (iii) contains both *A. niger (intermedius)* strains (CBS 117.32 produces a PL of the same MW as CBS 559.65, although it is detected only with polyclonal antibodies).

Of all other strains (Fig. 1a and 1b) both *A. foetidus* and one *A. carbonarius* strain produce a PL of approximately the same MW as PLII. *A. japonicus (aculeatus)* CBS 172.66 shows a PL of a lower molecular weight and both *A. japonicus* CBS 114.51 and *A. japonicus (aculeatus)* CBS 115.80 produce a PL reactive only with the polyclonal antibodies, and of an even lower molecular weight (not shown).

Restriction enzyme patterns and Southern blotting of chromosomal DNA.
Chromosomal DNA was isolated from all isolates listed in Table 2, except from *A. pulverulentus*. The DNA was digested with the restriction enzymes *Pvu*II and *Sal*I which both cleave in the coding region of the *pel*D gene from *A. niger* CBS 120.49 (Gysler *et al.*). It is important to note, that the pectin lyase shown in the Western blots is PLII, and thus not coded by the *pel*D gene which was used as a probe in the Southern blot experiments. Chromosomal DNA digests were separated on an agarose gel containing the intercalating agent ethidium bromide. When digested DNA was photographed under UV light illumination, the DNA was seen as a smear. Some pronounced bands are discernable, probably caused by highly repetitive, ribosomal DNA (Fig. 3). All *A.niger* isolates, as well as the two *A. foetidus* isolates show the same pattern. Both isolates of *A. japonicus* and *A. aculeatus* also have similar patterns, as did *A. ellipticus* and *A. helicothrix* (Fig. 3a).

After electrophoresis, the DNA was blotted on nitrocellulose and subsequently hybridized with the *Bam*HI/*Pst*I probe containing the whole structural part of the *pel*D gene from *A. niger* CBS 120.49 (Fig. 2). Thus, when *A. niger* CBS 120.49 chromosomal DNA digested with *Pvu*II and *Sal*I is hybridized with this probe, three hybridizing fragments are to be expected: a *Pvu*II-*Sal*I fragment of 800 bp (D in Fig. 2), and two other fragments of approximately 500 bp which partly contain *pel*D gene sequences and partly flanking, non-coding, DNA.

Autoradiograms of these blots are shown in Fig. 4 a-b. All *A.niger* and *A. foetidus* isolates with the exception of *A. intermedius* CBS 117.32 and *A. hennebergii* CBS 125.52, contain fragment D. The latter two isolates do, however, show a unique hybridizing band of a higher molecular weight, which probably contains the *pel*D gene.
The other hybridizing bands seen probably result from heterologous hybridization of *pel*D with other *pel* genes. With regard to these bands, the *A. niger* isolates can be divided into three groups: group I contains isolates characterized by two bands (a and b) indicated by arrows in Fig. 4 *i.e. A. niger* CBS 554.65, *A. awamori* CBS 563.65, *A. phoenicis* CBS 126.49 and both *A. usami* isolates; group II contains isolates which show just the upper band (a): *A. niger* 120.49, *A. awamori* CBS 557.65, *A. nanus* CBS 131.52, *A. intermedius* CBS 117.32, and both *A. hennebergii* isolates; and group III contains isolates with neither of these bands, but showing a band of a higher molecular weight instead (c): *A. phoenicis* CBS 135.48, *A. nanus* CBS 136.52, *A. intermedius* CBS 559.65 and *A. niger* NW756. Classification on the basis of

these results is clearly not in correspondence with the generally accepted taxonomic classification (Al-Musallam, 1980), *e.g.* neither the two *A niger*, nor the *A. awamori*, the *A. phoenicis*, the *A. nanus*, the *A. intermedius*, or the *A. hennebergii* isolates (the latter ones on basis of the presence of the 0.8 kb *pel*D fragment) would be classified as they are classified at present.

With respect to the other species of the *Aspergillus niger*-group, there is no clear homology with the *A.niger pel*D gene, except, as mentioned before, in the case of *A. foetidus* which clearly resembles *Aspergillus niger* CBS 554.65 and both *A. usami* isolates. Further, only *A. carbonarius* CBS 111.26 seems to contain the 0.8 kb *pel*D fragment. What can be seen, however, is that the patterns of isolates *A. ellipticus* and *A. helicothrix* are highly homologous, as was shown by the ribosomal banding patterns as well. *A. japonicus* CBS 114.51 and *A. aculeatus* CBS 115.80 seem to be more related to each other than the isolates of *A. japonicus* or *A. aculeatus*.

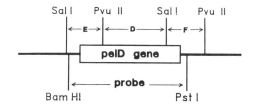

Figure 2. Restriction map of the *A. niger* CBS 120.49 *pel*D gene and the position of the probe used in the Southern blotting experiments. Fragments D, E and F, resulting from a *Pvu*II/*Sal*I digest, hybridize with the probe.

The results from both Western and Southern blotting are summarized in Table 3. In many cases the results obtained by Western and Southern blotting confirm each other. The *A. niger* isolates can easily be identified by both methods. However, isolates which are classified as one variety often have different Southern and Western blot patterns, *e.g.*, the *A. nanus* isolates are not similar, neither are the *A. intermedius* isolates nor the *A. hennebergii* isolates.

On the other hand, some isolates classified as different varieties of the same species show strong homology on the Southern blot, like *A. phoenicis* CBS 135.48 and *A. nanus* CBS 136.52 as well as *A. intermedius* CBS 117.32 and *A. hennebergii* CBS 118.35.

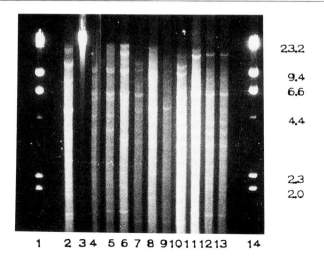

1. lambda DNA digested with *Hind*III, 2. *A. carbonarius* CBS 111.26, 3. *A. carbonarius* CBS 112.80, 4. *A. ellipticus* CBS 707.79, 5. *A. helicothrix* CBS 677.79, 6. *A. heteromorphus* CBS 117.55, 7. *A. japonicus* CBS 114.51, 8. *A. japonicus* CBS 621.78, 9. *A. aculeatus* CBS 172.66, 10. *A. aculeatus* CBS 115.80, 11. *A. hennebergii* CBS 125.52, 12. *A. foetidus* CBS 121.28, 13. *A. foetidus* CBS 618.78, 14. lambda DNA digested with *Hind*III

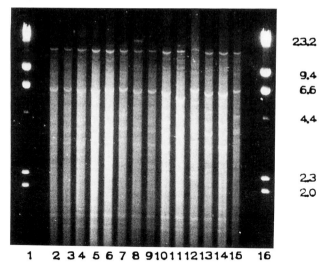

1. lambda DNA digested with *Hind*III, 2. *A. niger* CBS 554.65, 3. *A. awamori* CBS 557.65, 4. *A. awamori* CBS 563.65, 5. *A. phoenicis* CBS 126.49, 6. *A. phoenicis* CBS 135.48, 7. *A. nanus* CBS 136.52, 8. *A. nanus* CBS 131.52, 9. *A. usami* CBS 139.52, 10. *A. usami* CBS 553.65, 11. *A. intermedius* CBS 559.65, 12. *A. intermedius* CBS 117.32, 13. *A. hennebergii* CBS 118.35, 14. *A. niger* CBS 120.49, 15. *A. niger* NW756, 16. lambda DNA digested with *Hind*III

Figure 3. Ethidium bromide stained agarose gel electrophoresis patterns of chromosomal DNA digests. The lengths of the lambda fragments are shown in kb.

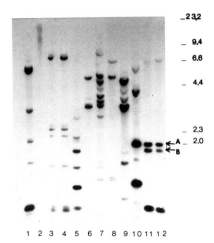

1.*A. carbonarius* CBS 111.26, 2.*A. carbonarius* CBS 112.80, 3.*A. ellipticus* CBS 707.79, 4.*A. helicothrix* CBS 677.79, 5.*A. heteromorphus* CBS 117.55, 6.*A. japonicus* CBS 114.51, 7.*A. japonicus* CBS 621.78, 8.*A. aculeatus* CBS 172.66, 9.*A. aculeatus* CBS 115.80, 10. *A. hennebergii* CBS 125.52, 11. *A. foetidus* CBS 121.28, 12. *A. foetidus* CBS 618.78

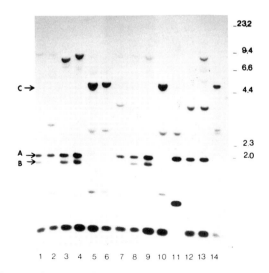

1. *A. niger* CBS 554.65, 2. *A. awamori* CBS 557.65, 3. *A. awamori* CBS 563.65, 4. *A. phoenicis* CBS 126.49, 5. *A. phoenicis* CBS 135.48, 6. *A. nanus* CBS 136.52, 7. *A. nanus* CBS 131.52, 8. *A. usami* CBS 139.52, 9. *A. usami* CBS 553.65, 10. *A. intermedius* CBS 559.65, 11. *A. intermedius* CBS 117.32, 12. *A. hennebergii* CBS 118.35, 13. *A. niger* CBS 120.49, 14. *A. niger* NW756

Figure 4. Southern analysis of genomic DNA of various isolates of *Aspergillus* sect. *Nigri*. DNA was digested with *Pvu*II and *Sal*I. After separation on a 0.6% agarose gel the DNA was transferred to nitrocellulose and hybridized with the 1.6 kb *Bam*HI/*Pst*I probe containing the whole structural *pel*D gene. V-banding is caused by trailing of non- or partially degraded RNA. Arrows indicate heterologous hybridizing bands A, B and C as mentioned in the text.The lengths of the lambda fragments are shown in kb.

DISCUSSION

In the past, classification of the black Aspergilli has largely been based on morphological features (Mosseray, 1934 a and b; Thom and Raper, 1945; Raper and Fennell, 1965; Al-Musallam, 1980). However, advanced methods in biochemical analysis, as well as the availability of cloned genes as molecular tools, now provide other independent criteria. In this paper we describe an initial attempt to use biochemical and genetic features to address existing problems in the classification of different Aspergilli. The degradation of pectin, a complex process shared by all the isolates examined, has been taken for further analysis. This heteropolysaccharide is present in the middle lamella and in the primary cell wall of higher plants. A large number of different enzymes and their corresponding structural genes are involved in catabolizing this substrate. We have limited ourselves in this study to an immunological analysis by Western blotting of only one particular enzyme, a pectin lyase which appears to be the major pectin lyase expressed under the growth conditions used. This enzyme corresponds with pectin lyase II described previously by van Houdenhoven (1975). We have also analyzed all isolates for the presence of a pectin lyase gene, *pel*D, which codes for another pectin lyase. This enzyme has recently been called PLD (Gysler *et al.*), but it is the same enzyme as the one described by van Houdenhoven (1975) as PLI. Due to the high degree of specificity which characterizes antibody-antigen interactions, immunological techniques are a powerful method to establish how different or how related proteins are. In this study our main interest was only to establish a rapid screening procedure using pectin lyase as a marker, enabling us to detect qualitative strain differences. Therefore, we have used besides polyclonal antibodies, a monoclonal antibody which recognizes a continuous epitope in both PLI and PLII, and reacts in Western blotting with the SDS-denatured protein bound to nitrocellulose. This approach by itself has some limitations. Apart from the fact that the conditions for staining are not such that they lead to quantitative data, it is also required that the gene one looks at becomes expressed. The final amount of pectin lyase which is detected in the culture medium is the result of a complex process which, amongst others, involves induction (depending on *e.g.* substrate, temperature, pH, oxygen levels, inoculation density, submerse or surface growth), glycosylation, secretion and stability of the secreted protein. It was possible, however, to classify the *A. niger* isolates which produce a pectin lyase into three groups on the basis of differences in molecular weight.

Analysis at the level of DNA by Southern blotting is a more direct and objective way to classify isolates. Small differences in nucleotide sequence, if present in the restriction enzyme sites used for digestion of the genomic DNA, lead to differences in restriction fragment patterns. Moreover, deletions or insertions larger than 50-100 bp, as well as differences in strength of hybridization can be detected. As can be seen from the results shown (Fig. 4b), the *pel*D gene is conserved throughout all the *Aspergillus niger* isolates. In the two isolates where the 0.8 kb band is absent, another band is seen, the length of which resembles the added lengths of fragments D and E or D and F (Fig. 2). This means that the non-coding region surrounding the gene, containg the next *Pvu*II or *Sal*I site, is conserved as well. In the 0.8 kb *Pvu*II/*Sal*I fragment three introns are present (Gysler *et al.*, submitted), the total length of which seems to be conserved as well. In all the other species, except *A. foetidus* and *A. carbonarius* CBS 111.26, the *pel*D band is absent; other bands are observed, but these clearly differ in strength of hybridization from the *A. niger* bands. In the *A. niger* isolates, two other bands were seen that hybridized strongly with the *pel*D gene. On the basis of presence or absence of these bands the *A. niger* isolates could be separated into three groups.

A number of similarities exist between the results obtained by Western and Southern blotting. The uniseriate species *A. japonicus* and *A. aculeatus* can be distinguished from the biseriate taxa. The examined isolates of *A. japonicus* and *A. aculeatus* could not be clearly distinguished from each other and this supports the classification of both taxa in one species. A distinction at varietal level as proposed by Al-Musallam (1980) based on differences of vesicle size and conidiophore-length cannot be made on the basis of the results of our investigation.

All taxa proposed as varieties of the *A. niger* by Al-Musallam (1980) show homology with respect to the two genes and the pectin lyase examined, indicating a strong relationship amongst these isolates. Strikingly, homology was also found between *A. ellipticus* and *A. helicothrix*. Al-Musallam (1980) described the latter as a new species after she isolated it as a distinct strain from the type culture of *A. ellipticus* and this might explain the homology between the two isolates. *A. helicothrix* differs from *A. ellipticus* by the production of typical cup-shaped sclerotia, a structure not found in other Aspergilli. Perhaps both isolates represent different stages of pleomorphic species, but such a phenomenon has not been observed in other anamorphic *Aspergillus* species.

The techniques applied in this paper are limited in their use to establish phylogenetic relationships. For that purpose, sequence determination would be necessary since differences between isolates can then be quantitated by scoring percentages of differing nucleotides after alignment of *e.g.* ribosomal RNA sequences (Logrieco *et al.*, 1990). The power of Southern and Western blotting, however, lies in the quick, yet reliable way by which isolates can be identified and classified.

ACKNOWLEDGEMENT

Margo Kusters acknowledges the financing of her project as part of a joint programme between the Biotechnology Department of Ciba-Geigy A.G., Basel, Switzerland and the Agricultural University, Wageningen, The Netherlands.

REFERENCES

AL-MUSALLAM, A. 1980. Revision of the black *Aspergillus* species. Thesis, State University Utrecht.

DE GRAAFF, L., VAN DEN BROEK, H. and VISSER, J. 1988. Isolation and expression of the *Aspergillus nidulans* pyruvate kinase gene. *Current Genetics* 13, 315-321.

GAMS, W., CHRISTENSEN, M., ONIONS, A.H.S., PITT, J.I. and SAMSON, R.A. 1985. Infrageneric taxa of *Aspergillus*. In Advances in *Penicillium* and *Aspergillus* systematics. eds., R.A.Samson and J.I. Pitt, pp. 55-62. New York and London: Plenum Press.

GODDARD, J.M., CAPUT, D., WILLIAMS, S.R. and MARTIN, D.W. 1983. Cloning of human purine-nucleoside phosphorylase cDNA sequences by complementation in *E.coli. Proceedings of the National Academy of Science, USA* 80, 4281-4285.

LAEMMLI U.K. 1970. Cleavage of structural proteins during the assembly of the head of bacteriophage T4. *Nature* 227: 680-685.

LOGRIECO, A., PETERSON, S.W. and WICKLOW, D.T. 1990. Ribosomal RNA comparisons among taxa of the terverticillate Penicillia. In. Modern Concepts in *Penicillium* and *Aspergillus* Systematics, eds. R.A. Samson and J.I. Pitt. pp. 343-354. New York and London; Plenum Press.

MANIATIS, T., FRITSCH, E.F. and SAMBROOK, J. 1982. Molecular cloning, a laboratory manual. New York: Cold Spring Harbor Laboratory Press.

MOSSERAY, R. 1934 a. Sur la systematique des *Aspergillus* de la section 'niger' Thom & Church. *Annals de Societé Science de Bruxelles Sér.* 2, 54: 72-85

—— 1934 b. Les *Aspergillus* de la section *niger* Thom et Church. *Cellule* 43: 203-285.

PONTECORVO, G., ROPER, J.A., HEMMONS, L.J., MACDONALD, K.D. and BUFTON, A.W.J. 1953. The genetics of *Aspergillus nidulans*. *Advances in Genetics* 5: 141-238.

RAPER, K.B. and FENNELL, D.I. 1965. The genus *Aspergillus*. Baltimore: Williams & Wilkins.

THOM, C. and RAPER, K.R. 1945. A manual of the Aspergilli. Baltimore: Williams & Wilkins.

VAN HOUDENHOVEN, F.E.A. 1975. Studies on pectin lyase. Thesis, Agricultural University Wageningen.

DIALOGUE FOLLOWING MRS KUSTERS PRESENTATION

PATERSON: You mentioned that three of the four isolates of *A. japonicus* produced a low molecular weight pectin lyase. How about the fourth one?

KUSTERS: Pectin lyase was not detected with the monoclonal or polyclonal antibodies in that isolate by our methods. I can't explain this. *Aspergillus japonicus* as it is now classified did not give us consistent results. We had two isolates each of *A. japonicus* var. *japonicus* and *A. japonicus* var. *aculeatus*. One of the isolates of *A. japonicus* and *A. aculeatus* gave identical results, so the varieties as they are defined are not consistent with our results. On the basis of the Southern blot, and based on some results not shown with the ethidium bromide stained gel of digested chromosomal DNA, the four isolates of *A. japonicus* are alike.

A MOLECULAR ASSESSMENT OF THE POSITION OF
STILBOTHAMNIUM IN THE GENUS *ASPERGILLUS*

J. Dupont[1], M. Dutertre[2], J.-F. Lafay[2], M.-F. Roquebert[1] and Y. Brygoo[2]

[1]*Laboratoire de Cryptogamie*
Museum National d'Histoire Naturelle
75 005 Paris, France

[2]*Laboratoire de Cryptogamie*
Université Paris Sud
91 405 Orsay Cedex, France

SUMMARY

Several *Stilbothamnium* and *Aspergillus* isolates were analysed by rapid rRNA sequencing of a variable region at the 5' end of the 28S-like rRNA molecule. This sequence permits evaluation of the genetic divergence between species and the construction of a phylogenetic tree computed from the number of nucleotide differences. Based on these data, the phylogenetic value of other taxonomical criteria used for the classification of *Stilbothamnium* and *Aspergillus* is discussed.

INTRODUCTION

Stilbothamnium Henn. is a genus of seminicolous fungi characterized by abundant, large, branched synnemata developping from seeds in tropical rain forests. These conidiomata are up to 50 mm high and brightly coloured, greenish or golden yellow. They are covered with *Aspergillus*-like heads bearing chains of echinulate conidia. Conidiomata sometimes develop together with grey-brown lobulate, stipitate sclerotia. These sclerotia often proved to remain sterile. Three other synnematous genera *Stilbodendron* H.& P.Sydow, *Penicilliopsis* Solms-Laubach, and *Pseudocordiceps* Haum., with the same ecological characteristics were described and discussed by Hauman (1936). Conidial structures of this group morphologically resemble those of *Penicillium* or *Aspergillus*.

Penicilliopsis was established by Solms-Laubach (1887) as an Ascomycete genus with a phialidic anamorph. For that reason, Samson and Seifert (1985) related the anamorph of *Penicilliopsis* sensu Hauman to the generic name *Sarophorum* Sydow. Examination of type material in herbaria allowed these authors to correlate the anamorphic genera *Stilbodendron* and *Sarophorum* with the teleomorph *Penicilliopsis*, but not *Stilbothamnium* even though its pedicellate sclerotia resemble *Penicilliopsis* ascostromata. The lack of evident correlation between *Stilbothamnium* and *Penicilliopis*, in spite of the morphological similarity of conidial apparatus, led Samson & Seifert (1985) to consider *Stilbothamnium* as a subgenus of *Aspergillus*. They proposed the new combination *Aspergillus togoensis* (Henn.) Samson and Seifert for *S. nudipes* Haum.(= *S. togoense* Henn. = *S. novoguineense* H.& P. Sydow = *Aspergillus coremiiformis* Bartoli & Maggi).

Gams *et al.* (1985) subdivided *Aspergillus* into six subgenera some with related teleomorphs. The subgenera were further divided into sections of similar taxonomic range to the "groups" established by Thom and Church (1926), Thom and Raper (1945) and Raper and Fennell (1965). *Aspergillus* subgenus *Stilbothamnium* (Henn.) Samson and Seifert appears closely related to the sect. *Circumdati* and *Flavi* of the subgen. *Circumdati*.

Modern Concepts in Penicillium and Aspergillus Classification
Edited by R. A. Samson and J. I. Pitt
Plenum Press, New York, 1990

Anamorphs of subgen. *Circumdati* are related to teleomorphs in*Petromyces* Malloch & Cain, *Chaetosartorya* Subr. and *Hemisartorya* Rai & Chowderi. *Penicilliopsis* is closely related to *Petromyces* Malloch & Cain and *Dichlaena* Mont. & Durrieu, both belonging to the family Trichocomaceae.

Morphological and structural analysis of *S. nudipes* shows clear relationships with the genus *Aspergillus*, and in particular with the sect. *Flavi* (Roquebert and Nicot, 1985). It differs from *A. tamarii* by the presence of coremia, septate phialides and by the mode of differentiation of conidial walls.

Christensen (1981) includes *A. coremiiformis* Bartoli & Maggi in sect. *Flavi* but suggested affinities with sect. *Circumdati* based on the pigmented, rough walls of the stipes and yellow conidial masses.

Wicklow *et al.* (1989) examined *A. togoensis* for the production of mycotoxins and did not find any mycotoxins characteristics of sect. *Flavi* but did find sterigmatocystin believed to be an intermediate in the biosynthesis of aflatoxins. These data support the linking of *A. togoensis* to sect. *Flavi* but not to species of sect. *Circumdati* which never produce sterigmatocystin. So the question is, where does *Stilbothamnium* fit, with regard to *Penicilliopsis* and *Aspergillus*, more precisely to sects *Flavi* and *Circumdati*. To study this problem we used the analysis of ribosomal RNA, which provides a powerful taxonomic indicator, because they are highly conserved and are universally found in living cells. The 5S rRNA was first used for this purpose (reviewed by Hori and Osawa ,1987). However, 5S rRNA is too short and too well conserved, to be suitable for studying closely related species. One has to look at the larger rRNA molecules, 18 S (Salim and Maden, 1981; Woese *et al.*, 1985) and 28S (Qu *et al.*, 1983). The development of a technique for rapid and simple sequencing of large stretches of 18 or 28S rRNA opened the way for systematic exploitation of the remarkable properties of these molecules as phylogenic indicators (Qu *et al.*, 1988). The larger rRNA subunits contain regions perfectly conserved adjacent to less conserved ones that are useful for phylogenetic evaluation (Hassouna *et al.*, 1984; Michot *et al.*,(1987). At the 5' end, two variable regions called D1 and D2, particularly D2 appear to be divergent enough to give informations on relationships between genera and possibly species (Baroin *et al.*, 1988; Guadet *et al.*, 1989).

In this paper, we present preliminary results on an evaluation of the relationships between *Stilbothamnium-Aspergillus* and *Penicillopsis*, using rapid rRNA sequencing methodology as a tool for classification. We show that this method is efficient for this purpose and may provide phylogenetic information as well.

MATERIAL AND METHODS

The names and origins of the isolates used in this study are listed in Table 1. Species which may be related to the genus *Stilbothamnium*, such as *A. flavus*, *A. tamarii*, *A. ochraceus* and *A. coremiiformis* were included for comparison. Less closely related species such as *A. clavatus*, *A. nidulellus*, *A. fumigatus*, *A. versicolor* and some *Penicillium* species were also examined. *Paecilomyces fumosoroseus* was added as an unrelated species for a relative estimation of distances.

RNA template isolation from two days old cultures was performed according to the method described by Maccecchini *et al.*, 1979; the method for RNA sequencing is that developed by Sanger *et al.* (1971) and modified by Qu *et al.*, (1983). The sequences were aligned manually (Fig.1). Divergence (or distance) between two sequences was estimated as the number of nucleotide differences, each difference counting for one. A computer

Table 1. List of sequenced isolates.

Aspergillus clavatus Desm.	U-Brest
A. coremiiformis Bartoli & Maggi	LCP 843396
A. flavus Link:Fr.	LCP 863450
	LCP 793244
	LCP 843365
	LCP 51508
	U-Brest
A. flavus var. *columnaris* Raper & Fennell	LCP 753053
A. fumigatus Fres.	U-Brest
A. itaconicus Kinoshita	LCP 611673
A. nidulellus Samson & Gams	U-Brest
A. niger van Tieghem	U-Brest
A. ochraceus Wilhelm	LCP 521064
	LCP 853389
	LCP 853498
	U-Brest
A. oryzae (Ahlburg) Cohn	U-Paris-Sud
A. parasiticus Speare	LCP 531525
	LCP 732236
	LCP 863455
A. tamarii Kita	LCP 853497
A. thomii G. Smith	LCP 561517
A. versicolor (Vuill.) Tiraboschi	U-Brest
Paecilomyces fumosoroseus (Wize) Brown & Smith	U-Paris-Sud
Penicilliopsis dichotomus Hauman	LCP 653802
Penicilliopsis sp.	LCP 653801
Penicillium canescens Sopp	LCP 51455
P. chrysogenum Thom	U-Paris-Sud
P. implicatum Biourge	LCP 763119
P. roqueforti Thom	U-Paris-Sud
P. spinulosum Thom	U-Paris-Sud
Stilbodendron sp.	LCP 653800
Stilbothamnium nudipes Hauman	LCP 651910
	LCP 673456

LCP: Laboratoire Cryptogamie Paris; U-Brest: Université de Brest; U-Paris-Sud: Université de Paris-Sud

comparison of all the sequences provided a distance matrix (Table 2) from which a phenetic tree was constructed by a Fitch Margoliash procedure using the Fitch program of Felsenstein's Philip package (version 2.9). The conversion of the distant matrix into a phylogeny is difficult and the phylogenetic tree will be subject to improvement as additional sequences and alternative interpretation of the data are used.

RESULTS

The technique used for our experiments provides sequences of about 130 nucleotides, which represents half of the length of the D2 domain. However, we found enough variability to separate representative species, keeping in mind that within an incomplete D2 domain, divergences are underestimated. More investigation is needed.

Table 2. Matrix of distances between the 17 representative strains

	B	C	D	E	F	G	H	I	J	K	L	M	N	O	P	Q
A *A. tamarii*	7	7	6	4	9	8	11	11	9	10	6	12	13	12	11	31
B *A. flavus*		2	1	11	4	5	6	12	9	12	7	12	13	14	15	30
C *A. oryzae*			1	11	4	5	6	12	9	12	7	12	13	14	15	29
D *Stilbothamnium nudipes*				10	3	4	5	11	8	11	6	11	12	13	14	30
E *Stilbodendron* sp.					13	12	13	11	9	10	8	14	13	12	10	28
F *A. ochraceus*						3	8	11	12	12	9	14	15	16	17	31
G *A. itaconicus*							8	13	12	13	9	14	15	16	17	31
H *A. niger*								9	13	12	8	14	15	15	16	31
I *A. fumigatus*									13	10	11	8	13	15	17	31
J *A. clavatus*										8	12	11	11	14	15	28
K *A. nidulellus*											11	8	11	12	12	31
L *A. versicolor*												6	15	18	17	32
M *P. chrysogenum*													12	13	13	30
N *P. canescens*														4	11	32
O *P. implicatum*															10	33
P *P. spinulosum*																34
Q *P. fumosoroseus*																32

Distances are expressed in absolute value; any differences (transition, transversion, each nucleotide deletion) counts for one, without correction for multiple changes at one given site.

Comparison of aligned sequences shows that different isolates of a single species are identical, so the sequence is considered as species specific. *A. flavus, A. flavus* var. *columnaris, A. parasiticus* and *A. coremiiformis* were very similar. Similarity can also be observed for *Stilbodendron* and *Penicilliopsis* spp as well as for *P. chrysogenum* and *P. roqueforti*. Only one sequence representative of each of these groups was kept for data processing.

From the phenetic tree, we can observe a separation of the 24 species into two clusters (1) *A. fumigatus, A. clavatus* and *Penicillium* spp., (2) all the other species. In this last branch *Aspergillus* and *Stilbothamnium* segregate clearly from *Penicilliopsis, Stilbodendron* and strikingly also from *A. tamarii*.

The presence of *Paecilomyces* in this study does not imply any phylogenetic significance; it is just an outer reference for "scaling".

DISCUSSION

Separations evidenced by partial rRNA sequences show a good correlation with morphological classification, particularly with infrageneric taxa of *Aspergillus* as established by Raper and Fennell (1965) and reviewed by Gams *et al.* (1985). The subgenera *Nidulantes, Fumigati* and *Clavati* are clearly distinct.

The position of *Stilbothamnium nudipes* close to *A. flavus* as well as to *A. ochraceus* supports its accomodation in the subgenus *Circumdati*. This finding agrees with previous suggestions (Christensen, 1981 and 1982; Roquebert and Nicot, 1985; Samson and Seifert, 1985 and Wicklow *et al.*, 1989). Consequently, the subgenus *Stilbothamnium* as proposed by Samson and Seifert (1985) seems to be superfluous. As suggested by Samson and Seifert (1985) we can confirm the identity of *S. nudipes* with *A. togoensis* (Henn.) Samson and Seifert.

```
                                                                                          **              *********  *
                  CGTTTACGCC OATTATGCCA GCGTCCGTGC COGAAGOCGC GTTCCTCGGT CCAGGCAGGC CGCATTGCCA ??CCTGGCTA TAAGGCGCCC CGAOOOGAOC OTACATTCCA GGGG
A. tamarii        CGTTTACGCC OATTATGCCA GCGTCCGTGC COGAAGOCGC GTTCCTCGGT CCAGGCAGGC CGCATTGCCA ??CCTGGCTA TAAGGCGCCC CGAOOOGAOC OTACATTCCA GGGG
A. coremiiformis  ------A--- ---------- ---------- ---------- ---------- ----T----- ---------- CT-C----  ----T----  --O-GA---O ----------  -A--
A. flavus         -----?-A-- ---------- ---------- ---------- ---------- ---T------ ---------- CT-C----  ----T----  --O-GA---O ----------  -A--
A. oryzae         ------A--- ---------- ---------- ---------- ---------- ---T------ --T------- CT-C----  ----T----  --O-----CG ----------  -A--
A. thomii         ------A--- ---------- ---------- ---------- ---------- ---T------ ---------- CT-C----  ----T----  --O-GA---O ----------  -A--
S. nudipes        ------A--- ---------- ---------- ---------- ---------- ---T------ ---------- CC------  --A-A---  -O-GG-GCA  ----------  -A--
Stilbodendron sp. --C------- ---------- ---------- ---------- ---------- ---------- ---------- CC------  --A-A---  ---AG-OTG- ----------  -A--
Penicilliopsis sp.--C------- ---------- ---------- ---------- ---------- ---------- ---------- CC------  --A-A---  -OO-GAAGTG C---------  -A--
P. dichotomus     --C------- ---------- ---------- ---------- ---------- ---T------ ---------- CC------  --A-A---  --AG-OTG-  ----------  -A--
A. ochraceus      ---G-AA- C ---------- ---------- ---------- ---------- ---T------ ---------- ---C----  ----T----  -O-TA---O  ----------  G---
A. itaconicus     ------AA- C ---------- ---------- ---------- ---------- ---T------ A--------- ---C----  ----T----  --AGAOO-O  ----------  ----
A. niger          ------A-- --O----- ---------- ---------- ---------- ---T------ ------A--- GC-C----  --AT-G--  -CG-GA---O -O--------  G---
A. fumigatus      -A------A- C----C--- ---------- ---------- ---------- ---------- ---------- CT-OO---  ---A-A--  ---GA-TGA  ----------  --A-
A. clavatus       ------A--- -----A--- ---------- ---------- --------T- ---------- ------?--- C---C---  ---A-A--  --OAGA---O ----------  --A-
A. nidulellus     -----G--G- -?--C----- ---------- ---------- ---------- ---T------ ---------- --+-A---  --A-AG--  -OO-GA-CGA ----------  -A--
A. versicolor     ------A--- ---------- ---------- ---------- ---------- ---T------ ---------- --+-A---  --A--T--  --OAGA---O ----------  --A-
P. chrysogenum    ------A--- -----C---- ---------- ---------- ---------- -T----T--- ---------- -C-TC---  --A-----  -T--G-OG-  ----C----G A-C-
P. canescens      --C-----A- ---------- A-----A--- ---------- ---------- -T------T- ---------- AC-TC---  --A-----  -OOTG-CGO  ----C----G A-?-
P. implicatum     --C-----?- ---------- ---------- ---------- --------G- ---------- -COO------ -C-TC---  --A-----  -O-G-OCGO  ----------G A---
P. spinulosum     --C------A- ---------A- ---------- ---------- ---T---T-- T----G---- AC-T----  --A-?---  -OO-GA-CTO ----------G A-?-
P. fumosoroseus   --C-G-T-A- C---------- --A--T--- O---T---- --A----A- --G-CGCA-O G-T-----G GAO-G---  --CA-T--  O---G--G-  -C---G----G -AA-
```

Figure 1. Aligment of 21 sequences. The sequence of *A. tamarii* is choosen as a model for the aligment. Identical bases are noted as dashes (-), undetermined as ? and deletion as O. The positions indicated by a * are not taken into account for the distance calculus.

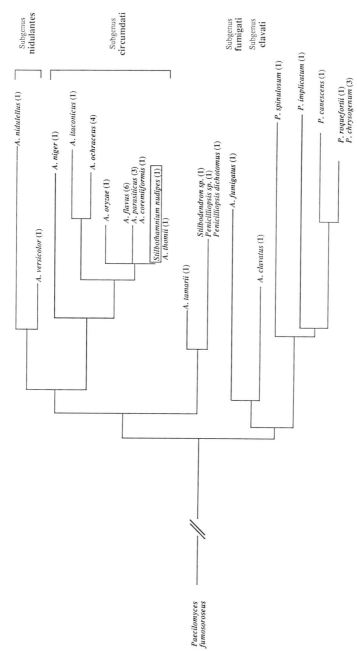

Figure 2. *Aspergillus* and *Penicillium* phylogenetic tree. Phylogenetic tree is constructed from table 2 data. Correspondence with *Aspergillus* subgenera is figured on the right. Horizontal distances are proportional to the number of different bases. Number of analysed strains is reported in brackets.

A. coremiiformis appears similar to *A. flavus*. Hence, it should be separated from *Stilbothamnium*, at variance with the synonymy established by Samson & Seifert (1985). Ambiguously, Christensen (1981) placed it in *Aspergillus* sect. *Flavi* and later in sect. *Circumdati*(Christensen, 1982). In our view, despite its resemblance to the latter, *A. coremiiformis* clearly belongs in sect.*Flavi*. Because some species of sect. *Flavi* produce mycotoxins, extensive research has been carried out by several different methods to differentiate species. Murakami (1971), Christensen (1981), Klich and Mullaney (1987) and Klich and Pitt (1988) concluded that *A. oryzae*, considered as a domesticated species, is distinct from *A. flavus*; our results confirm this distinction. In this case, it appears that ecological adaptations are correlated with hereditary changes. The distinct position of *A. oryzae* is not in agreement with the opinions of Vincent and Kulik (1970), Kulik and Brooks (1970) and Nasuno (1972). Based on nuclear DNA complementarity, Kurtzman *et al.* (1986) placed *A. flavus*, *A. oryzae*, *A. parasiticus* and *A. sojae* in a single species.

The close relationship of *A. thomii* with *A. flavus* was mentioned by Christensen (1981). Christensen and Tuthill (1985) proposed the transfer of *A. thomii* to sect. *Flavi* . Our results confirm this proximity.

The position of *A. tamarii* closer to *Penicilliopsis* than to *Aspergillus* is rather unexpected and disagrees with the analysis of mitochondrial DNA, which support the link between *A. tamarii* and *A. oryzae*.

The close correspondence of rRNA groupings with the morphological taxonomy confirms the phylogenetic validity of the morphological criteria used by mycologists and the usefulness of the molecular tool. This method can be used for a rapid determination and phylogenetic classification of isolates and will prove especially valuable for atypical isolates.

REFERENCES

BAROIN, A., PERASSO, R., QU, L.H., BRUGEROLLE, G., BACHELLERIE, J.P. and ADOUTTE, A. 1988. Partial phylogeny of the unicellular Eukaryotes based on rapid sequencing of a portion of 28S ribosomal RNA. *Proceedings of the National Academy of Sciences, USA* 85: 3474-3478.

CHRISTENSEN M. 1981. A synoptic key and evaluation of species in the *Aspergillus flavus* group. *Mycologia* 73: 1056-1084.

CHRISTENSEN M. 1982. The *Aspergillus ochraceus* group: two new species from Western soil and a synoptic key. *Mycologia* 74: 210-225.

CHRISTENSEN, M. and TUTHILL, D. 1985. *Aspergillus*: An overview. In Advances in *Penicillium* and *Aspergillus* systematics. eds., R.A.Samson and J.I. Pitt, pp. 195-209. New York and London: Plenum Press.

GAMS, W., CHRISTENSEN, M., ONIONS, A.H.S., PITT, J.I. and SAMSON, R.A. 1985. Infrageneric taxa of *Aspergillus*. In Advances in *Penicillium* and *Aspergillus* systematics. eds., R.A.Samson and J.I. Pitt, pp. 55-62. New York and London: Plenum Press.

GUADET, J., JULIEN, J., LAFAY, J.F. and BRYGOO, Y. , 1989. Phylogeny of some *Fusarium* species determined by large subunit rRNA sequence comparison. *Molecular Biology and Evolution* 6: 227-242.

HASSOUNA, N., MICHOT, B. and BACHELLERIE, J.P. 1984. The complete nucleotide sequence of the mouse 28S r RNA in higher Eukaryotes. *Nucleic Acids Research* 12: 3563-3583.

HAUMAN, L. 1936. Les champignons séminicoles des forêts tropicales. *Bulletin de la Societé royale Botanique de Belgique* 69: 96-129.

HORI, H. and OSAWA, S. 1987. Origin and evolution of organisms as deduced from 5S ribosomal RNA sequences. *Molecular Biology and Evolution* 4: 445-472.

KLICH, M.A. and MULLANEY, E.J. 1987. DNA restriction enzyme fragment polymorphism as a tool for rapid differenciation of *Aspergillus flavus* from *A. oryzae. Experimental Mycology* 11: 170-175.

KLICH, M.A. and PITT, J.I.. 1988. Differenciation of *Aspergillus flavus* from *A. parasiticus* and other closely related species. *Transactions of the British Mycological Society* 91: 99-108.

KULIK, M.M. and BROOKS, A.G. 1970. Electrophoretic studies of soluble proteins from *Aspergillus* spp. *Mycologia* 62: 365-376.

KURTZMAN, C.P., SMILEY, M.J., ROBNETT, C.J. and WICKLOW, D.T. 1986. DNA relatedness among wild and domesticated species in the *Aspergillus flavus* group. *Mycologia* 78: 955-959.

MACCECCHINI, M. L., RUDIN, Y., BLOBEL, G., and SCHATZ, G. 1979. Import of proteins into mitochondria: precursor forms of the extramitochondrially made F1-ATPase subunits in yeast. *Proceedings of the National Academy of Sciences, USA* 76: 343-347.

MICHOT, B. and BACHIELLERIE, J.P. 1987. Comparisons of large subunits RNA's reveal some eukaryotic specific elements of secondary structure. *Biochemie* 69: 11-23.

MURAKAMI, H. 1971. Classification of the koji moulds. *Journal of General and applied Microbiology, Tokyo* 17: 281-309.

NASUNO, S. 1972. Electrophorectic studies of alkaline 9 proteinases from strains of *Aspergillus flavus* group. *Agricultural and Biochemical Chemistry* 36: 684-689.

QU, T.H. , 1986. Dissertation 1273. Université Paul Sabatier, Toulouse, France.

QU, T.H. MICHOT B. and BACHELLERIE, J.P. 1983. Improved methods for structure probing in large RNAs: a rapid heterologous sequencing approach is coupled to the direct mapping of nuclease accessible sites. Application to the 5' terminal domain of eukaryotic 28S rRNA. *Nucleic Acid Research* 11: 5903-5920.

QU, T.H., NICOLOSO, M. and BACHELLERIE, J.P. 1988. Phylogenetic calibration of 5' terminal domain of large r RNA achieved by determining twenty eucaryotic sequences. *Journal of Molecular Evolution* 28: 113-124.

RAPER, K.B. and FENNELL, D.I. 1965. The genus *Aspergilus*. Baltimore: Williams & Wilkins.

ROQUEBERT, M.F. and NICOT, J.. 1985. Similarities between the genus *Stilbothamnium* and *Aspergillus*. In Advances in *Penicillium* and *Aspergillus* systematics. eds., R.A.Samson and J.I. Pitt, pp. 221-229. New York and London: Plenum Press.

SALIM, M. and MADEN, B.E.H. 1981. Nucleotide sequence of *Xenopus laevis* 18S ribosomal RNA inferred from gene sequence. *Nature, London* 291: 205-208.

SAMSON, R.A. and SEIFERT, K.. 1985. The Ascomycetes genus *Penicilliopsis* and its anamorphs. In Advances in *Penicillium* and *Aspergillus* systematics. eds., R.A.Samson and J.I. Pitt, pp. 397-428. New York and London: Plenum Press.

SOLMS-LAUBACH, H.G. 1887. *Penicilliopsis clavariaeformis*, ein neuer javanischer Ascomycet. *Annales Jardin Botanique, Buitenzorg* 6: 53-72.

THOM, C. and RAPER, K.B. 1945. A Manual of the *Aspergilli*. Baltimore: Williams and Wilkins.

THOM, C. and CHURCH, M.B. 1926. The *Aspergilli*. Baltimore: Williams and Wilkins Co.

VINCENT, P.G. and KULIK, M.M. 1970. Pyrolysis gas liquid chromatography of fungi: differentiation of species and strains of several members of the *Aspergillus flavus* group. *Applied Microbiology* 20: 957-963.

WICKLOW, D., VESONDER, R.F.,McALPIN, C.E., COLE, R.J. and ROQUEBERT, M.F. 1989. Examination of *Stilbothamnium togoense* for *Aspergillus flavus* group mycotoxins. *Mycotaxon* 34: 249-252.

WOESE, C.R., STACKBRANDT, E., MACKE, T.J. and FOX, G.E. 1985. A phylogenetic definition of the major eubacterial taxa. *Systematic Applied Microbiology* 6: 143-151.

RIBOSOMAL RNA COMPARISONS AMONG TAXA OF THE TERVERTICILLATE PENICILLIA

A. Logrieco[†], S.W. Peterson and D.T. Wicklow

Northern Regional Research Center
Agricultural Research Service
U.S. Department of Agriculture
Peoria, Illinois 61604, USA

SUMMARY

Ribosomal RNA sequences were determined for terverticillate Penicillia by the dideoxy nucleotide chain termination method and oligonucleotide primers. The sequences of individual isolates were compared for base differences upon proper alignment. Prior experience in other fungi suggests that a single nucleotide difference in the sequence of two *Penicillium* isolates may indicate that they are not the same species. A second baseline for data interpretation was provided by comparisons involving: *Penicillium, Saccharomyces,* and *Urnula.* These intergeneric comparisons revealed >100 base differences.

The maximum number of base differences between species classified in *Penicillium* subgenus *Penicillium* was 33 bases. Our results indicate that *Penicillium aurantiogriseum* NRRL 971, *P. viridicatum* NRRL 963, *P. verucosum* NRRL 965, *P. expansum* NRRL 976, *P. echinulatum* NRRL 1151, *P. hirsutum* NRRL 2032, *P. granulatum* NRRL 2036, and *P. puberulum* NRRL 845 are distinct species. *Penicillium claviforme* NRRL 2031 and *P. clavigerum* NRRL 1003 show a closer relationship to species in subgenus *Penicillium* than to *P. isariiforme* NRRL 2628. Morphological classification schemes that accommodate one or more of the above isolates into a single species are not supported by our results. Three isolates showed no base differences (i.e., *P. puberulum* NRRL 845, *P. resticulosum* NRRL 2021, and *P. camemberti* NRRL 877) and may represent variants of the same species. Ecological and physiological data, as well as secondary metabolite profiles, may be required if one is to distinguish *Penicillium* species by methods other than degree of nucleic acid relatedness.

INTRODUCTION

Considerable interest and controversy has surrounded taxonomic relationships among *Penicillium* species that produce terverticillate conidiophores (Samson *et al.,* 1976; Pitt, 1979; Frisvad and Filtenborg, 1983). This group includes important food and feed spoilage moulds, pathogens of mature fruits and cereal grains, and "domesticated" isolates used in the fermentation of cheeses or meats (Raper and Thom, 1949; Pitt, 1979; Leistner, 1984). In addition to causing deterioration and quality losses, these moulds may contaminate agricultural products with potent mycotoxins (Frisvad 1986). Correct identification is therefore essential to mycotoxicologists, plant pathologists and food microbiologists. There are considerable problems in attempting to identify terverticillate Penicillia because isolates commonly have characters found in more than one species. In approaching the taxonomy of this group one first must deal with the extensive variation among apparently "healthy isolates" while at the same time recognizing variation associated with strain deterioration in culture (Williams *et al.,* 1985). As with any group of organisms, the views

---

[†] Permanent address: Consiglio Nazionale delle Ricerche, Instituto Tossine e Micotossine da Parassiti Vegetali, Via Amendola, 197/F, 70126 Bari, Italy

Modern Concepts in Penicillium and Aspergillus Classification
Edited by R. A. Samson and J. I. Pitt
Plenum Press, New York, 1990

of taxonomists differ as to the importance of individual characters in delimiting species (Raper and Thom, 1949; Samson *et al.*, 1976; Pitt, 1979; Frisvad and Filtenborg, 1983). For the terverticillate Penicillia these taxonomic characters include: micromorphology (i.e., conidia, conidiogenous structures), macromorphology (i.e., colony texture and color), physiology (i.e., growth on different substrates at different temperatures and water activities), pathogenicity, and secondary metabolite profiles (SMPs). This research attempts to resolve some of the controversies by comparing of the ribosomal RNA (rRNA) sequences of selected species of terverticillate Penicillia by means of the dideoxy nucleotide chain termination method and oligonucleotide primers. We examined species that are the object of taxonomic disagreement as well as species accepted by all *Penicillium* taxonomists. To provide a baseline for RNA contrasts we examined teleomorph genera *Eupenicillium crustaceum* Ludwig and *Talaromyces helicus*
(Raper & Fennell) Benjamin, known to have a *Penicillium* anamorph state, and two unrelated ascomycetes *Saccharomyces cerevisiae* Hansen and *Urnula craterium* (Schw.) Fr., which presumably are only distantly related to either *Eupenicillium* or *Talaromyces*.

MATERIAL AND METHODS

The isolates we analyzed for rRNA base sequences are listed in Table 1. Isolates were predominantly from subgenus *Penicillium*. *Eupenicillium crustaceum* also produces terverticillate penicillia (anamorph state = *P. gladioli* McCulloch & Thom), while *Talaromyces helicus* was included as a species representing *Talaromyces*. It produces acerose phialides and typically biverticillate symmetrical penicilli (anamorph state = *P. spirillum* Pitt).

The isolates were grown at 25°C, in 100 ml of YM medium (Wickerham, 1951), on a rotary shaker (200 rpm) for 16-36 hours, until the cultures were in log phase growth. Ribosomal RNA isolation was according to Chirgwin *et al.* (1979), with the exceptions that cells were harvested by filtration, suspended in guanidinium thiocyanate reagent (10 ml/g), and broken in a Braun cell homogenizer with 0.5-mm glass beads. Intact undegraded rRNA, as assessed from denaturing agarose gel electrophoresis, was obtained by this method.

The base sequences of selected regions of the large (25S) and small (18S) subunit rRNA were determined, with specific oligonucleotide primers, by the dideoxy nucleotide chain termination method for RNA sequencing as described by Sanger *et al.* (1977) and Lane *et al.* (1985). Oligonucleotide primer C was purchased from Boehringer-Mannheim (Indianapolis, IN); the other primers were a gift from Carl Woese, University of Illinois. The first base synthesized from the small subunit primer, in relation to the *S. cerevisiae* primary structure (Rubstov *et al.*, 1980), is C, 1627. The first bases synthesized from the large subunit primers, based on *S. cerevisiae* primary structure (Georgiev *et al.*, 1981), are E, 1841 and F, 635. Sulfur-35 labeled nucleotide fragments generated in the chain extension reactions were separated by electrophoresis on 8% acrylamide-8 M urea gels. RNA base sequences were read from autoradiographs of the fixed and dried gels. Sequences with few differences or apparent insertions were rerun side by side on the same gel to verify differences. Some of the sequences were verified by repeating all steps from the beginning. Ribosomal RNA base sequences were aligned manually with a text editor. Alignment was necessary to compare homologous sequences. The data were evaluated with a set of programs that measure simple matching of aligned sequences.

Table 1. Isolates examined

P. atramentosum Thom NRRL 795 ex type,
P. puberulum Bainier NRRL 845 ex neotype,
P. roqueforti Thom NRRL 849 ex type,
P. camemberti Thom NRRL 877 ex type,
P. viridicatum Westling NRRL 963 ex neotype,
P. verrucosum Dierckx NRRL 965 ex neotype,
P. aurantiogriseum Dierckx NRRL 971 ex neotype,
P. expansum Link NRRL 976 ex neotype,
P. italicum Wehmer NRRL 983 ex neotype,
P. clavigerum Demelius NRRL 1003,
P. echinulatum Raper & Thom ex Fassatiová NRRL 1151 ex type,
P. brevi-compactum Dierckx NRRL 2011 ex neotype,
P. resticulosum Birkinshaw *et al.* NRRL 2021 ex type,
P. claviforme Bainier NRRL 2031 ex neotype,
P. hirsutum Dierckx NRRL 2032 ex neotype,
P. granulatum Bainier NRRL 2036 ex neotype,
T. helicus (Raper & Fennell) Benjamin NRRL 2106 ex type,
P. isariiforme Stolk & Meyer NRRL 2638 ex type,
E. crustaceum Ludwig NRRL 3332 ex type,
P. arenicola Chalabuda NRRL 3392 ex type,
P. fennelliae Stolk NRRL 3697 ex type,
P. olsonii Bainier & Sartory NRRL 13058 ex neotype,
U. craterium (Schw.) Fr. SWP-1,
S. cerevisiae Hansen NRRL Y-12632.

RESULTS AND DISCUSSION

Technical limitations, possible artefacts, and difficulties of dideoxy sequencing in ribosomal DNA have been thoroughly considered by Elwood *et al.* (1985), who estimated that 99% sequencing accuracy can be achieved with the dideoxy method and a double-stranded DNA template. Direct ribosomal RNA sequencing with dideoxy methods yields similar accuracy; however, a small percentage of the base positions are impossible to determine because a single stranded template is used. Therefore, we are probably underestimating the total genetic distance between the taxa we have examined. Even so, our sequences are representative of the complete sequences (Lane *et al.*, 1985). The ribosomal RNA base sequences are presented in Figures 1a-c. For *T. helicus* and *U. craterium*, we were unable to read approximately 30% of the sequences located in the region most distal from the primer. To accommodate these species a second matrix was generated based on the readable sequences located proximal to the primer (Fig. 2b).

A baseline for data interpretation was provided by comparisons between species of *Penicillium*, *Eupenicillium*, or *Talaromyces* and two outgroup species, *S. cerevisiae* and *U. craterium*. The relative rates of sequence change for all of the species in this study can be determined by the distance of each strain from the outgroup species. McCarrol *et al.* (1981)recorded an approximate 30% sequence difference between the 18s rRNA of the cellular slime mould *Dictyostelium discoideum* Raper and the ascomycetous yeast *Saccharomyces cerevisiae*. If all of the isolates have been mutating at a nearly constant rate since their divergence from a common ancestor, each strain should be nearly equally separated from the outgroup. Our results show that the outgroup species, *S. cerevisiae* and *U. craterium*, have approximately the same number of base differences with each of the isolates producing a *Penicillium* anamorph (Figs. 2 a-b).

```
(375-470)
P. puberulum          845   AGAGUU AAAAAGCNCG UGAAAUUGUU GAAAGGGAAG ---CGCUU-G CGACCNGNCU CG----CUCG CGGGG---UU CN-GGCAUUC GUGCC--NNN
P. camembertii        877   ...... N.....UN.. .......... .......... ---...-.. .....A.A. ..----... .......---. NG-C.....  .....--NNU
P. resticulosum      2021   ...... .....AA.. .......... .......... ---...-.. .....A.A. ..----... AA-...... .....--UNU
P. hirsutum          2032   ...... ......NN.. .......... .......... ---N...--- .....A.A. ..----...N .......---. .NGC...... N.....--NNU
P. echinulatum       1151   ...... ......NN.. .......... .......... ---...--- .....N.A. ..----... .......---. .NGC...... .....--NNU
P. granulatum        2036   ...... ......N.. .......... .......... ---...-.. .....A.A. ..----... .......---. .AGC...... U...N--NG-
P. olsonii          13058   ...... N.....N.. .......... .N........ ---...-.. .....A.A. ..----...U .......---. .NGC...... .....--NNN
P. roqueforti         849   ...... ......UN.. .......... .......... ---...-.. -....A.A. ..----... ....U--... N-G.C.... .....--NNG
P. verrucosum         965   ...... ......N.. .......... .......... ---...-.. .....A.A. ..----... .......---. .AGC...... .....--NNN
P. italicum           983   ...... ......NN.. .......... .......... ---N...--- .....N.A. ..----... .......---. .NGC...... N.....--NNN
P. expansum           976   ...... ......N.. .......... .......... ---...-.. .....A.A. ..----... .......---. .NGC...... N.....--NNN
P. brevi-compactum   2011   ...... ......N.. .......... .......... ---...-.. .....A.A. .N----...N .......---NG .N-NN.N..N N..N.--NNU
P. clavigerum        1003   ...... -.....C.. .......... .......... ---...-.. .....A.A. ..----...U .......---. NNGC.-... U.....--UU-
P. viridicatum        963   ...... ......A.. .......... .......... ---...-.. .....A.A. ..----...- .......---. NNGC....U U.....--GGN
P. fennelliae        3697   ...... ......N.. .......... .......... ---...-.. .....A.A. ..----... .......---. .UA....N..U UN...--NNN
P. claviforme        2031   ...... U.....UU.. .......... ..U...U... ---...-.. .....N.U. ..----...U .......---. .UGC..... U.....--NNN
P. atramentosum       795   ...... ......NA.. .......N.. N......... ---...-.. .....U.U. ..----...N. .......---. NAGC..... .....--NGN
P. aurantiogriseum    971   ...... ...C.U .--UUNAAA- .G....CC.. ---...-.. .....N.A. .....----...N ...NN----. .UGC.....N U.U.--NN-
E. crustaceum        3332   ...... N.....NU.. .......... .......... ----U...-- -....-.A. ..----... .......---. .UGC...... .....U-UNU
P. arenicola         3392   ...... ...C..A.. .......... .......... ---...-.. .....A.A. ..----... ......U--.C AGC.CNN... .N..N-NNU
P. isariforme        2638   ...... ......N.. .N........ .......... --U.U..-C .....A.A. ..----... .-....---. .AG...CA.U U...U-UGU
P. helicum           2106   ------ ---------- ---------- ---------- ---------- ---------- ---------- ...NN---. .UNNN.--. U....--NNN
S. cerevisiae    Y-12632   .....G ......UA.. .......... .......... GGCAUU.GAU .AGA.AUGG. GUUUUGUG.C .UCU.CUCC. UGU..G---U AG.GGA-AU-
U. craterium                ...... ......UA.. .......... .......N... ---UU...-. A....A.A.. UNGGUGA.A. GU.AUUAG.G GUU.CUU..N C.C.GU-AUA
```

```
(471-570)
P. puberulum          845   NUNUUUCCCN NNNUGGCCAG CGUCGG-UUU GGGCGGUCGG UCNAAGGCNC UCGGAAGGUA ------CNCUA GGGGNNUCUU AUAGCUNAGN G-UGC-AAUG
P. camembertii        877   N.N.......G GGNN..... ......-... .......... .NN....N. ........... -----.C... .NN..... ...U..G ....-N...
P. resticulosum      2021   N.N.......  GGU...... ......-... .......N... .NA....N. ........... -----.C... N...NN.... .N...U..N N-....-N...
P. hirsutum          2032   N.NNN.....G GGNN..... ......-... .......N... .N.....N. ........... -----.N.N. N....UN.... ...NG..N N-....-N...
P. echinulatum       1151   N.N.....GC GGNN..... ......-... .......N... .N.....U. ........... -----.C... .....UN.... .NN..N N-....-N...
P. granulatum        2036   -.A.......U GGU...... ......-... .......U... .N.....N. ........... -----.C... N....UU... .U..G .G....N.
P. olsonii          13058   N.NC.....G NGNG..... ......-... .......... ..A....N. ........N -----.C... N...NU...- .NN.CG..N N-...-N..
P. roqueforti         849   UNU..C.G.G GGU...... ......-... .......... .NN....N. ........... -----.N... .....UNN.. .N..NG..G ..-....N..
P. verrucosum         965   N.NC......G GGNN..... ......-... .......N... .N.....N. ........... -----.C.C N..NNN.... ...G..N N-N-.....
P. italicum           983   N.N.....GC GGUN..... ......-... .......N... .N.....N. ........... -----.C... N...NN.... .NN..N N-N-.....
P. expansum           976   UNU.....GC GGUN..... ......-... .......N... .A.....N. ........... -----.C... .....UN.... ...NG..N .-.-...-..
P. brevi-compactum   2011   U.NC..G..N NNA....... ......-... .......N... .N.....NN CU........ -----.C.NN .....UN.... .CAG.G N-......
P. clavigerum        1003   -.-......- GGU...... ......-... .......-... .A.....-. ........... -----.N... N....U..N ....CG..N N-......
P. viridicatum        963   G.UC.....GC GGCG..... ......-... .......... .UN....C. ........U ----A.C... .....NG.... .N..NG..G .-...N...
P. fennelliae        3697   N.NC......G GGN...... ......-... .......... .A.....C. CU........ ----C.CU.N .....NN.... .NU..G N-......
P. claviforme        2031   UNC.C.U..C GGU...... ......-... .......N... .UN....U. ........N ----C.C.C. .....UN.... ...NG..U U--.AU...
P. atramentosum       795   N.NC...N.N AGNG..... ......-... .......N...U CAA.G.C.U. .-........ -----.C N.....UU.... .U..G .G...N..
P. aurantiogriseum    971   -.NN.....- GGU...... ......-... .......... .A.G..U.. ........... -----.C..U N....UU.... N.NN..G..N N-....-NN..
E. crustaceum        3332   ACU..C.G.G GGU...... ......-... .......N... .NNN....U. ........... -----.C.C -...UU...... .....U..N .A-..AU...
P. arenicola         3392   NCU.NC.NUA GGU...... ......-... .......C... .A.....U. -.N...U... -----.C.CC N...UU.... ..N.NNGAG .-.N-....
P. isariforme        2638   NCA.A.A..C GGU...... ......-... .......C... .N.....N. ...A..U... ----A.C.CC .....NN.... ...CG..N N-...---...
P. helicum           2106   N.NC.CU..U UNU....... .N....-... ...N.CU.. .NA.....C. CG....U... -----.C..C NNNNUNC... .....NN.U U-..U-NN..
S. cerevisiae    Y-12632   C.CGCAUUUC ACUG...... .A.A.-... U..U..CA .AUA.U-.CA .A....U... GCUU-GC..C .UAAG.A... ....CUGUG .---.-..A
U. craterium                C.UGCCUNUU AUUN...N.. NA.NA.U.GC .UA...CG.. AUA...C.CN AG.....U.- GGCNCUCU.N N..AG.G.. .....CCNNN N--NU-...-
```

```
(571-614)
P. puberulum          845   CGACCUGCCU AGACCGAGGA -ACGCGC--U UCGGCUCGGA CGCU
P. camembertii        877   .......... .......... -.........- .......... ....
P. resticulosum      2021   .......... N......... -......-.. .......... ....
P. hirsutum          2032   N...N...N N.......... -......-.. ...NN..... ....
P. echinulatum       1151   .......N N......... -......-.. .......... ....
P. granulatum        2036   .......... ...NN..... -N........ .......... ....
P. olsonii          13058   .........N .......... -......-.. .......... ....
P. roqueforti         849   .......... .......... -......-.. .......... ....
P. verrucosum         965   ....N..N .......... -......--N .......... ....
P. italicum           983   ....N..N N......... -......-.. .......... ....
P. expansum           976   .......... .......... -......-.. .......... ....
P. brevi-compactum   2011   ..G......C G......... -......-.. .......... ....
P. clavigerum        1003   .......... .......... -......-.. .......... ....
P. viridicatum        963   .......... N......... -......-.. .......... ....
P. fennelliae        3697   ..G......N G......... -......-.. .......... ....
P. claviforme        2031   ....N...N .......... -......--N .......... ....
P. atramentosum       795   U.G......C G......... -N.....-.. .......... ....
P. aurantiogriseum    971   N......... .......... -......-.. .U....N..N ....
E. crustaceum        3332   .........N .N........ U......... .......... ....
P. arenicola         3392   ..G.NN...N N.....NN. -......-.. .....A.... ....
P. isariforme        2638   ..G......N G......... -U........ .......... ....
P. helicum           2106   ..G..N.... G.........C -C.....U-. ..N..GA... U...
S. cerevisiae    Y-12632   .UG..A..UG G...U..... -CU...ACG. AA.U.AA... U...
U. craterium                GCG.NN...G CN..N..... -CU....--N N.N.G.A... U...
```

Figure 1a. Aligned sequences obtained with the F (580r) primer. First base synthesized with this primer corresponds to position 635 of *S. cerevisiae* 25S rRNA. Dots indicate the same base as is found in the first line; dashes indicate missing data (in *U. craterium* and *T. helicum*) or gaps in the sequences; and N indicates that the correct base for that position could not be determined.

```
(1581-1680)
P. puberulum         845   AAAGGGA-AN CCGGUUUACU UUCCGGUNCC UAGAU-UGGA UUCUCCACGG CNACGUNACU GAACGCGGAG ACGUCGGCGG GNNUCCUGGG AAGAGUUCUC
P. camembertii       877   .......-.G .....A..N .......A.. .......-... .......... .N....N... .......... ........... .UG...... ..........
P. resticulosum     2021   .......-.- N......... .......U.. .N...-.... .......... .N....N... .......... ........... NUN...... ..........
P. hirsutum         2032   .......-.G .....N..A .......N.. .......-... .......... .N....N... .......... ..........N. .UN...... ..........
P. echinulatum      1151   .......-.G .....N..N .......N.. .......-... .......... .A....N... .......... ........... .UN...... ..........
P. granulatum       2036   .......-.G .....N..N .......U.. .......-... .......... .N....N... .......N.. ........... .UN...... ..........
P. olsonii         13058   .......-.- .....NN.. .......NU.. .N...-.... .......N.. NN....U... .......... .........N. .NG...... ..........
P. roqueforti        849   .......-VC .N...A.NN .......N.. .......-... .......... NA....N... .......... ......NN... .NG...... ..........
P. verrucosum        965   .......-.- .......N .......NN.. .N...-.... .......N.. .U..N.U... N......... ........... .UN...... ..........
P. italicum          983   .....N-.- .......N .......U.. .N...-.... .......... .N....U... .....NN... ..........N. .UN...... ..........
P. expansum          976   .......-.G .....N..N .......NA.. .......-... .......... .N....N... .......... ........... .NG...... ..........
P. brevi-compactum  2011   .......-.N N.....N..N .......N.. .......-... .......... .N....A... .......... ........... .NG...... ..........
P. clavigerum       1003   .......-.G .......N .......U.. .......-... .......... .NN...U... .....N... ........... .UU...... ..........
P. viridicatum       963   .......-.- .....A..A .......NA.. .......-... .......... .A....A... .......... ........... .NG...... ..........
P. fennelliae       3697   .......-.G .....N... .......NN.. .N...-.... N......... .N....N... .......... ........... .UG...... ..........
P. claviforme       2031   ......U.G .......N.A .......U.. .......-... .......... .N....N... .......N.. ........... UUU...... ..........
P. atramentosum      795   .......-.- .......N .......U.. .......-... .......... .N....U... .......... .........N. .UG...... ..........
P. aurantiogriseum   971   .......-.- .....NN.N .......N.. .N...-.... .......N.. UNN..UU... .......... ........... .NN...... ..........
E. crustaceum       3332   .......-.G .......N .......N.. .......--.. .......... .N....A... .......... ........... .UG...... ..........
P. arenicola        3392   .......-.G .....A..A .......U.. .G...G.... .......... .A....A... .......... ........... .NG...... ..........
F. isariforme       2638   .......-.G .....NNUN .......NN.. .G...G.... C......... .NNN..UN... .......... ......N.... .UG...... ..........
P. helicum          2106   .....U-.G .....NNUN .......U.. .G...G-... .......... .N....N... .......... ......A.... .UG...... ..........
S. cerevisiae     Y-12632   ....-.U ....N.GA .......NA.. .G...A.... .......U... UA....A... ...U.U... .......... C .AGC...... .G.....A..
U. craterium              ---------- ---------- ---------- ---------- ---------- ---------- ---------- ---------- --GC...... .G....U..

(1681-1780)
P. puberulum         845   UUUUCUUCUU GACAGCCUAU C-ACCCUGAA AUCGGUUUGU CCGGAGCUAG GGUUCUA-UG GCUN---GCA G-ACGCACUU UUGCGG-NNU CCGGUUGCGCC
P. camembertii       877   .......... .......... .-........ .......... .......... .......... ...N---... .-........ .......-NN. ..........
P. resticulosum     2021   .......... .......... .-........ .......... .......... .......... ...N-..... .-........ .......-NN. ..........
P. hirsutum         2032   .......... .......... .-........ .......... .......... .......... ...U----... .-........ .......-NN. ..........
P. echinulatum      1151   .......... .......... .-........ .......... .......... .......... ...N---... .-........ .......-NN. ..........
P. granulatum       2036   .......... .......... .-........ .......... .......... .......... ...N---... .-........ .......-NN. ..........
P. olsonii         13058   .......... .......G... .-........ .......... .......... .......... .NN--CAG. .--.NN..N. .....-NNN
P. roqueforti        849   .......... .......... .-........ .......... .......... .......... ...N---.G. .ACG,..N. .N....-NN. ..........
P. verrucosum        965   .......... .......... .-........ .......... .......... ...N-..... .U----..-. .-........ .......-UU. ..........
P. italicum          983   .......... .......... .-........ .......... .......... ...N-..... .U--GC-. .ACGCACU.. .GCG.NNUU. ..........
P. expansum          976   .......... .......... .-........ .......... .......... ...N---... .-........ .......-NN. ..........
P. brevi-compactum  2011   .......... .......... .-........ .......... .......... .......... .GC--AG. .CU.....C. .....-AGNN
P. clavigerum       1003   .......... .......... .-........ .......... .......... ...U----... .-........ .......-NN. ..........
P. viridicatum       963   .......... .......... .-........ .......... .......... ...N-..... .U---... .-........ .......-AA. ..........
P. fennelliae       3697   .......... .......G... .-........ .......... .......... ...N-..... .GC--N.AG CCU.....C. .....A-GG. ..........
P. claviforme       2031   .......... .......... .-........ .......... .......... .......... ...U----... .-........ .......-NN. ..........
P. atramentosum      795   .......... .......G... .-........ .......... .......... ...N-..... .G--CNG. .NNN.NN... .....-UU. .N......
P. aurantiogriseum   971   .......... .......... .-N......N N........ .......N... ...NN-... .U----..N .-...NN... ..N....-NN. ......N..
E. crustaceum       3332   .......... .......... .-........ .......... .......... .......... ...N---... .-........ .......-NN. ..........
P. arenicola        3392   .......... .......... .-........ .......... .......... ...C.-... .U----..N .-........ .......-NN. ..........
F. isariforme       2638   .......... N..G...... .-N......  .......... .......... ...C.CN. .----C..N .-.GCGNAC. ..UGC.GNNN
P. helicum          2106   .......... .......... .-........ .......N... ...N....N... .....N... .CCN. .NGCA.AG CUCG....C. N...C.-NG. ..........
S. cerevisiae     Y-12632   ....-.... A.....U.. .-....C.G. .U....A. .....A.G. .NNNU.-.. .AGAA.AG .CCA....CN N..U.-NN. .........U
U. craterium              -----..-.. A.......... .-........ ...A....... .U......... ...NUA.-.. .NGAA.AG CACU....C. ....A.-NN. .........GU

(1781-1824)
P. puberulum         845   CNNGACGACC CUUGAAAAUC CGCGGGAAGG AAUAGUUUUC ACGC
P. camembertii       877   .NN....... .......... .......... .......... ....
P. resticulosum     2021   .NNN...... .......... .......... .......... ....
P. hirsutum         2032   .NNN...... .......... .......... .......... ....
P. echinulatum      1151   .NNN...... .......... .......... .......... ....
P. granulatum       2036   .NNNN..... .......... .......... .......... ....
P. olsonii         13058   .NNN...... .......... .......... .......... ....
P. roqueforti        849   .NNCN..... .......... .......... .......... ....
P. verrucosum        965   .NNNN..... .......... .......... .......... ....
P. italicum          983   .NNNN..... .......... .......... .......... ....
P. expansum          976   .NN....... .......... .......... .......... ....
P. brevi-compactum  2011   .NNN...... .......... .......... .......... ....
P. clavigerum       1003   .NNNN..... .......... .......... .......... ....
P. viridicatum       963   .NNN...... .......... .......... .......... ....
P. fennelliae       3697   .CN....... .......... .......... ...N...... ....
P. claviforme       2031   .NNNN..... .......... .......... .......... ....
P. atramentosum      795   .NNN.....N .......... .......... .......... ....
P. aurantiogriseum   971   .NNNN....N .......... .......... .......... ...N
E. crustaceum       3332   .NN....... .......... .......... .......... ....
P. arenicola        3392   .NN.N..... .G........ .......... ...N...... ....
F. isariforme       2638   .NN.....U. .......... .......... .......... ....
P. helicum          2106   .NN.....U. ...N...... .......... ...NN..... ....
S. cerevisiae     Y-12632   UGU....G.. .G........ .A.A...... .......... .U..
U. craterium              .NNNN...NN .......... .G........ ...C...... .N..
```

Figure 1b. Aligned sequences obtained with the E (1611r) primer. The first base synthesized from the primer corresponds to position 1841 of *S. cerevisiae* 25S rRNA. Symbols are the same as Figure 1a.

```
1378-1470)
F. puberulum       845    NC UNUAGCCG-A UGGAAGUGCG CGGCAAUNAC NNGUCUGUGA UGCCCUUNGA UGUUCUGGGN C-NCGCGCGC UACNCUGA-C AGGGCCAGCG
F. camembertii     877    N.  .CN.....-. .......... .......A.. NG........ .......A.. ...N..N.-. ....N....
F. resticulosum   2021    G.  NC.....-. .......... ...N..NN.. NN........ .......... .NN....... ...U...-. .....N....
F. hirsutum       2032    N.  .C.....-. .......... ...N..N.. NN........ ...N..N.. .NN....... .N.U...-. ....N....
F. echinulatum    1151    N.  .C.....-. .......... ...N.. NN........ ...N..N.. ..UN...... .N.N...-. N.........
F. granulatum     2036    G.  .N......-. .......... ....U.. NU........ ...N..N.. ..NA...... .N.A...-. .....N....
F. olsonii       13058    -N  .N......-. .......... .......A.. AN........ ...U..A... ...UA..... ...A....-. ....N....
F. roqueforti      849    N.  .N......-. .......... ..NN..N.N NN........ ...U..A... ..-NN..... .N.N...-. .....U....
F. verrucosum      965    -.  NN......-. .......... .......NN........ ...U..N.. .NN....... .UNN...-. N........
F. italicum        983    -.  .NN.....-. .......... ...N..N.. NN........ ...N..N.. .UA....... .N.U...-. N....NN...
F. expansum        976    -.  .C......-. .......... .......A.. AG........ ...N..A... .NN....... .N.N...-. N...UN...
F. brevi-compactum 2011   N.  .CN.....-. .......... ...N.. AG........ ...N..NN.. ...UA..... .N.A...-. ....N....
F. clavigerum     1003    G.  .C......-. .......... ....U.. UN........ ...N..A... .NA....... .N.A...-. ....N....
F. viridicatum     963    G.  .NN.....-. .......... ...A.N UG........ ...N..A... ...UA..... .N.N...-. .....N....
F. fennelliae     3697    G.  .CN.....-. .......... .......A.. AG........ .......A.. ...UA..... .NN....-- UN..G....
F. claviforme     2031    G.  .UA.....-. .......... .......A.. AG........ ...N..A... ...UA..... .N.A...-. .....N....
F. atramentosum    795    -.  .N......-. .......... ...N..U.. NN........ ...N..A... ...UA..... .N.A...-. ....U....
F. aurantiogriseum 971    -.  NNNU....-. .......... ...U..U.U UN........ ...N..N.. ...NA..... .N.U...-. N....U...
E. crustaceum     3332    -.  .C......-. .......... ...N..U.. NN........ ...N..A... ...NA..... .N.U...-. N...U....
F. arenicola      3392    N.  .CN.....-. .......... .......N.. NG........ .......A.. .C NCA.... .N.N...-. .........
F. isariforme     2638    -.  .N.N....-. .......... .......N.. UN........ ...N..U.. .UA....... .N.N...-. N........
F. helicum        2106    -.  .C.U....N. .......U.. .U....U.. UN........ ...N..U.. ...UA..... .N.N...-. N........
S. cerevisiae   Y-12632   UU  .CA.....-. ...... UU. A......A.. AG........ ...N..A.. C........C NCA...... .A....--. -C..AGCCA.
U. craterium              UU  .CNN.A..-. ...NN..UU. A......A.. AN........ ...N..A.. C........C NCA...... .A....--. -CA.AGCCAA

(1471-1570)
F. puberulum       845    AGUA--CNUC ACCUUAACCG AGAGGUUUGG GUAAUCUUGU UAAACCCUNU NGUGCUGGGG AUAGAGCAUU GCNAUUAUUG CUCUUCNACN AGGAAUGCCU
F. camembertii     877    .......A.. .......... .......... .......... ........NN. N......... .......... ....A..... .......N..G .........
F. resticulosum   2021    .......N.. N......... .......... .......... .........NNN N......... .......... ....N..... ......N..G .........
F. hirsutum       2032    .......N.. .......... .......... .......... .........N. N......... .......... ....N..... .......N..G .........
F. echinulatum    1151    .......A.. .......... .......... .......... .........N. N......... .......... ....N..... ......N..N .........
F. granulatum     2036    .......A.. .......... .......... .......... .........N. N......... .......... ....A..... ......N..N NN.......
F. olsonii       13058    .......A.. .......... .......... .......... .N.....N.. N.     U. .......N... ....A..... ......N..N ..U......
F. roqueforti      849    .......A.. .......... .......... .......... ........NN. U......... .......N... ....N..... ......N..N .N.......
F. verrucosum      965    ...N--.N.. .......... .......... ...N...... ........N.. N......... .......N... ....N..... ......NN..N .........
F. italicum        983    ...N--.N.. .......... .......... .......... ........NN. N......... .......N... ....N..... ......NNNN .N....-..
F. expansum        976    .......A.. .......... .......... .......... ........NN. N......... .......... ....A..... ........A..G .........
F. brevi-compactum 2011   .......A.. .......... .......... .......... ........NN. N......... .......... ....A..... ........A..G .........
F. clavigerum     1003    .......A.. .......... .......... .......... ........NN. N......... .......... ....A..... ........A..G .........
F. viridicatum     963    ----NA.. .......... .......... .......... ........G. U......... .......... ....A..... ........A..G .N......
F. fennelliae     3697    .......A.. .......... .......... .......... .......G. U......... .......... ....U..... ......NU.NN .N......
F. claviforme     2031    .......A.. .......... .......... .......... ........G. U......... .......... ....A..... ......N..N .........
F. atramentosum    795    .......N.. .......... .......... .......... .....NNN N......... .......... ....A..... ........A..G ...CC....
F. aurantiogriseum 971    ...N--.N. N......... .......... ...N...... ........N. U......... .......... ....A..... ........A..G .........
E. crustaceum     3332    .......A.. .......... .......... .......... .N......NG. N......... .......... ....U...N.. ......N..G .........
F. arenicola      3392    .......A.. .....G.... ...N...C.. .......... ........G. N......... .......... ....A..... ......A..N .........
F. isariforme     2638    .......N.. .....-G... .N...CC.. .......... .N......N. N......... ...N...... ....A..... ......A..N NN......
F. helicum        2106    .......A.. .....GG... ......C.. .......... ........G. N......... ...N...... ....U..... ......NU.UU N.......
S. cerevisiae   Y-12632   C.AG--UC.A .....GG... ........C.U. .......... G....U.NNN N......... .......... ....N..... .......ANNN N.........
U. craterium              C.AGUU.A.. .....GG... GA....C... .......... U..G. GUGU. .......... .......... ....UA..... ......A..G ......U...

(1571-1597)
F. puberulum       845    AGU-AGCACG AGUCAUC-AG CUCGUGC
F. camembertii     877    ...-...... ........-.. .......
F. resticulosum   2021    ...-...... ......N.-N. .......
F. hirsutum       2032    ...-...... ........-.. .......
F. echinulatum    1151    ...-...... ........-.. .......
F. granulatum     2036    ...-...... .......... .......
F. olsonii       13058    ...-..N... .......... .......
F. roqueforti      849    ...-...... .......... .......
F. verrucosum      965    ..A-UA.G.. ......A... .......
F. italicum        983    ...-...... .......... .......
F. expansum        976    ...-...... ........-.. .......
F. brevi-compactum 2011   ...-...... ........-.. .......
F. clavigerum     1003    ...-...... .......... .......
F. viridicatum     963    ...-...... ..NNNN-N. ..U....
F. fennelliae     3697    ...-..N... ........-.. .......
F. claviforme     2031    ...-...... ........-.. .......
F. atramentosum    795    ...-..N... .......... .......
F. aurantiogriseum 971    ...-...... .....NNN. .......
E. crustaceum     3332    ...-.CG... .......... .......
F. arenicola      3392    ...-.CG... ......-NN .......
F. isariforme     2638    ...-...... ......NN. .......
F. helicum        2106    ...A-..... ........-.. .......
S. cerevisiae   Y-12632   ...A...G.A .........-.. ..U....
U. craterium              ...A...G.A .........-.. ..U.C.N
```

Figure 1c. Aligned sequences obtained with the C primer. The first base synthesized corresponds to position 1627 of the *S. cerevisiae* 18S rRNA. Symbols are the same as in Figure 1a.

	845	877	2021	2032	1151	2036	13058	849	965	983	976	2011	1003	963	3697	2031	795	971	3332	3392	2638
877	0																				
2021	0	0																			
2032	2	2	2																		
1151	3	3	3	1																	
2036	2	2	2	2	3																
13058	4	4	4	4	5	4															
849	5	5	5	5	6	5	5														
965	6	6	6	6	7	6	8	9													
983	7	7	7	5	4	7	8	9	11												
976	9	9	9	10	9	10	11	12	14	13											
2011	7	7	7	9	10	9	9	10	13	13	16										
1003	6	6	6	6	7	6	8	9	10	11	13	13									
963	7	7	7	7	6	7	9	10	11	10	13	14	11								
3697	8	8	8	10	11	10	9	12	14	14	17	8	14	14							
2031	10	10	10	10	11	10	12	13	14	15	18	16	14	14	16						
795	11	11	11	11	12	11	9	12	15	15	19	13	15	16	13	19					
971	13	13	13	13	14	13	15	16	17	18	20	20	15	18	21	21	22				
3332	11	11	11	9	10	11	13	10	14	14	19	18	14	15	19	19	19	22			
3392	19	19	19	20	20	20	20	17	23	23	24	19	24	24	23	28	25	31	25		
2638	24	24	24	23	24	24	24	25	28	25	31	28	25	27	27	29	29	35	28	30	
Y12632	115	115	115	115	115	116	117	116	117	118	116	114	117	115	116	119	116	125	120	109	116

Figure 2a. Matrix of base differences between the *Penicillium* species and *S. cerevisiae*. Gaps in sequences are counted as mismatches to any base present. Base positions for which the correct base could not be determined for one or more strains were excluded from the calculation. Total sequence length analysed is 708 positions.

	845	3332	3697	2638	2106	URNULA
3332	6					
3697	8	14				
2638	16	22	19			
2106	27	31	26	27		
URNULA	85	88	81	85	75	
Y-12632	80	84	83	79	72	75

Figure 2b. Matrix of base differences between *Eupenicillium, Talaromyces, Urnula,* and *Saccharomyces.* Results were calculated as in figure 2a. Total length of sequence examined, 555 bases.

It was recently proposed that the teleomorph genera *Eupenicillium* and *Talaromyces* with *Penicillium* anamorphs, represent separate lines of evolution involving cleistothecial Ascomycetes (Malloch, 1985). Malloch theorized that species in subgenus *Biverticillium* are more closely related to *Talaromyces* since they can degrade cellulose and were probably derived from species colonizing decayed wood such as *Trichoma* in the subfamily Trichocomoideae. Malloch (1985) classified *Penicillium* anamorphs with a marked affinity for starchy or oily substrates in the subfamily Dichlaenoideae. The latter would encompass those species in subgenus *Penicillium* that are commonly isolated from agricultural products (Pitt, 1979). *Eupenicillium crustaceum* and *T. helicum* differed by 31 bases in the abbreviated sequence length. Because the non-readable portion of the sequence contained numerous base differences in other species, we suggest that these genera could have as many as 40-45 different bases over the entire sequence length. *Urnula craterium* and *S. cerevisiae* differed from *E. crustaceum* and *T. helicum* by 72-88 bases in the abbreviated sequence length (Fig. 2b). At the same time, *U. craterium* and *S. cerevisiae* differ from each other by 75 bases (Fig. 2b). These results suggest that the two major teleomorph genera having *Penicillium* anamorphs can be traced to the same branch in Ascomycete evolution. If the two genera had entirely independent origins we would have expected a number of base differences equivalent to that recorded in contrasts involving *Urnula* and/or *Saccharomyces*.

All but three of the isolates we examined differed by one or more bases and may represent distinct species (Fig. 2a). Strains having no base differences (i.e., *P. puberulum* NRRL 845, *P. resticulosum* NRRL 2021, and *P. camemberti* NRRL 877) may represent variants of the same species. In heterothallic yeasts, isolates of a sexually reproducing species have an identical ribosomal RNA sequence, but isolates identified as siblings, on the basis of mating reactions and DNA complementarity, differ by as few as 2 and up to 7 base substitutions. Six isolates of *S. cerevisiae* representing isolates from different sources had identical base sequences (S. Peterson and C. P. Kurtzman, unpublished). If these data are representative of other fungi, a single nucleotide difference in the sequence of two *Penicillium* isolates suggests that they are not the same species. This information will aid in the resolution of several questions about taxonomic and evolutionary relationships among the isolates of terverticillate Penicillia that we sequenced. Our results indicate that *P. verrucosum* NRRL 965, *P. viridicatum* NRRL 963, *P. aurantiogriseum* NRRL 971, *P. hirsutum* NRRL 2032, and *P. puberulum* NRRL 845 (all ex neotype cultures) represent distinct species. Samson *et al.* (1976) accommodated these and several other species in *P. verrucosum* Dierckx. At that time, this was justified primarily on the basis of morphological characteristics of the conidiogenous structures (e.g., fasciculate Penicillia with two-staged, sometimes three-staged branched, rough-walled conidiophores and globose to subglobose, smooth to slightly rough-walled conidia). Samson *et al.* (1976) recognized strain NRRL 965 as the neotype culture of *P. verrucosum* and included this strain in *P. verrucosum* var. *verrucosum* Samson *et al.*, along with strain NRRL 963 (= *P. viridicatum* Westling). Pitt (1979) retained *P. verrucosum* as a species and distinguished it from *P. viridicatum*. Frisvad and Filtenborg (1983) used SMPs to place these and other isolates of terverticillate Penicillia into species and provisional nonbotanical subgroups. The authors proposed that SMPs, combined with recognizable microscopic and simple physiological criteria, should be one of the bases for the establishment of a new classification system of the terverticillate Penicillia. It was the authors' intent to allow mycologists time to consider these experimental groupings before formally erecting new varieties or species. Stolk and Samson (1985), citing "practical reasons" and the SMPs of Frisvad and Filtenborg (1983), decided to reverse their earlier classification scheme

(Samson *et al.*, 1976) and list these Penicillia as species. Our results provide evidence that these distinct *Penicillium* chemotypes represent distinct species.

P. *puberulum* NRRL 845 and *P. camemberti* NRRL 877, ex type, showed identical base sequences. At the same time, *P. camemberti* NRRL 877 and *P. aurantiogriseum* NRRL 971, ex neotype, differed by 15 bases. This result does not support the hypothesis that *P. aurantiogriseum* is the wild-type ancestor of the domesticated cheese mould *P. camemberti* as suggested by Samson (1985). Cruickshank and Pitt (1987) reported that *P. puberulum* (NRRL 2040, ex neotype) produced zymograms, suggesting synonomy with *P. aurantiogriseum*, but our data indicating 13 base substitutions argues strongly against this (Table 2a). The authors considered *P. commune* Thom NRRL 890a ex neotype to be incorrectly placed in *P. puberulum* by Pitt (1979). The rRNA base sequences of this *P. commune* strain were not examined and, therefore, cannot address the question of whether *P. commune*, like *P. puberulum*, should also be recognized as a synonym of *P. camemberti*.

"Domesticated" Penicillia used in food fermentations were derived from naturally occurring "wild" species (Samson, 1985) but *Penicillium* taxonomists may disagree as to which species represent the "wild" progenitor (Polonelli *et al.*, 1987). Frisvad and Filtenborg (1983) established the chemotype *P. camemberti* II to include species formerly classified in *P. commune* Thom. *Penicillium camemberti* was recognized as a domesticated form of *P. commune*, the wild form occurring in nature (Polonelli *et al.*, 1987). The search for a wild-type strain of *P. camemberti* is now answered with the type strain of *P. puberulum* isolated from corn. Because *P. camemberti* was described in 1906, while *P. puberulum* was described in 1907, the combination *P. puberulum* var. *camemberti* would be unacceptable according to the rules of nomenclature. Our data do not support placement of *P. puberulum* in synonymy with *P. aurantiogriseum* (Samson *et al.*, 1976) because the neotype isolates differed by 15 bases. Raper and Thom (1949) noted that *P. puberulum* NRRL 1889 and *P. puberulum* NRRL 845 came from the same original source, Thom No. 4876.20, a strain isolated from *Zea mays* L. and the basis of a classic paper on penicillic acid formation by Alsberg and Black (1913). Strain NRRL 845, received by C. Thom in 1935, had changed in cultural appearance, becoming more loose in texture and lighter sporing, and resembled *P. commune*. Thom and Raper (1949) were not certain of the taxonomic position of *P. puberulum*. The production of velvety colonies led Thom (1930) to place *P. puberulum* in the Asymmetrica-velutina section, but Thom and Raper (1949) noted the development of limited fasiculate structures in older colonies, and other characters suggested a relationship to *Penicillium cyclopium* series in the Asymmetrica-fasciculata section.

P. *resticulosum* was originally isolated as a culture contaminant in Birkinshaw's laboratory (Raper and Thom, 1949). *P. puberulum* NRRL 845 and *P. resticulosum* NRRL 2021 have identical base sequences. *P. puberulum* NRRL 845 is a loose-textured, lightly sporulating, cultural variant of isolate NRRL 1889. Both NRRL 845 and NRRL 1889 were extensively investigated in Birkinshaw's laboratory and it is interesting to speculate that NRRL 2021 represents another cultural variant of *P. puberulum* NRRL 1889. *P. puberulum* is reported to form limited fasiculate structures suggesting a relationship to the *P. cyclopium* series in the Asymmetrica-fasciculata section (Raper and Thom, 1949).

Samson *et al.* (1976) considered *P. resticulosum* to be a floccose variant of *P. expansum*. This is not supported by our results, which show that *P. expansum* and *P. resticulosum* differed by 9 bases. Pitt (1979) suggested that *P. resticulosum* was a distinct, rare species, but reduced it to synonymy with *P. expansum* (Cruickshank and Pitt, 1987). It is important to recognize that the three isolates with identical base sequences (i.e., *P. puberulum* NRRL 845, *P. resticulosum* NRRL 2021, *P. camemberti* NRRL 877) show identical numbers of different bases in contrast with other *Penicillium* isolates (Fig. 2a). The observation that some species assigned by Raper and Thom (1949) to the sections *Asymmetrica* subsect. *Funiculosa* and subsect. *Lanata* represent cultural variants of isolates classified in subsections *Velutina* or *Fasiculata* (Samson *et al.*, 1976; Pitt, 1979) is consistent with our findings.

Frisvad and Filtenborg (1983) proposed that *P. arenicola, P. fennelliae* and *P. olsonii,* species that Pitt (1979) included in subgen. *Penicillium*, were taxonomically distinct from "true species" of terverticillate Penicillia. We could not separate *P. fennelliae* or *P. olsonii* from the more typical species belonging to subgen. *Penicillium* on the basis of substantial differences in rRNA base sequences. *P. arenicola* showed a consistent pattern of higher numbers of base differences when contrasted with the other terverticillate Penicillia. Stolk and Samson (1985) noted that *P. arenicola* is not a typical *Penicillium*, but retained it in *Penicillium* in agreement with Pitt (1979). Pitt (1979) included *P. fennelliae* in subgen. *Penicillium* on the basis of the orginal illustrations, but noted that the isolates he examined produced predominantly biverticillate Penicillia. Our results indicate a closer relationship to species in subgen. *Penicillium* than to *P. isariiforme* in subgenus *Biverticillium*.

Raper and Thom (1949) classified *P. olsonii* in sect. *Biverticillata-Symmetrica*. Our results suggest that *P. olsonii* NRRL 13058 (ex neotype) is more closely aligned with species classified in subgen. *Penicillium* (Pitt, 1979).

P. claviforme NRRL 2031 (= *P. vulpinum* Cooke & Massee) Seifert & Samson) and *P. clavigerum* NRRL 1003 share more bases in common (14 base differences) than either taxon does with *P. isariiforme* NRRL 2628 (28 and 24 differences, respectively). Raper and Thom (1949) placed *P. claviforme* and *P. clavigerum* in subsection *Fasciculata* because they form coremia, but Pitt (1979) classified these species in subgen. *Biverticillium* with *P. clavigerum* being placed in synonymy with *P. duclauxii*. Frisvad and Filtenborg (1983) distinguished *P. isariiforme* on the basis of SMPs and strongly yellow-colored mycelium, agreeing with its placement in subgen. *Biverticillium*, with *P. claviforme* and *P. clavigerum* remaining in subgen. *Penicillium*. Our base se quence data supports their classification.

P. cyclopium var. *echinulatum* Raper & Thom was not validly published and Fassatiová (1977) validated and raised it to species status. Our results confirm that *P. echinulatum* NRRL 1151 and *P. aurantiogriseum* NRRL 971 (= *P. cyclopium*) are distinct species. *P. granulatum* Bainier (= *P. glandicola* (Oud.) Seifert & Samson) is recognized as sharing characteristics in common with *P. verrucosum* and *P. brevicompactum* (Pitt, 1979) but our results indicate that *P. granulatum* NRRL 2036 shares more bases with *P. puberulum, P. olsonii,* and *P. hirsutum.*

Classification schemes which rely on physiological characters (e.g., growth rates, toxin production) as well as morphological characters are supported by our results. Ecological and physiological data as well as SMPs are required if one is to distinguish *Penicillium* species by methods other than degree of nucleic acid relatedness. Wicklow (1985) observed that physiological attributes (Pitt, 1979), and SMPs (Frisvad *et al.*, 1983) are ecologically relevant characters that define the fungal niche. The fundamental niche of a fungus can be defined in the laboratory by careful control of climate, substrate chemistry, and interacting organisms (McNaughton, 1981). If the niche parameters of two isolates are distinct, it is likely they occupy different niches and would represent different species.

Williams *et al.* (1985) suggest that the considerable variation we find in subgen. *Penicillium* may result from the "rapid adaptation of a relatively few ancient species to take advantage of the many new nutritional niches provided by man during the few millennia of his agricultural activity." An example of this is demonstrated by our results showing that the domesticated white cheese mould *P. camemberti* has no base differences with the naturally occurring wild species *P. puberulum*. At the same time, those terverticillate *Penicillia* whose sequences differ by one or more bases represent species that predate human agriculture.

REFERENCES

ALSBERG, C. L. and BLACK, O.F. 1913. Contributions to the study of maize deterioration; biochemical and toxicological investigations of *Penicillium puberulum* and *Penicillium stoloniferum*. *U.S. Department of Agriculture Bureau of Plant Industry, Bulletin* 270, pp. 1-47.

CHIRGWIN, J. M., PRZYBYLA, A.E., MACDONALD, R.J. and RUTTER, W.J. 1979. Isolation of biologically active ribonucleic acid from sources enriched in ribonuclease. *Biochemistry* 18: 5294-5299.

CIEGLER, A., FENNELL, D.I., SANSING, G.A. , DETROY, R.W. and

BENNETT, J.A.1973. Mycotoxin-producing isolates of *Penicillium viridicatum*: classification into subgroups. *Applied Microbiology* 26: 271-278.

CRUICKSHANK, R. H. and PITT, J.I. 1987. Identification of species in *Penicillium* subgenus *Penicillium* by enzyme electrophoresis. *Mycologia* 79:614-620.

ELWOOD, H. J., OLSEN, G.L. and SOGIN. M.L. 1985. The small subunit ribosomal RNA gene sequences from the hypotrichous ciliates *Oxytricha nova* and *Stylonychia pustulata*. *Molecular Biological Evolution* 2: 399-410.

RAMIREZ, O. 1977. A taxonomic study of *Penicillium* series Expansa Thom emend. Fassatiová. *Acta Universitatis Carolinae Biologica* 12: 283-335.

FRISVAD, J. C. 1986. Taxonomic approaches to mycotoxin identification (taxonomic indication of mycotoxin content in foods). *In* Modern Methods in the Analysis and Structural Elucidation of Mycotoxins, ed. R. J. Cole, pp. 415-457. New York: Academic Press.

FRISVAD, J. C. and FILTENBORG, O. 1983. Classification of terverticillate *Penicillia* based on profiles of mycotoxins and other secondary metabolites. *Applied and Environmental Microbiology* 46: 1301-1310.

GEORGIEV, O. I., NIKOLAEV, N., HADJIOLOV, A.A., SKRYABIN, K.G., ZAKHARYEV, V.M. and BAYEV, A.A. 1981. The structure of the yeast ribosomal RNA genes. 4. Complete sequence of the 25S rRNA gene from *Saccharomyces cerevisiae*. *Nucleic Acids Research* 9: 6953-6958.

LANE, D. J., PACE, B., OLSEN, G.J., STAHL, D.A., SOGIN, M.L. and PACE, N.R. 1985. Rapid determination of 16s ribosomal RNA sequences for phylogenetic analyses. *Proceedings of the National Academy of Science, USA* 82: 6955-6959.

LEISTNER, L. 1984. Toxigenic Penicillia occurring in feeds and foods. *In* Toxigenic Fungi-Their Toxins and Health Hazard. eds. H. Kurata and Y. Uenopp. pp. 162-171. Tokyo: Kodansha.

MALLOCH, D. 1985. The Trichocomaceae: relationships with other Ascomycetes. *In* Advances in *Penicillium* and *Aspergillus* Systematics. eds. R.A. Samson and J. I. Pitt. pp. 365-382. New York and London: Plenum Press.

MCCARROLL, R., OLSEN, G.J., STAHL, Y.D., WOESE, C.R. and SOGIN, M.L. 1983. Nucleotide sequence of the *Dictyostelium discoideum* small-subunit ribosomal ribonucleic acid inferred from the gene sequence; evolutionary implications. *Biochemistry* 22: 5858-5968.

MCNAUGHTON, S. J. 1981. Niche: Definition and generalizations. *In*The Fungal Community: Its Organization and Role in the Ecosystem, eds. D. T. Wicklow and G. C. Carroll pp. 79-88. New York: Marcel Dekker.

PITT, J. L. 1979. The genus *Penicillium* and its teleomorphic states *Eupenicillium* and *Talaromyces*. London: Academic Press.

POLONELLI, L., MORACE, G., ROSA, R., CASTAGNOLA, M. and FRISVAD, J.C. 1987. Antigenic characterization of *Penicillium camemberti* and related common cheese contaminants. *Applied and Environmental Microbiology* 53: 872-878.

RAPER, K. B. and THOM, C. 1949. *A Manual of the Penicillia*. Baltimore: Williams & Wilkins.

RUBSTOV, P. M., MUSAKHANOV, M. M., ZAKHARYEV, V.M., KRAYEV, A.S., SKRYABIN, K.G. and BAYEV, A.A. 1980. The complete structure of yeast ribosomal RNA genes. I. The complete nucleotide sequence of the 18S ribosomal RNA gene from *Saccharomyces cerevisiae*. *Nucleic Acids Research* 8: 5779-5794.

SAMSON, R. A. 1985. Taxonomic considerations of harmful and beneficial moulds in food. *In* Filamentous Microorganisms: Biomedical Aspects, eds. T. Kuga, K. Terao, M. Yamazaki, M. Miyaji, and T. Unemoto pp. 157-163. Tokyo: Japanese Scientific Society Press.

SAMSON, R. A., STOLK, A.C. and HADLOK, R. 1976. Revision of the subsection fasciculata of *Penicillium* and some allied species. *Studies in Mycology, Baarn* 11:1-47.

STOLK, A.C. and SAMSON, R.A. 1983. A new taxonomic scheme for *Penicillium* anamorphs. *In* Advances in *Penicillium* and *Aspergillus* Systematics. eds. R.A. Samson and J. I. Pitt. pp. 163-192.New York and London: Plenum Press.

SANGER, F., NICKLEN, S. and COULSON, A.R. 1977. DNA sequencing with chain-terminating inhibitors. *Proceedings of the National Academy of Sciences, USA* 74: 5463-5467.

THOM, C. 1930. *The Penicillia*. Baltimore: Williams & Wilkins.

WICKERHAM, L. J. 1951. Taxonomy of yeasts. *United States Department of Agriculture Technical Bulletin* 1029: 1-56.

WICKLOW, D. T. 1985. Ecological adaptation and classification in *Aspergillus* and *Penicillium*. *In* Advances in Penicillium and Aspergillus Systematics. eds. R.A. Samson and J. I. Pitt. pp. 255-265. New York and London: Plenum Press.

WILLIAMS, A. P., J. I. PITT, and A. D. HOCKING. 1985. The closely related species of subgenus *Penicillium* a phylogenic exploration. *In* Advances in Penicillium and Aspergillus Systematics. eds. R.A. Samson and J. I. Pitt. pp. 121-128. New York and London: Plenum Press.

DIALOGUE FOLLOWING DR. PETERSON'S PRESENTATION

GAMS: I would like to ask Dr. Taylor and Dr. Peterson where they would draw the line to distinguish species using your techniques. Are your techniques sensitive enough to really distinguish species?

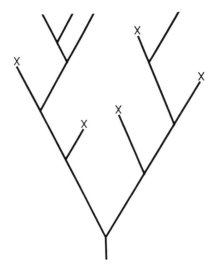

TAYLOR: This figure gives an indication of what molecular techniques can and cannot tell us. Molecular techniques can give us this whole story if we do enough work. In this diagram, we see species diverging and becoming extinct, diverging and becoming extinct, as we pass through time. Finally, at the bottom, we see the species on the left is quite distinct and no one has any problem recognizing it. The three species on the right, however, remain close together, and are difficult to distinguish, no matter what methods are used. With anamorphic genera there will always be the problem that closely related species are going to be difficult to distinguish. If these taxa are important, such as being mycotoxin producers, then they will be distinguished for practical reasons. If not, if nobody cares about them, they will be lumped together.

PETERSON: I agree with Dr. Taylor. Ribosomal RNA shows us the phylogeny but doesn't give us ability to assign a taxonomic level to a taxon. So, we're seeing a pattern of descent and it's still a philosophical decision whether something is a species or a variety.

GAMS: You said you could not distinguish some of the terverticillate Penicillia at all, but in your diagrams you show differences of two or three base changes. Is this not sufficient?

PETERSON: Our work with heterothallic species of yeasts, in which we do have a biological species concept, is the only way we have of calibrating what these base changes mean taxonomically. In sexually reproducing species, up to two base changes may exist in a single species. If there were fifteen bases differences, the case for considering these distinct species is overwhelming.

PITT: The work that is done with yeasts is fascinating, but it is irrelevant to the kind of fungi we are considering here. It's impossible to relate a yeast species to a *Penicillium* or *Aspergillus* species. The genome sizes are so different. We don't know anything about the mating patterns in these moulds, of course. I think you should ask quite a different question. Can you take ten isolates of *P. aurantiogriseum* and ten of *P. commune*, which people in this area consider to be separate species, and make the distinction between intraspecific variation in the parameter you are measuring, and the variation between species?

PETERSON: The point is well taken. We have been planning to take this approach in our laboratory. We had planned to use *P. chrysogenum* rather than *P. commune*. This needs to be done.

PITT: When you do this work, please have your isolates checked by at least one other taxonomist.

RIBOSOMAL DNA RESTRICTION STUDIES OF *TALAROMYCES* SPECIES WITH *PAECILOMYCES* AND *PENICILLIUM* ANAMORPHS

J.W. Taylor[1], J.I. Pitt[2] and A.D. Hocking[2]

[1]*Department of Plant Biology*
University of California
Berkeley, California 94720, USA

[2]*CSIRO, Division of Food Processing*
North Ryde 2113, Australia

SUMMARY

Ribosomal DNA (rDNA) in species of *Talaromyces* and related genera were examined in an initial attempt to understand their phylogenetic relationships. The variability in the nuclear rDNA repeat unit was studied by the restriction fragments of total DNA that hybridized to the rDNA repeat unit of *Neurospora crassa* (pMF2). Each fragment was treated as a taxonomic character with two states, present or absent. Pairwise comparisons of all taxa were used to produce a matrix of similarity coefficients which were subjected to UPGMA cluster analysis.

A comparison of 29 species with 21 having multiple isolates showed 14 with a majority of the isolates as closest neighbours. *Talaromyces* species with *Paecilomyces* anamorphic states cluster with *Byssochlamys* and *Thermoascus* species having *Paecilomyces* anamorphic states and not with *Talaromyces* species having *Penicillium* anamorphs. The results also indicate that the strictly anamorphic *Penicillium* species are not mixed in with the holomorphic species, but with one exception are clustered in a group that is well separated from most of the *Talaromyces* species.

INTRODUCTION

The evolutionary relationships of *Talaromyces* species are interesting because the morphology of both teleomorphs and anamorphs of these fungi indicate a close relationship to other genera of Trichocomaceae (Fig. 1). We are applying molecular methods to evolutionary studies of these fungi to add new data to the already well-described morphological features. At the present time, morphology is used to determine both the evolutionary history of these fungi and the evolution of their morphology. Morphology can be freed of this dual role by using molecular data to determine phylogenetic relationships and then studying the evolution of morphology against the background of genetic relatedness.

Talaromyces shares ascomata characters or anamorphic characters with the holomorphic fungi *Hamigera*, *Byssochlamys* and *Thermoascus* (Stolk and Samson, 1971, 1972; Subramanian, 1979; Malloch, 1981, 1985). Anamorphs of *Talaromyces* are assignable to *Penicillium* subgenus *Biverticillium*, *Paecilomyces* and *Geosmithia* (Pitt, 1979a, 1979b; Samson, 1974). Obviously, there are many possible evolutionary questions involving these taxa. For our initial study, we chose two topics: the first concerned the relative importance to systematics of teleomorphic and anamorphic reproductive morphology, and the second concerned the relationships of strictly anamorphic fungi to their presumed holomorphic relatives.

The first topic concerns *Talaromyces* species with *Paecilomyces* anamorphs: are their closest affinities with *Talaromyces* species with *Penicillium* anamorphs, as current

Modern Concepts in Penicillium and Aspergillus Classification
Edited by R. A. Samson and J. I. Pitt
Plenum Press, New York, 1990

taxonomies suggest, or are they more closely related to *Byssochlamys* or *Thermoascus* species which have *Paecilomyces* anamorphs? Stolk and Samson (1971) separated *Hamigera* from *Talaromyces* on a difference in teleomorph morphology (asci formed in chains vs. asci formed singly), but included species with *Penicillium* and *Paecilomyces* anamorphs. Pitt (1979a) retained *Hamigera* in *Talaromyces* but did not treat the species with *Paecilomyces* anamorphs. The main distinction between *Talaromyces* and *Byssochlamys*, the presence of hyphae surrounding the asci in *Talaromyces* and its absence in *Byssochlamys*, is not absolute. Stolk and Samson (1971) noted that some *Byssochlamys* cultures demonstrate a covering over the asci, while Pitt (1979a) emphasized the difficulty in distinguishing single asci from short chains, and of detecting croziers in the absence of detailed studies of nuclear behavior. *Talaromyces* remains heterogeneous and ill-defined: on the one hand, it includes species with asci formed both in chains and singly; on the other the separation from *Byssochlamys* is not clear.

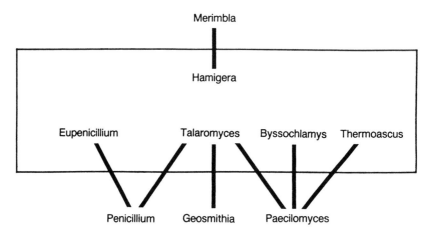

Figure 1. Chart of *Talaromyces* and relatives. Teleomorphic genera are inside the box, anamorphic genera outside. The links between teleomorphs and anamorphs are shown by the broad lines.

The second question concerns *Talaromyces* species with *Penicillium* anamorphs: are they closely interrelated with strictly anamorphic species in *Penicillium* subgen. *Biverticillium*? *Penicillium* subgen. *Biverticillium* species are morphologically similar to some *Penicillium* anamorphs of *Talaromyces* species, suggesting a close relationship. This similarity may be superficial, however, as Pitt (1988) has noted that, "...anamorphs of *Talaromyces* and species in subgenus *Biverticillium* appear to be quite distinct." Molecular evidence may help mycologists realize the long held but perhaps unrealistic goal of aligning strictly anamorphic species with the holomorphic taxa from which they presumably arose.

   To address these questions, variability in ribosomal DNA repeat regions was assessed by comparing restriction fragments of miniprep DNA showing sequence identity with a cloned rDNA repeat unit from *Neurospora crassa* (pMF2, Free *et al.*, 1979). Although rRNAs are evolutionarily conservative, the non-coding regions between the genes are known to be much more variable (Jorgenson and Cluster, 1988); our approach sampled variability in both the coding and non-coding regions. Distinct fragments were treated as characters with two states, present or absent, and subjected to UPGMA cluster analysis. This approach has been used with mitochondrial DNA (mtDNA) and rDNA in *Agaricus* (Anderson et al., 1986), with random nuclear DNA fragments in *Neurospora* (Natvig *et al.*, 1987), and with mtDNA in *Phytophthora* (Förster *et al.*, 1988).

MATERIAL AND METHODS

Fungi and cloned DNA.
The names, classification and culture collection accession numbers of the fungi studied are given in Table 1. Plasmid pMF2, which contains the nuclear rDNA repeat unit of *Neurospora crassa* (Free *et al.*, 1979), was obtained from Dr. Peter Russell of Reed College, Oregon through Dr. Michael Milgroom of Cornell University, New York.

Culture methods.
The fungi were maintained on Czapek Yeast Extract Agar (CYA, Pitt, 1988) or Malt Extract Agar (MEA, Pitt, 1979a) at 25°C or 30°C. To produce mycelium for DNA extraction, the fungi were grown in 250ml flasks containing 100 ml of broth (MEA or CYA without agar) on a rotary shaker at 25°C or 30°C for from two to seven days.

DNA extraction.
Mycelium was harvested from broth through Miracloth (Calbiochem, La Jolla, California) by vacuum filtration on a Buchner funnel, folded in the Miracloth, pressed briefly between pads of paper towelling, frozen in liquid N_2, and lyophilized. Lyophilized mycelium was ground in a mortar and pestle and stored in microcentrifuge tubes at -20°C. DNA was extracted from the ground mycelium by the miniprep methods of Lee et al. (1988) or Zolan and Pukkila (1986).

Digestion and electrophoresis.
From 5 μl to 10 μl of DNA isolated by miniprep was digested by restriction enzymes using the manufacturer's suggested conditions (New England Biolabs, Beverly, Massachusetts; Bethesda Research Labs, Gaithersburg, Maryland; Pharmacia, North Ryde, N.S.W.). Miniprep DNA that did not digest to completion was further purified by spermine precipitation (5 μl of miniprep DNA was incubated with 1 μl of 100mM spermine on ice for 15 minutes, microcentrifuged for 2 minutes, and the pellet resuspended in restriction buffer; Alan Brownlee, CSIRO Division of Animal Production, Prospect, N.S.W., personal communication; Hoopes and McClure, 1981). Digestion fragments were separated electrophoretically adjacent to a molecular size marker (1.0kb marker, Bethesda Research Labs, Bethesda, Maryland) in 1.0% agarose gels in 0.1M Tris, 125 mM NaAcetate, and 1.0 mM EDTA. DNA in gels was stained with ethidium bromide (0.5 μg per ml) and photographed with Polaroid film.

Southern hybridization.
Target DNA was transferred in alkali from agarose to a nylon membrane (ZetaProbe, BioRad) in alkali by the general method of Reed and Mann (1985). Probe DNA (pMF2) was labeled with ^{32}P dATP by the general method of Rigby et al. (1977) using a nick translation kit (BRESA, Adelaide, S. Australia). Hybridization and washing were carried out at 65°C by the general method of Reed and Mann (1985) using 7.0% lauryl sulphate,

Table 1. Taxa and isolates studied.

Taxon	Isolates
Genus *Talaromyces*	
Section *Talaromyces*	
Series *Flavi*	
+*T. flavus*	FRR 1265, 404, 629, 1019, 2098, 3380, 1976, 2268, 2386, 2417
–*T. striatus*	FRR 717, 2080
Series *Lutei*	
+*T. luteus*	FRR 1941, 1010, 2235, 1727
Series *Trachyspermi*	
T. trachyspermus	FRR 1792, 1026
T. mimosinus	FRR 1875
T. intermedius	FRR 3526 (CBS 152.65)
Section *Purpureus*	
T. purpureus	FRR 1731
Section *Thermophilus*	
+*T. thermophilus*	FRR 2155, 1791
Section *Paecilomyces*	
+*T. byssochlamyoides*	FRR 3523 (CBS 413.71), FRR 3524 (CBS 533.71)
+*T. leycettanus*	FRR 1655, FRR 3525 (CBS 398.68)
Section *Emersonii*	
+*T. bacillisporus*	FRR 947, 1025
+*T. emersonii*	FRR 1324, 1326, 3221
Genus *Byssochlamys*	
B. fulva	FRR 1125, 2785, 3493
+*B. nivea*	FRR 2205, 2231
Genus *Hamigera*	
H. avellanea	FRR 1938
Genus *Penicillium*	
Subgenus *Biverticillium*	
Section *Simplicium*	
Series *Miniolutea*	
+*P. minioluteum*	FRR 1095, 1741
+*P. funiculosum*	FRR 833, 1823, 1630
+*P. purpurogenum*	FRR 1061, 1147
Series *Islandica*	
P. islandicum	FRR 1036, 1399
+*P. variable*	FRR 1048, 1055
Genus *Geosmithia*	
G. putterilli	FRR 2037, 2024
G. cylindrospora	FRR 1366, 2673
Genus *Thermoascus*	
Th. crustaceus	FRR 1328, 1563
Genus *Paecilomyces*	
Section *Paecilomyces*	
Pa. variotii	FRR 1658, 3054
Section *Isarioidea*	
Pa. farinosus	FRR 2670
+*Pa. marquandii*	FRR 3583, 2015
+*Pa. lilacinus*	FRR 895, 1079

+ Majority of isolates clustered together in UPGMA analysis of all isolates
– Majority of isolates not clustered together.
FRR: CSIRO Division of Food Processing, N. Ryde, Australia.
CBS: Centraalbureau voor Schimmelcultures, Baarn.

Table 2. Number and molecular weight of restriction fragments hybridizing to *Neurospora* cloned nuclear rDNA (pMF2) from all isolates.

EcoRI		BamHI		BglII		DraI		HindIII	
#	kbp	#	kbp	#	kbp	#	kbp	#	kpb
1	9.0	1	20	1	11.3	1	17	1	20
2	8.6	2	15	2	10.0	2	13	2	18
3	8.0	3	12	3	9.3	3	10	3	15
4	7.6	4	10.5	4	8.7	4	9.5	4	14
5	7.0	5	9.25	5	8.0	5	9.1	5	12
6	6.7	6	8.25	6	7.6	6	8.5	6	11
7	6.4	7	7.3	7	7.1	7	8.0	7	9.3
8	5.7	8	6.5	8	6.0	8	7.5	8	8
9	5.4	9	5.9	9	5.1	9	6.2	9	7.5
10	5.2	10	5.2	10	4.0	10	5.5	10	7.0
11	5.0	11	4.5	11	2.9	11	4.7		
12	4.7	12	4.0	12	2.3	12	4.3		
13	4.4	13	3.5	13	1.6	13	2.8		
14	4.2	14	3.0						
15	4.0	15	2.7						
16	3.7	16	2.0						
17	3.4	17	1.8						
18	3.2	18	1.5						
19	3.0	19	1.3						
20	2.8								
21	2.2								
22	1.8								
23	1.2								
24	0.8								
25	0.4								

0.5% powdered milk and 1% polyethylene glycol (20M) in place of Denhardt's solution (cf. Maniatis *et al.*, 1982). Hybridization of probe DNA to target DNA was visualized by autoradiography (Fuji X-Ray film) over 3 to 24 hours.

Data analysis.
Molecular lengths of restriction fragments hybridizing to plasmid DNA were determined by superimposing the autoradiographs over enlargements of photographs showing the digested DNAs and the molecular length markers. For each enzyme, the molecular weight was determined for each hybridizing fragment; each fragment was assigned a number so that fragments of the same size from different isolates have the same number (Table 2). Numbered fragments were treated as characters with two states, present or absent, and their distribution was tabulated for all the fungi studied (Table 3). These discrete character state data were converted to Jaccard similarity coefficients (Sokal and Sneath, 1963; Simqual in Rholf, 1988) which emphasizes the positive matches over the negative matches as is appropriate for restriction fragment data. The similarity coefficients were subjected to UPGMA (unweighted pair-group method, arithmetic avearage clustering) with the "find" option to search for trees of equally good fit (SAHN in NTSYS; Rholf, 1988). How well the trees represent the similarity values was assessed by comparing the similarity coefficient

matrix with a matrix of cophenetic values synthesised from the similarities taken from each tree, using COPH in NTSYS, Sokal and Sneath (1963). If matrix correlations between the original similarity coefficients and the cophenetic values, asessed by MXCOMP in NTSYS, are greater than 0.9, the fit of tree to the similarity coefficient matrix is considered very good (Rholf, 1988). Numerical methods were carried out using an IBM PC/AT computer (registered) or its equivalent.

RESULTS

Selection of Restriction Enzymes.
Patterns of DNA hybridization of pMF2 to miniprep DNA from *T. flavus* 1265 were tested for digestions made with 16 enzymes: *Ssp*I, *Sca*I, *Hinf*I, *Bcl*I, *Dra*I, *Rsa*I, *Cla*I, *Hind*III, *Bgl*II, *EcoR*V, *Xba*I, *Taq*I, *Sal*I, *Nsi*I, *Bam*HI and *EcoR*I. Six enzymes were selected for further study: *EcoR*I, *Bam*HI, *Bgl*II, *Dra*I, *Hind*III, and *Taq*I.
Hybridizations of pMF2 to miniprep DNAs of all isolates showed that five of these enzymes gave from 10 to 25 distinct fragments in all isolates (Table 2) and from one to three distinct bands per isolate (Table 3). Digestions with *Taq*I, which recognizes only four nucleotides, produced too many small fragments hybridizing to pMF2 to be analyzed in 1.0% agarose.

Multiple isolates from a single species.
*EcoR*I fragments hybridizing to pMF2 were examined from at least two isolates in 21 of the 29 species studied (Table 1). UPGMA cluster analysis of all the isolates showed that 14 of the species with multiple isolates had a majority of the isolates as closest neighbors (marked "+" in Table 1), but 7 did not (marked "-" in Table 1). We consider this molecular evidence generally supportive of the current species concepts. In subsequent comparisons, often only one isolate was analysed from species where patterns were identical.

Talaromyces species with *Paecilomyces* anamorphs.
To assess relationships of *Talaromyces* species having *Paecilomyces* anamorphs with those having the more usual *Penicillium* anamorphs and with other taxa having *Paecilomyces* anamorphs, isolates of *T. byssochlamyoides* and *T. leycettanus*, both of which have *Paecilomyces* anamorphs, were compared with: (i) *Talaromyces* species having typical *Penicillium* anamorphs (*T. flavus*, *T. luteus*, and *T. trachyspermus*); (ii) *Talaromyces* species having *Penicillium* anamorphs that also resemble *Paecilomyces* (*T. intermedius* and *T. thermophilus*); (iii) *Talaromyces* species with the *Hamigera* type of ascus development having *Penicillium* anamorphs (*T. striatus* and *T. mimosinus*); and (iv) *Thermoascus* and *Byssochlamys* species having *Paecilomyces* anamorphs.
    UPGMA cluster analysis based on restriction fragments hybridizing to pMF2 gave one tree (Fig. 2) which showed that *Talaromyces* species with *Paecilomyces* anamorphs clustered closest to *Byssochlamys nivea* and *Thermoascus crustaceus*, fungi which also have *Paecilomyces* anamorphs. Of the three *Talaromyces* species with *Penicillium* anamorphs that show some resemblance to *Paecilomyces*, none clustered with the species having *Paecilomyces* anamorphs. The two *B. fulva* isolates did not cluster with the other *Byssochlamys* or *Thermoascus* species, nor did they cluster together.

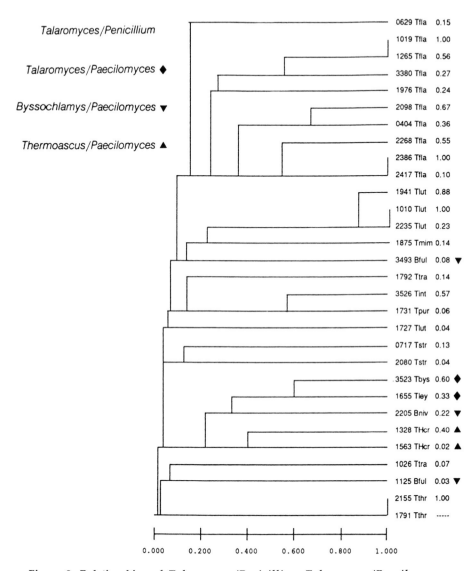

Figure 2. Relationships of *Talaromyces/Penicillium, Talaromyces/Paecilomyces, Thermoascus/Byssochlamys,* and *Byssochlamys/Paecilomyces* using UPGMA cluster analysis of restriction fragments hybridizing to cloned *Neurospora* nuclear rDNA (pMF2). The correlation between the similarity matrix supporting this tree and the matrix of cophenetic values extracted from the tree is 0.927 (see Data Analysis in Material and Methods).

Table 3. [1]Characters and [2]character states based on restriction fragments hybridizing *Neurospora* cloned nuclear rDNA (pMF2)

[3]Isolate		EcoRI Characters	BamHI Characters
		00000000001111111111222222	0000000001111111111
		1234567890123456789012345	1234567890123456789
0629	Tfla	0000000000000001100000001	0001000000000000000
1019	Tfla	0000001000000000100000000	0000000100000000100
1265	Tfla	0000001000000000100000000	0000000100000000100
2098	Tfla	0000000000000000101000001	0001000000000000000
3380	Tfla	0000001000000000100000000	0000000100000000100
0404	Tfla	0000000000000000101000001	0000000010001000000
1976	Tfla	0000001000000000100000000	0001000000000000000
2268	Tfla	0000000000000000101000001	0000000010110000000
2386	Tfla	0000000000000000110000001	0000000010001000000
2417	Tfla	0000000000000000110000001	0000000010001000000
1727	Tlut	0000100000000001000000000	0001000000000000000
1941	Tlut	0000000010000000001000000	0000100000000000000
1010	Tlut	0000000010000000001000000	0000100000000000000
2235	Tlut	0000000010000000001000000	0000100000000000000
1026	Ttra	0000000000000000010000010	xxxxxxxxxxxxxxxxxxx
1792	Ttra	0001000000000100000001000	0000010100000000000
3526	Tint	0000001000000000100000000	0000010000000000100
1731	Tpur	0000001000000000100000000	0001000000000000000
2155	Tthr	0000000000000000010010000	0000001000001000000
1791	Tthr	0000000000000000010010000	0000001000001000000
0717	Tstr	0100000000100000000001100	0000010000000000000
2080	Tstr	0000000000010010000000000	xxxxxxxxxxxxxxxxxxx
1875	Tmim	1000010100000000001010000	0001001001000010000
3523	Tbys	0000000000010000001000000	0000001001000100000
1655	Tley	0001000000100000001000000	0000000000100100000
1125	Bful	0000000001000000010000000	xxxxxxxxxxxxxxxxxxx
3493	Bful	1001001000010100000001000	0001000000000000000
2205	Bniv	0001000000100000000100000	0010001000000000000
1328	THcr	0000000010000000001000000	0000000010000000100
1563	THcr	0000000001000000001000000	0000000010000000100
1095	Pmin	0100000001000000010000000	0000010000000000000
0833	Pfun	0000000000000001010000000	0000000010000001000
1630	Pfun	0100001000000000010000000	0000010000000000000
1823	Pfun	0000001000000000010000000	0000000010000001000
1061	Ppur	0000000000000000010010000	0000010000000000000
1147	Ppur	0000000000000000010010000	0000010000000000000
1036	Pisl	0100000000100000010000000	0000001000000000000
1399	Pisl	0000000000100000010000000	0000001000000000000
1048	Pvar	0100000000000100000000000	0000000101000000000

[3]Isolate		BglII Characters	DraI Characters	HindIII Char.
		0000000001111	0000000001111	0000000001
		1234567890123	1234567890123	1234567890
0629	Tfla	0000001000101	0000100000000	0010001000
1019	Tfla	0001000000000	0000100000000	1000000000
1265	Tfla	0001000000000	0000100000000	1000000000
2098	Tfla	0001000000000	0000100000000	1000000000
3380	Tfla	0010000000000	0001000000000	1000000000
0404	Tfla	0001000000000	0000100000000	1000000000
1976	Tfla	0010000000000	0001000000000	0000001000
2268	Tfla	0010000000000	0001000000000	1000000000
2386	Tfla	0010000000000	0001000000000	1000000000
2417	Tfla	0010000000000	0001000000000	1000000000
1727	Tlut	0100000000000	0010000000000	1000000000
1941	Tlut	0001000000000	0000100000001	0000001000
1010	Tlut	0001000000000	0000100010001	0000001000
2235	Tlut	0001000000000	0000100010001	0000001000
1026	Ttra	1000000000000	xxxxxxxxxxxxx	xxxxxxxxxx
1792	Ttra	0001000000000	xxxxxxxxxxxxx	1000000000
3526	Tint	0010000000000	xxxxxxxxxxxxx	1000000000
1731	Tpur	0010000000000	xxxxxxxxxxxxx	1000000000
2155	Tthr	0010000000000	xxxxxxxxxxxxx	0000000100
1791	Tthr	0010000000000	xxxxxxxxxxxxx	0000000100
0717	Tstr	0000000010010	0000001001001	1000000000
2080	Tstr	0000100000000	0000001001001	1000000000
1875	Tmim	0001000000000	0000010001001	1000010000
3523	Tbys	0000001000000	0000001000000	0100000000
1655	Tley	0000001000000	0000001000000	0100000000
1125	Bful	0000100000000	xxxxxxxxxxxxx	0100000000
3493	Bful	0000001000100	0010000010010	0001001000
2205	Bniv	0000001000000	0000000100000	0100000000
1328	THcr	0000010000000	0000001000000	0100000000
1563	THcr	0000001000000	0000000100000	0100000000
1095	Pmin	0000001000000	0000010000000	0000010001
0833	Pfun	0000010000000	0000010000000	0000100010
1630	Pfun	0000000010100	0000001000000	0000010001
1823	Pfun	0000010000000	0000100000000	0000010010
1061	Ppur	0000010000000	0000100000000	0000010010
1147	Ppur	0000010000000	0100000100000	0000010010
1036	Pisl	0000001000000	1100000100000	0000010001
1399	Pisl	0000001000000	1100000100000	0000010001
1048	Pvar	0100000000000	xxxxxxxxxxxxx	0000001000

[1] Characters are based on restriction fragments are given in Table 2.
[2] Character states: 0=absent, 1=present, x=no data.
[3] Isolate names and classification are given in Table 1.

Penicillium **subgenus** *Biverticillium.*

To assess relationships between species of *Penicillium* subgen. *Biverticillium* and *Talaromyces*, isolates of the strictly anamorphic *Penicillium* subgen. *Biverticillium* were compared with: (i) *Talaromyces* species having typical *Penicillium* anamorphs; (ii) *Talaromyces* species having anamorphs resembling *Paecilomyces* (i.e. *T. intermedius* and *T. thermophilus*); and (iii) *Talaromyces* species having the *Hamigera* type of ascus development (i.e. *T. striatus* and *T. mimosinus*).

UPGMA cluster analysis based on restriction fragments hybridizing to pMF2 gave one tree (Fig. 3) which showed that *Penicillium* subgen. *Biverticillium* isolates group together and, with the exception of *P. variabile* 1048, did not closely cluster with *Talaromyces* species having typical biverticillate *Penicillium* anamorphs.

DISCUSSION

The size of the rDNA repeat unit of *N. crassa*, which we used as our probe (pMF2), is ca 10kbp. The sum of the sizes of the fragments hybridizing to pMF2 in particular digests was found to be ca 9 to 10 kbp (Tables 2 and 3). However, the sums for single isolates were not equal for different enzymes and for many isolates the sizes of the *Hind*III fragments were much larger (ca 20kbp). The inequality of sums for different enzymes was probably due to our inability to resolve restriction fragments smaller than 3-400 bp in 1% agarose. The very large *Hind*III fragments did not appear to be due to incomplete digestion and their presence is difficult to explain.

The variability in the sum of the sizes of the hybridizing fragments indicates that the approach used is relatively crude. Fragments of similar size are being compared, and some regions are not accounted for in each isolate because small fragments could not be analyzed. Also, the cluster analyses combined changes in fragment size due to restriction site changes (caused by nucleotide substitutions or small length mutations) with fragment size changes due to large length mutations (cf. Taylor, 1986). In spite of the simplicity of this technique, it has provided interesting information on the evolutionary relationships of *Talaromyces* and suggests questions that deserve intensive study by more laborious techniques such as DNA sequencing.

In 14 of the 21 cases where more than one isolate from a species was examined, conspecific isolates were closest neighbors in cluster analysis (Table 1). We consider this molecular evidence generally supportive of species concepts in these fungi. However, cases where isolates did not cluster should be studied further. Perhaps convergent evolution or unsuspected misidentifications have occurred.

The comparison (Fig. 2) of *Talaromyces* species having *Paecilomyces* anamorphs with *Talaromyces* species having *Penicillium* anamorphs and with *Byssochlamys* and *Thermoascus* species having *Paecilomyces* anamorphs shows that *Talaromyces* species with *Paecilomyces* anamorphic states cluster with *Byssochlamys* and *Thermoascus* and not with *Talaromyces* species having *Penicillium* anamorphs. Perhaps in these cases the anamorphs may be a better indicator of relatedness than the ascomata of the teleomorphs. This result, if confirmed by sequence analysis, will be important in Ascomycete systematics.

In the comparison of species in *Penicillium* subgen. *Biverticillium* with *Talaromyces* species having *Penicillium* anamorphs (Fig. 3), the strictly anamorphic *Penicillium* species, with one exception, have clustered in a group that is well separated from the *Talaromyces* species. It may be inferred that anamorphic *Penicillium* species in subgen. *Biverticillium* arose from holomorphs on rare occasions and that most of their interspecific diversity has

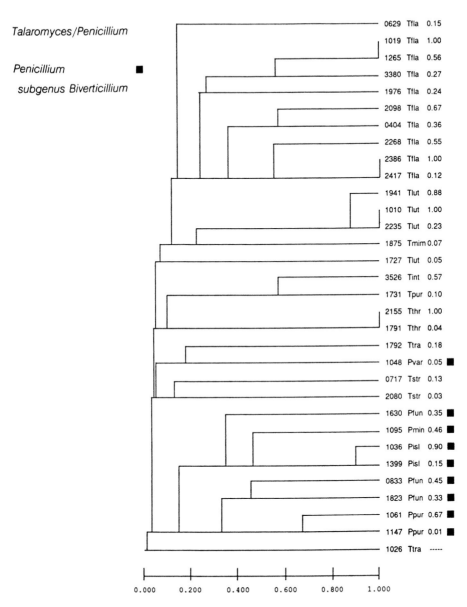

Figure 3. Relationships of *Talaromyces/Penicillium* and anamorphic *Penicillium* subgenus *Penicillium* using UPGMA cluster analysis of restriction fragments hybridizing to cloned *Neurospora* nuclear rDNA (pMF2). The matrix correlation between the similarity matrix supporting this tree and the matrix of cophenetic values extracted from the tree is 0.931 (see Data Analysis in Material and Methods).

developed since their divergence from the holomorph. If supported by further investigation, this result has application throughout the Deuteromycotina, where mycologists have hoped to link strict anamorphs with their nearest teleomorphic relative. This result indicates that in many cases a one-to-one relationship does not exist.

The questions addressed here are representative of the many that will be important to understanding the relationships of *Penicillium* and its anamorphic and teleomorphic relatives. Our results, although preliminary, indicate that currently accepted species are generally well defined, but that relationships at higher taxonomic levels, or between holomorphs and strict anamorphs, are less clear. This survey has shown that questions of interest to students of Trichocomaceae in particular and Ascomycotina in general can be approached by molecular methods: needed now is a thorough investigation with a more powerful technique. We are beginning to use DNA polymerase chain reaction amplification (PCR, Saiki *et al.* 1985, Mullis and Faloona 1987) of rDNA and direct sequencing of the product (Wrischnik *et al.* 1987, Gyllensten and Erlich 1988) to obtain better data on fungal evolution.

ACKNOWLEDGEMENTS

We thank N. Charley of CSIRO Division of Food Processing for assistance with fungal cultures, and J. Mattick, K. Finney, M. Bills, and S. Livingstone of CSIRO Division of Biotechnology for making the molecular work possible at North Ryde, and W. Stone of the Jepson Herbarium, University of California, Berkeley for helping to format the figures. Support for this research was provided by CSIRO and NSF (INT 8702240, BSR 8516513).

REFERENCES

ANDERSON, J. B., PETSCHE, D. M. and SMITH, M. L. 1986. Restriction polymorphisms in biological species of *Armillaria mellea. Mycologia* 79: 69-76.

FöRSTER, H., KINSCHERF, T. G., LEONG, S. A. and MAXWELL, D. P. 1988. Estimation of relatedness between *Phytophthora* species by analysis of mitochondrial DNA. *Mycologia* 80: 466-478.

FREE, S. J., RICE, P. W. and METZENBERG, R. L. 1979. Arrangement of the genes coding for ribosomal ribonucleic acids in *Neurospora crassa. Journal of Bacteriology* 137: 1219-1226.

GYLLENSTEN, U. B. and ERLICH, H. A. 1988. Generation of single stranded DNA by the polymerase chain reaction and its application to direct sequencing of the HLA DQ *alpha* locus. *Proceedings of the National Acadamy of Science, USA* 85: 7652-7656.

HOOPES, B. C. AND McCLURE, W. R. 1981. Studies on the selectivity of DNA precipitation by spermine. *Nucleic Acids Research* 9: 5493-5505.

JORGENSON, R. A. and CLUSTER, P. D. 1988. Modes and tempos in the evolution of nuclear ribosomal DNA: new characters for evolutionary studies and new markers for genetic and population studies. *Annals of the Missouri Botanical Garden* 75:1238-1247.

LEE, S. B., MILGROOM, M. G. and TAYLOR, J. W. 1988. A rapid, high yield mini-prep method for isolation of total genomic DNA from fungi. *Fungal Genetics Newsletter* 35: 23-24.

MALLOCH, D. 1981. The Plectomycete centrum. *In* "Ascomycete Systematics: The Luttrellian Concept". D. R. Reynolds, ed. Pp. 73-91. New York: Springer Verlag.

MALLOCH, D. 1985. The Trichocomaceae: relationships with other Ascomycetes. *In* Advances in Penicillium and Aspergillus Systematics. R. A. Samson and J. I. Pitt, eds. pp. 365-382. New York and London: Plenum Press.

MANIATIS, T., FRITSCH, E. F. and SAMBROOK, J. 1982. Molecular Cloning: A Laboratory Manual. Cold Spring Harbor Laboratory; Cold Spring Harbor, New York.

MULLIS, K. B. and FALOONA, F. A. 1987. Specific synthesis of DNA in vitro via a polymerase catalysed chain reaction. *Methods in Enzymology* 155: 335-350.

NATVIG, D. O., JACKSON, D. A. and TAYLOR, J. W. 1987. A random fragment hybridization analysis of evolution in the genus *Neurospora*: the status of four-spored strains. *Evolution* 41: 1003-1021.

PITT, J.I. 1979a. The Genus *Penicillium* and its teleomorphic states *Eupenicillium* and *Talaromyces*. Academic Press; London.

PITT, J. I. 1979b. *Geosmithia* gen. nov. for *Penicillium lavendulum* and related species. *Canadian Journal of Botany* 57: 2021-2030.

PITT, J.I. 1988. A laboratory guide to common *Penicillium* species. 2nd ed. North Ryde, N.S.W.: CSIRO Division of Food Processing.

REED, K. R. and MANN, D. A. 1985. Rapid transfer of DNA from agarose gels to nylon membranes. *Nucleic Acids Research* 13: 7207-7221.

RHOLF, F. J. 1988. NTSYS-pc: numerical taxonomy and multivariate analysis system. Version 1.4. Setauket, New York: Applied Biostatistics.

RIGBY, P. W. J., DIECKMANN, M., RHODES, C. and BERG, P. 1977. Labeling deoxyribonuclec acid to high specific activity in vitro by nick translation with DNA polymerase I. *Journal of Molecular Biology* 113: 237-251.

SAIKI, R. K., SCHARF, S., FALOONA, F., MULLIS, K. B., HORN, G. T., ERLICH, H. A. and ARNHEIM, N. 1985. Enzymatic amplification of ß-globin genomic sequences and restriction site analysis for diagnosis of sickle cell anemia. *Science, N.Y.* 230: 1350-1354.

SAMSON, R. A. 1974. *Paecilomyces* and some allied Hyphomycetes. *Studies in Mycology, Baarn* 6: 1-119.

SOKAL, R. R. and SNEATH, P.H.A. 1963. Principles of Numerical Taxonomy. Freeman; San Francisco.

STOLK, A. C. and SAMSON, R. A. 1971. Studies on *Talaromyces* and related genera. I. *Hamigera* gen. nov. and *Byssochlamys*. *Persoonia* 6: 341-357.

STOLK, A. C. and SAMSON, R. A. 1972. The Genus *Talaromyces*. Studies on *Talaromyces* and Related Genera II. *Studies in Mycology, Baarn* 2: 1-67.

SUBRAMANIAN, C. V. 1979. Phialidic Hyphomycetes and their teleomorphs - an analysis. *In* "The Whole Fungus". Vol. 1. W. B. Kendrick, ed. Pp. 125-151. Ottawa: National Museums of Canada.

TAYLOR, J. W. 1986. Fungal evolutionary biology and mitochondrial DNA. *Experimental Mycology* 10: 259-269.

WRISCHNIK, L. A., HIGUCHI, R. G., STONEKING, M., ERLICH, H. A., ARNHEIM, N. and WILSON, A. C. 1987. Length mutations in human mitochondrial DNA: direct sequencing of enzymatically amplified DNA. *Nucleic Acids Research* 15: 529-542.

ZOLAN, M. E. and PUKKILA, P. J. 1986. Inheritance of DNA methylation in *Coprinus cinereus*. *Molecular and Cellular Biology* 6: 195-200.

DIALOGUE FOLLOWING DR. TAYLOR'S PRESENTATION

SAMSON: Dr. Frisvad and I recently completed a chemotaxonomic study looking for a connection between the biverticillate Penicillia and *Talaromyces*. We also found no connection. There was some similarity between *P. variabile* and some *Talaromyces* species just as you found. *T. flavus* divided into two distinct groups, which correspond to the two varieties Dr. Stolk and I described in 1972. These varieties are chemically quite distinct. Strains of *T. flavus* var. *macrosporus* are now becoming a problem during food processing because of their heat resistant ascospores.

TAYLOR: We also have two general groups within *Talaromyces flavus* that more or less correspond with the two varieties. But there were a couple of isolates which, based on the ribosomal repeat region, fall into the wrong variety. It would be interesting to look at the secondary metabolites of those particular isolates. It may be that the intergenic region is variable and the same mutation has occurred in different isolates. This lack of correlation is not due to misidentification.

SAMSON: I'm not surprised to see *Talaromyces byssochlamydoides* is very close to *Byssochlamys*. That's why it has that name, of course. The ascomata of this species are more typical for *Talaromyces*, which is why we placed it there.

TAYLOR: We don't know *a priori* which of the characters we are looking at is evolutionarily more stable. By comparing morphology with the evolutionary story that we get from a completely different kind of character, we may be able to select the characters that are most stable.

SAMSON: What strikes me is that the *Paecilomyces* teleomorphs, *Talaromyces*, *Thermoascus* and *Byssochlamys*, are a relatively homogeneous group: they are thermophilous and have similar ascospores. Only the ascomata are different.

8

NEW APPROACHES FOR PENICILLIUM AND ASPERGILLUS SYSTEMATICS: BIOCHEMICAL AND IMMUNOLOGICAL TECHNIQUES

SECONDARY METABOLITES AS CONSISTENT CRITERIA IN *PENICILLIUM* TAXONOMY AND A SYNOPTIC KEY TO *PENICILLIUM* SUBGENUS *PENICILLIUM*

J.C. Frisvad and O. Filtenborg

Department of Biotechnology
The Technical University of Denmark
2800 Lyngby, Denmark

SUMMARY

As taxonomic characters in *Penicillium*, morphology and profiles of secondary metabolites have proved to be consistent and reproducible, provided they are recorded in a standardized and systematic way. Techniques for standardising profiles of secondary metabolites include the use of griseofulvin and preferably other commercially available secondary metabolites, and the use of standardised media, inoculation and incubation conditions and extraction techniques. Confirmation of results obtained using thin layer chromatography is strongly recommended, for example by high performance liquid chromatography and diode array detection, especially when new or usual results are found. Examples of consistent production of secondary metabolites in known culture collection strains of important *Penicillium* species are provided in this paper. A synoptic key to *Penicillium* subgen. *Penicillium*, based on secondary metabolites, is also given.

INTRODUCTION

Biochemical products such as citric acid (*Citromyces* Wehmer), oxalic acid (*Penicillium oxalicum* Currie et Thom) and citrinin (*P. citrinum* Thom), mycelial colours and diffusible pigments have been used in Penicillium taxonomy for a long time (Thom, 1930). Despite the monumental works of Raistrick and coworkers on the identification of a great number of *Penicillium* metabolites (Turner and Aldridge, 1983), individual secondary metabolites did not influence the systematics of *Penicillium* much before 1980. Ciegler *et al.* (1973) used penicillic acid, ochratoxin A and citrinin as criteria in their subdividing *P. viridicatum* Westling into subgroups, but did not draw any taxonomic conclusions. However, Frisvad (1981, 1983), emphasized secondary metabolites, backed up by physiological characters, when he proposed "*P. viridicatum* o-c" (= *P. verrucosum* Dierckx), "*P. crustosum* pA" (= *P. crustosum* Thom), "*P. cyclopium* p" (= *P. aurantiogriseum* Dierckx), "*P. melanochlorum*" and "*P. caseiphilum*" as species concepts. Later work (Frisvad, 1985; Cruickshank and Pitt, 1987) has validated these concepts, which are now recognised as *P. verrucosum* Dierckx, *P. crustosum* Thom, *P. aurantiogriseum* Dierckx, *P. solitum* Westling and *P. commune* Thom respectively. In introducing profiles of secondary metabolites into *Penicillium* taxonomy, Frisvad and Filtenborg (1983) provisionally used the subgroup concept of Ciegler *et al.* (1973), but these were later transferred species, varieties or chemotypes to correct names (Frisvad, 1985; Frisvad and Filtenborg, 1989).

Profiles of secondary metabolites has been proposed as objective and consistent taxonomic characters by Frisvad and coworkers (Filtenborg *et al.*, 1983; Frisvad and Filtenborg, 1983; Frisvad and Thrane, 1987; Frisvad, 1988, 1989; Frisvad *et al.*, 1989). Although some isolates in *Penicillium* may in time lose the ability to produce one or two secondary metabolites, the remaining profile of can be sufficient to identify an unknown

Modern Concepts in Penicillium and Aspergillus Classification
Edited by R. A. Samson and J. I. Pitt
Plenum Press, New York, 1990

(Frisvad, 1989), provided a standardized method is used for the analysis. An effective, standardized system with diode arrays detection high performance liquid chromatography (HPLC-DAD) method for analysis of secondary metabolites in fungi has been introduced by Frisvad and Thrane (1987). However, advanced liquid chromatographs are available to few mycologists, so thin layer chromatography (TLC) is the method of choice in the average mycological laboratory. Filtenborg and Frisvad (1980) and Filtenborg et al. (1983) introduced a simple TLC method for the analysis of mycotoxins and other secondary metabolites from growing fungi. The method, application of small agar plugs to TLC plates with or without extraction) results in a quite low sensitivity, so optimial inoculation techniques, media and incubation conditions, etc., are necessary.

Some authors have reported good results with the agar plug method (Blaser and Schmidt-Lorenz, 1981; Abarca et al., 1988; Klich and Pitt, 1988). Other authors have in reported inconsistent or irreproducible results with Penicillium species (Paterson et al., 1987; Bridge et al., 1986, 1987; Land and Hult, 1987; Stenwig, 1988). For example, Stenwig (1988) reported that clearest and most intense spots on the TLC plates were produced by all isolates in a species, but that less distinct spots showed poor reproducibility, caused by metabolite concentrations varying from above to below their detection limit in different chromatographic runs. He also stated that detection of the weak spots improved if several isolates of a species were compared on the same TLC plate, or if standards were used.

In this paper, we report on ways to consistently and reproducily detect secondary metabolites of Penicillium species by the agar plug method. A synoptic key for Penicillium subgen. Penicillium, based on secondary metabolites and other characters is also provided.

MATERIAL AND METHODS

Ex type and authentic cultures of Penicillium subgen. Penicillium species were examined for production of secondary metabolites using the TLC agar plug method (Filtenborg and Frisvad, 1980; Filtenborg et al., 1983, Frisvad et al., 1989). For TLC analysis agar plugs were taken from Czapek yeast extract agar (CYA), malt extract agar (MEA), yeast extract sucrose agar with Difco yeast extract (YES) and yeast extract sucrose agar with Sigma (Y-4000) yeast extract (SYES) (Frisvad and Filtenborg, 1983). Trace metals were added to all media (Frisvad and Filtenborg, 1983). Cultures were incubated for 5 to 21 days in the dark at an upright position in 9 cm plastic Petri dishes in polyethylene bags, with holes to avoid carbon dioxide accumulation.

TLC plates were eluted in CAP (chloroform/acetone/2-propanol, 85:15:20) and TEF (toluene/ethylacetate/90% formic acid, 5:4:1) with griseofulvin as external standard. Plates eluted in CAP were treated with 1% Cerium sulphate in 50% sulphuric acid. Plates eluted in TEF were treated in two ways: with cold 50% sulphuric acid for intracellular metabolites; and with anisaldehyde spray for extracellular metabolites. Anisaldehyde spray contains 0.5% anisaldehyde in methanol/acetic acid/sulphuric acid, 17:2:1 and visualisation is by heating for 8 min at 130° (Frisvad et al., 1989).

TLC plates were also scanned with a CAMAG TLC scanner and UV reflectance spectra recorded. High performance liquid chromatography (HPLC) with diode array detection (DAD) was also employed using the method of Frisvad and Thrane (1987). In that way UV spectra could be compared from HPLC and TLC to assure a good confirmation of the identity of the secondary metabolites by comparison of fungal metabolites with standards.

RESULTS AND DISCUSSION

Different ways of improving the consistency of detection of secondary metabolites in the terverticillate Penicillia were tested (Table 1).

Table 1. Methods for increasing production and/or reproducibility of detection of secondary metabolites in agar cultures.

Use of fresh cultures or cultures directly from freeze dried tubes or from silica gel, i.e. avoidance of repeatedly transferred cultures.

Avoidance of carbon dioxide accumulation during incubation.

Combined point and streak inoculation.

Use of agar plugs from different places in the fungal colony.

Repeated application of several plugs on the same spot.

Use of several effective media, such as CYA, YES, SYES, MEA). Special media may be necessary for particular secondary metabolites.

Use of more than one incubation temperature: 25° is generally best, but 20° or 30° may be much better in some cases.

Optimization of extraction time and extraction technique: usually chloroform/methanol, 2:1 is excellent, but chloroform/acetone or other mixtures may be significantly better in some cases.

Use of known standards (see Table 2).

Use of known cultures with good secondary metabolite production as standards (Frisvad, 1985).

Optimization of all the above conditions on good, authentic cultures from each taxonomic group.

Usually the standard conditions recommended by Filtenborg *et al.* (1983) were sufficient for reproducible recovery of important secondary metabolites, but each laboratory should optimise the methods, using fresh isolates with known profiles of secondary metabolites (Frisvad, 1985). Medium composition is of particular importance (Filtenborg *et al.*, 1990) In some cases a particular combination of medium and spray reagent is 10 to 100 times as sensitive as the standard method. For example, patulin is best produced on potato dextrose agar and visualized by MBTH spray (Frisvad *et al.*, 1989) and verrucosidin is produced optimally on Merck malt extract agar and visualized as a intracellular metabolite by anisaldehyde spray (El-Banna and Leistner, 1987). The use of commercially available standards (Table 2) greatly improves the TLC identification procedure and use is strongly recommended.

The intracellular alkaloids roquefortine C, meleagrin, oxaline and penitrem A are produced very consistently on CYA by all isolates in the relevant species, indicated in (the synoptic key). The production of these alkaloids on YES and SYES is usually limited, though better in strongly sporulating cultures. Analysis for intracellular mycotoxins produced on CYA and eluted in CAP using Cerium sulphate spray is therefore recommended as the first step in the identification of the terverticillate Penicillia by profiles of secondary metabolites.

Table 2. Commercially available standards of secondary metabolites produced by *Penicillium*.

Mycotoxin	Optimal production medium
Alternariol monomethylether	CYA
Chaetoglobosin C	CYA, SYES
Citreoviridin	YES, SYES
Citrinin	YES, SYES
Cyclopiazonic acid	CYA
Dipicolinic acid	YES
Gliotoxin	TGY (Frisvad *et al.*, 1987)
Griseofulvin	YES, SYES
Kojic acid	YES
Mycophenolic acid	YES
3-nitropropionic acid	YES
Ochratoxin A	YES, SYES
Patulin	YES, SYES, Potato dextrose agar
Penicillin	TGY (Frisvad *et al.*, 1987)
Penicillic acid	YES, Raulin-Thom (Betina, 1989)
PR-toxin	YES
Rubratoxin B	YES
Secalonic acid D	YES, CYA

Four important indol alkaloids detected consistently as intracellular metabolites on TLC plates eluated in CAP after visualization with Cerium sulphate spray:

Roquefortine C: Orange spot, Rf 0.46 (a) in *P. crustosum* (not fluorescing)

Penitrem A: Blue spot, Rf 1.16 in *P. crustosum* (not fluorescing)

Meleagrin: Yellow spot, Rf 0.65 in *P. chrysogenum* (a) (olive yellow fluorescence)

Oxaline: Olive yellow, Rf 0.71 in *P. atramentosum* (olive green flourescence)

(a) Rf value relative to griseofulvin; (b) *P. chrysogenum* also produce roquefortine C

The next recommended step is analysis for intracellular metabolites produced on CYA and YES, eluated in TEF. By this procedure a variety of mycotoxins can be detected: xanthomegnin, and brevianamide A, yellow and viomellein, yellow brown, in daylight; griseofulvin, a blue flourescence under UV light; citrinin, a yellow green fluorescing tail; ochratoxin A, a turquoise fluorescing spot, mycophenolic acid, a violet blue fluorescing spot, and cyclopiazonic acid, a brown fluorescing tail. Spraying with cold sulphuric acid visualizes cylopenin and viridicatin, which produce a violet blue fluorescence) and the characteristic series of 6 bluish flourescing metabolites in *P. echinulatum*, the penechins. Further spraying with anisaldehyde and heating visualizes chaetoglobosin C, verrucosidin and compactin.

The third recommended step in the identification procedure is analysis for extracellular metabolites produced on YES and SYES agar, eluted in TEF and visualized by anisaldehyde spray and heating. Predominantly, extracellular mycotoxins such as penicillic acid, patulin and terrestric acid, a yellow spot in daylight, are detected in this system.

Table 3. Consistent production of known secondary metabolites by *Penicillium coprophilum*

Isolate	Griseofulvin	Roquefortine C	Meleagrin	Oxaline
CBS 477.75	+	+	+	+
CBS 473.75	+	+	+	+
FRR 1403	+	+	+	+
NRRL 13627	+	+	+	+
IMI 285526	+	+	+	+
IMI 321500	+	+	+	+
IMI 293196	+	+	+	+
IMI 321501	+	+	+	+
ATCC 64629	+	+	+	+

Table 4. Consistent production of known secondary metabolites by *Penicillium griseofulvum*

Isolate	Griseofulvin	Cyclopiazonic acid	Patulin	Roquefortine C
NRRL 2300	+	+	+	+
NRRL 993	+	+	+	+
NRRL 2159A	+	+	+	+
NRRL A-23324	+	+	+	+
NRRL A-26914	+	+	+	+
IMI 296933	+	+	+	+
ATCC 9260	+	+	+	+
IMI 293195	+	+	+	+
IMI 285525	+	+	+	+
IFO 7010	+	+	+	+
CSIR 1399	+	+	+	+
CSIR 1082	+	+	+	+
FRR 1232	+	+	+	+

Table 5. Consistent production of secondary metabolites by *Penicillium brevicompactum*

Isolate	Mycophenolic acid	Raistrick phenols	Brevianamide A
IMI 40225	+	+	+
IMI 92044	+	+	+
IMI 94149	+	+	+
IMI 92219	+	+	+
IMI 17456	+	+	+
IMI 125546	+	+	+
IMI 285520	+	+	+
IMI 293191	+	+	+
NRRL 859	+	+	−
NRRL A-23329	+	+	+
CBS 317.59	+	+	−
CBS 256.74	+	+	+
CBS 210.28	+	+	−
NRRL 886	+	+	+

Many secondary metabolites are produced very consistently in *Penicillium* species as shown in Tables 3, 4 and 5, and Fig. 1. Fig. 1 shows HPLC traces of two *P. echinulatum* isolates. The traces are strikingly similar, and they are also similar to traces produced by the culture ex type (NRRL 1151) and several other isolates tested by this method. Cyclopenin, cyclopenol, viridicatin, viridicatol and the penechins were all consistently produced. UV spectra as determined by either HPLC DAD or UV reflectance spectra on TLC plates were valuable in designating the chromophore families in the different species (Fig. 2.).

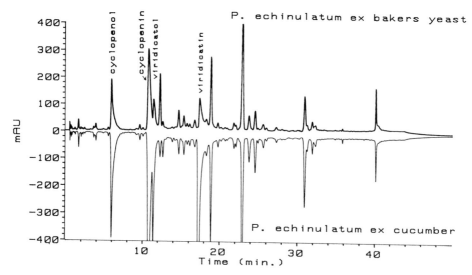

Figure 1. HPLC traces of two isolates of *P. echinulatum*. Note the very high qualitative similarity and the quantitative differences in individual compounds. All major peaks, except those marked, are members of the penechin chromophore family.

For example, *P. coprophilum* always produces griseofulvin, dechlorogriseofulvin, roquefortine C, meleagrin and oxaline irrespective of geographic source or habitat, including soil, dung, feedstuffs or foods). Study of *P. coprophilum* metabolites by HPLC DAD revealed a consistently produced series of unknown specific secondary metabolites as well. Similarily *P. griseofulvum* always produce griseofulvin, dechlorogriseofulvin, cyclopiazonic acid, roquefortine C and patulin (Table 4).

Except for brevianamide A, all known alkaloids are produced very consist ently by all isolates in each relevant terverticillate *Penicillium* species (Frisvad and Filtenborg, 1989). The insecticidal toxin brevianamide A (Paterson *et al.*, 1987) is produced by two species, *P. brevicompactum* and *P. viridicatum*, but only by a minor proportion of isolates. Other metabolites are consistent produced by *P. brevicompactum* (Table 5) and *P. viridicatum*, however, so this is not a major problem. Consistent production of mycophenolic acid and the Raistrick phenols clearly identify *P. brevicompactum*.

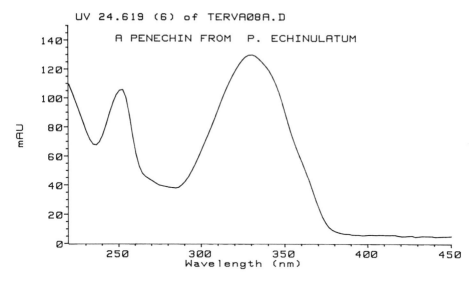

Fig. 2. UV spectrum of a major penechin obtained by diode array detection in a HPLC run of an extract of *P. echinulatum* NRRL 1151.

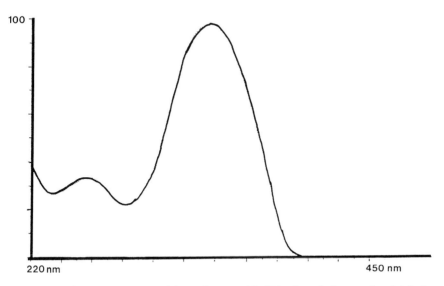

Fig. 3. UV reflectance spectrum of the major penechin (Rf value relative to griseofulvin in TEF: 1.0) obtained by a CAMAG TLC scanner. Note the similarity to the UV spectrum obtained by HPLC-DAD (see Fig. 2).

Some extracellular mycotoxins such as patulin, penicillic acid and citrinin are not consistently produced by some isolates of *P. expansum* or *P. aurantiogriseum* on the standard media, but are other media often assist toxins production. Land and Hult (1987), for example, could not detect patulin in some isolates of *P. expansum*. When some of those isolates were tested on oatmeal agar, MEA or potato dextrose agar, we found large amounts of patulin. The other *Penicillium* isolates from wood investigated by Land and Hult (1987), which were *P. roqueforti* and *P. commune*, also produced their typical mycotoxins in our hands (Frisvad and Filtenborg, 1989). Bridge *et al.* (1986, 1987) also reported inconsistent secondary metabolite production by some isolates of *P. viridicatum*, *P. glandicola*, *P. crustosum* and *P. commune*.

These fungi were all very efficient producers of their usual mycotoxins in our hands (Frisvad and Filtenborg, 1989).

SYNOPTIC KEY TO *PENICILLIUM* SUBGENUS *PENICILLIUM*

As secondary metabolite production by species in *Penicillium* subgen. *Penicillium* is very consistent when the recommended methods of analysis listed in Table 1 are followed, we offer the following synoptic key. It is based on individual secondary metabolites, physiological characters and some morphological characters.

List of included species and varieties.
Species bracketted (26-28) are not classified in subgen. *Penicillium*, but are included because of similarities in metabolites.

1.   *P. aethiopicum* Frisvad 1989
2.   *P. atramentosum* Thom 1910
3a.   *P. aurantiogriseum* Dierckx 1901 var. *aurantiogriseum*
3b.   *P. aurantiogriseum* var. *polonicum* (Zaleski) Frisvad 1989
3c.   *P. aurantiogriseum* var. *melanoconidium* Frisvad 1989
3d.   *P. aurantiogriseum* var. *neoechinulatum* Frisvad *et al.* 1987
3e.   *P. aurantiogriseum* var. *viridicatum* (Westling) Frisvad et Filtenborg 1989
4.   *P. brevicompactum* Dierckx 1901
5a.   *P. camemberti* Thom 1906
6a.   *P. chrysogenum* Thom 1910 var. *chrysogenum*
6b.   *P. chrysogenum* var. *dipodomyis* Frisvad *et al.* 1987
7.   *P. clavigerum* Demelius 1922
5b.   *P. commune* Thom 1910 chemotype I
5c.   *P. commune* chemotype II
8.   *P. confertum* (Frisvad *et al.*) Frisvad 1989
9.   *P. coprophilum* (Berk. et Curt.) Seifert et Samson 1985
10.   *P. coprobium* Frisvad 1989
11.   *P. crustosum* Thom 1930
12.   *P. digitatum* (Pers.:Fr.) Sacc. 1832
13a. *P. echinulatum* (Raper et Thom) Fassatiová1977 chemotype I
13b. *P. echinulatum* chemotype II
14.   *P. expansum* Link 1809
15.   *P. fennelliae* Stolk 1969
16a. *P. glandicola* (Oud.) Seifert et Samson 1985 var. *glandicola*
16b. *P. glandicola* var. *glaucovenetum* Frisvad 1989

17a. *P. griseofulvum* Dierckx 1901 var. *griseofulvum*
17b. *P. griseofulvum* var. *dipodomyicola* Frisvad *et al.* 1987
18a. *P. hirsutum* Dierkx 1901 var. *hirsutum*
18b. *P. hirsutum* var. *albocoremium* Frisvad 1989
18c. *P. hirsutum* var. *allii* Frisvad 1989
18d. *P. hirsutum* var. *hordei* (Stolk) Frisvad 1989
18e. *P. hirsutum* var. *venetum* Frisvad 1989
19.   *P. italicum* Wehmer 1894
20.   *P. solitum* Westling 1911
21.   *P. mononematosum* (Frisvad *et al.*) Frisvad 1989
22.   *P. olsonii* Bain. et Sart. 1912
23a. *P. roqueforti* Thom 1906 var. *roqueforti*
23b. *P. roqueforti* var. *carneum* Frisvad 1989
24a. *P. verrucosum* Dierckx 1901 chemotype I
24b. *P. verrucosum* chemotype II
25.   *P. vulpinum* (Cooke et Massee) Seifert et Samson 1985
(26. *P. sclerotigenum* Yamamoto 1955)
(27. *P. oxalicum* Currie et Thom 1915)
(28. *P. lanosum* Westling 1911)

Synoptic key.
Production of griseofulvin: 1, 9, 17a, 17b, 26, 28
Production of mycophenolic acid: 4, 23a, 23b
Production of ochratoxin A: 24a, 24b
Production of citrinin: 14, 18b, 24b
Production of cyclopiazonic acid: 5a, 5b, 5c, 17a, 17b
Production of chaetoglobosin C: 13b, 14
Production of kojic acid: 28
Production of secalonic acid D: 27
Production of PR-toxin: 23a
Production of patulin: 10, 14, 16a, 16b, 17a, 17b, 23b, 25
Production of penicillic acid: 3a, 3b, 3c, 3d, 3e15, 18b, (23b)
Production of penicillin: 6a, 6b
Production of terrestric acid: (3a), 11, 18a, 18b, 18d, 18e
Production of roquefortine C: 2, 6a, 8, 9, 11, 14, 16a, 16b, 17a, 18a, 18b, 18c, 18d, 18e, 23a, 23b, 25, 26, 27
Production of meleagrin: (2), 6a, 8, 9, 16a, 16b, 18b, 18c
Production of oxaline: 2, 3c, 9, 16a, 25, 27
Production of cyclopenin and viridicatin: 3a, 3b, 3c, 3d, 3e, (5c), 11, 13a, 13b, 18b, 18c, 18e, 20, 25
Production of palitantin: 5b, 5c, 13b, (20)
Production of rugulovasine A: 2, 5b
Production of viridicatumtoxin: 1
Production of tryptoquivalins: 1, 12
Production of aurantiamine: 3a, 3d
Production of viridamine: 3e
Production of verrucofortine: 3a, 3b, 3e
Production of verrucosidin: 3a, 3b, 3c
Production of penitrem A: (3a), 3c, 7, 11, 16a, (16b)

Production of xanthomegnin and viomellein: 3a, 3e
Production of brevianamide A: (3e), (4)
Production of botryodiploidin: (4), (23b)
Production of asperphenamate: 4
Production of Raistrick phenols: 4, 7, 14
Production of pebrolides: 4
Production of xanthocillin X: (6a)
Production of emodic acid: (6a)
Production of isofumigaclavine A: 7, (5b), (5c), 23a
Production of penechins: 13a
Production of mycelianamide: (17a)
Production of cyclopiamine: (17a)
Production of compactin: 18a, 20, 28
Production of carolic acid: 18d
Production of cyclopaldic acid: 5b, (5c), 21
Production of isochromantoxin: 21
Production of verrucolone: 24a, 24b
Production of marcfortin: 23a
Production of deoxybrevianamide E: 19
Production of 5,6-dihydro-4-methoxy-2H-pyran-2-one: 19
Production of puberulic acid: (3a)

Production of rot in apples: 11, 14, 20
Production of rot in citrus fruits: 12, 19
Production of rot in sweet potato: 26
Production of rot in garlic: 18c
Production of rot in onions and bulbs: 18a, 18b, 18e

Growth at 37°C: 1, (6a), (6b), 8, 21, 27

Good growth on creatine sucrose agar (and good base production): 2, (3b), 5a, 5b, 5c, 9, 10,
    11, 13a, 13b, 14, 16a, 16b, 20, 23a, 23b, 25
No acid production on creatine sucrose agar: 2, (4), (6a), (6b), 7, (9), 10, 12, 15, (17a), 18c,
    18e, 19, (20), 22, 23a, 23b, 24a, 24b, 26, 28
Good growth on nitrite sucrose agar: 2, 4, (5a), (5b), (5c), 7, 9, 10, (11), (16a), (16b), 18c, 18e,
    (19), (20), 22, 23a, 23b, 24a, 24b, 26, 27, 28
Growth on 0.5% acetic acid: 23a, 23b

Conidia very dark green: 2, 3c, 6b, (5c), 10, (11), 13a, 13b, 20, (27)
Conidia blue green: 3a, 3b, 3d, (6a), 16b, 18a, 18b, (18d), 18e, (20)
Conidia olive green: 12
Conidia grey: 17a
Conidia white or grey green late in the growth phase: 5a, (mutants)
Conidia cylindrical with rounded ends: 12, 15, 19
Conidia ellipsoidal: 1, (3), (4), (6a), (6b), 7, (5b), (5c), 8, 9, 10, 11, 12, (13b), 14, 15, 16a, 16b,
    19, (20), 21, 22, 25, 26, 27
Conidia echinulate: 3d, 13a, 13b
Conidia finely roughened: 4, 15, 18d, (20), 22, 28
Conidia > 3.5 µm long: 12, 19, 26, 27

Short phialides (< 6.5 μm): 17a, 17b
Quite short phialides with a thick-walled neck: 6a, 6b, 21
Rami divergent: 2, 6a, 6b, (8), (12), (13b), 17a, 17b, (18d), 21, 28
Penicilli multiramulate: ((2)), (4), ((6a & b)), (21), 22
Very few rami: 12, 15, 26, 27, (28)
Stipes rough on MEA: 1, (3a), (3b), (3c), 3d, 3e, (5a), 5b, 5c, (7), 11, 13a, 13b, ((14)), 16a, 16b,
    18 a,b,c,d,e, 20, 23a, 23b, 28
Production of sclerotia on MEA: 10, ((23a)), 26
Formation of conidial crusts: (3a), (5b), 11, 19, 27
Production of synnemata with distinct capitula: 9, 10, (13b), (14), 16a, 16b, (17a), (18a), 18b,
    18d, (18e), (19), 25
Synnemata yellow: 18a, 18d
Synnemata without sterile stalks: 7
Colonies floccose: 5a (repeatedly transferred strains)
Weak sporulation on YES agar: 3a, 3d, 3e, 5a, 5b, 7, ((11)), (14), 18b, 18d, ((19)), ((20)), 24a,
    24b, (25), 28
Violet brown reverse on YES: 19, 24b
Dark blackish green reverse on CYA: 23a
Colonies < 20mm MEA, 1 week, 25 C: ((2)), ((3a,c)), (4), ((5a)), (6b), (7), (8), (9), (10), 16a,
    16b, (17a), 21, ((22)), ((23a)), 24a, 24b, (25)
Colonies > 40 mm MEA, 1 week, 25 C: (3b), (6a), (11), (12), (14), (15), (18a,b,c,e), (19),
    (23a,b), 26, ((27))
Colonies > 40 mm YES, 1 week, 25 C: 1, (2), ((3a)), (3b), (3e), (5 b,c), 6a, ((9)), 11, 12, (13a,b),
    14, ((16a)), (18a,b,c,d,e), 19, (20), 22, (23a), 23b, (25), 26, 27

REFERENCES

ABARCA, M.L., BRAGULAT, M.R., BRUGUERA M.T. and CABANES, F.J. 1988. Comparison of some
    screening methods for aflatoxigenic moulds. *Mycopathologia* 104: 75-79.
BETINA, V. 1989. Mycotoxins. Amsterdam: Elsevier.
BLASER, P. and SCHMIDT-LORENZ, W. 1981. *Aspergillus flavus* contamination von Nüssen, Mandeln, mais
    mit bekannten Aflatoxin-gehalt. *Lebensmittel Wissenschaft und Technologie* 14: 252-259.
BRIDGE, P.D., HAWKSWORTH, D.L., KOZAKIEWICZ, Z., ONIONS, A.H.S. and PATERSON, R.R.M. 1986.
    Morphological and biochemical variation in single isolates of *Penicillium*. *Transactions of the British
    Mycological Society* 87: 389-396.
BRIDGE, P.D., HUDSON, L., KOZAKIEWICZ, Z., ONIONS, A.H.S. and PATERSON, R.R.M. 1987.
    Investigation of variation in phenotype and DNA content between single-conidium isolates of single
    Penicillium strains. *Journal of General Microbiolgy* 133: 995-1004.
CIEGLER, A., FENNELL, D.I., SANSING, G.A., DETROY, R.W. and BENNETT, G.A. 1973. Mycotoxin-
    producing strains of *Penicillium viridicatum*: classification into subgroups. *Applied Microbiology* 26: 271-278.
CRUICKSHANK, R.H. and PITT, J.I. 1987. Identification of species in *Penicillium* subgenus *Penicillium* by
    enzyme electrophoresis. *Mycologia* 79: 614-620.
EL-BANNA, A.A. and LEISTNER, L. 1987. Quantitative determination of verrucosidin produced by
    Penicillium aurantiogriseum. *Microbiologie Aliments Nutrition* 5: 191-195.
FILTENBORG, O. and FRISVAD, J.C. 1980. A simple screening-method for toxigenic moulds in pure
    cultures. *Lebensmittel Wissenschaft und Technologie* 13: 120-130.
FILTENBORG, O., FRISVAD, J.C. and SVENDSEN, J.A. 1983. Simple screening method for molds producing
    intracellular mycotoxins in pure culture. *Applied and Environmental Microbiology* 45: 581-585.
FILTENBORG, O., FRISVAD, J.C. and THRANE, U. 1990. The significance of yeast extract composition on
    metabolite production in *Penicillium*. *In* Modern Concepts in *Penicillium* and *Aspergillus* Classification,
    eds. R.A. Samson and J.I. Pitt, pp. 433-440. New York and London: Plenum Press.

FRISVAD, J.C. 1981. Physiological criteria and mycotoxin production as aids in identification of common asymmetric Penicillia. *Applied and Environmetal Microbiology* 41: 568-579.

FRISVAD, J.C. 1983. A selective and indicative medium for groups of *Penicillium viridicatum* producing different mycotoxins in cereals. *Journal of Applied Bacteriology* 54: 409-416.

FRISVAD, J.C. 1985. Classification of asymmetric Penicillia using expressions of differentiation. *In* Advances in Penicillium and Aspergillus Systematic, eds. R.A. Samson and J.I. Pitt, pp. 327-333. New York and London: Plenum Press.

FRISVAD, J.C. 1988. Fungal species and their specific production of mycotoxins. *In* Introduction to Foodborne Fungi, eds. R.A. Samson and E.S. van Reenen-Hoekstra, pp. 239-249. Baarn, Netherlands: Centraalbureau voor Schimmelcultures.

FRISVAD, J.C. 1989. The connection between the Penicillia and Aspergilli and mycotoxins with special emphasis on misidentified isolates. *Archives of Environmental Contamination and Toxicology* 18: 452-467.

FRISVAD, J.C. and FILTENBORG, O. 1983. Classification of terverticillate Penicillia based on profiles of mycotoxins and other secondary metabolites. *Applied and Environmental Microbiology* 46: 1301-1310.

FRISVAD, J.C. and FILTENBORG, O. 1989. Chemotaxonomy of and mycotoxin production by the terverticillate Penicillia. *Mycologia* 81 (in press).

FRISVAD, J.C. and THRANE, U. 1987. Standardized high-performance liquid chromatography of 182 mycotoxins and other fungal metabolites based on alkylphenone retention indices and UV-VIS spectra (diode array detection). *Journal of Chromatography* 404: 195-214.

FRISVAD, J.C., FILTENBORG, O. and THRANE, U. 1989. Analysis and screening for mycotoxins and other secondary metabolites in fungal cultures by thin-layer chromatography and high-performance liquid chromatography. *Archives of Environmental Contamination and Toxicology* 18: 331-335.

FRISVAD, J.C., FILTENBORG, O, and WICKLOW, D.T. 1987. Terverticillate Penicillia isolated from underground seed caches and cheek pouches of the banner-tailed kangaroo rat (*Dipodomys spectabilis*). *Canadian Journal of Botany* 65: 765-773.

KLICH, M.A. and PITT, J.I. 1988. Differentiation of *Aspergillus flavus* from *A. parasiticus* and other closely related species. *Transactions of the British Mycological Society* 91: 99-108.

KOZAKIEWICZ, Z. 1989. Ornamentation types of conidia and conidiogenous structures in fasciculate *Penicillium* species using scanning electron microscopy. *Botanical Journal of the Linnean Society* 99: 273-293.

LAND, C.J. and HULT, K. 1987. Mycotoxin production by some wood-associated *Penicillium* spp. *Letters in Applied Microbiology* 4: 41-44.

PATERSON, R.R.M., SIMMONDS, M.S.J. and BLANEY, W.M. 1987. Mycopesticidal effects of characterized extracts of *Penicillium* isolates and purified secondary metabolites (including mycotoxins) on *Drosophila melanogaster* and *Spodoptora littoralis*. *Journal of Invertebrate Pathology* 50: 124-133.

STENWIG, H. 1988. Thin-layer chromatography of plugs from agar cultures as an aid for identification of moulds in routine mycological examination of animal feeds. *Acta Agriculturae Scandinavica* 38: 215-222.

THOM, C. 1930. The Penicillia. Baltimore: Williams and Wilkins.

TURNER, W.B. and D.C. ALDRIDGE. 1983. Fungal metabolites II. London: Academic Press.

DIALOGUE FOLLOWING DR FRISVAD'S PRESENTATION

PITT: SPAS (Subcommission on *Penicillium* and *Aspergillus* Systematics) is now considering circulating morphologically "typical" isolates of selected Penicillium species, to allow people to calibrate their medium and incubation systems. Perhaps we could also consider doing the same thing with a selected set of producers of secondary metabolites.

ELECTROPHORETIC COMPARISON OF ENZYMES AS A CHEMOTAXONOMIC AID AMONG *ASPERGILLUS* TAXA: (1) *ASPERGILLUS* SECTS. *ORNATI* AND *CREMEI*

J. Sugiyama and K. Yamatoya

Institute of Applied Microbiology
University of Tokyo
Bunkyo-ku, Tokyo 113, Japan

SUMMARY

Polyacrylamide slab gel electrophoresis with specific staining for five enzymes was used as a chemotaxonomic criterion for comparing 35 isolates assigned to 12 species in *Aspergillus* subgen. *Ornati* sect. *Ornati* and three species in subgen. *Circumdati* sect. *Cremei*. These were glucose-6-phosphate dehydrogenase, malate dehydrogenase, fumarase, alcohol dehydrogenase, and glutamate dehydrogenase. A numerical analysis was performed on the basis of the similarity values in patterns of five enzymes. The results are presented as a dendrogram. Isolates from different geographical and ecological habitats within the same species were identical or very similar in the enzyme patterns. In the dendrogram, isolates were divided into five major clusters. Isolates within the same species were linked to each other with a high similarity value of 73% or more, whereas each species was linked to one another with a similarity value of 15 to 70 %. The three clusters were a heterogeneous assemblage in teleomorphs or ubiquinone systems, or both. Comparisons of enzyme patterns are considered to be useful tools at the species level in the classification of these *Aspergillus* taxa.

INTRODUCTION

Traditionally, circumscription and classification among *Aspergillus* taxa have been based on differences in cultural and morphological characteristics (Raper and Fennell, 1965), and a need for newer approaches has become apparent. Gams *et al.* (1985) classified *Aspergillus* into six subgenera and 18 sections: the complexity of the genus is emphasised by the fact that this single anamorph genus is associated with eleven teleomorph genera in the Trichocomaceae. In addition to the diversity of teleomorphs, the heterogeneity of *Aspergillus* ubiquinone systems has been reported by Kuraishi *et al.* (1985, 1990). It is time that the taxonomic systems in *Aspergillus* and its associated teleomorphs be revised on the basis of modern taxonomic criteria.

One of the six *Aspergillus* subgenera, *Aspergillus* subgen. *Ornati* sect. *Ornati* is associated with four different teleomorph genera, *Dichlaena*, *Hemicarpenteles*, *Sclerocleista*, and *Warcupiella*, and includes some strictly anamorphic species as well (Table 1). This subgenus is heterogeneous with respect to ubiquinone systems, as also shown in Table 1. Three major ubiquinone systems, Q-9, Q-10, and Q-10(H$_2$), occur in subgen. *Ornati* (Kuraishi *et al.*, 1985, 1990).

Enzyme electrophoresis has been made for the classification of *Aspergillus* isolates by Nealson and Garber (1967), Nasuno (1971, 1972a-b, 1974), and Kurzeja and Garber (1973). However, this technique has not been fully evaluated as an aid in *Aspergillus* taxonomy. In this paper, the electrophoretic comparison of enzymes has been studied as a chemotaxonomic tool for the classification of *Aspergillus* subgen. *Ornati*.

Modern Concepts in Penicillium and Aspergillus Classification
Edited by R. A. Samson and J. I. Pitt
Plenum Press, New York, 1990

Table 1. Genera, species and ubiquinone systems in the *Aspergillus ornatus* group and in *Aspergillus* subgen. *Ornati*.

A. ornatus group: Raper & Fennell (1965)	Aspergillus subgen. Ornati : Gams et al. (1985)	Teleomorph genus	Ubiquinone system[b]
	A. acanthosporus	Hemicarpenteles	Q-10
A. paradoxus	A. paradoxus	Hemicarpenteles	Q- 9
A. citrisporus	A. citrisporus	Sclerocleista	Q- 9
A. ornatus	A. ornatulus	Sclerocleista	Q- 9
A. spinulosus	A. warcupii	Warcupiella	Q-10
	Aspergillus sp.	Dichlaena	
A. brunneo-uniseriatus	A. brunneo-uniseriatus		Q- 9
A. raperi	[a]A. raperi		Q-10(H$_2$)
	A. apicalis		Q-10
	A. brunneo-uniseriatus var. nanus		Q-10(H$_2$)
	[a]A. ivoriensis		Q-10(H$_2$)

[a] *Aspergillus* spp. with Hülle cells are excluded from subgenus *Ornati*.
[b] From Kuraishi *et al.* (1985, 1990).

MATERIAL AND METHODS

Isolates examined.
Thirty isolates assigned to 12 species, both teleomorph and anamorph, in *Aspergillus* subgen. *Ornati* sect. *Ornati* (= the *A. ornatus* group of Raper and Fennell, 1965) were examined. Five isolates of three *Chaetosartorya* species in subgen. *Circumdati* sect. *Cremei* (= the *A. cremeus* group) were also included for comparison. Names, culture collection numbers and other relevant information are listed in Table 2. The species names shown in this table are as received from culture collections, but the appropriate teleomorph names have been used.

Cultivation and harvest of mycelia.
The isolates were cultivated in 500 ml flasks containing 200 ml of YM broth supplemented by 1 % glucose (pH 5.8) for 48 to 60 h at 27°C on a reciprocal shaker. The mycelia were harvested by filtration and were washed twice with 0.05 M Tris-HCl buffer, pH 7.8. The harvested mycelia were stored in a freezer at -20°C until used.

Preparation of cell-free extracts for electrophoresis.
Mycelia were distrupted by sonication for 6 to 10 min at 200 W (Tomy Seiko, Model UR-200P, Tokyo, Japan) and at 5°C in the same buffer solution. Usually 10 to 15 g of wet packed mycelia were employed. The homogenates were then centrifuged twice, for 30 min at 10,000 g and 5°C, and for 90 min at 100,000 g and 5°C. The resulting supernatant was used as an enzyme source for electrophoresis. It was stored in a freezer at -20°C until just before use.

Table 2. Isolates used

Species	Isolate[a]	Source
Hemicarpenteles acanthosporus	IFO 9490(T)	Soil, Papua New Guinea
Udagawa & Takada	IFO 9607	Soil, Papua new Guinea
	JCM 2268	Soil, Papua New Guinea
	JCM 2269	Soil, Papua New Guinea
Sclerocleista ornata	NRRL 2256(T)	Soil, Wisconsin, U.S.A.
(Raper *et al.*) Subram.	= JCM 2354	
	JCM 2263	Soil, Wisconsin, U.S.A.
	CBS 425.68	Forest soil, U.S.S.R.
Hemicarpenteles paradoxus	NRRL 2162(T)	Opossum dung, New Zealand
Sarbhoy & Elphick	= IFO 8172	
	= JCM 2355	
	IFO 9127	Dog dung, U.K.
Sclerocleista thaxteri	JCM 2264(T)	Caterpillar dung, U.S.A.
Subram.	JCM 2350	Soil, Massachusetts, U.S.A.
Warcupiella spinulosa	IFO 31800(T)	Jungle soil, Borneo
Subram.	= JCM 2358	
	JCM 2270	Soil, Singapore
Chaetosartorya chrysella	NRRL 5084(T)	Forest soil, Costa Rica
(Kwon & Fennell) Subram.	NRRL 5085	Forest soil, Costa Rica
Chaetosartorya cremea	NRRL 5081(T)	Forest soil, Costa Rica
(Kwon & Fennell) Subram.		
Chaetosartorya stromatoides	NRRL 4519(T)	Soil, Panama
Wiley & Simmons	= IFO 9652	
Aspergillus apicalis	CBS 236.81(T)	Wheat bean, India
B.S.Mehrotra & Basu		
Aspergillus brunneo-uniseriatus	IFO 6993(T)	Soil, India
Singh & Bakshi	= JCM 2348	
Aspergillus brunneo-uniseriatus var. *nanus*	JCM 2349(T)	Garden soil, India
Sankaran & Zachariah		
Aspergillus ivoriensis	JCM 2353(T)	Soil, Wisconsin, U.S.A.
Bartoli & Maggi		
Aspergillus raperi Stolk	IFO 6416(T)	Garden soil, Belgian Congo
	= JCM 2356	
	JCM 2266	Garden soil, Belgian Congo

[a](T): Strain derived from the type isolate. Culture collections: IFO, Insitute for Fermentation, Osaka; JCM, Japan Collection of Microorganisms, Wako; NRRL, USDA, Northern Regional Research Center Peoria, Illinois; CBS, Centraalbureau voor Schimmelcultures, Baarn.

Electrophoresis and staining of enzymes.

Electrophoresis and staining of enzymes were carried out as described by Yamazaki and Komagata (1981) with minor modifications. An apparatus (Rapidus Slab-gel Electrophoresis AE-6220, Atto Co. Ltd., Tokyo, Japan) was used for vertical gel electrophoresis. A 3.75 % polyacrylamide slab gel, 120 x 120 x 2 mm, with 12 slots, was prepared by the method of Yamazaki (1982). Extracts (0.06 to 0.1 ml) were applied to the slots. Enzymes were separated using Tris-HCl buffer, pH 8.9 in the gel and Tris-glycine buffer, pH 8.3 in the electrotrode vessels with 15 to 25 mA current for ca 5 h at 5°C. Bromophenol blue was used as the tracking dye. After electrophoresis, the gel was removed from the mould and stained to visualize the enzymes.

Enzymes.

The following five enzymes were examined: glucose-6-phosphate dehydrogenase (G6PDH; NADP-dependent, EC 1.1.1.49), malate dehydrogenase (MDH; NAD-dependent, EC 1.1.1.37), fumarase (Fmase; NAD-dependent, EC 4.2.1.2), alcohol dehydrogenase (ADH; NAD-dependent, EC 1.1.1.1), and glutamate dehydrogenase (GDH; NADP-dependent, EC 1.4.1.4). These enzymes were chosen because they play important roles in the metabolism of glucose by moulds.

Relative mobilities of enzyme bands and numerical analysis.

The relative mobilities (Rm) of the enzyme bands were calculated as the ratio of the distance that the enzyme moved from the origin to the distance that the tracking dye moved. The similarity values in enzymatic patterns were calculated by the following formula: S (%) = 2NAB / (NA + NB) x 100 (S, similarity value; NAB, the number of enzyme bands with identical Rms; NA, the number of enzyme bands of isolate A; NB, the number of enzyme bands of isolate B). These similarity values were averaged to find the overall similarity between isolates. Clustering was achieved by means of the unweighted-average linkage technique (Sneath and Sokal, 1973). Computations were performed on a FACOM M-380 computer in the Institute of Physical and Chemical Research (RIKEN), Wako-shi, Saitama Pref. using a FORTRAN program written by Dr. T. Kaneko. The results of the numerical analysis are presented as a dendrogram.

RESULTS AND DISCUSSION

Figure 1 shows representative examples of G6PDH and GDH electrophoretic patterns for the isolates of *Hemicarpenteles acanthosporus*, *Sclerocleista thaxteri*, *S. ornata*, *Chaetosartorya chrysella* and *Aspergillus apicalis*. Isolates from different geographical and ecological habitats within the same species showed identical or very similar enzyme patterns. GDH produced several enzyme bands which were characteristic in each species. Table 3 shows Rm's of five enzymes in 25 isolates of *Aspergillus* and three associated teleomorphs, *Hemicarpenteles*, *Sclerocleista*, and *Warcupiella* in subgen. *Ornati* sect. *Ornati*, and five isolates of three *Chaetosartorya* species with *Aspergillus* anamorphs in subgen. *Circumdati* sect. *Cremei*.

The G6PDH, MDH, Fmase, ADH, and GDH patterns were characteristic for each species, and clear differences were detected bewteen species. G6PDH and MDH produced one, two or three bands. Fmase and ADH produced one or two bands. GDH produced a single band. No enzyme band for Fmase and ADH was present in the isolates of *Hemicarpenteles acanthosporus*, *Warcupiella spinulosa*, *Chaetosartorya chrysella*, *C. cremea*,

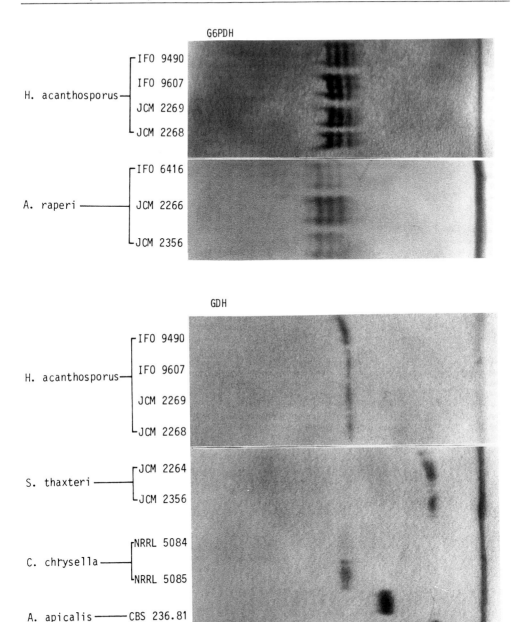

Figure 1. G6PDH and GDH electrophoretic patterns for 13 strains in *Aspergillus* sects. *Ornati* and *Cremei*.

Table 3. Comparison of electrophoretic Rm values of enzymes from 30 strains in *Aspergillus* sects. *Ornati* and *Cremei*.

		Enzymes (Rm x 100)				
		G6PDH	MDH	Fmase	ADH	GDH
Hemicarpenteles acanthosporus	IFO 9490	52,55,59	54,56,58	49	—[a]	54
	IFO 9607	52,55,59	54,56,58	49	–	54
	JCM 2268	52,55,59	54,56,58	49	–	54
	JCM 2269	52,55,59	54,56,58	49	–	54
Hemicarpenteles paradoxus	NRRL 2162	49,51	64,77	81	50,53	70
	IFO 8172	49,51	64,77	81	50,53	70
	IFO 9127	49,51	64,77	81	50,53	74
	JCM 2355	49,51	64,77	81	50,53	70
Sclerocleista ornata	NRRL 2256	57,61	72,76,79	75,80	78	87
	JCM 2263	57,61	72,76,79	75,80	78	87
	JCM 2354	57,61	72,76,79	75,80	78	87
	CBS 425.68	57	72,76,79	75,80	74	87
Sclerocleista thaxteri	JCM 2264	52,56	66,71,74	67,74	75	84
	JCM 2350	52,56	66,71,74	67,74	75	84
Warcupiella spinulosa	IFO 31800	64,68	43,49	50,58	–	67
	JCM 2270	64,68	43,49	61,66	–	67
	JCM 2358	64,68	43,49	50,58	–	67
Chaetosartorya chrysella	NRRL 5084	64	56,60	–	–	57
	NRRL 5085	64	56,60	–	–	57
Chaetosartorya cremea	NRRL 5180	63	55,58	–	–	61
Chaetosartorya stromatoides	NRRL 4519	68,72	60,63	50	–	55
	IFO 9652	68,72	60,63	50	–	55
Aspergillus apicalis	CBS 236.81	69,71	53,57	54	31,33	69
Aspergillus brunneo-uniseriatus	IFO 6993	61,63	66	53	–	72
	JCM 2348	55,57	66	53	–	72
Aspergillus brunneo-uniseriatus var. *nanus*	JCM 2349	54,57	48,52	58	–	84
Aspergillus ivoriensis	JCM 2353	55,59	56,69	–	–	65
Aspergillus raperi	IFO 6416	52,55,58	49,51	64	–	68
	JCM 2266	52,55,58	49,51	64	–	68
	JCM 2356	52,55,58	49,51	64	–	68

[a] Not detected.

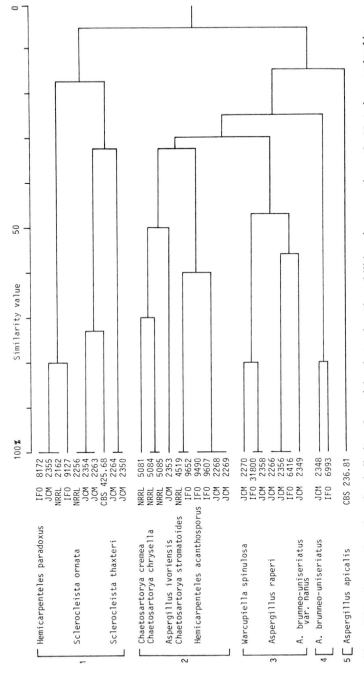

Figure 2. Dendrogram based on the similarity values of the electrophoretic mobilities of enzymes from the 30 strains examined in *Aspergillus* sects. *Ornati* and *Cremei*.

C. stromatoidea, Aspergillus brunneo-uniseriatus var. *nanus*, and *A. raperi*. The isolates within the same species produced identical or very similar patterns for each enzyme. *Sclerocleista ornata* CBS 425.68 had a single band for G6PDH and a different band (*Rm* 0.74) for ADH. Three other isolates of *S. ornata* showed identical patterns for all enzymes. *Warcupiella spinulosa* JCM 2270 showed a different Fmase pattern. The two isolates IFO 6993 and JCM 2348 of *A. brunneo-uniseriatus* produced different patterns for G6PDH.

The relative mobilities for five enzymes from the isolates tested are shown as a dendrogram to emphasize the similarities in enzyme patterns (Fig. 2). The isolates tested separated into five major clusters. Isolates in the same species were linked to each other with similarity values of 80 % or more except for *S. ornata* isolates (73% similarity). Species were linked with similarity values of 15 to 70%.

Cluster 1 included *Hemicarpenteles paradoxus, Sclerocleista ornata* and *S. thaxteri*. These showed little relationship in enzyme patterns, with similarity coefficients of only 18% and 32.5%. *H. acanthosporus*, the second *Hemicarpenteles* species examined, clustered quite separately from *H. paradoxus*. As *H. paradoxus* has the Q-9 ubiquinone system, and *H. acanthosporus* the Q-10 (Kuraishi *et al.*, 1985, 1990), a strong case exists for exclusion of *H. acanthosporus* from *Hemicarpenteles*. It showed a much higher relationship in enzyme similarity profiles with *S. stromatoides*, which, however, also has the Q-9 system.

Cluster 2 included isolates of three *Chaetosartorya* species, *Aspergillus ivoriensis* and *Hemicarpenteles acanthosporus*. *Chaetosartorya cremea* and *C. chrysella* were linked with a similarity value of 70%, while *C. stromatoides* was only distantly related, with a similarity value of 32.5%. *A. ivoriensis*, tentatively placed in sect. *Ornati* by Samson (1979) despite its production of Hülle cells, showed 50% similarity to *C. cremea* and *C. chrysella*, but less to other species. Although Gams *et al.* (1985) excluded species with Hülle cells from sect. *Ornati*, our data on this point are equivocal.

Cluster 3 included the isolates examined from *Warcupiella spinulosa, Aspergillus raperi* and *A. brunneo-uniseriatus* var. *nanus*. Intraspecific variation within this cluster was small, but the different species showed only a low level of similarity: 56% for *A. raperi* and *A. brunneo-uniseriatus* var. *nanus*, and 48% for *A. raperi* and *A. spinulosa*. *A. brunneo-uniseriatus* clustered alone (Cluster 4), and the two isolates examined, IFO 6693 and JCM 2348, were linked with a similarity value of only 80%. The two isolates were culturally and microscopically quite different, particularly in the shape and size of phialides and conida. The identity of JCM 2348 is therefore questionable. *A. brunneo-uniseriatus* clustered quite separately from *A. brunneo-uniseriatus* var. *nanus*, with a similarity value of only 25%. As these taxa also have different ubiquinone systems, Q-9 and Q-10 (H_2) respectively (Kuraishi *et al.*, 1990), their relationship is very doubtful.

The final cluster, 5, with *A. apicalis*, showed a very low level of relationship to any other species. Placement in sect. *Ornati* appears questionable.

Several conclusions may be drawn from the results shown here. Taxa in subgen. *Ornati* are heterogeneous in their enzyme patterns as well as their teleomorph-anamorph connections and ubiquinone systems. However, the enzyme patterns were characteristic for each species. Therefore, electrophoretic comparisons of enzymes will be of value in distinguishing species, and may provide valuable information for subgeneric taxonomy of *Aspergillus* and associated teleomorphs also.

ACKNOWLEDGEMENTS

We thank the curators of culture collections from which cultures were examined: CBS,

IFO, JCM, and NRRL. We also thank Dr. T. Kaneko, Tokyo University of Agriculture, Tokyo and Dr. K. Suzuki, Japan Collection of Microorganisms, The Institute of Physical and Chemical Research, Wako for the computing and Dr. M. Yamazaki for helpful comments on the electrophoresis. A Grant-in-Aid for Co-operative Research (A)(No. 61300005) from the Ministry of Education, Science and Culture, Japan to J.S., which partly supported this research is gratefully acknowledged.

REFERENCES

GAMS, W., CHRISTENSEN, M., ONIONS, A. H., PITT, J. I. and SAMSON, R. A. 1985. Infrageneric taxa of *Aspergillus*. In Advances in *Penicillium* and *Aspergillus* Systematics, eds. R. A. Samson and J. I. Pitt, pp. 55-62. New York and London: Plenum Press.

KURAISHI, H., KATAYAMA-FUJIMURA, Y., SUGIYAMA, J. and YOKOYAMA, T. 1985. Ubiquinone systems in fungi. I. Distribution of ubiquinones in the major families of Ascomycetes, Basidiomycetes, and Deuteromycetes, and their taxonomic implications. *Transactions of the Mycological Society of Japan* 26: 383-395.

KURAISHI, H., ITOH, M., TSUZAKI, N., KATAYAMA, Y. YOKOYAMA, T. and SUGIYAMA, J. 1990. Ubiquinone systems as a taxonomic tool in *Aspergillus* and its teleomorphs. *In* Modern Concepts in *Penicillium* and *Aspergillus* Classification, eds. R. A. Samson and J. I. Pitt, pp. 407-420. New York and London: Plenum Press.

KURZEJA, K.C. and GARBER, E. D. 1973. A genetic study of electrophoretically variant extracellular amylolytic enzymes of wild-type strains of *Aspergillus nidulans*. *Canadian Journal of Genetics and Cytology* 15: 275-287.

NASUNO, S. 1971. Polyacrylamide gel disc electrophoresis of alkaline proteinases from *Aspergillus* species. *Agricultural and Biological Chemistry* 35: 1147-1150.

NASUNO, S. 1972a. Differentiation of *Aspergillus sojae* from *Aspergillus oryzae* by polyacrylamide gel disc electrophoresis. *Journal of General Microbiology* 71: 29-33.

NASUNO, S. 1972b. Electrophoretic studies of alkaline proteinases from strains of *Aspergillus flavus* group. *Agricultural and Biological Chemistry* 36: 684-689.

NASUNO, S. 1974. Further evidence on differentiations of *Aspergillus sojae* from *Aspergillus oryzae* by electrophoretic patterns of cellulase, pectin-lyase, and acid proteinase. *Canadian Journal Microbiology* 20: 413-416.

NEALSON, K. H. and GARBER, E. D. 1967. An electrophoretic survey of esterases, phosphatases, and leucine amino-peptidases in mycelial extracts of species of *Aspergillus*. *Mycologia* 59: 330-336.

RAPER, K. B. and FENNELL, D. I. 1965. The Genus *Aspergillus*. Baltimore: Williams & Wilkins.

SAMSON, R. A. 1979. A compilation of the Aspergilli described since 1965. *Studies in Mycology, Baarn* 18: 1-38.

SNEATH, P. H. A. and SOKAL, R. R. 1973. Numerical Taxonomy. 2nd edition. San Francisco: W. H. Freeman.

YAMAZAKI, M. 1982. Electrophoretic patterns. *In* Biseibutu no Kagaku Bunrui Jikkenho (Chemotaxonomic Methods for Microorganisms), ed. K. Komagata, pp. 184-209. Tokyo: Gakkai Shuppan Center.

YAMAZAKI, M. and KOMAGATA, K. 1981. Taxonomic significance of electrophoretic comparison of enzymes in the genera *Rhodotorula* and *Rhodosporidium*. *International Journal of Systematic Bacteriology* 31: 361-381.

ELECTROPHORETIC COMPARISON OF ENZYMES AS A CHEMOTAXONOMIC AID AMONG *ASPERGILLUS* TAXA: (2) *ASPERGILLUS* SECT. *FLAVI*

K. Yamatoya[1], J. Sugiyama[1] and H. Kuraishi[2]

[1]*Institute of Applied Microbiology*
University of Tokyo
Bunkyo-ku, Tokyo 113, Japan

[2]*Faculty of Agriculture*
University of Agriculture and Technology
Fuchu, Tokyo 183, Japan

SUMMARY

Polyacrylamide slab gel electrophoresis with specific staining for eight enzymes was used as a chemotaxonomic aid in comparing 41 isolates in *Aspergillus* subgen. *Circumdati* sect. *Flavi*. The eight enzymes were 6-phosphogluconate dehydrogenase, malate dehydrogenase, phosphoglucomutase, glucose-6-phosphate dehydrogenase, fumarase, alcohol dehydrogenase, lactate dehydrogenase, and glutamate dehydrogenase. Numerical analysis was applied to similarity values of elctrophoretic relative mobilities of the eight enzymes. Five major clusters were obtained. *A. flavus*, *A. kambarensis*, *A. toxicarius*, and other very closely related *Aspergillus* taxa formed one major cluster at a 63% similarity level. This cluster could be divided into two subclusters, corresponding to *A. flavus* and *A. parasiticus*. *A. tamarii*, *A. leporis*, *A. nomius*, and *A. avenaceus* formed the separate major clusters at a similarity level of 38% or less; each cluster corresponded to a single species. The ubiquinone systems of 27 isolates were also determined. *A. avenaceus*, *A. flavus* var. *asper* and *A. leporis* possessed the Q-10 ubiquinone system, while all others had Q-10(H_2). The heterogeneity in the ubiquinone systems suggests that *Aspergillus* sect. *Flavi* is in need of revision.

INTRODUCTION

Aspergillus subgen. *Circumdati* sect. *Flavi* (Gams *et al.*, 1985), previously known as the '*A. flavus* group' (Raper and Fennell, 1965), includes industrially, agronomically and medically important species. Hence a great deal of interest persists in the classification of these species.

Recently Kurtzman *et al.* (1986) studied the DNA base composition and DNA-DNA homology of 17 isolates of *A. flavus*, *A. parasiticus*, *A. oryzae* and *A. sojae*. All four species had high (69-100 %) nuclear DNA complementarity and a similar genome size. They interpreted their data as indicating that these taxa represented a single species. They proposed the following taxonomic designations, taking into account morphological differences: *A. flavus* subsp. *flavus* var. *flavus*, *A. flavus* subsp. *flavus* var. *oryzae*, *A. flavus* subsp. *parasiticus* var. *parasiticus* and *A. flavus* subsp. *parasiticus* var. *sojae*. Klich and Mullaney (1987) and Klich and Pitt (1988) have disagreed with this taxonomic approach.

In this paper, we present data on electrophoretic relative mobility values of eight enzymes in 41 isolates in *Aspergillus* sect. *Flavi*, and attempt to evaluate these in relation to the conflicting concepts of speciation mentioned above, and our zymogram studies in *Aspergillus* subgen. *Ornati* (Sugiyama and Yamatoya, 1990). We also wished to compare our data to modern chemotaxonomic approaches such as DNA base composition, DNA-DNA homology, and ubiquinone systems.

Modern Concepts in Penicillium and Aspergillus Classification
Edited by R. A. Samson and J. I. Pitt
Plenum Press, New York, 1990

Table 1. Isolates studied.

Species	Isolate	Source
A. avenaceus	IAM 13120	Seed peas, London, U.K.
	IAM 13121[T]	Seed peas, London, U.K.
A. flavus var. *flavus*	NRRL 482	Wehmer's *A. flavus* isolate
	NRRL 1957[NT]	Mouldy cellophane, South Pacific
	NRRL 3251	Walnuts, U.S.A.
	NRRL 13135	Mouldy peanuts, U.S.A.
	NRRL 13136	Kangaroo rat cheek pouch, U.S.A.
	IAM 2670	Shoyu-koji, Chiba, Japan
A. flavus var. *asper*	IAM 13119[T]	Butter, Hokkaido, Japan
A. flavus var. *oryzae*	NRRL 447[NT]	Sake-koji, Japan
	NRRL 451	Soy Sauce, China
A. oryzae	IAM 2630	Miso-koji, Korea
	IAM 2735	Sake-koji, Kyoto, Japan
	IAM 12166	Okinawa, Japan
	IAM 13833	Cassava, South Africa
A. oryzae var. *magnasporus*	IAM 2660	Amazake-koji, Tokyo, Japan
	IAM 2719	Shoyu-koji, Tokyo, Japan
A. oryzae var. *microsporus*	IAM 2683	Shoyu-koji, Tokyo, Japan
	IAM 2699	Tamari-koji, Japan
A. oryzae var. *micro-vesiculosus*	IAM 2955	Shoyu-koji, Kumamoto, Japan
A. oryzae var. *pseudoflavus*	IAM 2956	K. Saito (IFO 4083)
A. oryzae var. sporofallus	IAM 2957	Miso-koji, Nagoya, Japan
A. oryzae var. tenuis	IAM 2958	T. Takahasi (IFO 4134)
A. oryzae var. *viridis*	IAM 2750	Shoyu-koji, Tottori, Japan
A. flavus var. *parasiticus*	NRRL 465	J. Takamine's Lab., Japan
	NRRL 502[T]	Sugarcane, Hawaii
A. parasiticus	IAM 2150	ATU A-8-3 = CLMR
A. flavus var. *sojae*	NRRL 5595[T]	Koji, Japan ?
	NRRL 5594	Koji, Manchuria ?
A. sojae	IAM 2631	P Shoyu-koji, Tokyo, Japan
	IAM 2669	Shoyu-koji, Chiba, Japan
A. kambarensis	IAM 12768[T]	Core sample, Niugata, Japan
A. leporis	NRRL 3216[T]	Dung of rabbit, U.S.A.
A. nomius	NRRL 3161	*Cycas circinalis*, Polynesia
	NRRL 5919	Diseased alkalibes, U.S.A.
	NRRL 13137[T]	Mouldy wheat, U.S.A.
A. tamarii	NRRL 429	Soy sauce, China
	NRRL 13139	Kangaroo rat cheek pouch, U.S.A.
	IAM 2138	Amazake-koji, Japan
A. toxicarius	JCM 2252	ATCC 15517 (as *A. parasiticus*)
	JCM 2253[T]	Groundnuts, Uganda

[T] Strain derived drom the type; [NT] strain derived from neotype.

Culture collections: IAM, Institute for Applied Microbiology, University of Tokyo, NRRL, USDA, Northern Regional Research Center, Peoria, Illinois; JCM, Japan Collection of Microorganisms, Wako.

MATERIAL AND METHODS

Isolates examined.

Forty-one isolates assigned to *Aspergillus* sect. *Flavi* were examined, including 17 NRRL isolates used by Kurtzman *et al.* (1986, 1987) to obtain data on DNA base composition and DNA relatedness values. Names, culture collection numbers, and ecological and geographical sources are listed in Table 1. Most species names were used according to their designation when they were received from culture collections.

Electrophoretic comparison of enzymes.

The cultivation, harvest of mycelia, electrophoresis and staining procedures of eight enzymes used in this study were as described previously (Sugiyama and Yamatoya, 1990). The numerical analysis based on the enzyme patterns was also made as described previously (Sugiyama and Yamatoya, 1990).

In addition to the five enzymes examined previously in isolates of *Aspergillus* subgen. *Ornati* (Sugiyama and Yamatoya, 1990), three further enzymes were studied. These were 6-phosphogluconate dehydrogenase (6PGDH; NADP-dependent, EC 2.7.5.1), phosphoglucomutase (PGm; NADP-dependent, EC 1.1.1.41) and lactate dehydrogenase (LDH; NAD-dependent, EC 1.1.1.27).

Extraction and analysis of ubiquinones.

Ubiquinone systems for 27 *Aspergillus* isolates were determined as described previously (Kuraishi *et al.*, 1985). The abbreviations used for ubiquinones were indicated as described previously (Sugiyama and Yamatoya, 1990).

RESULTS AND DISCUSSION

Summarized in Table 2 are the ubiquinone systems in 27 isolates assigned to of *Aspergillus* subgen. *Circumdati* sect. *Flavi* obtained in this study together with some data reported by Kuraishi *et al.* (1990). The four isolates of *A. avenaceus, A. flavus* var. *asper* and *A. leporis* examined possessed the Q-10 ubiquinone system. All others examined possessed the Q-10(H_2) system.

Figure 1 shows representative examples of G6PDH and 6PGDH patterns for seven *Aspergillus* isolates from the IAM, JCM and NRRL culture collections. The isolates of five taxa produced identical enzyme patterns for G6PDH, whereas these isolates were clearly divided into two groups for 6PGDH.

Table 3 presents data on the *Rm* values for eight enzymes from 41 isolates tested in *Aspergillus* sect. *Flavi*. On the whole, isolates within the same taxon were identical or showed very similar patterns for all enzymes. G6PDH produced three enzyme bands for each isolate. MDH produced one, two or three bands. Fmase produced two bands in most isolates; *A. nomius* produced a single band, while Fmase was not detected in *A. leporis*. ADH and LDH produced one or two bands, except that LDH was present in *A. leporis*. 6PGDH, PGm and GDH produced a single band in most isolates; however, no 6PGDH bands were seen in *A. avenaceus* or *A. leporis*.

A dendrogram based on calculated similarity values of eight enzymes in 41 isolates, examined by the unweighted-average linkage method, is shown in Fig. 2. This dendrogram was drawn from all enzyme bands detected. Figure 3 shows a dendrogram

Table 2. Ubiquinone in *Aspergillus* isolates examined.

Species	Isolate	Ubiquinone system
A. avenaceus	IAM 13120	Q-10*
	IAM 13121(T)	Q-10*
A. flavus var. flavus	NRRL 482	Q-10(H$_2$)
	NRRL 1957(NT)	Q-10(H$_2$)
	NRRL 3251	Q-10(H$_2$)
	NRRL 13135	Q-10(H$_2$)
	NRRL 13136	Q-10(H$_2$)
A. flavus var. asper	IAM 13119(T)	Q-10*
A. flavus var. oryzae	NRRL 447(NT)	Q-10(H$_2$)
	NRRL 451	Q-10(H$_2$)
A. oryzae var. magnasporus	IAM 2660	Q-10(H$_2$)
	IAM 2719	Q-10(H$_2$)
A. oryzae var. microsporus	IAM 2683	Q-10(H$_2$)
	IAM 2699	Q-10(H$_2$)
A. oryzae var. micro-vesiculosus	IAM 2955	Q-10(H$_2$)
A. oryzae var. pseudoflavus	IAM 2956	Q-10(H$_2$)
A. oryzae var. sporofallus	IAM 2957	Q-10(H$_2$)
A. oryzae var. tenuis	IAM 2958	Q-10(H$_2$)
A. oryzae var. viridis	IAM 2750	Q-10(H$_2$)
A. flavus var. parasiticus	NRRL 465	Q-10(H$_2$)
	NRRL 502(T)	Q-10(H$_2$)
A. flavus var. sojae	NRRL 5594	Q-10(H$_2$)
	NRRL 5595(T)	Q-10(H$_2$)
A. kambarensis	IAM 12768(T)	Q-10(H$_2$)
A. leporis	NRRL 3216(T)	Q-10
A. nomius	NRRL 3161	Q-10(H$_2$)
	NRRL 5919	Q-10(H$_2$)
	NRRL 13137(T)	Q-10(H$_2$)
A. tamarii	NRRL 429	Q-10(H$_2$)
	NRRL 13139	Q-10(H$_2$)
A. toxicarius	JCM 2252	Q-10(H$_2$)*

Footnotes: see Table 1.
*: Data from Kuraishi et al. (1990).

based on only the deeply stained bands for eight enzymes; those with underlines in Table 3 were omitted as weakly staining or unstable. No significant differences were found between the two dendrograms.

The isolates tested separated into five major clusters, as shown in Figs. 2 and 3. Thirty-two isolates assigned to *A. flavus* and other closely related taxa formed one major cluster (Cluster 1) at a 63% similarity level (Fig. 2). This cluster was further divided into two subclusters, 1-a and 1-b. Subcluster 1-a contained twenty-two isolates of *A. flavus*, *A. flavus* var. *asper*, *A. flavus* var. *flavus*, *A. flavus* var. *oryzae*, *A. oryzae* and most of its varieties, and *A. kambarensis*. These isolates in this sub-cluster were linked to each other with a high similarity values of 73 % or more. Also, Subcluster 1-a was composed of the isolates having only the Q-10(H$_2$) system except *A. flavus* var. *asper* IAM 13119 (Table 2).

Subcluster 1-b contained 10 isolates of *A. flavus* var. *parasiticus*, *A. flavus* var. *sojae*, *A. oryzae* var. *magnasporus* IAM 2719, *A. parasiticus*, *A. sojae*, and *A. toxicarius*. These isolates were linked to each other with a similarity value of 86% or more. All isolates in this

G6PDH

A. oryzae v. magnasporus	IAM 2260	
	IAM 2719	
A. oryzae	IAM 2735	
A. toxicarius	JCM 2252	
	JCM 2253	
A. flavus v. parasiticus	NRRL 502	
A. flavus v. flavus	NRRL 1957	

6PGDH

A. oryzae v. magnasporus	IAM 2260	
	IAM 2719	
A. oryzae	IAM 2735	
A. toxicarius	JCM 2252	
	JCM 2253	
A. flavus v. parasiticus	NRRL 502	
A. flavus v. flavus	NRRL 1957	

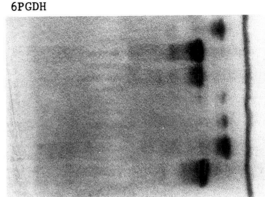

Figure 1. G6PDH and 6PGDH electrophoretic pattern for seven isolates in *Aspergillus* subgen. *Circumdati* sect. *Flavi*.

subcluster were of the same Q-10(H$_2$) system as in Subcluster 1-a (Table 2). Therefore, the isolates in these subclusters can be regarded as an electrophoretically homogeneous taxonomic group which corresponds to a single species. However, further taxonomic investigations will be required to determine the taxonomic position of *A. flavus* var. *asper* having the Q-10(H$_2$) system.

Cluster 2 contained three isolates of *A. tamarii* having the Q-10(H$_2$) system. Enzyme patterns for these isolates showed more than more than 94% similarity. This cluster was linked to Cluster 1 with a relatively low similarity value of 37%.

Cluster 3 contained only *A. leporis* NRRL 3216, which was characterized by us as having the Q-10 system (Table 2). This cluster was linked to Cluster 2 with a low similarity value of 27%.

Cluster 4 contained three isolates of *A. nomius*, of which two isolates (NRRL 13137 and NRRL 3161) had identical enzyme patterns (100% similarity). Isolate NRRL 5919 was linked to these two isolates with a similarity value of 73%. The three isolates of *A. nomius* were characterized by the same Q-10(H$_2$) system as Clusters 1 and 2. *A. nomius* appears to be a distinct species.

Table 3. Comparison of electrophoretic Rm values of enzymes from 41 isolates in *Aspergillus* subgen. *Circumdati* sect. *Flavi*

Species	Strains	Enzymes (*Rm* x100)							
		G6PDH	MDH	Fmase	ADH	LDH	6PGDH	PGm	GDH
A. avenaceus	IAM 13120	55,59,62	61,65,68	65,67	77,81	80,83	—	78	80
	IAM 13121	55,59,62	61,65,68	65,67	77,81	80,83	—	78	80
A. flavus v. flavus	NRRL 482	62,65,67	40,45,48	76,79	59,72	73,83	80	74	71
	NRRL 1957	62,65,67	40,45,48	76,79	59,72	73,83	80	74	71
	NRRL 3251	62,65,67	40,45,48	76,79	72	83	80	74	71
	NRRL 13135	62,65,67	40,45,48	76,79	72	83	80	74	71
	NRRL 13136	62,65,67	40,45,48	76,79	72	83	80	74	71
A. flavus	IAM 2670	62,65,67	40,45,48	76,79	72	83	80	74	71
A. flavus v. asper	IAM 13119	62,65,67	40,43	76,79	73	81,83	80	74	71
A. flavus v. oryzae	NRRL 447	62,65,67	40,45,48	76,79	57,72	72,83	80	74	71
	NRRL 451	62,65,67	40,45,48	76,79	57,72	83	82	74	71
A. oryzae	IAM 2630	62,65,67	40,45,48	76,79	72	83	80	74	71
	IAM 2735	62,65,67	40,45,48	76,79	72	83	80	74	71
	IAM 12166	62,65,67	40,44,48	76,79	57,72	73,83	80	74	71
	IAM 13883	62,65,67	40,44,48	76,79	57,72	73,80	91	74	71
A. oryzae v. magnasporus	IAM 2660	62,65,67	40,44,48	76,79	57,72	73,83	78	74	71
	IAM 2719	62,65,67	40,44	76,79	69,72	73,83	91	69	71
A. oryzae v. microsporus	IAM 2683	62,65,67	40,44,48	76,79	69,72	73,83	80	74	71
	IAM 2699	62,65,67	40,44,48	76,79	72	73,83	80	74	71
A. oryzae v. micro-vesiculosus	IAM 2955	62,65,67	40,44	76,79	79	83	80	74	71
A. oryzae v. pseudoflavus	IAM 2956	62,65,67	40,45,48	76,79	72	83	80	74	71
A. oryzae v. soprofallus	IAM 2957	62,65,67	40,45,48	76,79	72	83	80	74	71
A. oryzae v. tenuis	IAM 2958	62,65,67	40,45,48	76,79	72	83	78	74	71
A. oryzae v. viridis	IAM 2750	62,65,67	40,45	76,79	72	83	80	74	71
A. flavus v. parasiticus	NRRL 465	62,65,67	38,42	76,79	72	84	91	69	71
	NRRL 502	62,65,67	38,42	76,79	72	84	91	69	71
A. parasiticus	IAM 2150	62,65,67	38,42	76,79	72	84	91	69	71
A. flavus v. sojae	NRRL 5595	62,65,67	39,42	76,79	59,72	82,84	91	69	71
	NRRL 5594	62,65,67	39,42	76,79	72	82,84	91	69	71
A. sojae	IAM 2631	62,65,67	40,44	76,79	72	82,84	91	69	71
	IAM 2669	62,65,67	39,42	76,79	72	82,84	91	69	71
A. kambarensis	IAM 12768	62,65,67	40,45,48	76,79	73	83	80	74	71
A. leporis	NRRL 3216	59,61,65	59	—	72	—	—	83	70,74
A. nomius	NRRL 3161	56,59,61	40,49,57	73	70	69	87	62,66	70
	NRRL 5919	56,59,61	40,49	73	56	57	87	62,66	70
	NRRL 13137	56,59,61	40,49,57	73	70	69	87	62,66	70
A. tamarii	NRRL 429	62,65,67	50,54,63	76,78	56,65	75,86	72	72	71
	NRRL 13139	62,65,67	50,54,63	76,78	64	86	72	72	71
	IAM 2138	62,65,67	50,54,63	76,78	64	75,86	72	72	71
A. toxicarius	JCM 2252	62,65,67	40,44	76,79	72	83	91	69	71
	JCM 2253	62,65,67	40,44	76,79	72	83	91	69	71

— : Not detected.

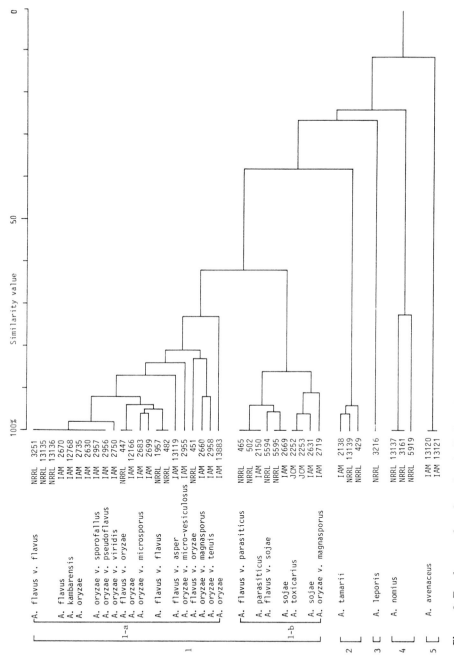

Figure 2. Dendrogram based on the similarity values of the electrophoretic mobilities of enzymes from 41 isolates examined in *Aspergillus* subg. *Circumdati* secti *Flavi*; this dendrogramn was drawn from all enzyme bands detected.

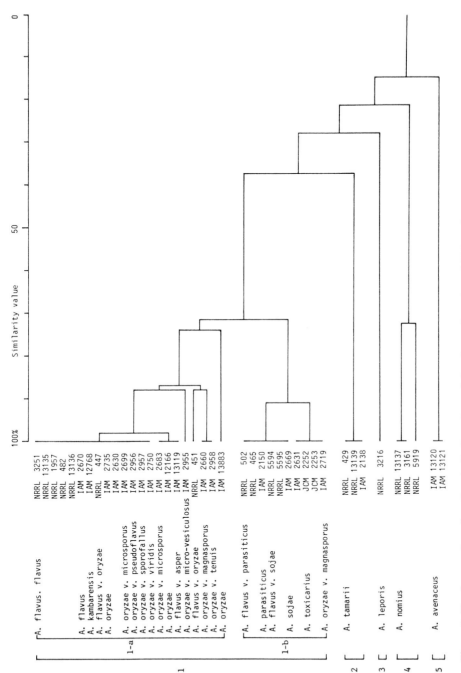

Figure 3. Dendrogram based on the similarity values of the electrophoretic mobilities of enzymes from 41 isolates examined in *Aspergillus* subgen. *Circumdati* sect. *Flavi*; this dendrogram was drawn from deeply stained bands for the respective enzyme.

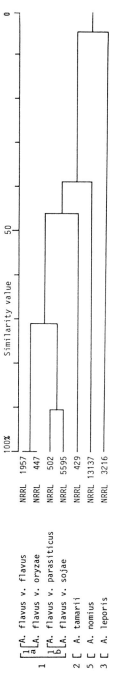

Figure 4. Dendrogram based on the data of DNA-DNA homology between *A. flavus* and other phenotypically similar species reported by Kurtzman *et al.* (1987).

Cluster 5 contained two isolates of *A. avenaceus*, with identical enzyme patterns and the same Q-10 system as in *A. flavus* var. *asper*, and *A. leporis* (Table 2). This cluster was linked to Cluster 4 with a low similarity value of 12% (Fig. 2).

As already mentioned, recently Kurtzman *et al.* (1986) studied taxonomic relationship between the isolates of *A. flavus* and six phenotypically similar species in the *A. flavus* group by DNA base composition and DNA-DNA homology. G+C ratios in the DNAs of these isolates ranged from 48.6 to 50.4 mol%. Their data showed no significant difference in the DNA base composition. However, DNA relatedness values reported by Kurtzman *et al.* (1986, 1987), divided the seven *Aspergillus* taxa into four major clusters, and we present this data as a dendrogram here (Fig. 4). A good correlation was found between the two dendrograms (Figs. 2 and 3) based on the similarity values in enzyme patterns obtained in this study and the dendrogram (Fig. 4) based on the DNA-DNA homology values reported by Kurtzman *et al.* (1986, 1987). These correlations support a conclusion that the respective major cluster in Figs. 2 and 3 correponds to a single species. We suggest that *A. flavus*, *A. oryzae*, *A. parasiticus*, *A. sojae*, and other closely related taxa in Cluster 1 are accommodated in a single species, *A. flavus* and *A. parasiticus*. From our data, the three taxa *A. flavus* var. *asper*, *A. leporis* and *A. avenaceus*, characterized by the Q-10 system, should be excluded from *Aspergillus* sect. *Flavi*.

Combinations of chemotaxonomic criteria such as electrophoretic comparison of enzymes, ubiquinone systems, DNA base composition, and DNA-DNA homology can support or question taxonomic decisions in recent years based mainly on morphological species concepts (e.g., Murakami, 1971; Christensen, 1981; Murakami *et al.*, 1982; Klich and Pitt, 1985, 1988). Such chemotaxonomic methods will be a valuable tool for clarification of interrelation among *Aspergillus* taxa lacking teleomorphs.

In conclusion, we emphasise the usefulness of electrophoretic comparisons of enzymes and the ubiquinone system for classifying taxa of *Aspergillus* and associated teleomorph genera.

ACKNOWLEDGEMENTS

We thank Dr. T. Nakase (JCM) and Dr. S. W. Peterson (NRRL) for supplying the cultures. We also thank Dr. T. Kaneko, Tokyo University of Agriculture, Tokyo and Dr. K. Suzuki, Japan Collection of Microorganisms, The Institute of Physical and Chemical Research, Wako for their helpful suggestions with respect to the computation of data and Dr. M. Yamazaki for helpful comments on the electrophoresis. Grateful acknowledgement is made of a Grant-in-Aid for Co-operative Research (A)(No. 61300005) from the Ministry of Education, Science and Culture, Japan to J. S., which partly supported this research.

REFERENCES

CHRISTENSEN, M. 1981. A synoptic key and evaluation of species in the *Aspergillus flavus* group. *Mycologia* 73: 1056-1084.

GAMS, W., CHRISTENSEN, M., ONIONS, A. H., PITT, J. I. and SAMSON, R. A. 1985. Infrageneric taxa of *Aspergillus*. *In* Advances in *Penicillium* and *Aspergillus* Systematics, eds. R. A. Samson and J. I. Pitt, pp. 55-62. New York and London: Plenum Press.

KLICH, M.A. and MULLANEY, E.J. 1987. DNA restriction enzyme fragment polymorphism as a tool for rapid differentiation of *Aspergiillus flavus* from *Aspergillus oryzae*. *Experimental Mycology* 11: 170-175.

KLICH, M. A. and PITT, J. I. 1985. The theory and practice of distinguishing species of the *Aspergillus flavus* group. *In* Advances in *Penicillium* and *Aspergillus* Systematics, eds. R. A. Samson and J. I. Pitt, pp. 211-220. New York and London: Plenum Press.

KLICH, M. A. and PITT, J. I. 1988. Differentiation of *Aspergillus flavus* from *A. parasiticus* and other closely related species. *Transactions of the British Mycological Society* 91: 99-108.

KURAISHI, H., KATAYAMA-FUJIMURA, Y., SUGIYAMA, J. and YOKOYAMA, T. 1985. Ubiquinone systems in fungi. I. Distribution of ubiquinones in the major families of Ascomycetes, Basidiomycetes and Deuteromycetes and their taxonomic implications. *Transactions of the Mycological Society of Japan* 26: 383-395.

KURAISHI, H., ITOH, M., TSUZAKI, N., KATAYAMA, Y., YOKOYAMA, T. and SUGIYAMA, J. 1990. Ubiquinone systems as a taxonomic tool in *Aspergillus* and its teleomorphs. *In* Modern Concepts in *Penicillium* and *Aspergillus* Classification, eds R. A. Samson and J. I. Pitt, pp. 407-420. New York and London: Plenum Press.

KURTZMAN, C. P., SMILEY, M. J., ROBNETT, C. J. and WICKLOW, D. T. 1986. DNA relatedness among wild and domesticated species in the *Aspergillus flavus* group. *Mycologia* 78: 955-959.

KURTZMAN, C. P., HORN, B. W., and HESSELTINE, C. W. 1987. *Aspergillus nomius*, a new aflatoxin-producing species related to *Aspergillus flavus* and *Aspergillus tamarii*. *Antonie van Leeuwenhoek* 53: 147-158.

MURAKAMI, H. 1971. Classification of the koji mold. *Journal of General and Applied Microbiology* 17: 281-309.

MURAKAMI, H. HAYASHI, K. and USHIJIMA, S. 1982. Useful key characters separating three *Aspergillus* taxa: *A. sojae, A. parasiticus,* and *A. toxicarius. Journal of General and Applied Microbiology* 28: 55-60.

RAPER, K. B. and FENNELL, D. I. 1965. The Genus *Aspergillus*. Baltimore: Williams & Wilkins.

SUGIYAMA, J. and YAMATOYA, K. 1990. Electrophoretic comparison of enzymes as a taxonomic aid among *Aspergillus* taxa. I. *Aspergillus* sects. *Ornati* and *Cremei. In* Modern Concepts in *Penicillium* and *Aspergillus* Classification, eds. R. A. Samson and J. I. Pitt, pp. 385-393. New York and London: Plenum Press.

DIALOGUE FOLLOWING DR. SUGIYAMA'S PRESENTATION

TAYLOR: It looks to me as if *Aspergillus flavus* and *A. oryzae*, identified on morphological characters, are very close and perhaps do not split completely. Maybe these really should be considered a single species, the only differences being the ability to produce certain toxins or enzymes.

SUGIYAMA: Taxonomically, these comprise one single species, I think, based on the enzyme pattern.

PITT: On the other hand, one could argue that there is a split between your groups 1A and 1B. This would still allow us to distinguish *A. flavus* and *A. parasiticus.*

SUGIYAMA: At the moment, I think that Cluster 1 consists of two species, *A. flavus* and *A. parasiticus.*

PITT: Comparing your two papers, it is interesting to see that the divergence in section *Ornati* is so much greater than you have shown in section *Flavi.* It's nice to see a modern technique that correlates so well with what earlier observers have seen in studying morphology.

THE UBIQUINONE SYSTEM AS A TAXONOMIC AID IN *ASPERGILLUS* AND ITS TELEOMORPHS*

H. Kuraishi[1], M. Itoh[1], N. Tsuzaki[1], Y. Katayama[1], T. Yokoyama[2] and J. Sugiyama[3]

[1]*Faculty of Agriculture*
Tokyo University of Agriculture and Technology
Fuchu, Tokyo 183, Japan

[2]*Institute for Fermentation*
OsakaYodogawa-ku Osaka 532, Japan

[3]*Institute of Applied Microbiology*
University of Tokyo
Bunkyo-ku, Tokyo 113, Japan

SUMMARY

Ubiquinones (Q) are important carriers in the electron transport chain of respiratory systems. It has been shown that the number of isoprene units of the ubiquinone side chain is an excellent aid in identification of genera or subgeneric taxa in microbial taxonomy.

In *Aspergillus*, 9 or 10 isoprene units (Q-9 and Q-10) occur, together with a hydrogenated form Q-10(H2). We have studied the ubiquinone systems in *Aspergillus* in relation to the taxonomy of Raper and Fennell, who subdivided *Aspergillus* into species without metulae, species with or without metulae and species with metulae. In species without metulae, Q-9 occurs in sects *Aspergillus* and *Restricti* and most species in sect. *Cervini*; and Q-10 in sects *Fumigati* and *Clavati*. Subgen. *Ornati* was heterogeneous in ubiquinone systems, in agreement with the observations of morphological heterogeneity. In the sections which may or may not produce metulae, belonging to sects *Nigri* and *Cremei* possessed Q-9 ubiquinone, as did *A. wentii*. Nearly all species in sects *Wentii*, *Flavi*, *Circumdati*, *Candidi and Sparsi* possessed the Q-10(H2) system. Species in sects *Nidulantes*, *Versicolores*, *Usti*, *Terrei* and *Flavipedes*, which include species always producing metulae, all possessed Q-10(H2) ubiquinone without exception.

Xerophilic species had Q-9 or Q-10, while nearly all species having Hülle cells possessed only the Q-10(H2) system. The isoprene units of ubiquinone were highly correlated with morphological and physiological characters in the infrageneric taxa of *Aspergillus*.

INTRODUCTION

The isoprene side chains of ubiquinone molecules are easily characterised from mycelia and fruiting structures of fungi by high performance liquid chromatography after extraction. Previously, Kuraishi *et al.* (1985) determined the general distribution of ubiquinone systems in representative species in the major families of the Ascomycotina, Basidiomycotina and Deuteromycotina. We demonstrated the usefulness of the ubiquinone system as an aid in the classification of fungal taxa in the major families of these subkingdoms, and for the elucidation of their phylogeny. Sugiyama *et al.* (1988)

---

* This paper constitutes Part III of a series entitled "Ubiquinone systems in fungi"

Modern Concepts in Penicillium and Aspergillus Classification
Edited by R. A. Samson and J. I. Pitt
Plenum Press, New York, 1990

showed that rust and smut fungi possessed Q-9 and Q-10, respectively, and the value of the ubiquinone system in their systematics was evaluated.

This paper describes the distribution of ubiquinone systems in *Aspergillus* and its teleomorphs, in relation to the 18 group organisation of Raper and Fennell (1965) or the six subgenera and 18 sections presented by Gams *et al.* (1985). The relationship between morphology and ubiquinone systems is also discussed.

MATERIAL AND METHODS

Isolates studied.

The isoprene side chains of ubiquinones were analysed on 397 isolates from 131 species of *Aspergillus*, representing both teleomorphs and anamorphs. These isolates are listed in Table 1.

Table 1. Principle ubiquinone systems in Aspergillus and its teleomorphs

Subgenus *Aspergillus*
  Section *Aspergillus*
    Eurotium
    E. amstelodami Mangin
      IAM 2035, IAM 13826(FRR 548), IAM 13827(FRR 2792), IFO 5721, IFO 6667,
      JCM 1565(NT), JCM 2605   Q-9
    E. chevalieri Mangin
      IAM 2001, IAM 13859(FRR 2773), IAM 13860(FRR 2795), IFO 5883, IFO 30570,
      JCM 1568(NT)   Q-9
    E. chevalieri Mangin var. *intermedium* (Thom *et* Raper) Malloch *et* Cain
      IFO 5322   Q-9
    E. echinulatum Delacroix
      IFO 5862, JCM 1570(NT)   Q-9
    E. halophilicum Christensen *et al.*
      IFO 7054, IFO 8156   Q-9
    E. herbariorum (Wiggers) Link : Fries
      JCM 2602, JCM 2603   Q-9
    E. leucocarpum Hadlok *et* Stolk
      JCM 1574(NT)   Q-9
    E. pseudoglaucum Blochwitz
      IAM 2578, JCM 1579   Q-9
    E. repens de Bary
      IFO 4332, IFO 7463, JCM 2600   Q-9
    E. rubrum Konig *et al.*
      IAM 13896(FRR 1968), IAM 13897(FRR 2887), IAM 13898(FRR 321)
      IAM 13899(FRR 2970), IFO 4089, IFO 7712   Q-9
    E. tonophilum Ohtsuki
      IFO 4089, IFO 7712   Q-9
    E. umbrosum (Bainier *et* Sartory) Malloch *et* Cain
      IFO 8207   Q-9
    E. xerophilum Samson *et* Moucchacca
      JCM 1583(T)   Q-9
    Edyuillia
    E. athecia (Raper *et* Fennell) Subramanian
      JCM 1850(T)   Q-9

Aspergillus
A. proliferans Smith
    JCM 1729   Q-9
Section *Restricti*
A. caesiellus Saito
    IFO 4882, JCM 1852   Q-9
A. conicus Blochwitz
    IFO 4047, IFO 6399   Q-9
A. gracilis Bainier
    JCM 1726   Q-9
A. penicilloides Spegazzini
    IAM 13890(FRR 2177), IAM 13891(FRR 2178), IFO 30615, NRRL 4550   Q-9
    IFO 8155   Q-(8(49%)+9(51%))
A. restrictus Smith
    IAM 13844(FRR 2973), IAM 13845(FRR 2176), IFO 7101, IFO 7683, JCM 1727(T) Q-9

Subgenus *Fumigati*
  Section *Fumigati*
    Neosartorya
N. aurata Malloch *et* Cain
    IFO 8783, IFO 9817   Q-10
N. aureola (Fennell *et* Raper) Malloch *et* Cain
    IFO 8105   Q-10
N. fennelliae Kwon-Chung
    JCM 1946 mate A, JCM 1947 mate a   Q-10
N. fischeri (Wehmer) Malloch et Cain
    IAM 2024, IAM 13863(NRRL 4161), IAM 13864(FRR 181),   Q-10
    JCM 1740(T), IFO 5866, IFO 8790
N. fischeri var. *glabra* (Fennell *et* Raper) Malloch *et* Cain
    IFO 8789, IFO 9857, IFO 30571, IFO 30572
N. fischeri var. *spinosa* (Fennell *et* Raper) Malloch *et* Cain
    IFO 5955, IFO 8007, IFO 8780, IFO 8782, IFO 30573
N. fischeri var. *verrucosa* (Udagawa *et* Kawasaki) Malloch *et* Cain
    IFO 8779   Q-10
N. quadricincta (Yuill) Malloch *et* Cain
    IFO 8778, IFO 9842, JCM 1855(T)
N. stramenia (Novak et Raper) Malloch *et* Cain
    IFO 9611, IFO 31358, JCM 1856(T)   Q-10
Aspergillus
A. brevipes Smith
    IFO 5821, JCM 1734(T)   Q-10
A. fumigatus Fresenius
    IAM 3006, IAM 13869(FRR 163), IAM 13870(SRRC 43), IFO 8868,   Q-10
    IFO 9733
A. unilateralis Thrower
    IFO 8136, JCM 1737(T)   Q-10
Section *Cervini*
A. cervinus (Massee) Neill
    IAM 2752, IAM 13857 (FRR 3531), IAM 13858(FRR 3532), JCM 1722   Q-9
A. kanagawaensis Nehira
    IAM 13874 (FRR 3520), IAM 13875(FRR 3528), JCM 1723(T)   Q-9
A. nutans McLennan *et* Ducker
    IFO 7869, IFO 8134, JCM 1730(T)   Q-9
A. bisporus Kwon-Chung *et* Fennell
    JCM 1721(T)   Q-10(H2)

(cont.)

Table 1 (cont.)

Subgenus *Ornati*
  Hemicarpenteles
    H. acanthosporus Udagawa *et* Takada
      IFO 9490(T), IFO 9607   Q-10
    H. paradoxus Sarbhou et Elphick
      IAM 13886(FRR 2162), IAM 13887(NRRL 5162), IFO 8172(T), IFO 9127,
      JCM 2355(T)   Q-9
  Sclerocleista
    S. ornatus Raper *et al.*
      IAM 13880(FRR 2291), IFO 4042, IFO 8130, JCM 2263, JCM 2354(T)   Q-9
    S. thaxteri (Subramanian) von Arx
      JCM 2264(T), JGM 2350(T)   Q-9
  Warcupiella
    W. spinulosa (Warcup) Subramanian
      IFO 31800(T), IFO 32023, JCM 2358(T), NHL 2824   Q-10
  Aspergillus
    A. brunneo-uniseriatus Singh *et* Bakshi
      IFO 6993, JCM 2348(T)   Q-9
    A. brunneo-uniseriatus var. *nanus* Singh *et* Bakshi
      IFO 32018, JCM 2349   Q-10(H2)
    A. apicalis Mehrotra *et* Basu
      CBS 236.81   Q-10
    A. ivoriensis Bartoli et Maggi
      JCM 2353(T)   Q-10(H2)
    A. raperi Stolk
      JCM 2265, JCM 2266, JCM 2356(T)   Q-10(H2)

Subgenus *Clavati*
  Section *Clavati*
    A. clavatus Desmazieres
      IAM 2002, IAM 13832(FRR 3029), IAM 13833(SRRC 19), IAM 13834(SRRC 52),
      IFO 5837, IFO 8605, IFO 8606, IFO 8865   Q-10
    A. clavato-nanica Batista *et al.*
      JCM 1858(T)   Q-10
    A. giganteus Wehmer
      IFO 5818, JCM 1719   Q-10
    A. longivesica Huang *et* Raper
      JCM 1720(T)   Q-(9(49%)+10(46%))

Subgenus *Nidulantes*
  Section *Nidulantes*
    Emericella
      E. aurantiobrunnea (Atkins *et al.*) Malloch *et* Cain
        IFO 30837   Q-(10(64%)+10(H2)(36%))
      E. bicolor Christensen *et* States
        IFO 30830   Q-10(H2)
      E. cleistominuta Mehrotra *et* Prasad
        IFO 30839   Q-10(H2)
      E. desertorum Samson *et* Mouchacca
        IFO 30840   Q-10(H2)
      E. foveolata Horie
        IFO 30559, IFO 30560   Q-10(H2)
      E. fruticulosa (Raper *et* Fennell) Malloch *et* Cain
        IFO 30841   Q-(10(52%)+10(H2)(48%))

E. heterothallica (Kwon *et al.*) Malloch *et* Cain
    IFO 30842, IFO 30843   Q-10(H2)
E. navahoensis Christensen *et* States
    IFO 30836   Q-10(H2)
E. nidulans (Eidam) Vuillemin
    IAM 2006, IAM 13876(NRRL 189), IAM 13877(NRRL 1092),
    IFO 9852, JCM 2728, IFO 30872   Q-10(H2)
E. nidulans var. *acristata* (Fennell *et* Raper) Subramanian
    IFO 30063, IFO 30844   Q-10(H2)
E. nidulans var. *dentata* (Sandhu *et* Sandhu) Subramanian
    IFO 30845, IFO 30907   Q-10(H2)
E. nidulans var. *echinulata* (Fennell *et* Raper) Godeas
    IFO 30248, IFO 30412, IFO 30896   Q-10(H2)
E. nidulans var. *lata* (Thom *et* Raper) Subramanian
    IFO 30847   Q-10(H2)
E. nivea Wiley *et* Simmons
    IFO 4112, IFO 9653(T), IFO 9654, IFO 32020   Q-10(H2)
E. parvathecia (Raper *et* Fennell) Malloch *et* Cain
    IFO 30848   Q-(10(85%)+10(H2)(15%))
E. purpurea Samson *et* Mouchacca
    IFO 30849   Q-10(H2)
E. quadrilineata (Thom *et* Raper) Benjamin
    IAM 13895(FRR 3539),   Q-(10(34%)+10(H2)(66%))
    IFO 5859, IFO 30850, IFO 30911, JCM 2730   Q-10(H2)
E. rugulosa (Thom *et* Raper) Benjamin
    IAM 13900(FRR 1587), IAM 13901(SRRC 93), IFO 5863,
    IFO 8626, IFO 8629, IFO 30853, IFO 30913, JCM 2729   Q-10(H2)
    IFO 30852   Q-(10(12%)+10(H2)(88%))
E. spectabilis Christensen
    IFO 30854   Q-10(H2)
E. striata (Rai *et al.*) Malloch *et* Cain
    IFO 30856   Q-10(H2)
E. sublata Horie
    IFO 30906   Q-(10(12%)+10(H2)(88%))
E. unguis Malloch *et* Cain
    IAM 13911(NRRL 216), IAM 13912(FRR 220), IFO 8087, IFO 30857, JCM 2726,
    JCM 2727   Q-10(H2)
E. variecolor Berkeley *et* Broome
    CBS 119.37, CBS 134.55, CBS 135.55, CBS 136.55, CBS 597.65, IFO 30855,
    IFO 31666, IFO 32051, IFO 32052   Q-10(H2)
E. violacea (Fennell *et* Raper) Malloch *et* Cain
    IFO 8106, JCM 2725(T)   Q-10(H2)

Aspergillus
A. multicolor Sappa
    IFO 8133   Q-10(H2)
Section *Versicolores*
A. asperescens Stolk
    IFO 5996   Q-10(H2)
A. caespitosus Raper *et* Thom
    IAM 13853(FRR 1929), IAM 13854(FRR 3521), IFO 8086   Q-10(H2)
A. janus Raper et Thom
    IFO 7627   Q-10(H2)
A. silvaticus Fennell *et* Raper
    IFO 8173   Q-10(H2)

(cont.)

Table 1 (cont.)

A. *sydowi* (Bainier *et* Sartory) Thom *et* Church
    IAM 13905(FRR 546), IAM 13906(FRR 2991)   Q-10(H2)
A. *sydowi* var. *achlamidosporus* Nakazawa *et al.*
    IFO 4096   Q-10(H2)
A. *sydowi* var. *inaequalis* Nakazawa *et al.*
    IFO 4097   Q-10(H2)
A. *varians* (Wehmer) Raper *et* Fennell
    IFO 4114   Q-10(H2)
A. *versicolor* (Vuillemin) Tiraboschi
    IAM 12156, IAM 13848(FRR 658), IAM 13849(SRRC 109), IFO 4105, IFO 4411,
    IFO 8004, IFO 30338, IFO 31223, JCM 2608, JCM 2609   Q-10(H2)
Section *Usti*
A. *deflectus* Fennell *et* Raper
    IFO 6357   Q-10(H2)
A. *puniceus* Kwon *et* Fennell
    IAM 13892(FRR 1852), IAM 13893(FRR 3537)   Q-10(H2)
A. *ustus* (Bainier) Thom et Church
    IAM 2057, IAM 13913(NRRL 279), IAM 13914(FRR 664), IFO 7013, IFO 8878
    Q-10(H2)
Section *Terrei*
A. *terreus* Thom
    IAM 3004, IAM 13909(FRR 1910), IAM 13910(SRRC 100), IFO 6123, IFO 7078,
    IFO 30537   Q-10(H2)
A. *terreus* var. *africanus* Fennell *et* Raper
    IFO 8835   Q-10(H2)
A. *terreus* var. *aureus* Fennell *et* Raper
    IFO 30536(T), IFO 31217   Q-10(H2)
Section *Flavipedes*
Fennellia
F. *flavipes* Wiley *et* Simmons
    IAM 2523, IAM 13865(SRRC 22), IAM 13866(FRR295), IFO 9655(T)   Q-10(H2)
F. *nivea* (Wiley *et* Simmons) Samson
    JCM 2731, JCM 2732(T)   Q-10(H2)
Aspergillus
A. *carneus* (van Tieghem) Blochwitz
    IAM 13855(NRRL 527), IAM 13856(FRR 2977), IFO 5861, IFO 30898, IFO 30899
    Q-10(H2)
A. *iizukae* Sugiyama
    IFO 8869   Q-10(H2)
A. *niveus* Blochwitz
    IAM 13878(NRRL5505), IAM 13879(FRR3516)   Q-10(H2)

Subgenus *Circumdati*
Section *Wentii*
A. *terricola* Marchal
    IFO 5867   Q-10(H2)
A. *terricola* var. *americana* Marchal
    IFO 5446   Q-10(H2)
A. *thomii* Smith
    IAM 2759(T)   Q-10(H2)
A. *wentii* Wehmer
    IAM 13915(FRR 1778), IAM 13916(FRR 2627), IFO 4108, IFO 6578, IAM 7126,
    IFO 8864, IFO 8879, JCM 2724   Q-9
A. *wentii* var. *minimus* Nakazawa *et al.*
    IFO 4110   Q-9

Section *Flavi*
  A. avenaceus Smith
    IFO 7539, IFO 8129   Q-10
  A. flavus Link : Fries
    IAM 13835(SRRC 100A), IAM 13836(SRRC 167), IFO 7600, IFO 30107, IFO 30180,
    JCM 2604   Q-10(H2)
  A. flavus var. *asper* Sasaki
    AHU B-8, IFO 5324   Q-10
  A. flavus var. *columnaris* Raper *et* Fennell
    JCM 2605   Q-10(H2)
  A. oryzae (Ahlburg) Cohn
    IAM 2609, IAM 13881(SRRC 266), IAM 13882(SRRC 2103),
    IAM 13883(SRRC 2104), JCM 2606   Q-10(H2)
  A. parasiticus Speare
    IAM 13888(SRRC 1039), IAM 13889(IMI 26862), IFO 30179, JCM 2720   Q-10(H2)
  A. sojae Sakaguchi *et* Yamada
    IAM 13902(SRRC 299), IAM 13903(SRRC 1124), IFO 5241   Q-10(H2)
  A. tamarii Kita
    IAM 13907(ATCC 26950), IAM 13908(VDR 33), IFO 7465   Q-10(H2)
  A. toxicarius Murakami
    IFO 30108   Q-10(H2)
  A. zonatus Kwon et Fennell
    IFO 8817, JCM 2254   Q-10(H2)
Section *Nigri*
  A. aculeatus Iizuka
    IAM 2907(T), IFO 31348   Q-9
  A. awamori Nakazawa
    IAM 2387, IFO 4033   Q-9
  A. awamori var. *fumeus* Nakazawa *et al.*
    IFO 4122   Q-9
  A. awamori var. *fuscus* Nakazawa *et al.*
    IFO 4119   Q-9
  A. awamori var. *minimus* Nakazawa *et al.*
    IFO 4115   Q-9
  A. awamori var. *piceus* Nakazawa *et al.*
    IFO 4116   Q-9
  A. carbonarius (Bainier) Thom
    IAM 13830(SRRC 16), IAM 13831(FRR 369), IFO 4039, IFO 5864   Q-9
  A. ficuum (Reichardt) Hennings
    IFO 4280, IFO 4320   Q-9
  A. foetidus (Nakazawa) Thom *et* Raper
    IAM 13867(NRRL 337), IAM 13868(NRRL 341), IFO 4312   Q-9
  A. helicothrix Al-Musallan
    JCM 1861(T)   Q-9
  A. heteromorphus Batista et Maia
    JCM 1862(T)   Q-9
  A. japonicus Saito
    IFO 4060, IFO 4337, IFO 4403   Q-9
  A. japonicus var. *aculeatus*
    IAM 13871(FRR 5118), IAM 13872(SRRC 168), IAM 13873(NRRL 1782)   Q-9
  A. niger van Tieghem
    IAM 3001, JCM 1864(NT), IAM 13840(SRRC 60), IAM 13841(SRRC 61)   Q-9
  A. niger var. *awamori* (Nakazawa) Al-Musallam
    IAM 13838(FRR 4795), IAM 13839(FRR 3534)   Q-9

(cont.)

Table 1 (cont.)

A. *niger* var. *phoenicis* (Corda) Al-Musallam
    JCM 2610   Q-9
A. *phoenicis* (Corda) Thom
    IFO 8873, IFO 8874   Q-9
A. *pulverulentus* (McAlp.) Thom
    IFO 4282, IFO 4353   Q-9
A. *saitoi* Sakaguchi *et al.*
    IAM 2209(T)   Q-9
A. *saitoi* var. *kagoshimaensis* Sakaguchi *et al.*
    IFO 2190   Q-9
A. *usamii* Sakaguchi *et al.*
    IAM 2185(T), IFO 4388   Q-9
Section *Circumdati*
Petromyces
P. *alliaceus* Malloch et Cain
    IAM 13824(SRRC 8), IAM 13825(FRR 3518), IFO 5320, IFO 7538, JCM 1948(T) Q-10
Aspergillus
A. *dimorphicus* Mehrotra *et* Prasad
    JCM 1952(T)   Q-9
A. *lanosus* Kamal et Bhargava
    JCM 1955(T)   Q-10
A. *auricomus* (Guéguen) Saito
    IAM 13851(NRRL 391), IAM 13852(SRRC 1031), JCM 1949(T)   Q-10(H2)
A. *bridgeri* Christensen
    JCM 1950(T)   Q-10(H2)
A. *campestris* Christensen
    JCM 1951(T)   Q-10(H2)
A. *elegans* Gasperini
    JCM 1953   Q-10(H2)
A. *insulicola* Montemayor *et* Santiago
    JCM 1954(T)   Q-10(H2)
A. *melleus* Yukawa
    IAM 2066, JCM 1956, JCM 2607   Q-10(H2)
A. *ochraceus* Wilhelm
    ATCC 1008, IAM 2022, IAM 13842(SRRC 65), IAM 13843(FRR 400), IFO 4344,
    IFO 4345, IFO 4346, IFO 8559, JCM 1958   Q-10(H2)
A. *ochraceus* var. *microsporus* Tiraboschi
    IFO 4073, IFO 4409, IFO 4410   Q-10(H2)
A. *ochraceoroseus* Bartoli et Maggi
    JCM 1957(T)   Q-10(H2)
A. *ostianus* Wehmer
    IAM 13884(FRR 2115), IAM 13885(FRR 3504), JCM 1959   Q-10(H2)
A. *petrakii* Voros
    JCM 1960(T)   Q-10(H2)
A. *robustus* Christensen *et* Raper
    JGM 1961(T)   Q-10(H2)
A. *sclerotiorum* Huber
    IAM 13846(FRR 3505), IAM 13847(FRR 3507), JCM 1962(T)   Q-10(H2)
A. *sulphureus* (Fresenius) Thom *et* Church
    JCM 1963   Q-10(H2)
Section *Candidi*
A. *candidus* Link : Fries
    IAM 2018, IAM 13828(NRRL 303), IAM 13829(NRRL 2297),
    IAM 13850(NRRL 312), IFO 4322, JCM 1867   Q-10(H2)

Section *Cremei*	
Chaetosartorya	
C. chrysella Kwon *et* Fennell	
IFO 32022	Q-9
C. cremea Kwon *et* Fennell	
IAM 13861(NRRL 5081), IAM 13862(FRR 3534), IFO 32021	Q-9
C. stromatoides Wiley *et* Simmons	
IFO 9652(T), IFO 9656, IFO 9657, IFO 9658, IFO 9659	Q-9
A. itaconicus Kinoshita	
IFO 4419	Q-9
Section *Sparsi*	
Aspergillus	
A. gorakhpurensis Kamal *et* Bhargava	
JCM 2267(T)	Q-9
A. biplanus Raper *et* Fennell	
JCM 2347(ST)	Q-10(H2)
A. diversus Raper *et* Fennell	
JCM 2351(ST)	Q-10(H2)
A. funiculosus Smith	
JCM 2352(T)	Q-10(H2)
A. sparsus Raper *et* Thom	
IAM 13904(FRR 1933), JCM 2357(T)	Q-10(H2)

Extraction and analysis of ubiquinones.

Mycelia grown on potato sucrose agar, or malt yeast extract agar were used for the extraction of ubiquinones. Xerophilic species were grown on potato sucrose agar containing 20% sucrose. Extraction, purification and characterisation of ubiquinones were carried out as described previously (Kuraishi *et al.*, 1985). Where one type of ubiquinone molecule constituted more than 90% of the types found in a particular species, it was considered to be the major quinone type.

RESULTS AND DISCUSSION

Influence of growth conditions.

Emericella fruticulosa IFO 30841 and *E. parvathecia* IFO 30848, which have two molecular species, Q-10 and Q-10(H2) as major ubiquinones, were grown on potato sucrose agar adjusted to two pH values, incubated for two weeks, harvested and ubiquinones analysed. Under these conditions, Q-10 and Q-10(H2) were equally produced by *E. fruticulosa*, while in *E. parvathecia* most of the ubiquinone molecules were Q-10 with a small proportion of Q-10(H2). As shown in Table 2, pH had no effect on ubiquinone composition.

    A. longivesica JCM 1720 produces both short and long conidiophores, the latter only under the light, and it normally produces equal levels of Q-9 and Q-10. As shown in Table 3, no differences were seen in the ubiquinone composition of hyphae cultivated under lights, and harvested at various ages. It is concluded that the molecular composition of the ubiquinone system is a stable character in fungal mycelia.

Ubiquinone systems in *Aspergillus* and its teleomorphs.

Major ubiquinone systems in the isolates studied are shown in Table 1.

Table 2. Ubiquinone compositions of *Emericella fruticulosa* IFO 30841 and *Emericella parvathecia* IFO 30848 grown on the different pH media.

pH	E. fructiculosa IFO 30841		E. parvathecia IFO 30848	
	Q-10	Q-10(H2)	Q-10	Q-10(H2)
5.6	52%	48%	88%	12%
7.0	46%	54%	89%	11%

Table 3. Ubiquinone compositions of *Aspergillus longivesica* JCM 1720 obtained at various ages and cultivated under the light.

		1 week	2 weeks	4 weeks
Dark	Q-8	8%	8%	8%
	Q-9	59%	57%	57%
	Q-10	34%	35%	36%
Light	Q-8	8%	7%	7%
	Q-9	59%	59%	59%
	Q-10	34%	34%	34%

Subgenus Aspergillus

The group includes species which are xerophilic and have no metulae. It is divided into two sections, *Aspergillus* and *Restricti*.

In sect. *Aspergillus*, Q-9 was the major ubiquinone system found in 38 isolates. Species in sect. *Restricti* also predominantly possessed the Q-9 system. Four isolates of *A. penicillioides* had the Q-9 system, but *A. penicillioides* IFO 8155 produced equal amounts of Q-8 and Q-9. The occurrence of Q-8 in this isolate was very unusual.

A large number of Ascomycetes produced ubiquinones with hydrogenated isoprene side chains such as Q-10(H2) (Kuraishi *et al.*, 1985), but hydrogenated ubiquinones have not been found in xerophilic fungi.

Subgenus Fumigati

Subgen. *Fumigati* includes sects *Fumigati* and *Cervini*; metulae are not produced. In sect. *Fumigati*, all 36 isolates examined from *Neosartorya* (six species and three varieties) and *Aspergillus* (three species) produced Q-10. In sect. *Cervini*, four species were examined: three species produced the Q-9 system, but *A. bisporus* JCM 1721 was exceptional, having the Q-10(H2) system.

Subgenus Ornati

Subgen. *Ornati*, corresponding to the *A. ornatus* group of Raper and Fennell (1965) is an admittedly artificial grouping of species without metulae (Raper and Fennell, 1965; Christensen and Tuthill, 1985). The subgenus is clearly heterogeneous, as it includes four teleomorph genera (Fennell, 1977; Samson, 1979; Gams *et al.*, 1985; Christensen and Tuthill, 1985).

In the teleomorph genus *Hemicarpenteles*, *H. acanthosporus* (two isolates) possessed the Q-10 system, while *H. paradoxus* (five isolates) had Q-9. Different ubiquinone systems in one teleomorph genus are most unusual. *Sclerocleista ornata* and *S. thaxteri* (seven isolates) possessed the Q-9 system, while four isolates of *Warcupiella spinulosa* had Q-10. The *Aspergillus* species were also heterogeneous, with Q-9, Q-10 and Q-10(H2) all being produced. Both Q-9 and Q-10 were produced by *A. brunneo-uniseriatus* and *A. apicalis*.

In bacteria, ubiquinone systems are species specific, so it appears to be unacceptable for varieties in one species to have different systems. As *A. brunneo-uniseratus* had the Q-9 system, while *A. brunneo-uniseriatus* var. *nanus* had Q-10(H2), further taxonomic study seems indicated.

A relationship may exist between the production of the Q-10(H2) system and the production of Hülle cells. *A. ivoriensis* produces Q-10(H2) and has Hülle cells, and appears unrelated to other species of this subgenus (Samson, 1979). *A. raperi*, also a Q-10(H2) species, was described as producing abundant Hülle cells, and *A. bisporus* in sect. *Cervini*, has Hülle cells and the Q-10(H2) system. Species which have uniseriate conidial heads, Hülle cells and the Q-10(H2) system should probably be excluded from sect. *Cervini* and subgen. *Ornati*.

Here, we tentatively propose to divide subgen. *Ornati* into two groups with Q-9 or Q-10.

Subgenus Clavati

Subgen. *Clavati* sect. *Clavati* is morphologically well defined and produces phialides only. Species examined (11 isolates) possessed Q-10 except *A. longivesica*, the single isolate of which produced about 50% of Q-9 in addition to Q-10.

Subgenus Nidulantes

This subgenus has five sections, sects *Nidulantes, Versicolores, Usti, Terrei* and *Flavipedes*: metulae are always present and Hülle cells are often produced. In sect. *Nidulantes*, which includes a number of *Emericella* species, most isolates possessed the Q-10(H2) system, but five isolates had Q-10 as well. In sects *Versicolores, Usti, Terrei* and *Flavipedes*, only the Q-10(H2) system was found.

Subgenus Circumdati

In subgen. *Circumdati*, metulae may or may not be present, and this large subgenus is divided into seven sections. In sect. *Wentii*, *A. wentii* (nine isolates) had the Q-9 system, but *A. terricola* and *A. thomii* had Q-10(H2). The result for *A. wentii* was surprising, as Kozlowski and Stepien (1982) reported that mitochondrial DNA fragments of *A. wentii* linked to those of *A. tamarii* and *A. oryzae*, which belong to sect. *Flavi* and have Q-10(H2). Christensen and Tuthill (1985) described that the *A. wentii* group is heterogenous like the *A. ornatus*group. Our results suggest that *A. wentii* (subgroup I) is clearly distinguished from the group which includes *A. thomii* and *A. terricola* (subgroup II) based on the ubiquinone isoprenologue.

In sect. *Flavi*, the Q-10(H2) system was almost universally found, but the Q-10 system was found in *A. avenaceus* and *A. flavus* var. *asper*. It appears likely that such Q-10 species are incorrectly classified in sect. *Flavi* and that detailed characterization of Q-10 strains are necessary for their correct taxonomic placement.

All forty isolates examined from sect. *Nigri*, comprising 13 species, have Q-9 ubiquinones. In sect. *Circumdati*, 13 anamorphic species had the Q-10(H2), while *A. dimorphicus* and *A. lanosus* were exceptional. *A. dimorphicus* had Q-9 system, which was morphologically characterized by Mehrotra and Prasad (1969) by branched conidiophores. According to Samson (1979) this species might be close to *A. petrakii*, but this taxon has Q-10(H2). *A. lanosus* had the Q-10 system, and shows some similarity with *A. fresenii* (= *A. sulphureus*), which has Q-10(H2), on the basis of the long conidiophores. Since these two species were analyzed on the ubiquinone system using one single isolate, their infrageneric position can be elucidated after the investigation of further isolates. *Petromyces* is the only teleomorph in this section; Q-10 ubiquinones are produced. Therefore, subsect. *Circumdati* can be divided into two subgroups I and II consisting of species with the teleomorph *Petromycetes* (Q-10) and other taxa (Q-10(H2)).

In sect. *Candidi*, with a single species, *A. candidus*, six isolates tested produced Q-10(H2). In sect. *Cremei*, where we examined three species of *Chaetosartorya* and one *Aspergillus*, all isolates were of Q-9. Four species in sect. *Sparsi* produced Q-10(H2), but only one isolate of *A. gorakhpurensis* (group *A. sparsus*) had a Q-9 system which have usually metulae, whereas other species in this group had Q-10(H2). According to Samson (1979), *A. gorakhpurensis* is biseriate, and considered as related to *A. pulvinus* of the *A. versicolor* group. Unfortunately, the quinone system of *A. pulvinus* was not analysed and detailed re-examination on the morphological characteristics might be worthwhile for this species.

In summary, species belonging in subgenera *Aspergillus*, *Fumigati*, *Ornati* and *Clavati*, where species do not form metulae, mainly produced Q-9 or Q-10, although a few species had Q-10(H2). Species belonging to subgen. *Nidulantes*, which always produce metulae, and often have Hülle cells, produce the Q-10(H2) system without exception. In subgen. *Circumdati*, a large and undoubtedly taxonomically heterogeneous subgenus, the ubiquinone systems produced were rather more complex than in the other subgenera. Molecular species of ubiquinone differed from section to section, or sometimes, as in sects Wentii and *Circumdati*, within sections as well.

The species of *Aspergillus* and its teleomorphs fall into 18 groups, which have been widely accepted. Figure 1 shows the main ubiquinone molecules of groups in the schematic presentation according to the order of the groups described by Raper and Fennell (1965). Among the groups, each of *A. ornatus*, *A. ochraceus* and *A. wentii* groups was further divided into two subgroups. In Fig. 2, the intrageneric subgenera and sections as proposed by Gams *et al.*(1985) are presented with their ubiquinone species. Like Fig. 1, the two sections and a subgenus were separated into two distinct clusters on the basis of ubiquinone isoprenologue. The schematic presentation of groups or sections with subgenera shown in Figs. 1 and 2 is modified and rearranged as shown in Fig. 3 on the basis of ubiquinone system, metulae and Hülle cells.

Morphological characters are considered to originate in the chromosomal DNA. On the other hand, ubiquinone would be distributed in the membrane of mitochondria to act as a carrier in electron transport. Although it is not clear whether biosynthesis of ubiquinone molecules is coded only on chromosomal DNA, mitochondrial DNA or both, it was found that there is some correlation between the presence or absence of metulae and ubiquinone isoprenologue. In the species belonging to subgenera which have no metulae, Q-9 or Q-10 was distributed as a major ubiquinone except for a few species such as *A. bisporus*and a member of subgenera Ornati, which have a dihydrogenated ubiquinone.

Further examinations including other biochemical approaches are required for the strains which have ubiquinone systems contradictious in the infrageneric taxa.

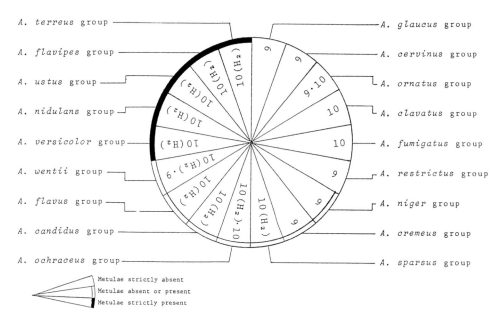

Figure 1. Schematic presentation of the main ubiquinone systems of the groups in the genus *Aspergillus* and its teleomorphs according to Raper and Fennell (1965).

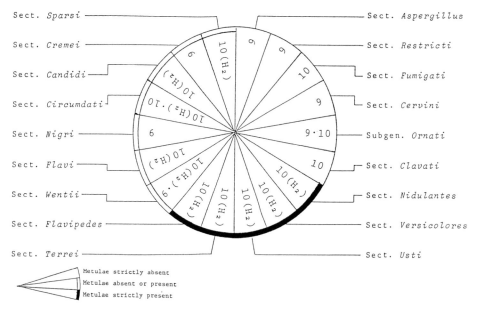

Figure 2. Schematic presentation of the main ubiquinone systems of the subgenera and sections in the genus *Aspergillus* and its teleomorphs according to Gams *et al.* (1985).

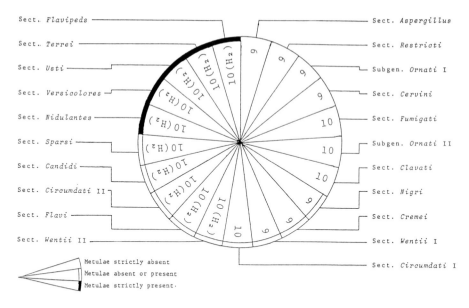

Figure 3. Rearrangment of the subgenera and section of *Aspergillus* and its teleomorphs based on the ubiquinone systems.

REFERENCES

CHRISTENSEN, M. and TUTHILL, D.E. 1985. *Aspergillus*: an overview. *In* Advances in *Penicillium* and *Aspergillus* Systematics, eds. R.A. Samson and J.I. Pitt, pp. 195-209. New York and London: Plenum Press.

FENNELL, D.T. 1977. *Aspergillus* taxonomy. *In* Genetics and Physiology of *Aspergillus*, eds. J.E. Smith and J.A. Pateman, pp. 1-21. London: Academic Press.

GAMS, W., CHRISTENSEN, M., ONIONS, A.H., PITT, J.I. and SAMSON, R.A. 1985. Infrageneric Taxa of *Aspergillus*. *In* Advances in *Penicillium* and *Aspergillus* Systematics, eds. R.A. Samson and J.I. Pitt, pp. 55-62. New York and London: Plenum Press.

KOZLOWSKI, M. and STEPIEN, P.P. 1982. Restriction enzyme analysis of mitochondrial DNA of members of the genus *Aspergillus* as an aid in taxonomy. *Journal of General Microbiology* 128: 471-476.

KURAISHI, H., KATAYAMA-FUJIMURA, Y., SUGIYAMA, J. and YOKOYAMA, T. 1985. Ubiquinone systems in fungi. I. Distribution of ubiquinones in the major families of Ascomycetes, Basidiomycetes and Deuteromycetes, and their taxonomic implications. *Transactions of the Mycological Society of Japan* 26: 383-395.

MEHROTRA, B.S. and PRASAD, R. 1969. *Aspergillus dimorphicus* and *Emericella cleisto-minuta* spp. nov. from Indian soils. *Transactions of the British Mycological Society* 52: 331-336.

RAPER, K.B. and FENNELL, D.I. 1965. The Genus *Aspergillus*. Baltimore: Williams and Wilkins.

SAMSON, R.A. 1979. A compilation of the Aspergilli described since 1965. *Studies in Mycology, Baarn* 18: 1-38.

SUGIYAMA, J., ITOH, M., KATAYAMA, Y., YAMAOKA, Y., ANDO, K., KAKISHIMA, M. and KURAISHI, H. 1988. Ubiquinone systems in fungi. II. Distribution of ubiquinones in smut and rust fungi. *Mycologia* 80: 115-120.

DIALOGUE FOLLOWING DR. KURAISHI'S PRESENTATION

SAMSON: This is a very impressive study. None of the isolates of *Hemicarpenteles paradoxus* that I have seen produced ascospores. So, I tend to agree that the placement of this species is doubtful.

CHRISTENSEN: As I recall, *A. thomii* is in the *A. wentii* group, but has been suggested to be a member of section *Flavi*. Do you have an opinion on this?

KURAISHI: Although *A. thomii* and nearly all isolates in sect. *Flavi* have the Q-10(H2) system, we need to study this problem more carefully.

IMMUNOLOGICAL DIFFERENTIATION BETWEEN *PENICILLIUM* AND *ASPERGILLUS* TAXA

B. Fuhrmann[1], M.F. Roquebert[2], V. Lebreton[1] and M. van Hoegaerden[1]

[1]*Chemunex S.A.*
94700 Maisons Alfort

[2]*Laboratoire de Cryptogamie*
Museum National d'Histoire Naturelle
75005 Paris, France

SUMMARY

Enzyme-linked immunosorbent assay (ELISA) and immunofluorescence have been used to detect species of *Penicillium*, *Aspergillus* and other moulds. Thirteen species of *Penicillium* subgenera *Aspergilloides*, *Furcatum* and *Penicillium* and five species of *Aspergillus* were tested for their immunological reaction with monoclonal antibodies directed to *P.glabrum* (Wehmer) Westling, *A.versicolor* (Vuill.) Tiraboschi, *Geotrichum candidum* Link and *Mucor racemosus* Fres. Cross reactions showed that monoclonal antibodies directed to *M. racemosus*, *G. candidum* and *A.versicolor* are specific. Similar reactions with monoclonal antibodies from *P. glabrum* or *A.versicolor* show that one antigenic determinant is shared by all species of moulds tested. Other monoclonal antibodies show that at least one epitope is shared by *Penicillium* and *Aspergillus* but not other genera listed above. At least one epitope is shared by all *Penicillium* subgen. *Aspergilloides* and *Aspergillus*, and at least one by all tested species of subgen. *Aspergilloides*. With the monoclonals tested, no distinction was found between subgen. *Furcatum* and subgen. *Penicillium*. Antigenicity appears to correlate well with morphological data. ELISA tests are faster than microscopical or biological examination. The method seems appropriate for defining precise relationships between taxa.

INTRODUCTION

Mould identification is based primarily on colony characteristics and micromorphology. Morphological and biochemical variations in cultures are frequent, even in single spores (Bridge *et al.*, 1986). Thus these criteria are not always reliable. The problem is particularly acute for *Penicillium* and *Aspergillus* where the large number of species and varieties make identifications difficult even for an experienced mycologist. Moreover, standard methods require examination of at least one week old cultures. For medical or industrial purposes, this length of time is often too long. Mycologists are attempting to propose other criteria for identifying moulds rapidly and reliably. Immunological methods may be the most efficient of all for this purpose. They were first investigated for medically important fungi (Pepys and Longbottom, 1978; Polonelli and Morace, 1985, 1986; Morace *et al.*, 1984; Sekhon *et al.*, 1982; Standard and Kaufman, 1977; Kaufman and Standard, 1987). The specificity of the tests showed them to be of value not only for medical identifications but also for resolving taxonomic problems in other fields of mycology (Hingand *et al.*, 1983; Morace *et al.*, 1984; Polonelli *et al.*, 1984, 1985, 1987; Lin *et al.*, 1986; Notermans *et al.*, 1986; van der Heide and Kauffman, 1987) and for characterization of an atypical isolate among typical ones (Dupont *et al.*, 1988). Our objective was to identify reliably fungal contaminants present in dairy products, in the fastest way possible. The principle of the method was to detect antigenic sites of the cell wall by using enzyme linked

Modern Concepts in Penicillium and Aspergillus Classification
Edited by R. A. Samson and J. I. Pitt
Plenum Press, New York, 1990

immunosorbent assays (ELISA) and immunofluorescent methods, with specific mould antibodies. A first experiment used *Penicillium viridicatum* Westling antiserum (Fuhrmann *et al.*, 1989). Results show that this antiserum contained antibodies reacting with all tested isolates from *Penicillium* subgen. *Penicillium* and *Furcatum* but not with the subgen. *Aspergilloides*, nor other tested moulds. This discrimination between subgenera permitted a first step in identification. On the other hand, adsorption experiments provided evidence similar biochemical structures between *P. glabrum* (Wehmer) Westling, *Aspergillus versicolor* (Vuill.) Tiraboschi and *A. fumigatus* Fres. In this case, antigenicity seemed to correlate with morphology,i.e. apical enlargment of the stipes, a common feature in *Aspergillus* and *Penicillium* subgen. *Aspergilloides*. More specific and precise recognition with monoclonal antibodies (MAB) was next attempted. This is reported here. Reactions with MAB have the advantage of reacting with one and only one antigenic site on cell walls. Each taxon being characterized by a set of antigenic determinants, the knowledge of the composition of this set may help in identification.

MATERIAL AND METHODS

Antigen production.
Fungal strains (Table 1) were cultured and prepared by the protocol described by Fuhrmann *et al.* (1989). The mycelium of *P. glabrum* (LCP 88.2494), *A. versicolor* (LCP 88.2500), *Mucor racemosus* (LCP 88.3095), *Geotrichum candidum* (LCP 88.2502) and *P. viridicatum* (LCP 88.2485) was used to immunize mice and rabbits. The other strains were tested for cross reaction with antibodies obtained from immunization.

Antibodies production.
Rabbit immunization protocol with *P. viridicatum* was described by Fuhrmann *et al.* (1989). For monoclonal antibody production, Balb/c mice were injected intravenously or intraperitoneally with 200 or 400 g of mycelium. Three days after the last boost injection, the mouse spleens were removed for fusions. Splenocytes were pooled with azaguanine resistant murine myeloma P3-X63-Ag8-65-3 or NS1 cells 3 and fused according to the procedure of Köler et Milstein (1975). Positive hybridomas were cloned by limiting dilution in presence of macrophages or thymocytes and were amplified as ascitic fluids in mice primed with pristane. The different antibodies obtained by these protocols and proposed by Chemunex, are grouped in Table 2.

ELISA assays.
In ELISA tests, antigens (in this work, from mycelium) are bound to the plastic surface of a microtiter plate. Incubation with polyclonal or monoclonal antibodies then allows binding with those specific for the immobilized antigens from the mycelium. The fixed antibodies are detected when incubated with a second specific antigen labeled with an enzyme. After adding the enzyme substrate, the optical density of the stained product is measured (van der Heide and Kauffman,1987). The protocol used here was described by Fuhrmann *et al.* (1989). The percentage of cross reaction was expressed as: O.D. (test fungal strain with antibody)- O.D. (test fungal strain with the preimune/ O.D.(test immunogen with antibody)- O.D.(test immunogen with preimune). This percentage indicates binding of the antibodies to the antigen of the mycelium. Cross reaction is considered not significant if under 12%.

Table 1. Fungal mycelium investigated by ELISA and immunofluorescence.

Penicillium subgen. *Aspergilloides*

P. glabrum	882494, 8930, MUCL 11337, MUCL 158571, 51162, 282, 8931, MUCL 134661, 8932
P. spinulosum	783183, 53798, 722169
P. decumbens	581377
P. implicatum	803265, 763119
P. montanense	49459
P. thomii	51417, 488
P. purpurescens	65412

Penicillium subgen. *Furcatum*

P. raistrickii	76803, 853436
P. jensenii	1389
P. citrinum	682011, 793276
P. waksmanii	692046
P. janthinellum	54268, 53105
P. oxalicum	571533, 51782
P. melinii	773173

Penicillium subgen. *Penicillium*

P. aurantiogriseum	882485, 50212, 611628, 753045, 863439, 873488, 882486, LMUB 8736, LMUB 8737, LMUB 8738, LMUB 8739
P. roqueforti	882492, 75146, 75148, 55352
P. camemberti	882493
P. chrysogenum	561177, 72284
P. brevicompactum:	611633, 52699
P. expansum	873492, 843384

Aspergillus

A. versicolor	882500, IP 1187-79, 62141, 642270, 763140, 833361, 863444
A. fumigatus	882499, IP 864, 65516, 722155
A. parasiticus	IP 1142-76, 531525
A. niger	853502, 853501, 742280
A. flavus	IP 855, 753053, 863450, 853495
A. nidulans	681992

Eurotium

E. repens	882498
E. rubrum	50115
E. amstelodami	50142
E. chevalieri	51991

Mucorales

M. racemosus	60427, 863459, 631816, 763095
M. plumbeus	843350, 49435
M. circinelloides	601609
M. lamprosporus	72151
M. dispersus	61629
M. fuscus	833343
M. ambiguus	732209
M. genevensis	783178
M. hiemalis	637
M. mucedo	305

Other genera

Absidia corymbifera	IP 1129
Geotrichum candidum	882502, 51590, IP 287, IP 285, IP 1447-83, 2288, 53577
Geotrichum capitatum	IP 1647, IP 1640-86
Ulocladium chartarum	882504

(cont.)

Table 1 (cont.)

Trichoderma longibrachiatum	882505, 823394
Trichoderma harzianum	823408, 803308
Trichoderma reesei	813296
Fusarium solani	8717
Fusarium decemcellulare	49119
Fusarium sp.	882507
Alternaria tenuissima	882503, 681988, 681989
Alternaria alternata	843390, 853387, 863465
Phialophora mustea	793219, 833332
Phialophora malorum	882506
Botrytis cinerea	831835, 83555, 823392

All isolates are from Laboratoire de Cryptogamie (LCP), except those indicated with IP: Institut Pasteur, Paris, LMUB: Laboratoire de Microbiologie de l'Universite Libre de Bruxelles and MUCL: Mycotheque Universite Catholique de Louvain, Belgium.

Immunofluorescence tests.
In the immunofluorescence tests, the antigen-antibody reaction is used for confirmation of binding. The principle is similar to ELISA, but the second antibody (specific for the first) is labelled with a fluorochrome and detected with an epifluorescence microscope. The protocol was described by Fuhrmann *et al.* (1989).

Table 2. Antibody production

Immunogen	Antigen	Monoclonal antibody
Penicillium glabrum	8824941	PF1-46/5
		PF3-93
		PF3-139
Aspergillus versicolor	8825001	AV1-15/37
		AV1-17/58
		AV1-32/70
Mucor racemosus	883095	MR3-20/12
		MR3-3/10
Geotrichum candidum	8825021	GC1-17/1

Immunogen	Antigen	Polyclonal antibody
Penicillium viridicatum	8824851	B3

RESULTS

A fungus or group of fungi may be characterized by the presence or absence of a specific antigenic site, in comparison with other antigenic profiles (Table 3). All Hyphomycetes and Mucorales tested (94 strains) have a common antigenic determinant (1) not shared

Table 3. Cross reaction results

FUNGAL MYCELIUM	NUMBER	A1	A2	A3	A4	A5	A6	A7	A8	A9	A10
Penicillium glabrum (A)	882494	+ 85	- 0	- 0	- 0	+ 100	- 6	+ 100	- 7	+ 100	- 8
	8930	+ 55	- 5	- 0	- 0	+ 93	- 1		- 12		- 7
	11337	+ 40		- 0	- 0	+ 62	- 2	+ 58	- 13	+ 72	- 8
	158571		- 3	- 0	- 0	+ 89	- 1		- 8		- 1
	51162		- 0	- 0	- 0	+ 93	- 4	+ 68	- 12	+ 93	- 2
	282		- 5	- 0	- 0	+ 58	- 1		- 6		- 2
	8921			- 0	- 0	+ 85	- 1		- 9		- 2
	134661		- 0			+ 89			- 6		
	8932		- 0	- 0	- 0	+ 55	- 3		- 9		- 3
P. spinulosum (A)	783183	+ 50	- 2	- 0	- 0	+ 102	- 0	+ 100	- 8		- 0
	53798		- 3	- 0	- 0	+ 78	- 0	+ 100	- 5	+ 36	- 0
	722169		- 2	- 0	- 0	+ 103	- 0	+ 100	- 9	+ 91	- 0
P. decumbens (A)	581377	+ 60				+ 24				- 0	
P. montanense (A)	49459	+ 80				+ 63		+ 32		- 8	
P. thomii (A)	51417		- 0	- 0	- 0		- 0	+ 94	- 11	+ 24	- 0
	488					+ 100			- 14		
P. purpurescens (A)	65412		- 0	- 0	- 0	+ 96	- 0	+ 95	- 10	+ 100	- 0
P. implicatum (A)	803265	+ 93		- 0	- 0	+ 39	- 0				- 0
	763119	+ 77							- 13		
P. raistrickii (F)	76803	+ 75				+ 37		+ 88		- 0	
	853436	+ 43				+ 48				- 0	
P. jenseni (F)	1389	+ 66	- 5	- 0	- 0	+ 58		+ 110		- 0	- 8
P. citrinum (F)	682011	+ 46				+ 67		- 10		- 0	
	793276	+ 69						- 0		- 0	
P. waksmanii (F)	692046	+ 68				+ 65		- 0		- 0	
P. janthinellum (F)	54268		- 2	- 0	- 0	+ 51			+ 51		- 5
	53105					+ 60			+ 75		
P. oxalicum (F)	571533		- 5	- 0	- 0	+ 36	- 3	- 0	+ 61	- 0	- 0
	51782		- 6							- 0	
P. melinii (F)	773173		- 0	- 0	- 0		- 10	- 0	+ 67	- 0	- 0
P. verrucosum (P)	882485	+ 128		- 0	- 0	+ 100	- 5	- 11	+ 100	- 0	- 4
	50212	+ 39				+ 21		- 0		- 0	
P. aurantiogriseum(P)	611628	+ 21				+ 63	- 2		+ 73	- 0	
	753045	+ 46				+ 53	- 0	- 0	+ 68	- 0	
	863439	+ 85				+ 73	- 2		+ 96	- 4	
	873488	+ 56				+ 78	- 0	- 2	+ 72	- 0	
	882486		- 0			+ 31					
	8736			- 0	- 0	+ 36	- 3		+ 76		- 0
	8737			- 0	- 0	+ 30	- 2		+ 100		- 0
	8738					+ 90	- 2		+ 100		- 0
	8739					+ 75	- 3		+ 100		- 0
P. roquefortii (P)	882492	+ 65	- 5	- 4	- 0	+ 93	- 11	- 1	+ 72	- 11	- 0
	75146	+ 21	- 0	- 0	- 0			- 5	+ 100	- 0	- 2
	75148		- 2	- 0	- 0	+ 21	- 10	- 0	+ 53	- 0	- 0
	55352		- 0	- 0	- 0	+ 41		- 10	17	- 0	
P. camembertii (P)	882493	+ 106	- 1	- 0	- 0	+ 87		- 5	+ 103	- 0	- 7
P. chrysogenum (P)	561177	+ 71				+ 82				- 0	
	72284	+ 84				+ 88		- 0		- 0	
P. brevicompactum (P)	611633	+ 65				+ 43		- 0		- 0	
	52699	+ 72				+ 51		- 0		- 0	
P. expansum (P)	873492	+ 66				+ 45		- 0		- 0	
	843384	+ 47				+ 51		- 0		- 0	
Aspergillus versicolor	882500	+ 100	- 3	- 0	- 0	+ 47	+ 100	+ 25	- 14	- 0	+ 100
	1187-79	+ 93	- 0	- 0		+ 92	+ 216		- 15		+ 100
	62141	+ 100				+ 66	+ 100	+ 100		- 0	
	642270	+ 100					+ 100				
	763140	+ 100				+ 44	+ 100			- 0	
	833361	+ 100				+ 73	+ 100	+ 63		- 0	
	863444	+ 100				+ 89	+ 100			- 0	
A. fumigatus	882499	+ 96	- 0	- 0	- 0	+ 42	+ 38	+ 84	- 5		- 0
	864	+ 84	- 2	- 0	- 0	+ 132	+ 210		- 6		- 0
	65516	+ 100				+ 92	+ 100	+ 88	- 11	- 0	
	722155	+ 67				+ 58	+ 100	+ 67	- 12	- 5	

(A). Subgenus Aspergilloides; (F). Subgenus Furcatum; (P) Subgenus Penicillium
+. Antigen present; -. no Antigen; Following numbers correspond to the percentage of cross reaction.

(cont.)

Table 3. Continued.

FUNGAL MYCELIUM	NUMBER	A1	A2	A3	A4	A5	A6	A7	A8	A9	A10
A. parasiticus	1142-76	+ 56	- 1	- 0	- 0	+ 94	+ 44		- 12		- 4
	531525	+ 100				+ 76	+ 100	+ 100		- 0	
A. niger	853502	+ 100				+ 100		+ 70		- 0	
	853501	+ 72				+ 24		+ 39		- 1	
	742280	+ 45				+ 59		+ 50		- 0	
A. flavus	855	+ 100	- 11	- 0	- 0	+ 151	+ 227		- 12		- 0
	753053	+ 42						+ 100		- 0	
	863450	+ 66								- 0	
	853495	+ 74				+ 21		+ 100		- 3	
A. nidulleius	681992	+ 54				+ 39		+ 100		- 0	
Eurotium repens	882498	+ 75	- 3	- 11	- 0	+ 48		- 0	- 11	- 3	- 0
E. rubrum	50115	+ 72					+ 100	- 0		- 0	
E. amstelodami	50142	+ 25					+ 100	- 0		- 0	
E. chevalieri	51991	+ 100				+ 89	+ 100				
Mucor racemosus	883095	+ 31	- 4	+ 100	+ 100	- 0		- 3	- 12	- 0	- 0
	60427	+ 47		+ 100	+ 75	- 0		- 0	- 0	- 0	- 0
	863459	+ 42		+ 88		- 0	- 0	- 0	- 0	- 0	- 0
	631816	+ 84		+ 67		- 0		- 0	- 0	- 0	- 0
M. plumbeus	843350	+ 100		+ 91		- 0		- 0	- 0	- 0	- 0
	49435	+ 51		+ 98	- 5	- 0	- 11	- 0	- 0	- 0	- 0
M. circinelloides	601609	+ 44		+ 100	- 2	- 0	- 0	- 0	- 0	- 0	- 0
M. lamprosporus	722151	+ 100		+ 100	+ 80	- 0		- 0	- 0		- 0
M. dispersus	611629	+ 100		+ 94	17	- 0		- 0	- 0	- 0	- 0
M. fuscus	833343	+ 67		+ 100	+ 24	- 0		- 0	- 0	- 0	- 0
M.ambiguus	732209	+ 68		+ 44	+ 30	- 0		- 0	- 0	- 0	- 0
M.genevensis	783178	+ 84		+ 92	+ 30	- 0		- 0	- 0	- 11	- 6
M. hiemalis	637	+ 92			+ 96	- 0		- 0	- 0	- 0	- 0
M. mucedo	305					- 0					
Absidia corymbifera	1129	+ 107	- 6				- 0	- 0		- 0	- 0
Geotrichum candidum	882502	- 9	+ 100	- 11	- 7	- 0	- 7		- 5		- 2
	51590	- 0	+ 82	- 0	- 0	- 4	- 0				- 0
	287	- 0	+ 79			- 0	- 10				- 2
	285	- 6	+ 75				- 11				- 0
	1447-83	- 0	+ 78			- 0	- 7				
	2288		+ 80				- 0		- 0		
	53577		+ 78								- 7
G. capitatum	1647	- 0	+ 75								
	1640-86	- 0	+ 73			- 0	- 4				- 0
Ulocladium chartarum	882504	+ 70	- 0	- 0	- 0	- 0		- 0	- 8	- 11	- 0
Trichoderma longibrachiatum	882505	+ 70	- 0	- 6	- 2	- 2	+ 28	- 0	- 10	- 0	- 0
	823394	+ 100					+ 100	- 9			
T. harzianum	803308	+ 100				- 10	+ 100	- 10	- 4	- 0	
	823408	+ 69				- 11	+ 100	- 11	- 0	- 0	
T. reesei	813296	+ 89					+ 100				
Fusarium solani	17	+ 108	- 1			- 0	+ 211	- 5		- 0	- 11
F. decemcellulare	49119	+ 68				- 0		- 0	- 0	- 0	
F. sp	882507	+ 87	- 0	- 0	- 0	- 0	+ 53	- 10	- 0	- 3	- 0
Alternaria tenuissima	882503	+ 37	- 0	- 0	- 0		- 8		- 11		- 1
	681988	+ 96						- 0			
	681989	+ 78				- 5	- 8	- 10		- 11	
A. alternata	843390	+ 55					- 0	- 2		- 0	
	853387	+ 34				- 0			- 7	- 0	
	863465	+ 88				- 8	- 8	- 3	- 4	- 0	
Phialophora mustea	793219					- 2		- 0	- 5	- 0	
	833332					- 8		- 0	- 7	- 0	
P. malorum	882506		- 8	- 0	- 0	- 0	- 6	- 1	- 0	- 0	- 0
Botrytis cinerea	1835		- 6			- 0					
	555		- 0								- 0
	3392							- 8		- 0	

+. Antigen present; -. no Antigen; Following numbers correspond to the percentage of cross reaction.

with *Geotrichum*. This genus is characterized not only by the absence of any common antigen with other fungi, but also by the presence of a specific one (2). These results were obtained with nine isolates of *G. candidum* and *G. capitatum*, and are in agreement with the distance of this genus with Endomycetous affinities from the other genera studied here. The 14 strains of *Mucor* have one antigen different from Hyphomycetes; cross reaction with the corresponding MAB (3) is 0%.

Aspergillus versicolor, *A. fumigatus*, *A. flavus* and *A. parasiticus* and *Eurotium rubrum*, *E. amstelodami* and *E. chevalieri* have in common three antigenic determinants (1, 5 and 6). The conjugation of the three is representative of *Aspergillus*, but not specific to it. Antigen 6 is produced in common with *Fusarium* and *Trichoderma*, 5 with *Penicillium* and 1 with each of these genera. *Eurotium* species are distinct from anamorphic *Aspergilli* by the absence of one antigen (7).

In our experiments, *Aspergilli* are the best characterized by four antigenic determinants: 1, 5, 6, 7. In this genus, *A. versicolor* species can be recognized by one species specific antigen (10).

For comparison, we tested representatives of *Penicillium* subgen. *Aspergilloides, Furcatum* and *Penicillium*. It appears that tested species in subgen. *Aspergilloides* have three antigens in common with *Aspergillus* and only two with subgen. *Penicillium*: 1 and 5. In these experiments, no differences were detected between subgen. *Furcatum* and *Penicillium*, which have two antigens in common with *Aspergillus* (11 and 5) and a specific one (8, polyclonal).

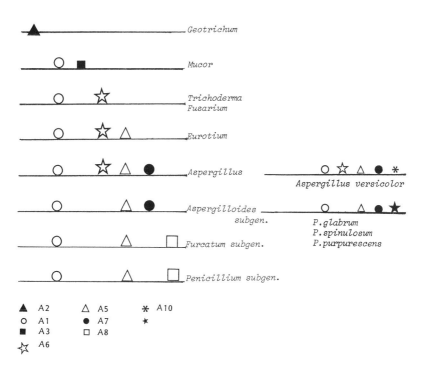

Figure 1. Schematic representation of antigenic sites on the cell wall.

DISCUSSION

Antibodies produced against specific antigenic structures of fungal cells have been shown to be important tools in classification and identification tests (Van der Heide and Kauffman, 1987). Production of species specific MAB is needed to use the corresponding species as an immunogen. Heterologous crossing with a given MAB permits establishment of clusters based on the presence or absence of one antigenic site (Fig.1). This study showed that the method allows distinctions at different levels of classification: generic, subgeneric and specific, even if the number of tested species is limited. *Aspergillus* and *Penicillium* have two antigens in common both genera may be separated from other tested fungi by a specific antigen (5).

Penicillium subgen. *Aspergilloides* appears to be closer to *Aspergillus* (3 antigenic sites in common) than to subgen.*Penicillium* (2 antigenic sites in common). These results confirm those obtained with antisera against *P. verrucosum* (Fuhrmann *et al.*, 1989). In this case, morphological features like apical enlargment of the stipes are correlated with antigenic similarities. This suggests a relative coherence of *Aspergillus* and *Penicillium* subgen. *Aspergilloides* and the intermediate position of subgen. *Aspergilloides* between *Penicillium* and *Aspergillus*. *Aspergillus* is well characterized with four antigens, of which three are common with *Eurotium* and three with *Penicillium* subgen. *Aspergilloides*. Subgenus *Aspergilloides* does not share antigen 6 with *Aspergillus* and *Eurotium* and, by this fact, and of course many other features, is not completely included in *Aspergillus*.

P. raistrickii G. Smith has antigen (7) in common with all tested species of subgen. *Aspergilloides*. This species is included by Pitt (1979) in subgen. *Furcatum*, who considered that it resembled *P. thomii* in subgen. *Aspergilloides* by production of sclerotia and rough walled vesiculate stipes, giving the appearance of being a "metulate *P. thomii*". Raper and Thom (1949) also previously noted a similarity between the two species. Our results provide further evidence of proximity between *P. raistrickii* and *P. thomii*, as both species reacted very strongly with MAB PF3.93 (88 and 94%).

P. glabrum (= *P.frequentans*), *P. purpurescens* and *P. spinulosum* possess a common antigen, A9, absent in other species of subgen.*Aspergilloides*. Morphologically these species are very close (Raper and Thom, 1949; Pitt, 1979), and immunogenic reactions confirm this relationship. Species tested of subgen. *Penicillium* and subgen. *Furcatum* do not react with the common *Aspergillus* subgen. *Aspergilloides* antigen and show a common reaction with antiserum A8. Differences in species reactions with this antiserum have been discussed earlier (Fuhrmann *et al.*, 1989). These reactions, however, are not so specific as MAB, as they detect more than one antigenic determinant in the same experiment. The highest level of identification in these experiments was obtained with *A. versicolor* who has four *Aspergillus* antigens and one species specific antigen. *Trichoderma* and *Fusarium*, both of which have teleomorphs in the family Nectriaceae, have a common antigenic determinant, shared with *Eurotium* and *Aspergillus* which are both in the family Trichocomaceae. These two families are grouped in the order Hypocreales by Malloch (1979). A6 antigen may be representative of this community but more strains need to be tested. *Mucor* (Zygomycotina) has an antigen in common with the Ascomycetes (A1) and Hyphomycetes with ascomycetous affinities. Surprisingly, this antigen does not react with *Geotrichum*, whose teleomorph *Dipodascus* is an Ascomycete. *Geotrichum* is antigenically segregated from other filamentous fungi with which it does not share any antigens. It is characterized by A2 antigen, which is specific to the genus.

The MAB ELISA method has shown to be very efficient for systematic purposes and to identify species, genera and groups. Results are in accordance overall with accepted

Table 4. Specificity of antibodies

Monoclonal antibody cross reaction with antigen specificity

AV1-15/37	91 strains	A1	Hyphomycetes and Mucorales
GC1-17/11	48 strains	A2	Genus *Geotrichum*
MR3-20/12	55 strains	A3	Genus *Mucor*
MR3-3/10	50 strains	A4	Specific for *Mucor* species
PF1-46/5	101 strains	A5	Genera *Penicillium* and *Aspergillus*
AV1-17/58	69 strains	A6	Genera *Aspergillus*, *Trichoderma* and *Fusarium*
PF3-93	77 strains	A7	Genus *Aspergillus* (except *Eurotium*) and *Penicillium* subgenus *Aspergilloides*
PF3-139	83 strains	A9	Specific for some *Penicillium* spp of subgenus *Aspergilloides*
AV1-32/70	62 strains	A10	*A. versicolor*

Polyclonal antibody cross reaction with antigen specificity

B3	73 strains	A8	*Penicillium* subgenus *Furcatum*

classifications. They have proved to be useful to pinpoint the intermediate position of some taxa. MAB has the advantage of using standard reagents produced indefinitely in vivo or in vitro. They bind to one, and only one, epitope and are quite specific. Using ELISA, many isolates may be tested and identified rapidly.

REFERENCES

BRIDGE, P.D., HAWKSWORTH, D.L., KOZAKIEWICZ, Z., ONIONS, A.H.S and PATERSON, R.R.M. 1986. Morphological and biochemical variations in single isolates of *Penicillium*. *Transactions of the British Mycological Society* 87: 389-396.

DUPONT, J., POLONELLI, L. and MORACE, G. 1988. Application de l'immunologie (anticorps monoclonaux) a la caractérisation d'une souche de *P. camemberti* Thom. *Le Lait* 68: 435-442.

FUHRMANN, B., ROQUEBERT, M.F., van HOEGAERDEN, M. and STROSBERG, A.D. 1989. Immunological differentation of *Penicillium* species. *Canadian Journal of Microbiology* 35 (in press).

HEIDE, van der S. and KAUFFMAN, F. 1987. Serological methods for taxonomic diagnostic research of yeasts. *Studies in Mycology, Baarn* 30: 351-360.

HINGAND L., LE COZ, S., KERLAN, C. and JOUAN, B. 1983. Application de l'immunofluorescence la détection de *Phoma exigua* var. *foveata*, agent de la gangrène de la pomme de terre. *Agronomie* 3: 51-56.

KAUFMAN L. and STANDARD P.G. 1987. Specific and rapid identification of medically important fungi by exoantigen detection. *Annual Review of Microbiology* 41: 209-225.

KOHLER, G. and MILSTEIN,C. 1975. Continuous cultures of fused cells secreting antibody of predefined specificity. Nature 256: 495.

LIN, H.H., LISTER R.M. and COUSIN, M.A. 1986. Enzyme linked immunoabsorbent assay for detection of mold in tomato puree. *Journal of Food Science* 51: 180.

MALLOCH, D. 1979. Plectomycetes and their anamorphs. In The Whole Fungus, Vol.1, ed. W.B. Kendrick, pp. 153-165. Ottawa: National Museums of Canada.

MORACE, G., ORSINI, D., CASTAGNOLA, M. and POLONELLI, L. 1984. Exoantigen studies of *Phanerochaete chrysosporium* and *Sporotrichum pruinosum* cultures. *Igiene Moderna* 81: 314-321.

NOTERMANS, S., HEUVELMAN, C.J., van EGMOND, H.P., PAULSH, W.E. and BESLING, J.R. 1986. Detection of mold in food by enzyme-linked immunosorbent assay. *Journal of Food Protection* 49: 135-142.

PEPYS, J. and LONGBOTTOM, J.L. 1978. Immunological methods in mycology. In Handbook of Experimental Immunology. Third edition, ed. D. Weir, pp. 1-27. Oxford: Blackwell Scientific Publications.

PITT J.I. 1979 The genus *Penicillium* and its teleomorphic states *Eupenicillium* and *Talaromyces*. Academic Press, London.

POLONELLI, L., ORSINI D., CASTAGNOLA, M. and MORACE, G. 1984. Serological identification of aflatoxin potential producing *Aspergillus* species. *Igiene Moderna* 81: 1128-1136.

POLONELLI, L. and MORACE, G. 1985. Antigenic characterization of *Micrsoporum canis, M. distortum, M. equinum, M. ferrugineum* and *Trichophyton soudanense* cultures. *Mycopathologia* 92: 7-10.

POLONELLI, L., CASTAGNOLA, M., D'URSO, C. and MORACE, G. 1985. Serological approaches for identification of *Aspergillus* and *Penicillium* species. *In* Advances in *Penicillium* and *Aspergillus* Systematics, eds. R.A. Samson and J.I. Pitt, pp. 267-280 New York and London: Plenum Press.

POLONELLI, L. and MORACE, G. 1986. Specific and common antigenic determinants of *C.albicans* isolates detected by monoclonal antibodies. *Journal of Clinical Microbiology* 8: 366-368.

POLONELLI, L., MORACE, G., ROSA, R., CASTAGNOLA, M. and FRISVAD, J.C. 1987. Antigenic characterization of *Penicillium camemberti* and related common cheese contaminants. *Applied and Environmental Microbiology* 53: 872-878.

RAPER, K.B. and THOM, C. 1949. A Manual of the *Penicillia*. Baltimore: Williams and Wilkins.

SEKHON, A.S., LI, J.S.K. and GARG, A.K. 1982. *Penicillium marneffei*: serological and exoantigen studies. *Mycopathologia* 77: 51-54.

STANDARD, P.G. and KAUFMAN, L. 1977. Immunological procedure for the rapid and specific identification of *Coccidioides immitis* cultures. *Journal of Clinical Microbiology* 5: 149-153.

DIALOGUE FOLLOWING DR. ROQUEBERT'S PRESENTATION

LATGÉ: Do you have any idea of the class of immunoglobulins or the chemical nature of your wall antigens?

ROQUEBERT: The immunoglobulins are IgM. We believe that the antigens are poly-saccharides.

PITT: I'd be intrigued to see how *Paecilomyces* fits in with *Aspergillus* and *Penicillium* using the techniques that you have described today.

THE SIGNIFICANCE OF YEAST EXTRACT COMPOSITION ON METABOLITE PRODUCTION IN *PENICILLIUM*

O. Filtenborg, J.C. Frisvad and U. Thrane

Department of Biotechnology
The Technical University of Denmark
2800 Lyngby, Denmark

SUMMARY

In the literature and in our own experience, significant variations occur in morphological characteristics and production of secondary metabolites by cultures grown on YES (2% yeast extract and 15% sucrose) agar, a substrate which is often used in screening for mycotoxins in moulds. In this investigation we have demonstrated a very significant influence of yeast extract brand (Difco, Sigma Y4000 and Y0325, Oxoid, Merck, Lab M and Gibco) on the production of mycotoxins in YES by some important Penicillia. Using a TLC screening method the variation in mycotoxin production due to the use of different brands of yeast extract ranged between detection in 5 days and none detected in 4 weeks. The difference in mycotoxin production was often accompanied by differences in several other characteristics like pH changes of the substrate, sporulation, colony diameter and reverse colour. We have been unable to find which components in the yeast extracts were responsible for the observed changes, but the addition of $MgSO_4$ appeared to be a satisfactory compensation in most respects. So it is suggested that this compound in general is added to the YES formula, along with previously suggested compounds like $ZnSO_4$ and $CuSO_4$, thus making this substrate a very valuable and reliable tool in screening for production of secondary metabolites and in mould taxonomy. Further it is suggested to use pH registration monitoring in the cultures parallel to screening for secondary metabolites, since pH differences proved to be a useful indication of significant changes in the detected profiles.

INTRODUCTION

It is generally recognized that substrate composition has a major impact on morphology and production of secondary metabolites of mould cultures (Aharonowitz, 1980; Betina, 1984; Orvehed *et al.*, 1988; Constantinescu, 1990). The importance of trace metals e.g. Zn, Cu, and magnesium on mycotoxin production and pigmentation has been the subject of several investigations (Weinberg, 1970; Coupland and Niehaus, 1987a, b; Mashaly *et al.*, 1988; Jackson *et al.*, 1989) and so has the influence of C and N sources (Krumphanzl *et al.*, 1982; Haggblom and Ghosh, 1985; Coupland and Niehaus, 1987b; Orvehed *et al.*, 1988).

During our taxonomic studies of moulds we have now and then experienced surprisingly large morphological differences between cultures of the same isolate on substrates differing only in the origin of the ingredients. The substrates which exhibit these differences are CYA, made with Czapek Dox Broth (Difco) and yeast extract 5 g/l, and YES (15% sucrose + 2% yeast extract) agar, which have a worldwide use in studies of morphology and production of secondary metabolites in moulds (Pitt, 1979; Frisvad and Filtenborg, 1983; Betina, 1984; Samson and Pitt, 1985). Yeast extract and peptones are important components in substrates used for mycotoxin production and it is known that considerable variations occurs in its composition due to variations in raw materials and production conditions (Odds *et al.*, 1978; Bird *et al.*, 1985; Bridge *et al.*, 1985). We have tested (unpublished data) the influence of different brands of yeast extract on morphology

Modern Concepts in Penicillium and Aspergillus Classification
Edited by R. A. Samson and J. I. Pitt
Plenum Press, New York, 1990

and secondary metabolite profiles of mould cultures on CYA and YES. We found that cultures on CYA were only slightly influenced by the choise of yeast extract brand, whereas cultures on YES were very much so. Compared to Difco, the yeast extract from Oxoid, Sigma Y-0375, Lab M, Gibco and Merck in general supported a significantly lower production of secondary metabolites, if any, as well as a reduced sporulation and less colourful reverses. On the other hand, the yeast extract Sigma Y-4000 supported increased production of certain secondary metabolites compared to Difco, while other metabolites were produced in equal or slightly smaller amounts. Similarly Sigma Y-4000 yeast extract supported greater sporulation and more colourful reverses than Difco. All tested brands of yeast extract produced a significant variation in colony diameters of cultures on YES (Filtenborg, Frisvad and Thrane, unpublished results).

This investigation aimed to establish the differences in yeast extract composition which were responsible for the significant variations in morphology and production of secondary metabolites of cultures on YES. It is important to clarify these differences as they may be the cause of disagreements concerning specific secondary metabolite profiles.

As obtaining the exact composition of all brands of yeast extract from the manufacturers has been impossible, this investigation is based on the lack of certain salts and trace metals in YES, inspired by the above mentioned results with CYA which implies the importance of adding Czapek Dox Broth.

MATERIAL AND METHODS

Fungi.

Isolates used were *Penicillium expansum* Link (Frank 597 = IBT 3034), *Penicillium roqueforti* Thom IBT 3035, *Penicillium verrucosum* Dierckx chemotype II IBT 3038 and *Penicillium crustosum* Thom IBT 3036.

Substrates.

A number of YES substrates were used all with 2% yeast extract, 15% sucrose and 1.5% agar, differing only in the brand of yeast extract: Difco (DYES), Oxoid (OYES), Sigma Y-4000 (S1YES), Sigma Y-0375 (S2YES), Lab M (LYES), Merck (MYES) and Gibco (GYES). The substrates were prepared with and without the addition of $MgSO_4 \cdot 7H_2O$ (0.5 g/l) or Czapek Dox Broth (Bacto) (35 g/l). The pH of S1YES was adjusted to 6.5, as it was pH 4.5 without adjustment. Other media were close to pH 6.5 unadjusted.

Inoculation.

Inoculation was in triplicate. Observed variations in triplicate colony diameters did not exceed 3 mm. The agar plug TLC method (Filtenborg *et al.*, 1983) was used in metabolite screening, with samples taken from all three colonies. Variation in metabolite concentrations between triplicates was never significant. Measurement of pH was performed by applying the agar plugs, used for the metabolite screening, onto pH indicator strips (Acilit pH 0-6, Merck art. 9531 and Neutralit pH 5-10, Merck art. 9533). Measurements were made on all three colonies, and variation did not exceed 0.5 pH units.

RESULTS AND DISCUSSION

Preliminary investigations performed with *Penicillium verrucosum* demonstrated that the addition of Czapek Dox Broth to YES substrates containing different brands of yeast extract almost completely eliminated the observed differences in morphological characteristics and ability to produce secondary metabolites. On the other hand, addition of trace metals, Cu and Zn, did not change the culture characteristics. Each component in the Czapek Dox Broth was then added separately to the YES substrates. Only the addition of $MgSO_4$ appeared to have an effect, and the effect was very much the same as observed with Czapek solution. Further investigations then aimed to see if other important moulds responded to the addition of $MgSO_4$ like *Penicillium verrucosum*. The importance of $MgSO_4$ concentration was also checked.

Penicillium expansum

When *Penicillium expansum* was grown all expected secondary metabolites were detected after 7 days in YES with Difco or Sigma Y-4000 yeast extract, but none were detected when Oxoid, Sigma-0375, Gibco, Lab M or Merck were used (Table 1).

Table 1. Influence of $MgSO_4$ addition on growth and metabolite production by *Penicillium expansum* on various YES formations after 7 days incubation.

Medium	MGSO$_4$ added	Colony Ø (mm)	Sporu-lation	pH	Roque-fortine C	Citrinin	Patulin
DYES (a)	–	64	weak	5.5	+3(b)	+4	+2
OYES	–	60	weak	3.5	–	–	–
S1YES	–	60	some	3.5	(+)	+1	+4
S2YES	–	NT(c)	NT	3.0	–	–	–
LYES	–	NT	NT	3.5	–	–	–
MYES	–	NT	NT	3.0	–	–	–
GYES	–	NT	NT	3.0	–	–	–
DYES	+	62	weak	4.0	+4	+4	+4
OYES	+	59	weak	3.5	+3	+3	+4
S1YES	+	61	some	3.0	(+)	+2	+6
GYES	+	NT	NT	5.5	–	+5	+3
MYES	+	NT	NT	5.5	–	+5	+3
LYES	+	NT	NT	3.5	–	+2	+3
S2YES	+	NT	NT	3.0	–	+1	+5
DYES	Cz(d)	61	some	7.0	+4	+4	+3
OYES	Cz	62	some	5.0	+6	+4	+5
S1YES	Cz	60	some	4.0	+3	+2	+5

(a) for formulae see text; (b) - no metabolite was detected, +1: detection limit, +2, +3 etc., indicates arbitraly chosen values for increasing metabolite-concentrations; (c) NT, not tested; (d) addition of Czapek Dox broth (35g/l).

By adding $MgSO_4$, alone or as a component of Czapek Dox Broth, citrinin and patulin could be detected in every YES substrate tested, with citrinin dominating at pH values above 5.5 and patulin at low pH. As YES is not the optimal substrate for roquefortine C production, detection can be variable, as seen in Table 1. Addition of Czapek Dox Broth, which stabilises the final pH at a slightly higher level than does $MgSO_4$ alone, appeared to increase detection of roquefortine C significantly.

Increasing or decreasing the $MgSO_4$ concentration to 2.5 or 0.1 g/l produced only a minor effect, with metabolite production tending to increase with increasing concentration. Increasing concentration of Czapek Dox Broth to 90 g/l significantly decreased patulin and citrinin production, but increased the roquefortine C concentration. This effect may be due to nitrate inhibition of production of polyketides (Ward and Packter, 1974; Grootwassink *et al.*, 1980; Orvehed *et al.*, 1988).

After 14 days incubation, citrinin concentration was higher or remained constant and detection was possible on every subtrate, although very weak on OYES. The level of roquefortine C did not change significantly, wheras the level of patulin significantly decreased and detection was only possible in S1YES, with and without added $MgSO_4$. The pH values in general increased to 6-7 between 7 and 14 days incubation, but the pH of S1YES rose only to about 4, and this may account for the detection of patulin in that medium. This is in agreement with results obtained for canescin production of *P. canescens* Sopp by Brian *et al.*, (1953), who stated that "a variety of media were suitable for its production, the main requirements being that the pH shall not rise above 6.5". It was of particular interest to note that media which did not support production of citrinin supported accumulation of one of its suggested precursors, 2-carboxy-3,5-dihydroxy-benzylmethylketone (Turner, 1971).

Variations in colony diameter were only of minor importance. Differences in sporulation, reverse colour and other morphological characteristics were observed, but reduced significantly by the addition of $MgSO_4$.

Table 2. Influence of $MgSO_4$ addition on growth and metabolite production by *Penicillium roqueforti* on various YES formations after 7 days incubation.

Medium	GSO$_4$ added	Colony Ø (mm)	Sporu- lation	pH	PR-toxin	Roque- fortine C
DYES(a)	–	60	heavy	4.5	+3(b)	+4
OYES	–	46	some	6.0	–	+5
S1YES	–	57	heavy	4.5	+1	+5
S2YES	–	NT(c)	NT	6.5	–	+4
LYES	–	NT	NT	6.0	–	+4
MYES	–	NT	NT	5.5	–	+1
GYES	–	NT	NT	6.0	–	+4
DYES	+	63	heavy	4.0	+1	+2
OYES	+	58	heavy	4.5	+1	+2
S1YES	+	60	heavy	4.0	–	+2
GYES	+	NT	NT	5.0	+1	+1
MYES	+	NT	NT	5.0	+3	+2
LYES	+	NT	NT	5.0	+4	+1
S2YES	+	NT	NT	5.0	–	+1
DYES	Cz(d)	53	heavy	5.0	–	+2
OYES	Cz	49	heavy	5.5	–	+2
S1YES	Cz	53	heavy	4.5	–	+3

Footnotes: see Table 1.

Penicillium roqueforti

PR-toxin was only detected when *P. roqueforti* was grown on DYES and S1YES. PR-toxin is more frequently, but not invariably, detected when MgSO₄ was added alone, but not as component of Czapek Dox Broth (Table 2). However, increasing the MgSO₄ concentration to 2.5 g/l (data not shown) significantly improved the detection frequency. When Cz broth was added PR-toxin was not produced. Again, this may be due to nitrate inhibition, or it may be an effect of pH, or both. After 14 days of incubation PR-toxin concentration decreased in substrates without MgSO₄, but increased in substrates with MgSO₄ added. Only minor pH changes were observed during prolonged incubation (14 days), except in DYES with added Czapek Dox Broth, where the pH rose to 7, and here it was not possible to detect PR-toxin.

Roquefortine was detected in every case, at higher concentrations in substrates without added MgSO₄. This differs from the detection of this toxin in *P. expansum* (see Table 1), but the pH values were higher in the *P.roqueforti* cultures.

Differences were observed in colony diameters, sporulation and reverse colours on the various media, especially OYES, but again addition of MgSO₄ greatly reduced these.

Penicillium crustosum

Terrestric acid was detected when *P. crustosum* was grown in each of the substrates, but little was found in OYES without MgSO₄ (Table 3). CYA is a much better substrate than YES for production of penitrem A and roquefortine C (Filtenborg *et al.*, 1983), but MgSO₄ and Czapek Dox Broth increased production of both alkaloids quite significantly.

Table 3. Influence of MgSO₄ addition on growth and metabolite production by *Penicillium crustosum* on various YES formations after 7 days incubation.

Medium	MGSO₄ added	Colony Ø (mm)	Sporu- lation	pH	Terrestric acid	Penitrem A	Roque- fortine
DYES(a)	–	51	heavy	3.5	+7(b)	+3	+2
OYES	–	50	weak	4.0	(+)	–	–
S1YES	–	55	heavy	3.0	+4	+5	+2
S2YES	–	NT(c)	NT	3.5	+3	–	NT
LYES	–	NT	NT	4.0	+3	+1	NT
MYES	–	NT	NT	7.0	+3	–	NT
GYES	–	NT	NT	4.0	+3	+1	NT
DYES	+	53	heavy	3.5	+5	+4	+3
OYES	+	50	heavy	3.5	+6	+4	+3
S1YES	+	53	heavy	3.0	+2	+2	+2
GYES	+	NT	NT	4.0	+3	+2	NT
MYES	+	NT	NT	3.5	+5	+3	NT
LYES	+	NT	NT	3.0	+4	+1	NT
S2YES	+	NT	NT	3.0	+3	–	NT
DYES	Cz(d)	52	weak	4.5	+7	+6	+5
OYES	Cz	53	weak	5.0	+6	+1	+4
S1YES	Cz	54	heavy	3.5	+5	+6	+6

Footnotes: see Table 1

Table 4. Influence of MgSO$_4$ addition on growth and metabolite production by *Penicillium verrucosum* on various YES formations after 7 days incubation.

Medium	MGSO$_4$ added	Colony Ø (mm)	Sporu-lation	Reverse colour	pH	Citrinin	Ochra-toxin A
DYES(a)	–	34	–	VB (e)	4.5	+2(b)	+3
OYES	–	28	–	C	5.0	–	–
S1YES	–	34	–	VB	4.5	+3	+2
S2YES	–	NT(c)	NT	NT	5.0	–	–
LYES	–	NT	NT	NT	4.0	+1	+1
MYES	–	NT	NT	NT	4.5	+1	–
GYES	–	NT	NT	NT	5.0	–	+1
DYES	Cz(d)	34	–	C	5.5	+5	+5
OYES	Cz	30	–	C	6.0	+6	+6
S1YES	Cz	34	–	LVB	5.5	+6	+6
DYES	+	32	–	VB	5.0	+4	+3
OYES	+	32	weak	VB	4.0	+5	+2
S1YES	+	32	–	VB	4.5	+6	+3
GYES	+	NT	NT	NT	4.5	+5	+4
MYES	+	NT	NT	NT	3.5	+5	+2
LYES	+	NT	NT	NT	5.0	+5	+3
S2YES	+	NT	NT	NT	3.5	+3	+3

Footnotes (a) to (d) see Table 1; (e) VB, violet brown; C, cream-coulored, LVB, light violet brown.

Penicillium verrucosum

After 7 days incubation, *P. verrucosum* produced citrinin and ochratoxin A in significant amounts only on DYES and S1YES (Table 4). The addition of MgSO$_4$ resulted in good detection of both toxins on every substrate tested. After 14 days incubation, *P. verrucosum* produced citrinin on every substrate, but ochratoxin A production was still variable. The violet brown reverse colour which is diagnostic for *P.verrucosum* chemotype II when grown on YES (Frisvad, 1983), did not appear on OYES. However, the characteristic colour appeared on every substrate when MgSO$_4$ was added.

CONCLUSIONS

The detection of the mycotoxins citrinin, patulin, roquefortine C, PR-toxin, terrestric acid, penitrem A and ochratoxin A on YES, using a simple TLC screening procedure (Filtenborg *et al.*, 1983, Frisvad *et al.*, 1989) has been shown here to be dependent on the brand of yeast extract used in the substrate. In several cases detection was not possible at all, even after 14 days of incubation. Only Difco and Sigma Y-4000 yeast extracts provided reasonably reliable secondary metabolite production by different Penicillia.

This is of course a very serious and unacceptable problem, as YES is a very valuable substrate for taxonomy based on secondary metabolites. This may explain reports in the literature (Bridge *et al.*, 1986, 1987; Land and Hult, 1987; Stenwig, 1988) that the significant and consistent profile of secondary metabolites observed by us for at great number of *Penicillium* species (Frisvad and Filtenborg, 1983, Frisvad and Filtenborg, 1989), may be difficult to reproduce everywhere. Perhaps it was pure luck that there was Difco yeast extract on our shelves when we started to work on mycotoxin producing moulds using the agar plug method some 13 years ago.

However, according to the results in this investigation, the problem can be overcome by including $MgSO_4$ (0.5 g/l) in YES, in addition to the $ZnSO_4$ and $CuSO_4$ recommended previously (Smith, 1949; Frisvad and Filtenborg, 1983). $MgSO_4$ obviously compensates for differences in yeast extract composition between brands, and perhaps between batches, which we have shown significantly affects production of secondary metabolites, and gross morphology and colours of mould cultures on YES. However, clarifying the exact nature of these differences has not yet been possible. Further investigations will be carried out to solve these problems.

Sigma Y-4000 yeast extract appears to be unique. The final pH of S1YES, around 4.5 makes pH regulation necessary to avoid softening of the agar gel. During growth of cultures on S1YES, pH initially decreases significantly, as in other substrates, but the usual subsequent pH increase is minimal in S1YES compared to the other media. Sporulation was often significantly heavier on S1YES, at the expense of mycelium production, and often the secondary metabolite profile was increased both qualitatively and quantitatively on S1YES compared to the other media.

Based on the results reported here it is suggested that culture pH always be measured at the time of screening for secondary metabolite profiles. In our experience knowledge of pH level may assist in detection of culture changes or problems with the screening procedure. Significant pH deviations mean that the screening must be carried out after a different incubation time, on a different substrate or perhaps that the culture is contaminated.

REFERENCES

AHARONOWITZ, Y. 1980. Nitrogen metabolite regulation of antibiotic biosynthesis. *Annual Review of Microbiology* 34: 209-233.

BETINA, V., ed. 1984. Mycotoxins: production, isolation, separation and purification. Amsterdam: Elsevier.

BIRD, N.P., CHAMBERS, J.G., LEECH, R.W. and CUMMINS, D. 1985. A note on the use of metal species in microbiological tests involving growth media. *Journal of Applied Bacteriology* 59: 353-355.

BRIAN, P.V., HEMMING, H.G., MOFFATT, J.S. and UNWIN, C.H. 1953. Canescin, an antibiotic produced by *Penicillium canescens. Transactions of the British Mycological Society* 36: 243-247.

BRIDGE, P.D., HAWKSWORTH, Z., KOZAKIEWICZ, Z., ONIONS, A.H.S., PATERSON, R.R.M. and SACKIN, M.J. 1985. An integrated approach to Penicillium systematics. *In* Advances in *Penicillium* and *Aspergillus* Systematics, eds. R.A. Samson and J.I. Pitt, pp. 281-309. New York and London: Plenum Press.

BRIDGE, P.D., Hawksworth, D.L., KOZAKIEWICZ, Z., ONIONS, A.H.S. and PATERSON, R.R.M. 1986. Morphological and biochemical variation in single isolates of *Penicillium. Transactions of the British Mycological Society* 87: 389-396.

BRIDGE, P.D., HUDSON, L., KOZAKIEWICZ, Z., ONIONS, A.H.S. and PATERSON, R.R.M. 1987. Investigation of variation in phenotype and DNA content between single conidium isolates of single *Penicillium* strains. *Journal of General Microbiology* 133: 995-1004.

CONSTANTINESCU, O. 1990. Standardisation of methods in *Penicillium* identification. *In* Modern Concepts in *Penicillium* and *Aspergillus* Classification, eds. R.A. Samson and J.I. Pitt, pp. 17-25 New York and London: Plenum Press.

COUPLAND, K. and NIEHAUS, W.G. 1987a. Stimulation of alternariol biosynthesis by zinc and magnesium ions. *Experimental Mycology* 11: 60-64.

COUPLAND, K. and NIEHAUS, W.G. 1987b. Effect of nitrogen supply, Zn++, and salt concentration on kojic acid and versicolorin biosynthesis by *Aspergillus parasiticus. Experimental Mycology* 11: 206-213.

FILTENBORG, O. and FRISVAD, J.C. 1980. A simple screening method for toxinogenic moulds in pure culture. *Lebensmittel Wissenschaft und Technologie* 13: 128-130.

FILTENBORG, O., FRISVAD, J.C. and SVENDSEN, J.A. 1983. Simple screening method for molds producing intracellular mycotoxins in pure culture. *Applied and Environmental Microbiology* 45: 581-585.

FRISVAD, J.C. 1983. A selective and indicative medium for groups of *Penicillium viridicatum* producing different mycotoxins in cereals. *Journal of Applied Bacteriology* 54: 409-416.

FRISVAD, J.C. and FILTENBORG, O. 1983. Classification of terverticillate Penicillia based on profiles of mycotoxins and other secondary metabolites. *Applied and Environmental Microbiology* 46: 1301-1310.

FRISVAD, J.C. and FILTENBORG, O. 1989. Terverticillate Penicillia: Chemotaxonomy and mycotoxin production. *Mycologia* 81 (in press).

FRISVAD, J.C. FILTENBORG, O. and THRANE, U. 1989. Analysis and screening for mycotoxins and other secondary metabolites in fungal cultures by thin-layer chromatography and high-performance liquid chromatography. *Archives of Environmental Contamination and Toxicology* 18: 331-335.

GROOTWASSINK, J.W.D. and GAUCHER, G.M. 1980. De novo biosynthesis of secondary metabolism enzymes in homogenous cultures of *Penicillium urticae*. *Journal of Bacteriology* 141: 443-455.

HAGGBLOM, P.E. and GHOSH, J. 1985. Postharvest production of ochratoxin A by *Aspergillus ochraceus* and *Penicillium viridicatum* in barley with different pH levels. *Applied and Environmental Microbiology* 49: 787-790.

JACKSON, M.A., SLINNINGER, P.J. and BOTHAST, R.J. 1989. Effect of zinc, iron, cobalt and manganese on *Fusarium moniliforme* NRRL 13616 growth and fusarin C biosynthesis in submerged cultures. *Applied and Environmental Microbiology* 55: 649-655.

KRUMPHANZL, V., SIKYTA, B. and VANEK, Z. eds. 1982. Overproduction of microbial products. London: Academic Press.

LAND, C.J. and HULT, K. 1987. Mycotoxin production by some wood-associated *Penicillium* spp. *Letters in Applied Microbiology* 4: 41-44.

MASHALY, R.I., HABIB, S.L., EL-DEEB, S.H., SALEM, M.H. and SAFWAT, M.M. 1988. Isolation, purification and characterization of enzyme(s) responsible for conversion of sterigmatocystin to aflatoxin B1. *Zeitschrift fur Lebensmittel-Untersuchung und -Forschung* 186: 118-124.

ODDS, F.C., HALL, C.A. and ABBOTT, A.B. 1978. Peptones and mycological reproducibility. *Sabouraudia* 16: 237-246.

ORVEHED, M., HAGGBLOM, P. and SODERHALL, K. 1987. Activity of NADPH-generating pathways in relation to polyketide synthesis in the fungus *Alternaria alternata*. *Experimental Mycology* 11: 187-196.

PITT, J.I. 1979. The genus *Penicillium* and its teleomorphic states *Eupenicilium* and *Talaromyces*. London: Academic Press.

SAMSON, R.A. and PITT, J.I., eds. 1985. Advances in *Penicillium* and *Aspergillus* Systematics. New York en London: Plenum Press.

SMITH, G. 1949. The effect of adding trace elements to Czapek-Dox medium. *Transactions of the British Mycological Society* 32: 280-283.

STENWIG, H. 1988. Thin-layer chromatography of plugs from agar cultures as an aid for identification of moulds in routine mycological examination of animal feeds. *Acta Agriculture Scandinavia* 38: 215-222.

TURNER, W.B. 1971. Fungal metabolites. London: Academic Press.

WARD, A.C. and PACKTER, N.M. 1974. Relationships between fatty acid and phenol synthesis in *Aspergillus fumigatus*. *European Journal of Biochemistry* 46: 323-333.

WEINBERG, E.D. 1970. Biosynthesis of secondary metabolites: roles of trace elements. *Advances in Microbial Physiology* 4: 1-44.

DIALOGUE FOLLOWING DR. FILTENBORG'S PRESENTATION

PITT: We've recently been noticing problems with yeast extracts. I devised CYA as a successor to Raper's corn steep liquor because yeast extract is much more widely available. The yeast extract that was available at the time had the advantage of containing all the trace elements and accessory nutrients that the fungi might require for growth. It compensated for differences in water supplies, grades of chemicals, including agar, and so on. The addition of Smith's trace elements, at that time, was either unnecessary or counterproductive. Later, it became apparent in some European labs that

CYA was no longer producing good sporulation. It is now clear that copper is the problem. Too little copper inhibits sporulation because it is a cofactor for the enzymes that produce the pigmentation of the *Penicillium* conidia. Too much copper is toxic. The problem is that the medium makers are now using a more purified yeast extract that is less suitable for fungal work. The cure for this is to add 0.05% $CuSO_4.5H_2O$ and 0.1 % $FeSO_4.5H_2O$ to CYA.

FRISVAD: The problem may not be as simple as the absolute concentration of copper. Metals might be chelated by some yeast extracts.

PITT: There is a fair amount of flexibility in copper concentration.

FRISVAD: You may get sporulation, but poor colour production also. The conidia may be brown if insufficient copper is present. Magnesium sulfate, as we mention in the paper, helps get the proper colouration with YES. It is already present in CYA.

ONIONS: Is there any problem with malt extracts? Malt extracts are highly variable.

PITT: This is true, but it doesn't seem to cause any problems.

FRISVAD: We have considered this in the Subcommission on *Penicillium* and *Aspergillus* Systematics, and we have a lot of data on colony diameters and so on using different malt extracts in different laboratories. Eventually, this data should be analyzed.

9

TAXONOMIC STUDIES ON THE TELEOMORPHS OF PENICILLIUM AND ASPERGILLUS

CHEMOTAXONOMY OF *EUPENICILLIUM JAVANICUM* AND RELATED SPECIES

J.C. Frisvad[1], R.A. Samson[2] and A.C. Stolk[2]

[1]*Department of Biotechnology*
The Technical University of Denmark
2800 Lyngby, Denmark

[2]*Centraalbureau voor Schimmelcultures*
3740 AG Baarn, The Netherlands

SUMMARY

The secondary metabolites of isolates of *Eupenicillium javanicum* and related species were examined by thin layer chromatography and high-performance liquid chromatography. *E. javanicum* produces xanthomegnin and viomellein, while *E. ehrlichii* differs by the production of brefeldin A, palitantin and penicillic acid. *E. zonatum* produce janthitrems, xanthomegnin and brefeldin A, indicating an intermediary position between the two former taxa. Isolates of *P. janthinellum* also produces xanthomegnin and related compounds and this support its identity as the anamorph of *E. javanicum*. Chemically *P. cremeogriseum* fits very well with *E. ehrlichii* and it could be considered as the anamorph of this ascomycete. Production of janthitrems and brefeldin A indicates that P. piscarium is the first available name for the anamorph of E. zonatum. The slow growing species *E. meloforme, E. lineolatum, E. angustiporcatum* and *E. cryptum* all differ from *E.javanicum* and should be recognized as distinct species.

INTRODUCTION

In their monograph of the ascomycetes genus *Eupenicillium* and related *Penicillium* anamorphs, Stolk and Samson (1983) accommodated several species of *Eupenicillium* as synonyms of *E. javanicum*. They were regarded as synonyms on the basis of the production of single asci, broadly lenticular ascopores with spinulose valves and identical phialide shape. *E. levitum* (Raper and Fennell) Stolk and Scott, *E. lineolatum* Udagawa and Horie and *E. meloforme* Udagawa and Horie also proved to be closely related, but differed by the ascospore ornamentation. These three taxa were therefore recognized as varieties. Stolk and Samson (1983) also found, that *P. janthinellum* as the anamorph of *E. javanicum*. Based on an original drawing of in Oudemans' herbarium, they also considered that the neotypification of *P. simplicissimum* (Oud.) Thom by Pitt (1979) was incorrect and that *P. janthinellum* was a synonym of *P. simplicissimum*.

Only a few connections are known to exist between anamorphic nonsclerotial Penicillia and *Eupenicillium* species. A connection between *P. janthinellum* and *E. javanicum* and related species would be of great theoretical and practical importance, especially because strains of the teleomorphic state are also important in biotechnology (Hamlyn *et al.*, 1987).

Chemotaxonomy, especially based on profiles of secondary metabolites and extracellular enzymes, has been a succesfull independent taxonomic criterion in *Penicillium* and *Aspergillus* taxonomy (Frisvad, 1989a). In this paper we evaluate the species related to *E. javanicum* based on chemotaxonomic data.

Modern Concepts in Penicillium and Aspergillus Classification
Edited by R. A. Samson and J. I. Pitt
Plenum Press, New York, 1990

Table 1. Production of mycotoxins by isolates in species related to *Eupenicillium javanicum*.

Species	Number	Former name	Mycotoxins produced
E. javanicum	NRRL 707		xanthomegnin
			palitantin (weak)
	CBS 341.48		xanthomegnin
			palitantin (weak)
	CBS 251.66		xanthomegnin
			palitantin (weak)
	NRRL 2078		xanthomegnin
			palitantin (weak)
	NRRL 2079		xanthomegnin
	CCM F-374		xanthomegnin
	CBS 349.51	*P. oligosporum*	palitantin
	NRRL 2016	*P. janthinellum*	xanthomegnin
	NRRL 2674	*P. raperi*	xanthomegnin
	CBS 191.67	*P. janthinellum*	xanthomegnin
	CBS 346.68	*P. janthinellum*	xanthomegnin
	IMI 108033	*P. janthinellum*	xanthomegnin
	NRRL 904	*P. janthinellum*	xanthomegnin
	ATCC 42743	*P. citreoviride*	xanthomegnin
	IBT NIPB10		xanthomegnin
	IBT JMRS2		xanthomegnin
E. zonatum	CBS 992.72		xanthomegnin
			brefeldin A (weak)
			janthitrems
	79-4L-61	*P. janthinellum*	brefeldin A
			janthitrems
	NRRL 2022	*P. piscarium*	brefeldin A
			paspaline
	IBT LOJO3		xanthomegnin
			brefeldin A
			janthitrems
E. ehrlichii	NRRL 708		brefeldin A
			palitantin
			penicillic acid
	NRRL 2083	*E. brefeldianum*	brefeldin A
			penicillic acid
			palitantin
			fulvic acid
	NRRL 710	*E. brefeldianum*	brefeldin A
E. ehrlichii	CBS 235.81	*E. brefeldianum*	brefeldin A
	CBS 233.81	*E. brefeldianum*	brefeldin A
	CBS 234.81	*E. brefeldianum*	brefeldin A
			paspaline
			paspalinine
	NRRL 2093	*E. brefeldianum*	brefeldin A
	CBS 577.70	*E. javanicum*	brefeldin A
			palitantin
			fulvic acid

Species	Number	Former name	Mycotoxins produced
	CBS 682.77	E. brefeldianum	brefeldin A penicillic acid palitantin
	CBS 421.66	E. brefeldianum	brefeldin A
	CBS 291.62	E. brefeldianum	brefeldin A
	NRRL 3389	P. cremeogriseum	brefeldin A
	CBS 277.83	P. sajarovii	brefeldin A
P. onobense	IMI 253737		brefeldin A
P. pulvillorum	NRRL 2026		penicillic acid
	IMI 177905	P. simplicissimum	penicillic acid
	IMI 190029	P. simplicissimum	penicillic acid
P. simplicissimum sensu Raper & Thom	CBS 372.48		–
P. species	FRR 1893	P. janthinellum	verrucologen penicillic acid
E. levitum	CBS 345.48		–
	CBS 228.81		–
E. ludwigii	CBS 417.63		–
E. lineolatum	CBS 188.77		–
E. angustiporcatum	CBS 202.84		–
E. meloforme	CBS 445.74		–
	CBS 446.74		–
	CBS 447.74		–
E. cryptum	ATCC 60138		–

Note: Several secondary, but specific metabolites were found in all isolates but they could not be identified.

MATERIAL AND METHODS

Isolates of *E. javanicum* and related species (Table 1) were grown on CZ, CYA, MEA, YES, and OAT agars at 25 C and on CYA at 37 C and examined after one, two and three weeks for morphological, physiological and chemical characters. For secondary metabolite production all isolates were examined using the agar plug method (Filtenborg *et al.*, 1983), while ex type cultures and some other representative isolates of all species were examined by high performance liquid chromatography (HPLC) with diode array detection (Frisvad and Thrane, 1987).

RESULTS AND DISCUSSION

All isolates of *E. javanicum* and all strongly coloured isolates of *P. janthinellum* produced xanthomegnin and related metabolites (Table 1) supporting the viewpoint of Samson and Stolk (1983) that this species is the anamorph of *E. javanicum*. Similar growth rates and other physiological features, including growth at 37°C and weak growth on creatine-sucrose agar also supported that conclusion. Production of palitantin in some isolates indicated relationship to *E. brefeldianum* and *E. ehrlichii*, however. *E. zonatum* produced xanthomegnin, brefeldin A and janthitrems, so from a chemotaxonomic point of view this species is intermediate between *E. javanicum* and *E. ehrlichii*. The anamorph of *E. zonatum* differs from *P. janthinellum* by the very rough conidia, but the size and shape of the ascospores resemble those of *E. javanicum* (Fig. 1). *E. zonatum* shares secondary metabolites with *E. javanicum* and *E. ehrlichii*, but the production of janthitrems seems to be unique to the former species. The production of janthitrems and very rough conidia by *P. piscarium* suggests that the latter is the anamorph of *E. zonatum*.

The less strongly coloured isolates in *E. ehrlichii* and *E. brefeldianum* were very good producers of brefeldin A and palitantin and some of them produced penicillic acid and fulvic acid. Pitt (1979) and Stolk and Samson (1983) emphasized the intermediate position of *E. ehrlichii* between *E. javanicum* and *E. brefeldianum*. The strong micromorphological resemblance between *E. ehrlichii* and *E. brefeldianum* was fully supported by their identical profiles of secondary metabolites. The connection between *E. ehrlichii* and weakly coloured strains of *P. janthinellum* is also substantiated by their common production of palitantin and brefeldin A.

P. cremeogriseum Chalabuda (Fig. 2) would fit the weakly coloured isolates and this name appear to be the first available for the anamorph of *E. ehrlichii*.

Among the anamorph species examined *P. piscarium* and *P. onobense* both produce brefeldin A suggesting relationship to *P. cremeogriseum*. However, *P. piscarium* has distinctly rough conidia and stipes and the conidia are globose to subglobose and clearly related to *E. zonatum*. *P. onobense* seems to be closely related to *P. brasilianum* because of its rough stipes and ellipsoidal spirally roughened conidia, but the profiles of secondary metabolites are very different: *P. brasilianum* produces penicillic acid, verrucologen, viridicatumtoxin and verruculotoxin (Frisvad, 1989b).

Because *P. simplicissimum* has been used for several taxa, the name should maybe be restricted to isolates related to CBS 372.48 = NRRL 902, the isolate on which Raper and Thom (1949) based their concept of that species. Chemotaxonomically this isolate produced unique secondary metabolites different from those of *P. pulvillorum, P. brasilianum, P. piscarium, P. cremeogriseum, P. ochrochloron, P. onobense* or *P. janthinellum*. The situation is further complicated by strains such as *P. janthinellum* FRR 1893 and "*P. solitum*" CBS 288.36, both producers of verrucologen and penicillic acid. These strains may be related to *P. brasilianum*, but they do not produce verruculotoxin, viridicatumtoxin or other metabolites unique for *P. brasilianum*.

We did not find any known secondary metabolites in *E. levitum, E. lineolatum* and *E. meloforme* and since we have not observed similarities in the chemical profiles of these taxa with *E. javanicum*, we tentatively place them as separate species (Fig. 3).

Several attempts to induce the teleomorph in the ex type culture of *E. cryptum* Gochenaur & Cochrane (1986) failed, but the description, as well as the micrographs and drawings strongly suggest a close relationship with the other ascosporic species of *Eupenicillium* series *Javanica*. The type culture of *E. angustiporcatum* Takada & Udagawa (1983) produces only a few, reduced conidiophores (Fig. 3m-n), agreeing with those of the series *Javanica*. No teleomorph was observed. According to Takada and Udagawa's

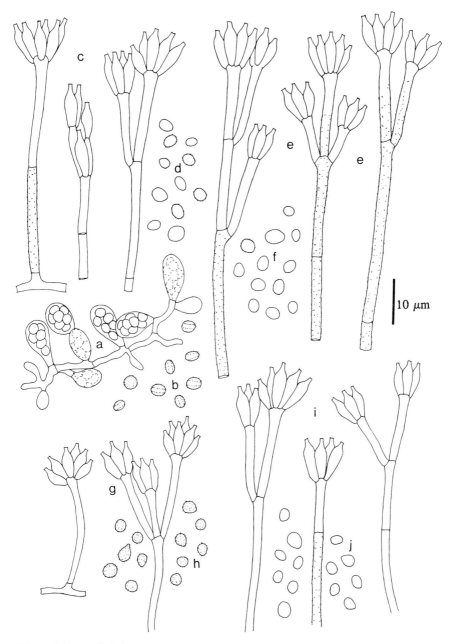

Figure 1. *P. simplicissimum* - teleomorph: *E. javanicum* CBS 310.48, a. ascus-development; b. ascospores; c. conidiophores; d. conidia CBS 283.36 e. conidiophores; f. conidia, CBS 992.72 g. conidiophores; h. conidia, CBS 340.48 i. conidiophores; j. conidia

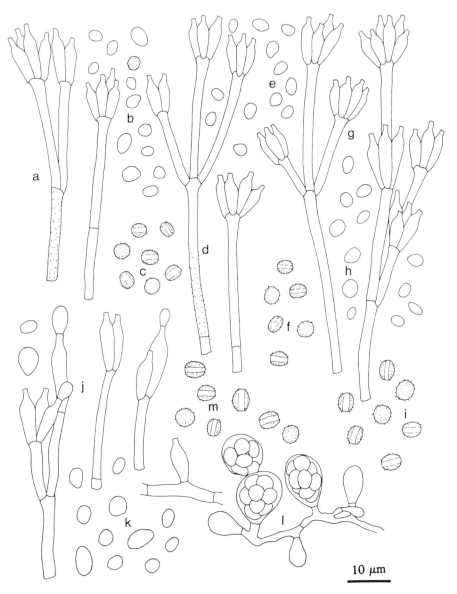

Figure 2. *P. cremeogriseum* - teleomorph: *E. ehrlichii* CBS 235.81 a. conidiophores; b. conidia; c. ascospores CBS 417.16 g. conidiophores; h. conidia; i. ascospores CBS 324.68; j. conidiophores; k. conidia; l. ascus-development; m. ascospores

Figure 3. *P. meloforme* - teleomorph: *E. meloforme* CBS 443.75 a. conidiophores; b. conidia CBS 447.74 c. ascus-development; d. ascospores *P. rasile* - telemorph: *E. levitum*, CBS 345.48 e. conidiophores; f. conidia; g. ascus-development; h. ascospores *P. lineolatum* - teleomorph: *E. lineolatum*, CBS 187.77 i. conidiophores; j. conidia; k. ascus-development; l. ascospores *P. angustisporcatum*, CBS 202.84 m. conidiophores; n. conidia

Table 2. Production of secondary metabolites in fast growing species related to *Eupenicillium javanicum.*

Species	1	2	3	4	5	6	7
E. javanicum	++	–	–	–	w	–	–
P. janthinellum	++	–	–	–	w	–	–
E. zonatum	++	–	++	++	w	–	–
P. piscarium	–	–	++	++	w	–	–
E. ehrlichii	–	++	++	–	++	–	–
P. cremeogriseum	–	++	++	–	++	–	–
P. simplicissimum sensu Raper & Thom	–	–	–	–	–	–	–
P. pulvillorum	–	++	–	–	–	–	–
P. brasilianum	–	++	–	–	–	++	++
P. onobense	–	–	++	–	–	–	–
P. ochrochloron	–	–	–	–	–	–	–
E. levitum	–	–	–	–	–	–	–
E. ludwigii	–	–	–	–	–	–	–

1: Xanthomegnin	4: Janthitrem	7: Viricatumtoxin
2: Penicillic acid	5: Palitantin	w: weak production
3: Brefeldin A	6: Verrucologen	++: strong production

description the ascospores were ornamented with two prominent well-separated equatorial ridges and the valves show several low ribs. The species is probably related to *E. lineolatum.*

The production of secondary metabolites in *E. javanicum, E. ehrlichii* and other species was apparently not dependent on the production of ascomata. Cultures grown on OA or MEA, where ascomata production was usually prolific, did not contain extra secondary metabolites compared to those produced on other media where ascomata were absent. The profiles of secondary metabolites of ascomatal isolates were identical to those produced by conidial isolates obtained from sectors in some isolates of *E. javanicum* or *E. ehrlichii.* Furthermore the profiles of secondary metabolites in *E. javanicum* isolates were alike those produced by strictly conidial strains of *P. janthinellum* and the same was the case for *E. ehrlichii* isolates compared to *P. cremeogriseum,* and *E. zonatum* as compared to *P. piscarium.* An overview of known mycotoxins produced by species previously regarded as closely related to *E. javanicum, P. janthinellum* or *P. simplicissimum* (Table 2) shows that several closely related taxa could be recognized

This means that comparison of profiles of secondary metabolites could assist to elucidate relationships between anamorphs and teleomorphs. However it is still possible that some secondary metabolites are morph related, as Wicklow and Shotwell (1983) and

Wicklow and Cole (1982) found that some fungal tremorgens were only found in the sclerotia (and not in the conidia) of *Aspergillus flavus*.

ACKNOWLEDGEMENTS

The authors thank the NATO Scientific Affairs Division (Brussel, Belgium) for the research grant (0216/86) for international collaboration.

REFERENCES

DOMSCH, K.H., W. GAMS and T.-H. ANDERSON. 1980. Compendium of soil fungi. Academic press, London.

FILTENBORG, O., FRISVAD, J.C. and SVENDSEN, J.A. 1983. Simple screening method for molds producing intracellular mycotoxins in pure culture. *Applied and Environmental Microbiology* 45: 581-585.

FRISVAD, J.C. 1989a. The use of high-performance liquid chromatography and diode array detection in fungal chemotaxonomy based on profiles of secondary metabolites. *Botanical Journal of the Linnean Society* 99: 81-95.

—— 1989b. The connection between the Penicillia and Aspergilli and mycotoxins with special emphasis on misidentified isolates. *Archives of Environmental Contamination and Toxicology* 18: 452-467.

FRISVAD, J.C. and THRANE, U. 1987. Standardized high performance liquid chromatography of 182 mycotoxins and other fungal metabolites based on alkylphenone retention indices and UV-VIS spectra (diode array detection). *Journal of Chromatography* 404: 195-214.

GOCHENAUR, S.E. and E. COCHRANE. 1986. *Eupenicillium cryptum* sp. nov., a fungus with self-limiting growth and restricted carbon nutrition. *Mycotaxon* 26: 345-360.

HAMLYN, P.F., D.S. WALES and B.F. SAGAR. 1987. Extracellular enzymes of *Penicillium*. *In* J.F. PEBERDY (ed.): *Penicillium* and *Acremonium*, pp. 245-286. New York and London: Plenum Press.

PITT, J.I. 1979. *The genus Penicillium and its teleomorphic states Eupenicillium and Talaromyces*. London and New York: Academic Press.

RAPER, K.B. and C. THOM. 1949. Manual of the Penicillia. Baltimore: Williams and Wilkins.

STOLK, A.C. and SAMSON, R.A. 1983. The ascomycete genus *Eupenicillium* and related *Penicillium* anamorphs. *Studies in Mycology, Baarn* 23: 1-149.

TAKADA, M. and S.-I. UDAGAWA. 1983. Two new species of *Eupenicillium* from Nepalese soil. *Transactions of the Mycological Society of Japan* 24: 143-150.

WICKLOW, D.T. and COLE, R.J. 1982. Tremorgenic indol metabolites and aflatoxins in sclerotia of *Aspergillus flavus*: an evolutionary perspective. *Canadian Journal of Botany* 60: 525-528.

WICKLOW, D.T. and SHOTWELL, O.L. 1983. Intrafungal distribution of aflatoxins among conidia and sclerotia of *Aspergillus flavus* and *Aspergillus parasiticus*. *Canadian Journal of Microbiology* 29: 1-5.

DIALOGUE FOLLOWING DR. FRISVAD'S PRESENTATION

PITT: In *Penicillium*, especially subgen. *Penicillium*, your approach to taxonomy using secondary metabolites tends to split species more than my morphological concepts. However, secondary metabolites appear to be more conservative in what I would consider to be older species and genera, such as the soil Penicillia and species of *Aspergillus*. Here, you tend to be seeing fewer species using your techniques than some of the morphologists do. In the species you have just discussed, I would tend to be more conservative and maintain species that you are tending to lump together. I have tended to maintain *Eupenicillium javanicum* as a monoverticillate species and am rather surprised to see the well developed metulate forms you showed included in that species. Dr.

Samson and Stolk's concept of this species is undoubtedly broader than mine, and we must do more work before we will come to a consensus.

TAYLOR: Is there any variability in the wall of the ascocarp of any of the *Eupenicillium* species you have looked at?

SAMSON: Yes. The wall can be either pseudoparenchymatous or sclerenchymatous.

PITT: How many other species have you actually looked at, and how many possible connections have you seen? When Dr. Taylor looked at *Talaromyces*, there were very few connections. I would expect to see more in *Eupenicillium* but still not very many.

FRISVAD: Perhaps *P. decumbens* and *Eupenicillium meridianum* might be very closely related. Maybe there are a few other connections. In my opinion, the *Eupenicillium euglaucum* complex may have a lot of species in it.

PATERSON: In some of the HPLC traces, some of the peaks have quite similar retention times. Is the diode ray detector completely satisfactory for distinguishing between peaks with similar retention times?

FRISVAD: You sometimes see that the peaks aren't very well separated at all or maybe completely together. So then we can take ultraviolet spectra from the up slope and the down slope and distinguish two compounds that way. If they are really close, then you get a mystical ultraviolet spectrum that you cannot interpret.

THE GENUS *NEOSARTORYA*: DIFFERENTIATION BY SCANNING
ELECTRON MICROSCOPY AND MYCOTOXIN PROFILES

R.A. Samson[1], P.V. Nielsen[2] and J.C. Frisvad[2]

[1]Centraalbureau voor Schimmelcultures
3740 AG Baarn, The Netherlands

[2]Department of Biotechnology
Technical University of Denmark
2800 Lyngby, Denmark

SUMMARY

Isolates of *Neosartorya* including extype cultures were examined morphologically and chemotaxonomically and allocated to nine taxa (*N. aurata, N. aureola, N. fennelliae, N. fischeri, N. fischeri* var. *glabra, N. fischeri* var. *spinosa, N. quadricincta, N. spathulata,* and *N. stramenia*), three chemotypes and three undescribed species. All the species were clearly distinguished by their ascospore morphology and their characteristic profile of secondary metabolites. Tremorgens were produced by many species including *N. fischeri, N. aureola* and *N. fischeri* var. *spinosa. N. fennelliae* produced vidiricatumtoxin, found for the first time outside the genus *Penicillium*. Trypacidin, was found in *N. aureola, N. fennelliae* and *N. fischeri* var. *glabra.*

INTRODUCTION

The genus *Neosartorya* was established by Malloch and Cain (1972) for members of "*the Aspergillus fischeri*" series, a series in the "*Aspergillus fumigatus* group" of Raper and Fennell (1965) (= subgen. *Fumigati* sect. *Fumigati;* Gams *et al.,* 1985). Raper and Fennell accepted 5 species and 3 varieties in the *N. fischeri* series, separated by ascospore ornamentation and colour of the ascomata. Malloch and Cain (1972) listed 10 taxa, including three teleomorphs in *Aspergillus,* later re classified by Samson (1979) and von Arx (1981) in the teleomorphic genera *Hemicarpenteles* and *Chaetosartorya.* In additon of the taxa accepted by Raper and Fennell (1965), Kwon-Chung and Kim (1974) and Takada and Udagawa (1985) described two heterothallic species, *N. fennelliae* and *N. spathulata.*

Table 1. Reported production of mycotoxins by Neosartorya species

Taxa	Mycotoxins and antibiotics
N. aurata	Helvolic acid (Turner and Aldridge, 1983)
N. fischeri var. *fischeri*	Fumitremorgin A, B, C, verruculogen (Horie and Yamazaki, 1981; Patterson et al., 1981; Nielsen et al., 1988), Avenaciolide (Turner and Aldridge, 1983), Terrein (Mizawa *et al.,* 1962)
N. fischeri var *glabra*	Canescin (Turner and Aldridge, 1983)
N. quadricincta	Cyclopaldic acid (Turner and Aldridge, 1983)
N. stramenia	Canadensolide (Turner, 1971)

Modern Concepts in Penicillium and Aspergillus Classification
Edited by R. A. Samson and J. I. Pitt
Plenum Press, New York, 1990

Table 2. Examined isolates of *Neosartorya*

N. aurata Warcup:
CBS 466.65 = WB 4378, T, ex soil, Brunei

N. aureola Fennell & Raper:
CBS 105.55 = NRRL 2244, ex soil, Gold Coast; CBS 106.55 = NRRL 2391, ex soil Liberia;
CBS 651.73 A and B, ex soil, Surinam

N. fennelliae Kwon-Chung & Kim:
CBS 598.74 = NRRL 5534, mating a and CBS 599.74 = NRRL 5535, mating A, T ex rabbit eye ball,
USA; CBS 410.89 = NHL 2951 ("A") and CBS 411.89 = NHL 2952 ("a"), ex marine sludge, Japan

N. fischeri var. *fischeri* :
CBS 111.51 ex Camellia sinensis; CBS 113.64 ex man; CBS 525.65 = WB 4161; CBS 544.65 = WB
181, T, original C. Wehmer ex Dahlia tuber which were preserved in alcohol, Switserland
isolated by E. Fischer; CBS 404.67 T, Aspergillus fischeri var. thermomutatus, ex mouldy
cardboard, Canada; CBS 681.77, ex soil, Baarn; CBS 832.88 = IBT 3023, ex soil Lyngby, Denmark;
IBT 3003, 3012 and 3013, ex soil Lyngby, Denmark; IBT 3005, ex soil Urmston, UK; IBT 3007,
3008, 3009, ex soil Holte, Denmark; WB 4075 = IMI 16143 ex garden Soil, UK

N. fischeri var. *glabra* :
CBS 111.55 = WB 2163, T. ex old tires, USA; CBS 112.55 = NRRL 2392, ex garden soil, South
Australia; CBS 165.63, ex soil, the Netherlands; IBT 3004, 3006, ex soil Urmston, UK; IBT 3010,
3011 ex soil Lyngby; IBT 3014 and 3015 ex soil, Sweden; WB 4179, soil Australia; WB 2392 = CBS
112.55, garden soil, Australia; IMI 144207, ex canned strawberries; RO 343; RO 27-3

N. fischeri var. *spinosa*:
CBS 483.65 = WB 5034, T, ex soil Nicaragua; CBS 297.67 ex soil, Pakistan; CBS 865.70 from
unrecorded source, India; CBS 448.75 = NHL 5083, T, Aspergillus fischeri var. verrucosus ex
Japan; CBS 161.88 = TRTC 50994, ex soil Sudan; IBT 3001, Anatto seeds, Brazil; IBT 3002, ex soil,
Lyngby, Denmark; WB 4076 = IMI 16061, ex hay, UK; WB 5034 = CBS 483.65, T, ex soil,
Nicaragua

N. quadricincta Yuill
CBS 135.52 = WB 2154, T. ex cardboard, UK; CBS 941.73, ex soil, Surinam; IMI 58374 ex soil, NT,
Australia; WB 4175 ex soil Australia; WB 2221 ex fabric, Florida, USA; WB 2154 = CBS 135.52, T,
ex cardbord, UK

N. spathulata Takada & Udagawa
CBS 408.89 = NHL 2948 ("A") and CBS 409.89 = NHL 2949 ("a"), T ex soil, Formosa

Neosartorya spp.
CBS 652.73 A and 652.73 B, ex soil, Surinam; CBS 290.74 ex *Acer pseudoplatanus*, The
Netherlands; IBT 3016, ex soil, Lyngby, Denmark; RO 342; IBT 3017

Neosartorya stramenia
CBS 498.65 = WB 4652, T. ex forest soil, Wisconsin

Isolates of *Neosartorya fischeri* have frequently been reported as heat resistant spoilage
fungi in foods and beverages (Kavanagh *et al.*, Beuchat, 1986; 1963; Scott and Bernard,
1987; Conner and Beuchat, 1987; Baggerman and Samson, 1988; Nielsen *et al.*, 1988;
Samson, 1989). In addition, several mycotoxins have been reported from *Neosartorya*
species (Table 1).

Correct identification of *Neosartorya* taxa is therefore important. This paper reassesses the taxonomy of this genus, primarily on the basis of ascospore morphology and profiles of secondary metabolites.

MATERIAL AND METHODS

The isolates examined are listed in Table 2. They were grown on CYA, MEA, YES, and OA (Samson and van Reenen-Hoekstra, 1988) at 25°C and on CYA and MEA at 37°C. Examination was made after one and two weeks for morphological, physiological and chemical characters.

For scanning electron micropscopy (SEM), mature cleistothecia were transferred to aluminium stubs which were covered with double sided adhesive tape. A small drop of water containing some Tween-80 has added and the cleistothecia crushed. The suspension was air dried, coated with gold and examined by SEM JEOL. For comparison of ornamentation patterns, cleisthothecia were also fixed in glutaraldehyde and chemically dehydrated prior to critical point drying as described by Samson *et al.* (1979).

For secondary metabolite production, all isolates were examined using the agar plug method (Filtenborg *et al.*, 1983), Representatives of all species, including extype cultures, were examined by high performance liquid chromatography (HPLC) with diode array detection (Frisvad and Thrane, 1987). For this purpose, isolates were grown on 3 plates each of CYA, YES and Sigma Y-4000 Yeast extract sucrose agar (SYES) for two weeks. After extraction with 150 ml chloroform/methanol (2:1) for 2 min in a Colworth Stomacher, filtration and evaporation of the solvent the procedure was as described by Frisvad and Thrane (1987).

RESULTS AND DISCUSSION

Ascospore morphology.

Light microscopical examination of the taxa accepted by Raper and Fennell (1965) showed that the species can be readily recognized, on the basis of the colour of ascomata and ascsopore morphology. However, for the varieties of *N. fischeri* where the differentation is only based on the ornamentation of the ascospores, light microscopy proved to be more difficult. SEM, however, provided a much clearer picture of ascospore ornamentation, and proved to be consistent and typical for each taxon. Specimens fixed in glutaraldehyde and critical point dried showed similar surface structures to ascospores which were unfixed and airdried. The latter simple preparation technique was therefore chosen to examine all species.

Raper and Fennell (1965) described the ascospores of *N. fischeri* var. *fischeri* as having convex surfaces bearing anastomising ridges. Under SEM the ascospores are typically reticulate (Fig. 1 A-B). Ascospores of *N. fischeri* var. *glabra* observed by Fennell and Raper (1955) to be smooth or nearly so, were finely roughened under the SEM (Fig. 1 C-D). Ascospores *N. fischer* of var. *spinosa* are spinulose, although some gradation exists from rough to distinctly spinulose (Fig. 1 C-F, 2 A-B). Ascospores of *N. fischeri* var. *verrucosa* (Fig. 2 B), are identical with *N. fischeri* var. *spinosa*, and so the two taxon are synonymous. Spinulose ascospores were also observed in *N. aureola* (fig. 2 E-F), but this species can be separated from *N. fischeri* var *spinosa* by yellow rather than white ascomata. Ascospores of *N. stramenia* (Fig. 4 A-B) and *N. aurata* (Fig. 2 C-D) resemble those of *N. fischeri* var. *glabra*, but again the ascomata are yellow. The type culture of *N. quadricincta* produces only very limited and small ascomata, but the ascospores have the four crests characteristic of

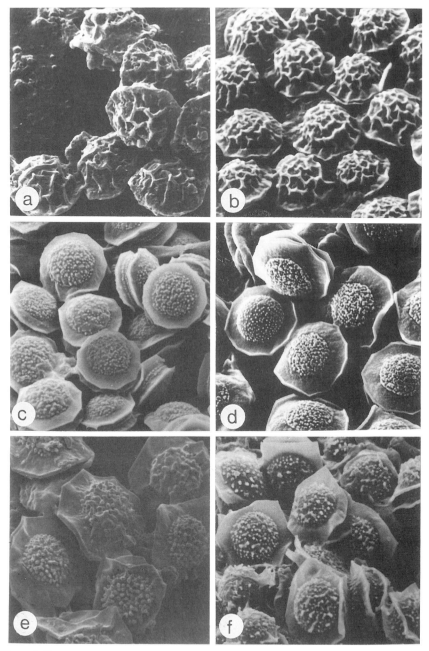

Figure 1. Scanning electronmicrographs of ascospores. A-B. *N. fischeri* A. CBS 544.65 = WB 181, B. CBS 832.88 = IBT 3023, C-D. *N. fischeri* var. *glabra* C. CBS 112.55 = NRRL 2392, D. IBT 3004, E-F. *N. fischeri* var.*spinosa*, E. CBS 483.65 = Wb 5034, F. WB 4076 = IMI 16061 (all 3500 x).

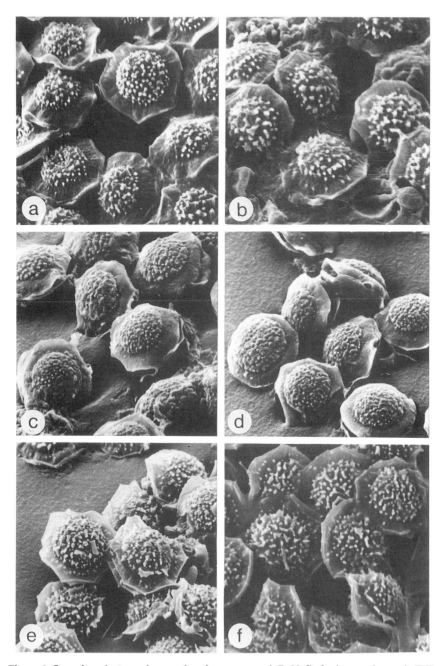

Figure 2. Scanning electronmicrographs of ascospores. A-B. *N. fischeri* var. *spinosa,* A. IBT 3001. B. CBS 448.75 (T of *A. fischeri* var. *verrucosus*) C-D. *N. aurata* CBS 466.65 = WB 4378, E-F. *N. aureola,* E. CBS 105.55 = NRRL 2244, F. CBS 651.73 A (all 3500 x).

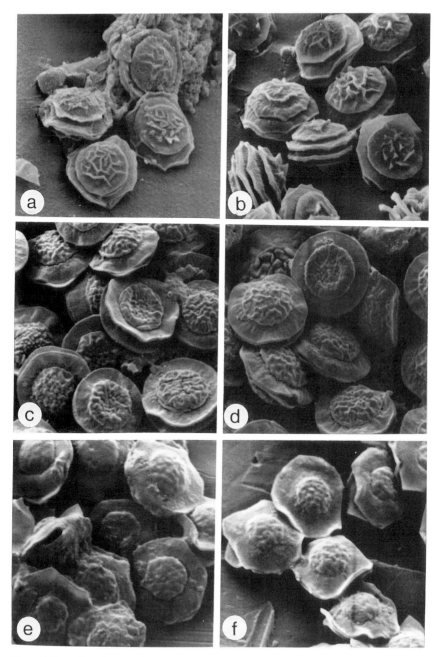

Figure 3. Scanning electronmicrographs of ascospores. A-B. *N. quadricincta* CBS 135.52 = WB 2154, B. WB 4175, C-D. *N. fennelliae*, C.CBS 598.74 x CBS 599.74, D. CBS 410.89 x CBS 411.89, E-F. *N. spathulata* CBS 408.89 x CBS 409.89 (all 3500 x).

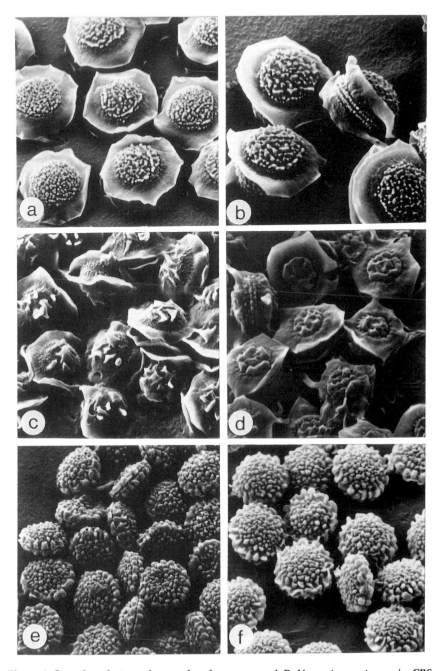

Figure 4. Scanning electronmicrographs of ascospores. A-B. *Neosartorya stramenia*, CBS 498.65 = WB 4652, C-F. *Neosartorya* spp. C. IBT 3002, D. IBT 3016, E. CBS 652.73 A, F. 652.73 B (all 3500 x).

this species, while the convex surface is slightly reticulate (Fig. 3 A-B). Ascospores of the heterothallic *N. fennelliae*, produced after are mated in pairs, have a distinct cerebriform surface structure (Fig. 3 C-D). *N. spathulata* isolates produce almost smooth-walled ascospores with distinct crests (Fig. 3 E-F). This species also produces a quite different *Aspergillus* anamorph as described by Takada and Udagawa (1985).

Amongst the material examined, several isolates were found with ascospores different from any described species (Fig. 4 C-F). Their secondary metabolites were also distinct, so these isolates may represent three undescribed species.

Secondary metabolites profiles.
Secondary metabolite data correlated well with the morphological data. All species studied produced a characteristic profile of secondary metabolites as evaluated by HPLC-DAD. Two examples of consistent secondary metabolite production by isolates of the same taxon are depicted: in Fig. 5 and in Fig.6. Nearly all isolates of *N. fischeri* var. *fischeri* produced fumitremorgin A, B, C and verrucologen, Horie and Yamazaki (1981), found fumitremorgin production in only a small proportion of isolates they examinated, but the differences in results may be explained by differences in media used for mycotoxin production. We have found CYA, YES and Sigma Y-4000 YES are a very effective combination of media for mycotoxin production.

Isolates identified as *N. fischeri* var. *glabra* and *N. fischeri* var. *spinosa* produced on morphological criteria different profiles of secondary metabolites (Table 3). However, morphologically similar isolates of *N. fischeri* var. *glabra* produced two quite distinct secondary metabolite profiles. Two of them produced very large amounts of mevinolins (CBS 112.55 and RO 27-3) while three others, including the ex type culture of *N. fischeri* var. *thermomutatus*, produced a somewhat heterogeneous range of metabolites.

The potentially tremorgenic tryptoquivalins were found for the first time in *N. aureola*, *N. fischeri* var. *fischeri*, *N. fischeri* var. *glabra* chemotype III and *N. fischeri* var. *spinosa* (Table 1, 3). Tryptoquivalins have previously earlier been found in *A. fumigatus* and *A. clavatus* (Buchi *et al.*, 1977). Two tryptoquivalins were produced in very high amounts by *N. aureola* CBS 651.73 A and B.

N. fennelliae isolates were found to produce viridicatumtoxin which was previously only found in *Penicillium* (Frisvad, 1989). Viriditoxin was found in *N. aureola* (CBS 106.55 and 671.73B) and *Neosartorya* sp. (CBS 652.73A & B). Xanthocillin derivatives were produced by *N. spathulata* and *Neosartorya* sp. (IBT 3002). Trypacidin, another metabolite characteristic of *Aspergillus fumigatus* was also produced by *N. fennelliae* and the ex type culture of *N. fischeri* var. *thermomutatus*. Two families of secondary metabolites with characteristic UV spectra (FUA and FUB), found in *A. fumigatus*, were also produced by several *Neosartorya* species (Table 1). However, although some secondary metabolites were common to several species in *Neosartorya* and *Aspergillus* sect. *Fumigati*, each *Neosartorya* taxon we investigated produced a specific distinct profile of secondary metabolites.

CONCLUSIONS

On the basis of differences in ascospore ornamentation and profiles of secondary metabolites the described taxa in *Neosartorya* can be well separated. The two varieties *glabra* and *spinosa* of *N. fischeri* differ from var. *fischeri* that they could be raised to species rank (also compare Kozakiewicz, 1989). Since there are some biochemical and morphological similarties between isolates assigned to *N. fischeri* var. *spinosa* and *N. fischeri*

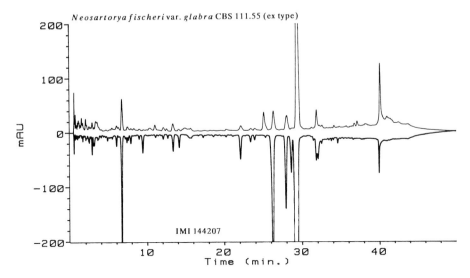

Figure 5. Comparison of HPLC traces of two isolates of *N. fischeri* var. *glabra*, CBS 111.55 (ex type) and IMI 144207. The prominent peak at 29.3 min. was unique to var. *glabra* chemotype I.

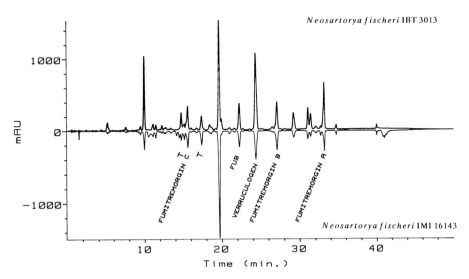

Figure 6. Comparison of HPLC traces of two strains of *N. fischeri*, IBT 3013 (soil, Denmark) and IMI 16143 (soil, Great Britain). Both strains produce fumitremorgins A, B, and C and verrucologen two tryptoquivalins (T) and the unknown metabolite FUB.

Table 3. Secondary metabolites determined in the Neosartorya isolates investigation

	Mycotoxins produced								
	1	2	3	4	5	6	7	8	9
N. aurata									
CBS 466.65	–	–	–	–	–	–	–	–	–
N. aureola									
CBS 105.55	+	+	–	–	–	–	–	–	–
CBS 106.55	+	+	–	–	–	–	–	–	–
CBS 651.73	+	+	–	–	–	–	–	–	–
CBS 651.73	–	+	–	–	–	–	–	–	–
N. fennelliae									
CBS 598.74	–	–	+	+	–	–	–	–	–
CBS 599.74	–	–	+	+	–	–	–	–	–
NHL 2951	–	–	–	+	–	–	–	–	–
NHL 2952	–	–	–	+	–	–	–	–	–
N. fischeri var. *fischeri*									
CBS 544.65	–	–	–	–	–	+	–	–	–
WB 4075	–	–	–	–	+	+	–	–	–
IMI 16143	–	+	–	–	+	+	–	–	–
IBT 3005	–	–	–	–	+	+	–	–	–
IBT 3007 (NT)	–	–	–	–	–	–	–	–	–
IBT 3008 (NT)	–	–	–	–	–	–	–	–	–
IBT 3009	–	+	–	–	+	+	–	–	–
IBT 3012 (NT)	–	–	–	–	–	–	–	–	–
IBT 3013	–	+	–	–	+	+	–	–	–
IBT 3023	–	+	–	–	+	+	–	–	–
CBS 448.75 (NT)	–	–	–	–	–	–	–	–	–
N. fischeri var. *glabra* chemotype I									
CBS 111.55	–	–	–	–	–	–	–	–	–
IMI 144207	–	–	–	–	–	–	–	–	–
IMI 102173	–	–	–	–	–	–	–	–	–
IBT 3004	–	–	–	–	–	–	–	–	–
IBT 3006 (NT)	–	–	–	–	–	–	–	–	–
IBT 3010	–	–	–	–	–	–	–	–	–
IBT 3011 (NT)	–	–	–	–	–	–	–	–	–
IBT 3014 (NT)	–	–	–	–	–	–.	–	–	–
IBT 3015 (NT)	–	–	–	–	–	–	–	–	–
N. fischeri var. *glabra* chemotype II									
WB 2392	–	–	–	–	–	–	+	–	–
CBS 112.55	–	–	–	–	–	–	+	–	–
RO 27–3	–	–	–	–	–	–	+	–	–
N. fischeri var. *glabra* chemotype III									
CBS 404.67	–	+	+	–	–	+	–	–	–
CBS 941.73	–	–	–	–	–	–	–	–	–
IMI 16061	+	+	–	–	–	–	–	–	–
RO 343	–	+	–	–	–	+	–	–	–
WB 4076	+	+	–	–	–	–	–	–	–
WB 4179	–	–	–	–	–	–	–	–	–
N. fischer var. *spinosa*									
CBS 483.65	+	–	–	–	–	–	–	–	–
CBS 297.67	+	+	–	–	–	–	–	+	–
IBT 3001	+	+	–	–	–	+	–	–	–
CBS 681.77 (NT)	–	–	–	–	–	–	–	–	–
WB 5034	+	–	–	–	–	–	–	–	–

	Mycotoxins produced								
	1	*2*	*3*	*4*	*5*	*6*	*7*	*8*	*9*
N. quadricincta									
CBS 135.52	–	–	–	–	–	–	–	–	–
WB 2154	–	–	–	–	–	–	–	–	–
IMI 58374	–	–	–	–	–	–	–	–	–
WB 4175	–	–	–	–	–	–	–	–	–
WB 2221	–	–	–	–	–	–	–	–	–
N. spathulata									
NHL 2948	–	–	–	–	–	–	–	–	–
NHL 2949	–	–	–	–	–	–	–	–	–
N. stramenia									
CBS 498.65	–	–	–	–	–	–	–	–	–
N. species 1									
CBS 652.73	–	–	–	–	–	–	–	–	–
CBS 652.73	–	–	–	–	–	–	–	–	–
CBS 290.74	–	–	–	–	–	–	–	–	–
N. species 2									
IBT 3016	–	–	–	–	–	–	–	–	+
N. species 3									
RO 342	–	+	–	–	–	–	–	–	–
N. species 4									
I BT 3002	–	+	–	–	–	+	–	–	–

1 = FUA	6 = FUB
2 = tryptoquivaline	7 = mevinolins
3 = trypacidin	8 = terrein
4 = viridicatumtoxin	9 = kotanin
5 = Fumitremorgins A,B,C and verrucologen	NT = Not tested by HPLC-DAD

var. *glabra* chemotype III a more detailed investigation is required to solve the taxonomic classification of these taxa.

It is important to note that both varieties *glabra* and *spinosa* are the only two species found on food products. IMI 144207 and IMI 102173 were both found in canned strawberries from UK and IBT 3001 was found on anatto seeds from Brazil. The two isolates asigned to var. *glabra* examined did not produce any known mycotoxins, but var. *spinosa* IBT 3001 produced tryptoquivalins. Kozakiewicz (1989) incorrectly stated that the type of *Neosartorya fischeri* CBS 544.65 = WB181 = IMI 211391 was isolated from canned apples. This culture was received by Wehmer (1907) from Prof. E. Fischer and isolated from tubers of *Dahlia* flowers, which were preserved in alcohol. In our analysis we did not find any known mycotoxins in this isolate.

The significance of the food-borne taxa will be further examined. The isolates representing the new species will be described elsewhere.

ACKNOWLEDGEMENTS

The authors thank the NATO Scientific Affairs Division (Brussel, Belgium) for the research grant (0216/86) for international collaboration.

REFERENCES

ARX, von J.A. 1981. Genera of fungi sporulating in pure culture. 3rd. ed. Vaduz: J. Cramer.

BAGGERMAN, W.I. and SAMSON, R.A. 1988. Heat resistance of fungal spores. *In* Introduction to food-borne fungi, eds. R.A. Samson and E.S. van Reenen-Hoekstra, pp. 262-267. Netherlands: Centraalbureau voor Schimmelcultures, Baarn.

BEUCHAT, L.R. 1986. Extraordinary heat resistance of *Talaromyces flavus* and *Neosartorya fischeri* ascospores in fruit products. *Journal of Food Science* 51: 1506-1510.

BÜCHI, G., LUK, K.C., KOBBE, B. and TOWNSEND, J.M. 1977. Four new mycotoxins of *Aspergillus clavatus* related to tryptoquivaline. *Journal of Organical Chemistry* 42: 244-246.

COLE, R.J. 1977. Tremorgenic mycotoxins. *In* Mycotoxins in human and animal health. eds. Rodricks, J.V., Hesseltine, C.W. and M.A. Mehlman, pp. 583-595. Pathotox Publishers, Park Forest South.

CONNER, D.E. and BEUCHAT, L.R. 1987. Heat resistance of ascospores of *Neosartorya fischeri* as affected by sporulation and heating medium. *International Journal of Food Microbiology* 4: 303-312.

DE JESUS, A.E., HULL, W.E., STEYN, P.S., VAN HEERDEN,F.R. and VLEGGAAR, R. 1982. Biosynthesis of viridicatumtoxin, a mycotoxin from *Penicillium expansum. Journal of the Chemical Society Chemical Communications* 1982: 902-904.

FENNELL, D.I. and RAPER, K.B. 1955. New species and varities of *Aspergillus*. Mycologia 47:68-79.

FILTENBORG, O., FRISVAD, J.C. and SVENDSEN, J.A.. 1983. Simple screening method for molds producing intracellular mycotoxins in pure culture. *Applied and Environmental Microbiology* 45: 581-585.

FRISVAD., J.C. 1989. The connection between the Penicillia and Aspergilli and mycotoxins with special emphasis on misidentified isolates. *Archives of Environmental Contamination and Toxicology* 18: 452-467.

FRISVAD, J.C. and THRANE, U. 1987. Standardized high-performance liquid chromatography of 182 mycotoxins and other fungal metabolites based on alkylphenone retention indices and UV-VIS spectra (diode array detection). *Journal of Chromatography* 404: 195-214.

GAMS, W., CHRISTENSEN, M., ONIONS, A.H.S., PITT, J.I. and SAMSON, R.A. 1985. Infrageneric taxa of *Aspergillus. In* Advances in *Penicillium* and *Aspergillus* Systematics, eds. R.A. Samson and J.I. Pitt, pp. 55-62. New York and London: Plenum Press. o7 3

HOCKING, A. D. and PITT, J.I. 1984. Food spoilage fungi. II. Heat-resistant fungi. *CSIRO Food Research Quarterly* 44: 73-82.

HORIE, Y. and YAMAZAKI, M. 1981. Productivity of the tremorgenic mycotoxins, fumitremorgins A and B in *Aspergillus fumigatus* and allied species. *Nippon Kingakkai Kaiko* 22: 113-119.

KAVANAGH, J., LARCHET, N., and STUART, M. 1963. Occurrence of a heat resistant species of *Aspergillus* in canned strawberries. *Nature, London* 198: 1322.

KOZAKIEWICZ, Z. 1989. *Aspergillus* species on stored products. *Mycological Papers* 161: 1-188.

KWON-CHUNG, K.J. and KIM, S.J. 1974. A second heterothallic *Aspergillus*. Mycologia 66: 628-638.

MALLOCH, D. and CAIN, R.F. 1972. The Trichocomataceae: Ascomycetes with *Aspergillus, Paecilomyces and Penicillium* imperfect states. *Canadian Journal of Botany* 50: 2613-2628.

MISAWA, M., NARA, T., NAKAYAMA, K. and KINOSHITA, S. 1962. Formation of terrein by *Aspergillus fischeri* Wehmer. *Nippon Nogei Kagaku Kaishi* 36: 699-703.

NIELSEN, P.V., BEUCHAT, L.R. and FRISVAD J.C. 1988. Growth and fumitremorgin production by *Neosartorya fischeri* as affected by temperature, light and water activity. *Applied and Environmental Microbiology* 54: 1504-1510.

PATTERSON, D.S., SHREEVE, B.J., ROBERTS, B.A. and MACDONALD, S.M. 1981. Verruculogen produced by soil fungi in England and Wales. *Applied and Environmental Microbiology* 42: 916-917.

RAPER K.B. and FENNELL, D.I. 1965. The genus *Aspergillus*. Baltimore: Wiliams and Wilkins.

SAMSON, R.A. 1979. A compilation of the Aspergilli described since 1965. *Studies in Mycology, Baarn* 18: 1-38.

SAMSON, R.A. 1989. Filamentous fungi in food and feed. *Journal of Applied Bacteriology Symposium Supplement* 1989: 27S-35S.

SAMSON, R.A., STALPERS, J.A. and VERKERKE, W. 1979. A simplified technique to prepare fungal specimens for scanning electron microscopy. *Cytobios* 24: 7-11

SAMSON, R.A. and VAN REENEN-HOEKSTRA, E.S. 1988. Introduction to Food-borne Fungi. Baarn, Netherlands: Centraalbureau voor Schimmelcultures.

SCOTT, V. N. and BERNARD, D.T. 1987. Heat resistance of *Talaromyces flavus* and *Neosartorya fischeri* isolated for fruit juice. *Journal of Food Protection* 50: 18-20.

TAKADA, M. and UDAGAWA, S. 1985. A new species of heterothallic *Neosartorya*. *Mycotaxon* 24: 395-402.

TAKADA, M., UDAGAWA, S. and NORISUKI, K. 1986. Isolation of *Neosartorya fennelliae* and interspecific pairings between *N. fennelliae*, *N. spathulata* and *Aspergillus fumigatus*. *Transactions of the Mycological Society of Japan* 27: 415-423.

TURNER, W.B. 1971. Fungal Metabolites. London: Academic Press.

TURNER, W.B. and D.C. ALDRIDGE. 1983. Fungal metabolites II. London: Academic Press.

UDAGAWA, S. 1986. Current topics in the taxonomy of thermophilous fungi, as potential pathogens of opportunistic infections. *Japanese Journal of Medical Mycology* 27: 5-14.

WEHMER, C. 1907. Zur Kenntnis einiger *Aspergillus*-Arten. *Centralblatt für Bakteriologie und Parasitenkunde*, Abt. II, 18: 385-395.

DIALOGUE FOLLOWING DR. SAMSON'S PRESENTATION

TAYLOR: Does anyone know how these ascospores are dispersed in nature? Or is the whole cleistothecium dispersed?

SAMSON: This is only speculation, but I think that the whole ascoma is dispersed. Insects might eat them and then excrete the ascospores.

TAYLOR: It is possible that the ascospore ornamentation in these fungi is involved in dispersal, and so the ascospore morphology may be evolving quite rapidly compared to other characters.

PITT: Regarding the dispersal of the teleomorph in nature, perhaps the real answer is that these things don't occur in nature! *Neosartorya* species are found in processed foods because of their heat resistance. Have you looked at other *Aspergillus* teleomorph genera that appear to be closely related, such as *Sclerocleista*. The number of known isolates of these genera is very low. They all have the same ascocarp wall and appear to me to be very closely related.

SAMSON: I would like to combine some of these genera. But we have to consider the quite distinct anamorphs. I'm very interested in the studies that Dr. Taylor and Dr. Kuraishi have done, which might help us understand the relationships among these teleomorph genera. Perhaps we should consider erecting a teleomorph genus that has different anamorphs in section *Ornati*.

TAYLOR: Has anyone crossed isolates of *Neosartorya fennelliae* from different parts of the world to see if it is really all one species, or if there might be vegetative compatability groups?

SAMSON: Takada and Udagawa published a study in Japanese. Judging from their tables, they saw lots of isolates and tried to establish the mating pattern.

CHRISTENSEN: We should be looking in the tropics a lot more for *Aspergillus* species and their teleomorphs.

SAMSON: I agree. The heat shock technique, in which you heat the sample at 60-70°C, or microwave a sample in a petri dish for one minute, is very effective.

TAYLOR: Bob Metzenburg at Wisconsin has developed a chemical method that simulates heat shock. Using it, he has found a number of homothallic *Neurospora* species from tropical Mexican and Central American soils.

PARTICIPANTS

Australia
Dr. J.I. Pitt
C.S.I.R.O.
Division of Food Research
P.O. Box 52,
NORTH RYDE, N.S.W. 2113

Czechoslovakia
Dr. O. Fassatiova
Dept. of Cryptogamic Botany
Benatska 2,
128 01 PRAHA 2

Belgium
Prof. Dr. G.L. Hennebert
Laboratoire de Mycologie Systématique et
Appliquée
MUCL, Université Catholique de Louvain
Place Croix du Sud 3,
1348 LOUVAIN-LA-NEUVE

Dr. J.F. Berny
Laboratoire de Mycologie Systématique et
Appliquée
MUCL, Université Catholique de Louvain
Place Croix du Sud 3,
1348 LOUVAIN-LA-NEUVE

Dr. D. Stynen
ECO/BIO, Diagnostics Pasteur
Woudstraat 24
3600 GENK

Canada
Dr. K.A. Seifert
Forintek Canada Corp.
800 chemin Montreal Road,
OTTAWA, Ontario K1G 3Z5

Denmark
Dr J.C. Frisvad
The Technical University of Denmark
Food Technology Lab., Building 221
2800 LYNGBY

Dr O. Filtenborg
The Technical University of Denmark
Food Technology Lab., Building 221
2800 LYNGBY

Federal Republic of Germany
Dr. Helgard I. Nirenberg
Institut für Mikrobiologie
Biologische Bundesanstalt für Land- und
Forstwirtschaft
Konigin Louise Str. 19,
1000 BERLIN 33

France
Dr. J.-P. Latgé
Institut Pasteur
28, Rue du Dr. Roux
75724 PARIS

Dr. J.-P. Debeaupuis
Institut Pasteur
28, Rue du Dr. Roux
75724 PARIS

Dr. Eveline Gueho
Institut Pasteur
28, Rue du Dr. Roux
75724 PARIS

Dr. Marie-France Roquebert
Laboratoire de Cryptogamie
M.N.H.N.
12, Rue de Buffon,
75005 PARIS

Mad. Joële Dupont
Laboratoire de Cryptogamie
M.N.H.N.
12, Rue de Buffon,
75005 PARIS

Italy
Prof.L. Polonelli
Istituto di Microbiologia
Facolta di Medicina e Chirurgia
Universita degli Studi di Parma
Via A. Gramsci 14,
I-43100 PARMA

Japan
Dr. Junta Sugiyama
Institute of Applied Microbiology
Univ. of Tokyo
Bunkyo-ku, TOKYO

Prof. Hiroshi Kuraishi
Faculty of Agriculture
Tokyo Univ. of Agriculture and Technology
5-8 Saiwaicho 3-chome
Fuchu, TOKYO 183 Japan

Netherlands
Prof. Dr. K.W. Gams
Centraalbureau voor Schimmelcultures
P.O. Box 273,
3740 AG BAARN

Drs. Ellen S. van Reenen-Hoekstra
Centraalbureau voor Schimmelcultures
P.O. Box 273,
3740 AG BAARN

Dr. R.A. Samson
Centraalbureau voor Schimmelcultures
P.O. Box 273,
3740 AG BAARN

Drs. Amelia C. Stolk
Wittelaan 19,
3743 CP BAARN

Dr. J. Visser
Section Molecular Genetics
Dept. of Genetics, Agricultural University
Dreyenlaan 2,
6703 HA WAGENINGEN

Turkey
Dr. N. Aran
Dept. Food Science, Tubitak
Marmara Scientific and Industrial
Research Institute
GEBZE/ISTANBUL

United Kingdom
Dr. P.D. Bridge
C.A.B. Mycological Institute
Ferry Lane,
KEW, Surrey TW9 3AF

Dr. Jim Croft
Department of Genetics
University of Birmingham
P. O. Box 363,
BIRMINGHAM B15 2TT

Prof. Dr. D.L. Hawksworth
C.A.B. Mycological Institute
Ferry Lane,
KEW, Surrey TW9 3AF

Dr. Z. Lawrence
C.A.B. Mycological Institute
Ferry Lane,
KEW, Surrey TW9 3AF

Dr. Agnes H. S. Onions
The nineteenth, North Road
Dunbar,
EAST LOTHIAN EH42 1 AY

Dr. R.R. Paterson
C.A.B. Mycological Institute
Ferry Lane,
KEW, Surrey TW9 3AF

Mr. A.P. Williams
Leatherhead Food Research Association
Randalls Road,
LEATHERHEAD, Surrey KT 22 7RY

Dr Veronia Hearn
Public Health Laboratory Serivce
Mycological Reference Laboratory
61 Colindale Avenue,
LONDON NW9 5HT

United States of America
Prof. Dr. M. Christensen
Dept. of Botany
Aven Nelson Building, University of Wyoming
LARAMIE, Wyoming 82071

Dr. Maren Klich
USDA Southern Regional Research Center
P.O. Box 19687,
NEW ORLEANS, LA 70179

United States of America

Dr. S. Peterson
United States Department of Agriculture,
NRRL
1815 North University Street,
PEORIA, Illinois 61604

Dr. Ed. Mullaney
USDA Southern Regional Research Center
P.O. Box 19687,
NEW ORLEANS, LA 70179

Dr. John Taylor
Dept. of Botany, University of California
2017 Life Sciences Building,
BERKELEY, California 94720

Sweden

Dr. O. Constantinescu
The Herbarium, University of Uppsala,
P.O.Box 541,
S-75121 UPPSALA

Switserland

Mrs Nadine Braedlin
NESTEC Ltd, Quality Assurance Dept.
55 Avenue Nestle,
1800 VEVEY

INDEX